GALACTIC AND EXTRA-GALACTIC
RADIO ASTRONOMY

Contributors

Robert L. Brown

W. Butler Burton

Edward B. Fomalont

M. A. Gordon

Robert M. Hjellming

Sebastian von Hoerner

David E. Hogg

Kenneth I. Kellermann

Richard N. Manchester

Barry E. Turner

Gerrit L. Verschuur

Melvyn C. H. Wright

GALACTIC AND EXTRA-GALACTIC
RADIO ASTRONOMY

By the Staff of
The National Radio Astronomy Observatory

Edited by
Gerrit L. Verschuur
and
Kenneth I. Kellermann

With the Assistance of
Virginia Van Brunt

Springer-Verlag New York Heidelberg Berlin

1974

The National Radio Astronomy Observatory, Green Bank, West
Virginia, is operated by Associated Universities, Inc., under con-
tract with the National Science Foundation.

Library of Congress Cataloging in Publication Data

United States. National Radio Astronomy Observatory,
  Green Bank, W. Va.
  Galactic and extra-galactic radio astronomy.
  Includes bibliographies.
  1. Radio astronomy. I. Verschuur, Gerrit L.,
  1937–   ed. II. Kellermann, Kenneth I., 1937–   ed.
  III. Title.
  QB475.U54 1974   522'.682   72–97680
  ISBN 0–387–06504–0

ISBN 0–387–06504–0 Springer-Verlag New York Heidelberg Berlin
ISBN 3–540–06504–0 Springer-Verlag Berlin Heidelberg New York

PREFACE

The present set of chapters by members of the staff of the National Radio Astronomy Observatory deals with the basic fields of research concerned with radio astronomy outside the solar system. The emphasis in this volume is on the type of data available and its interpretation. Basic theory is considered only where absolutely necessary, and little discussion of receivers or techniques is entered into in most of the chapters. The book is intended to take over where most textbooks on radio astronomy leave off, that is, in the discussion of what is actually known from the research done. In addition there is a chapter on the technical aspects of interferometry and aperture synthesis, since so much of modern radio astronomy depends, and will depend in an ever increasing manner, on such tools.

The editors want to stress that the chapters were not necessarily expected to be comprehensive reviews of any of the fields being covered, but rather, overall outlines which the individual authors felt would be suitable for graduate students and interested workers in other fields. As a result, the lists of references are not complete. This only reflects the preferences of the individual authors and not the relative merit of those references included or omitted.

It was also thought worthwhile to maintain the style of presentation of the individual authors rather than to make the various chapters of like format and style. Again this is indicative of the preferences and interests of the authors as well as the state of the subject being described. As a result, some of the chapters are necessarily more quantitative, others more qualitative.

The book would not have been possible without the cooperation of Dr. David S. Heeschen, Director of the National Radio Astronomy Observatory, to whom we express our grateful thanks.

In the preparation of the final versions of each chapter we have employed the assistance of many voluntary referees to whom we are also most grateful. For their assistance in this arduous task we sincerely wish to thank the following: A. H. Barrett, D. C. Backer, P. Baker, B. J. Bok, A. H. Bridle, E. Churchwell, J. R. Dickel, R. H. Gammon, C. Heiles, J. A. Högbom, W. E. Howard, G. K. Miley, I. I. K. Pauliny-Toth, A. Sandage, G. A. Seielstad, S. C. Simonson, J. H. Taylor, C. M. Wade, and M. Walmsley. One of us (B. E. T.) wishes to gratefully acknowledge the extensive assistance of R. H. Gammon.

The editors also want to thank Donna Beemer and Beaty Sheets for their patient cooperation in typing the manuscripts, and Peggy Weems and Gene Crist and his staff for the production of the diagrams.

Gerrit L. Verschuur
Kenneth I. Kellermann

CONTENTS

GALACTIC AND EXTRA-GALACTIC
RADIO ASTRONOMY

CHAPTER 1

GALACTIC NONTHERMAL CONTINUUM EMISSION

Robert L. Brown

The third group [of radio static] *is composed of a very steady hiss type static the origin of which is not yet known* . . . [however] *the direction of arrival changes gradually throughout the day going almost completely around the compass in 24 hours.* Karl Jansky, 1932.

1.1 Introduction

1.1.1 Historical Preface

In the course of an investigation of short-wave interference for the Bell Telephone Laboratories, Karl Jansky in 1932 built a rotating antenna array operable at 14.6-m wavelength and used this instrument to study local radio disturbances. He found that the static that was received usually could be categorized by one of the following descriptions: intermittent and strong, intermittent and weak, or very steady and weak. The first two types of static resulted from thunderstorms which were either quite close or distant from the antenna, respectively; but the third type was conspicuous as it never became intense nor did it ever completely vanish. Moreover Jansky (1932) noted that the apparent direction to this disturbance rotated through nearly 360° in 24 hours. Such periodic behavior indicated to him that the phenomenon was associated with the Sun, either directly or at least causally related through some process occurring at the subsolar point. These inferences were sharply revised when, after an entire year of observation, it was noted that the direction to the disturbance was not correlated with the position of the Sun, that it rather referred to a fixed direction in space. Jansky (1933) established this direction to be near right ascension 18^h and declination $-10°$, *i.e.*, very near the galactic center. Before it was concluded that a cosmic source of radio radiation existed at the galactic center, other possibilities, such as the presence of a cosmic ray source in that direction whose particles interacted with the Earth's atmosphere producing radio photons, had to be evaluated. Even these hypotheses, however, had to be dropped two years later when Jansky (1935), with increased sensitivity, was able to establish that radiation could be received continuously as the antenna beam swept along the galactic equator: A relative intensity maximum appeared in directions toward the galactic center while a minimum was noted near the anti-center. Pressured by the force of these observations, the conclusion that the Galaxy itself was a source of intense radiation at frequencies of a few tens of megahertz was inescapable.

The origin of this radiation remained obscure. Jansky suggested that since the nearby stars seemed to be largely confined to the galactic plane—a distribution reflecting the concentration of radio emission—perhaps merely long-wavelength stellar emission was being measured. In this model the radio

1

intensity was simply proportional to the number of stars in the beam. Such an explanation was only partially satisfying because at least one representative star, the Sun, appeared to be radio-quiet. Alternatively, interstellar dust—which optical extinction measurements showed is also concentrated toward the plane—served as a vehicle for two other early theories of the galactic radio emission: One involved radiation emitted by atoms recombining on the surface of grains and the other suggested that the observations refer to the Rayleigh-Jeans tail of a blackbody distribution of 30°K dust (Whipple and Greenstein, 1937).

In 1940 Grote Reber operating a 9.4-m filled parabolic reflector, attempted to observe the "cosmic noise" at frequencies of 3300 and 900 MHz which, if the Whipple-Greenstein blackbody theory had been applicable, should have been extremely intense. He failed to find any radiation at all (to his limits) at these frequencies. At 162 MHz, however, emission was detected along the galactic equator; these new measurements confirmed Jansky's impression that the intensity was in fact a function of longitude and that it appeared most pronounced at the galactic center (Reber, 1940a). Moreover, Reber (1940b) interpreted his fluxes as arising from thermal bremsstrahlung (free-free radiation) in a hot ($T_e = 10^{4\circ}$K), dense ($n_e = 1$ cm^{-3}) interstellar gas. This interpretation, while certainly consistent with the 162-MHz data, was not in agreement with Jansky's lower-frequency measurements that would have required a thermal medium with $T_e = 1.5 \times 10^{5\circ}$K. This discrepancy aside, Reber (1944) went on to interpret secondary maxima in the longitude distribution of the 160-MHz continuum flux as indicating the presence of spiral arms seen tangentially along particular lines of sight. In this manner he concluded that spiral arms should exist in the directions of Cygnus, Cassiopeia, and Canis Major. Moreover, the Sun was also detected weakly at 160 MHz, a result which confirmed that if stellar radiation produced the galactic radio background, then a population of "radio stars" quite distinct from the Sun was needed.

1.1.2 Early Galactic Surveys and Interpretations

a) Galactic Distribution of Radio-Frequency Radiation

In the 10 years immediately following the early work of Jansky and Reber several additional galactic surveys were undertaken at frequencies from 9.5 to 3000 MHz, with increasingly greater sensitivity and angular resolution. Taken together, these early surveys pointed to a roughly consistent picture of the galactic distribution of radio-frequency emission: Regions of greatest intensity were closely confined to the galactic plane, with the maximum brightness temperature seen at all frequencies occurring in directions toward the galactic center. On nearly all the maps several secondary maxima were seen, one in a region in Cygnus and another in the anti-center (Taurus) direction. A sample of these early observations at several frequencies is shown in Figure 1.1, where the results for the galactic center and anti-center regions are presented as Figures 1.1(a) and 1.1(b), respectively (in *old* galactic coordinates, l^I, b^I).

The principal impression to be drawn from these maps, aside from the similarity of their general features, is that the overall concentration to the plane increases and the effect of discrete sources is more pronounced as the frequency of observation increases—a result that reflects the fact that in most instruments the angular resolution increases with ν. More specific analysis reveals that in the galactic center region the radio plane of symmetry is displaced slightly south of the galactic plane $b^I = 0°$, while in the anti-center direction the 480-MHz map indicates that the relative maximum seen at all frequencies is largely due to the presence of a small discrete source.

With the recognition in the late 1940's that discrete sources of intense radio emission existed, the question as to what proportion of the disk radiation was in fact due to such sources was considered. The obvious correlation between the concentration of radio emission to the galactic plane and the limited

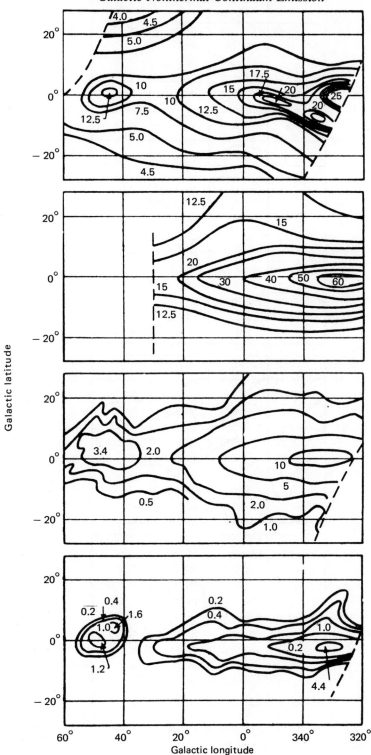

Figure 1.1 (a) Early surveys of the galactic center region at frequencies of 64, 100, 160 and 480 MHz, reading top to bottom. Old galactic co-ordinates are indicated. (Bolton and Westfold, 1951. *Australian J. Sci. Res.* A3:251).

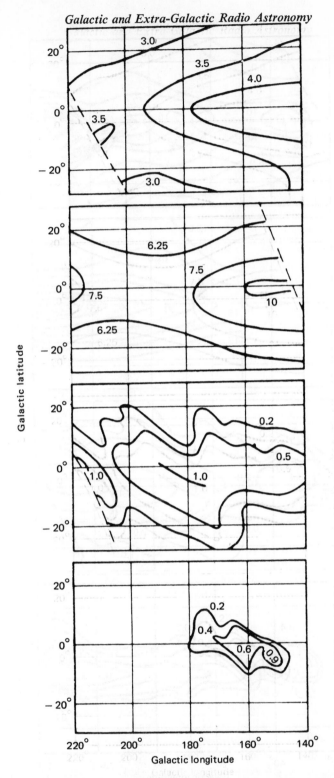

Figure 1.1 (b) Early surveys of the galactic anti-center region at frequencies of 64, 100, 160, and 480 MHz, reading top to bottom. Old galactic co-ordinates are indicated. (Bolton and Westfold, 1951. *Australian J. Sci. Res.* A3:251.)

z-extent of normal-type stars suggested to many that the radio background originated entirely in the coronae of "radio stars," *i.e.*, stars that were distributed like faint G and M dwarfs but whose ratio of radio to optical luminosities $L_r/L_o \sim 10^7$ times greater than that of the Sun. This explanation was supported by the observation that the density of early-type stars and novae had been long recognized to be greatest in the Cygnus and Sagittarius regions, and these were precisely the same regions exhibiting radio maxima. However, to account for two observational results—the intensity very near the galactic center greatly exceeds the intensity elsewhere, and the intensity contrast between the galactic poles and the galactic equator away from the Sgr-Cyg regions is not large—two distributions of radio stars are required. One must be highly concentrated toward the galactic center and the other must be essentially isotropic, *i.e.*, result from a much more extended, perhaps spherical, "corona" which is either concentric with the galaxy (Shklovsky, 1952) or due to a superposition of extra-galactic sources (Bolton and Westfold, 1951). Moreover, the brightness temperatures needed by the radio stars in such models must be enormous, $T \sim 10^{16}{}^\circ\mathrm{K}$. Alternatively, models were considered in which "radio stars" contributed to the galactic emission at only the lowest frequencies, but the high-frequency radiation was almost entirely due to thermal bremsstrahlung in the interstellar gas. Scheuer and Ryle (1953) recognized that these latter two-component star-gas models were observationally verifiable since they predict that the hot interstellar gas should show up in emission at $v \gtrsim 100$ MHz and in absorption at lower frequencies. To test this hypothesis they mapped the galactic plane with an interferometer operating at 81.5 and 210 MHz and were able to show that not only is the emission even more concentrated toward the galactic plane than ever thought before ($< 2°$, compare Figure 1.1), but the predicted interstellar absorption is not present. The latter result implies that the electron temperature of the emitting gas must exceed 15,000°K, a value inconsistent with the usual thermal explanation.

b) *Spectral Distribution of the Radio-Frequency Radiation*

While an analysis of the spatial distribution of galactic radio emission is one way to establish the origin and nature of the radiation, a study of the frequency spectrum is another. Table 1.1 presents a sample of the spectral data available in the early 1950's: T_{GC} and T_{AC} refer to the brightness temperatures seen in the galactic center and anticenter regions, respectively; α_{GC} and α_{AC} refer to the slopes of the corresponding T versus v curves.

Note that the shape of the spectrum is roughly similar in both regions and that the slope of the spectrum at high frequencies is

Table 1.1 Spectral Properties of the Galactic Emission from Early Surveys*

Frequency (MHz)	Brightness temperature (°K)		Spectral index	
	T_{GC}	T_{AC}	α_{GC}	α_{AC}
18.3	2.0×10^5	7.5×10^4	-1.28	-2.13
40	6.7×10^4	1.19×10^4	-1.88	-2.55
64	2.1×10^4	3100	-2.24	-3.08
100	6000	720	-2.32	-2.94
200	1190	120	-2.41	-2.32
480	145	16.6	-2.23	-2.04
1200	18	—	-2.17	—

*Adapted from Piddington (1951).

relatively constant and close to -2, as expected from a gas emitting thermal bremsstrahlung radiation, whereas at lower frequencies it begins to flatten. The trend shown in this table led Piddington (1951) to conclude that radio stars determine the intensity at $\nu \lesssim 400$ MHz but that bremsstrahlung in the interstellar gas is responsible for the emission at $\nu \gtrsim 400$ MHz. Scheuer and Ryle's observations, which showed that the hypothetical radio-emitting gas must be prohibitively hot (see p. 5), also revealed that in the galactic plane the slope α of T with frequency was $\alpha < 2$, a result arguing strongly against any thermal interpretation—even at short wavelengths. Consequently, to explain these results one is forced to hypothesize still a third distribution of radio stars, this one highly confined ($< 2°$) to the galactic plane and most intense at high frequencies ($\nu \gtrsim 250$ MHz).

As a result of all the early surveys one can express the following important questions that later were to direct the course of study into the galactic radio continuum; interestingly many of these puzzles remain unanswered today.

(1) How is the galactic radio background produced?
(2) What contribution do discrete sources make?
(3) What proportion of the galactic continuum is thermal radiation?
(4) What is the relation of extended regions such as that in Cygnus to the rest of the galactic radio background?
(5) How is the radio background attenuated at low frequencies?
(6) Is there an isotropic distribution of radio emission extended about the galaxy?

It is largely the study of these and related questions that constitutes the principal concern of this chapter.

1.2 Physical Mechanisms

1.2.1 Synchrotron Radiation

In the nascent years of the radio-star hypothesis Alfven and Herlofson (1950) considered models of ordinary stars which would produce extremely high radio brightness temperatures. They concluded that the most likely process was by synchrotron radiation, *i.e.*, radiation emitted by an electron spiraling in a magnetic field. Their suggestion was that perhaps many stars are surrounded by a shell of intense magnetic field into which highly energetic electrons are continuously injected. In the same year Kiepenheuer (1950) generalized this model by supposing instead that the interstellar gas itself was laced by a magnetic field and that streams of relativistic electrons gyrating in this magnetoplasma emitted synchrotron radiation and were in fact responsible for the galactic radio background. This model however attracted little attention. Even after polarization measurements verified Shklovsky's (1952) prediction that synchrotron radiation was responsible for the radio flux from the Crab Nebula, the mechanism was still not considered to be a likely explanation for the galactic continuous emission. Bolton (1956), in a summary of the most widely held ideas of the time, continued to adopt the view that radiation from population I stars superimposed on emission from a halo coextensive with the galaxy provided a reasonable interpretation of the available data. It was not until the high-resolution work of Mills (1959), which indicated that the radio continuum was correlated with the spiral arms, that theories for the galactic background involving synchrotron emission in the ambient interstellar medium were widely discussed—and this was nearly nine years after the model was originally proposed.

a) Principal Formulae

Since many excellent reviews of the synchrotron process as applied to radio astronomy already exist (Ginzburg and Syrovatsky, 1964, 1965; Pacholczyk, 1970), the details are not repeated here; rather only those basic results necessary for the interpretations which follow are summarized.

A relativistic electron which passes through a region of magnetic field will experience the $\mathbf{v} \times \mathbf{B}$ Lorentz force, spiral around the field lines, and emit radio-frequency electromagnetic radiation. The expression for the energy radiated in unit time by a single electron into a particular solid angle Ω,

$$P(\Omega,t) = -\int \mathbf{j}\cdot\mathbf{E}\,dV = \int \mathbf{j}\cdot\left(\frac{1}{c}\frac{\partial\mathbf{A}}{\partial t} + \nabla\phi\right)dV \quad (1.1)$$

can be used to determine the mean angle of emission θ,

$$\langle \sin^2\theta \rangle = \int \sin\theta\, P(\Omega,t)\frac{d\Omega}{\int P(\Omega,t)}\, d\Omega \quad (1.2)$$

where θ is the angle between the electron's direction of motion and the direction of the emitted photon. Relativistically,

$$\langle \theta^2 \rangle^{1/2} \sim 1 - \beta = \frac{1}{\gamma^2} \quad (1.3)$$

with $\gamma = (1 - v^2/c^2)^{1/2}$ (see Jackson, 1962, for details) so that the radiation is confined to a very narrow cone. As a consequence of this restricted angle, an observer in the orbit plane of the electron will receive a pulse of radiation every time the electron completes an orbit, *i.e.*, at intervals $\tau = 2\pi\gamma/\omega_o$; the duration of the pulse will be $\Delta t \sim \gamma^{-2}\omega_o^{-1}$, where $\omega_o = eB/mc$ is the electron gyrofrequency. The emitted photon spectrum is a continuum formed by harmonics of the fundamental

frequency ω_o, and this spectrum extends up to a maximum frequency defined by the condition $\omega_{max} \sim 1/\Delta t \sim \gamma^2\omega_o$. We expect, therefore, that a single electron will radiate very little energy at frequencies $\omega > \omega_{max}$. Detailed computations bear this out; the instantaneous spectral power radiated is

$$P(\nu) = \frac{\sqrt{3}e^2\,B}{mc^2}\left[\left(\frac{\nu}{\nu_c}\right)\int_{\nu/\nu_c}^{\infty} K_{5/3}(\xi)\,d\xi\right] \quad (1.4)$$

where $K_{5/3}(\xi)$ is a modified Bessel function and the shape of the quantity in square brackets is shown in Figure 1.2. Here it can be seen that virtually no power is emitted at frequencies in excess of ν_c, where

$$\nu_c = \frac{3}{2}\gamma^2\left(\frac{\omega_o}{2\pi}\right) = \frac{3e}{4\pi mc}B_\perp\gamma^2$$

$$= 4.21\,B_\perp\,\gamma^2 \text{ MHz} \quad (1.5)$$

and $B_\perp = B\sin\theta$ gauss is the component of the magnetic field perpendicular to the instantaneous velocity of the electron. For radio-frequency emission in typical astronomical fields, $B \sim 10^{-5} - 10^{-6}$ G, radiating electrons must have Lorentz factors $\gamma \sim 10^3 - 10^5$, which are very energetic particles indeed! Moreover, the radiation from an ensemble of such monoenergetic particles in a homogeneous magnetic field will be polarized to a substantial degree (Ginzburg and Syrovatsky, 1964):

$$\Pi = \frac{K_{5/3}(\nu/\nu_c)}{\int_{\nu/\nu_c}^{\infty} K_{5/3}(\xi)d\xi}$$

$$= \begin{cases} \frac{1}{2} & \nu \ll \nu_c \\ 1 - 2/3(\nu/\nu_c) & \nu \gg \nu_c \end{cases} \quad (1.6)$$

In summary, the fundamental results of single-particle synchrotron radiation in a homogeneous magnetic field are (1) the emission received at a particular frequency can be related directly to the energy of the electron producing this radiation and the strength of the magnetic field, and (2) the radiation is strongly polarized.

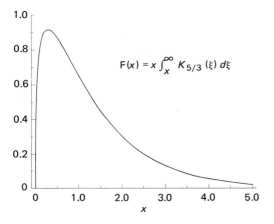

$$F(x) = x\int_x^\infty K_{5/3}(\xi)\,d\xi$$

Figure 1.2 The synchrotron spectrum from a single electron as a function of $x = \nu/\nu_c$. (Blumenthal and Gould, 1970. *Rev. Mod. Phys.* 42:237.)

b) *Radiation from an Ensemble of Particles*

The specific intensity emitted by a homogeneous and isotropic ensemble of electrons having an energy distribution $N(E)dE$ in the interval between E and $E + dE$ when the particles are moving in a uniform magnetic field B of spatial extent L is obtained from Equation (1.4):

$$I(\nu) = \frac{\sqrt{3}e^2\, BL}{mc^2} \int N(E)dE \left(\frac{\nu}{\nu_c}\right) \int\limits_{\nu/\nu_c}^{\infty} K_{5/3}(\xi)d\xi$$

(1.7)

Adopting the approximation (to be discussed below) that the distribution of electrons is describable by a power law of spectral index γ between particular limits E_1 and E_2, i.e.,

$$N(E)dE = KE^{-\gamma}\, dE \qquad (E_1 \le E \le E_2) \quad (1.8)$$

then the specific intensity can be integrated in closed form

$$I(\nu) = \frac{e^3 L}{mc^2} \left[\frac{3e}{4\pi mc}\right]^{(\gamma-1)/2}$$
$$\times\, a(\gamma)\, K\, B^{(\gamma+1)/2}\nu^{-(\gamma-1)/2}$$
$$\times\, \text{erg cm}^{-2}\, \text{s}^{-1}\, \text{sr}^{-1}\, \text{Hz}^{-1} \quad (1.9)$$

where $a(\gamma)$ is a slowly varying function of γ, in general $a(\gamma) \sim 0.1$. The important result here is that a power law distribution of electrons of spectral index γ reflects itself as a power law photon spectrum characterized by an intensity $I(\nu) \propto \nu^{-\alpha}$; the two indices α, γ are related by

$$\alpha = \frac{-(\gamma-1)}{2} \qquad (1.10)$$

The assumption introduced above that the distribution of relativistic electrons is a power law finds considerable application in radio astronomy because many radio spectra are observed to exhibit a power law dependence. However, this is not the sole justification for its use. From Figure 1.2 it is readily apparent that the range of electron energies which contribute to the emission spectrum at a particular frequency is strongly dependent on the electron spectrum. In fact, it can be shown that 90 % of the emission at a particular frequency comes from a range of electron energies $(E_2/E_1)_{90\%}$ that becomes increasingly narrow as the electron spectral index γ increases. This behavior is made explicit in Table 1.2. Consequently, the power law approximation amounts to an assumption that over an interval of ~ 5 to ~ 100 in energy the distribution of relativistic electrons is a power law: Its shape outside this restricted interval may deviate enormously from such a description without significantly influencing the resultant synchrotron emission spectrum.

The degree of polarization similarly depends on the electron spectral index. In a homogeneous magnetic field this is

$$\Pi_H = \frac{\gamma + 1}{\gamma + \frac{7}{3}} \qquad (1.11)$$

whereas in a totally random field $\Pi = 0$. A more realistic magnetic field configuration—perhaps applicable to the galactic plane—is composed of a weak homogeneous field B onto which is superimposed a random field B_r. If $B_\perp = B \sin\theta$ is the projected component of the homogeneous field and if $B_\perp \ll B_r$ (*i.e.*, the homogeneous field is a weak perturbation on the random field), then the degree of polarization becomes

$$\Pi = \Pi_H \frac{(\gamma + 3)(\gamma + 5)}{32}\left(\frac{B_\perp}{B_r}\right)^2 \quad (1.12)$$

so that the polarization, while present, is not in general very large.

Table 1.2 Relative Energies of the Synchrotron Electrons

Electron spectral index $\equiv \gamma$	1.0	1.5	2.0	2.5	3.0	4.0	5.0
$(E_2/E_1)_{90\%}$	1620	117	56	22	15	8.9	6.1

c) Influence of the Medium

All the results obtained above refer to processes occurring in a vacuum. When these same processes are considered in the presence of a thermal gas, two effects become manifest: Both the synchrotron intensity and its degree of polarization may be suppressed.

In a manner precisely analogous to the emission of free-free radiation, an ionized thermal gas may also *absorb* electromagnetic radiation. An unbound (free) electron in passing in the Coulomb field of a proton may encounter a radio-frequency photon, absorb it, and make a transition to another hyperbolic orbit of slightly greater energy. The absorption coefficient μ per centimeter for this process is

$$\mu = 9.75 \times 10^{-3} \, n_e^2 \, T_e^{-3/2} \, \nu^{-2}$$
$$\times \ln \left(4.97 \times 10^7 \, T_e^{3/2} \, \nu^{-1} \right) \quad (1.13)$$

so that it is of greatest import in a dense cool ionized gas, and it attenuates most effectively those protons at the lowest frequencies. Neutral atoms also contribute to the free-free absorption process, but their effect is diminished by the fact that the absorption coefficient for neutrals, $\mu_n \propto T^{1/2}$, becomes most important at high temperatures, where one expects preferentially ions, not neutrals.

The presence of the plasma also means that the index of refraction of the medium is less than unity; as a result the synchrotron radiation is suppressed at frequencies below those where the phase velocity of light is greater than c ("Razin-Tsytovich" effect). There is a consequent low-frequency cut-off near

$$\nu_R = \frac{2\nu_P^2}{3\nu_0} \sim \frac{20 n_e}{B} \quad (1.14)$$

where ν_P is the plasma frequency and ν_0 the gyrofrequency. At frequencies $\nu < \nu_R$ the synchrotron spectrum has the form $\nu^{3/2} \exp(-\nu_R/\nu)$.

Electromagnetic radiation, in addition to being attentuated in passing through a plasma-filled medium, also may have its plane of polarization rotated. This *Faraday rotation* arises because of the difference in effective refractive indices for left-handed and right-handed circularly polarized waves, *i.e.*, for the ordinary and extraordinary rays. The angle of rotation

$$\chi_F = \frac{\omega}{2c} \int (n_{\text{ord}} - n_{\text{ext}}) dl$$
$$= 2.36 \times 10^4 \, \nu^{-2} \int n_e \, B_{11} dl \quad (1.15)$$

depends on the integral electron density along the line of sight and the component of the magnetic field in the direction of the emitted photon. When $\chi_F \geq 2\pi$ across an extended source, the radiation received will be largely unpolarized, independent of the local intrinsic polarization of the emission (see Chapter 8 for a more complete discussion).

1.2.2 Collective Processes

The study of plasma radiation mechanisms was largely motivated by the observation of sporadic solar radio bursts in which the measured intensities proved difficult to explain by the conventional bremsstrahlung or synchrotron explanations. As it is precisely this characteristic—intense short-lived emission—that is expected from the conversion of plasma oscillations into electromagnetic waves, such an interpretation for solar flares is widely embraced.

Many types of collective waves may be excited by means of some instability in a magnetoactive plasma: (1) electron waves with an angular frequency $\omega_e = (4\pi e^2 n_e / m_e)^{1/2}$, where m_e is the electron mass; (2) electron plasma waves with a frequency near the gyrofrequency $\omega_o = eB/m_e c$, where $\omega_o > \omega_e$ must be satisfied for this mode to be excited; (3) ion plasma waves, $\omega_i = \omega_e (m_e/m_i)^{1/2}$, where m_i is the ion mass; (4) ion-acoustic waves, $w_{ia} = k(T_e/m_i)^{1/2}$, where $\omega_{ia} \ll \omega i$ and the electron temperature must greatly exceed the ion temperature; (5) Alfven and magnetosonic waves having frequencies $\omega_A < \omega_o (m_e/m_i)$; (6) whistlers with frequencies ω_W satisfying $\omega_o \ll \omega_W \ll \omega_A$; and (7) longitudinal ion waves with $\omega_L \simeq \omega_o (m_e/m_i)$. In a homogeneous medium in the linear approximation these waves do not interact, but in a nonlinear approximation or in an inhomogeneous plasma, interactions occur. These interactions are accompanied by an exchange of energy, either between waves of the same type with different wave numbers or between waves of different types. Consequently if one particular mode is preferentially excited, the interaction with the other modes may lead to an intense emission of radio-frequency electromagnetic radiation.

In considering the application of these collective processes to radio astronomy it is necessary to include both an investigation of the excitation mechanisms of various waves and also the conversion of the plasma-turbulent spectrum into electromagnetic waves. Unfortunately, very little work of this nature has been done, with most of it being applied to intrinsically small sources of high-energy density which are known to be variable at short wavelengths (cf. the review of Kaplan and Tsytovich, 1969). An interesting exception in which collective processes are suggested for the production of the low-frequency galactic background is considered in Section 1.3.6.

1.3 Observations of Galactic Nonthermal Radio Radiation

Although surveys of the galactic radio-frequency background have now been carried out for nearly 30 years and specific features have been studied in considerable detail, many of the most fundamental characteristics of the observed radiation remain subjects of uncertainty and debate. A representative map of the whole sky taken from the 150-MHz work of Landecker and Wielebinski (1970) with $2° \times 2°$ resolution is presented as Figure 1.3; it reveals most of the gross properties of present interest. Here the strong concentration of the emission toward the galactic plane is readily apparent, as is the fact that the galactic center seems to serve, approximately, as a center of symmetry for the entire sky (recall that both of these features were recog-

nizable even in the earliest low-resolution maps). Substantial emission is present all over the sky, but its distribution is far from uniform: Large-scale prominences that distort the high-latitude uniformity can be seen arising from rather low latitudes and extending nearly to the poles, and indeed, even in the galactic disk the emission is highly irregular. Taken by themselves, these properties of the observed galactic background are of considerable interest, but in order to draw reliable conclusions about the physical nature and morphology of our Galaxy it is necessary to incorporate the data into an appropriate model for the distribution of the emission. Presently the observed radiation is often described in terms of a superposition of four components: The galactic disk, symmetric about the galactic center; local features, mainly confined to the co-ambient spiral arm; the galactic halo, symmetric about the galactic center; and an isotropic contribution due to all extragalactic systems. Observations pertaining to the relative importance of these systems and their ramifications are discussed below.

1.3.1 Continuous Emission from the Disk

a) *Distribution in Longitude*

The distribution of emission in galactic longitude very near the galactic plane, as seen in Figure 1.3, exhibits a rather irregular behavior that is not completely unexpected; it is in fact a consequence of the rather large angular resolution employed in which discrete sources of angular size $\lesssim 2°$, particularly HII regions and supernova remnants, cannot be excluded from the more extended continuum of interest. The emission toward the galactic center appears relatively smooth because the path lengths involved extend through the whole Galaxy so that discrete variations are smoothed out to the point that they become nearly invisible against the integrated background. In the anti-center direction the emission path length is considerably shorter, and local features con-

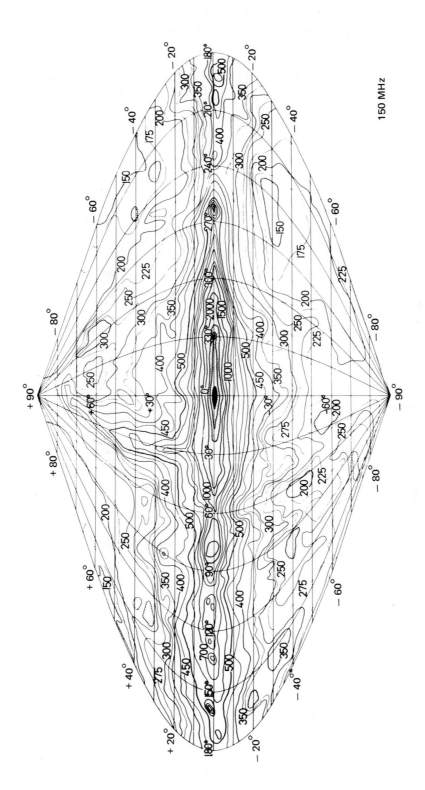

Figure 1.3 150-MHz sky brightness map in galactic coordinates. (Landecker and Wielebinski, 1970. *Australian J. Phys. Astrophys., Suppl.* 16.)

sequently appear more prominent. Interestingly, the minimum in the disk component does *not* occur in the direction of the anticenter, as would be expected from a planar distribution of radiation concentric about the galactic center, but rather the minimum brightness temperature ($T_B = 305°$K at 150 MHz) occurs near longitude $l = 240°$. The assumption of galactic symmetry about $l = 0°$, for this disk component at least, is evidently unjustified.

In addition to a study of the spatial distribution of the disk radiation, further insight can be gained by adopting a model in which the emission is due to the synchrotron process operating in a uniform interstellar magnetic field. Then from a comparison of the brightness temperatures from particular spatial regions at well-separated frequencies it is possible to obtain the slope of the resultant temperature versus temperature plot. In the disk this is $\alpha = -0.4 \pm 0.1$ (Bridle, 1967) at $\nu \sim 178$ MHz, steepening to $\alpha = -0.6 \pm 0.1$ at higher frequencies (see Section 1.4). An interesting anti-correlation is found in which α is largest when T_B is small and α assumes its lowest values for the very brightest continuum regions. This behavior seems to imply that in the brightest regions an additional source of radiation is present which has a characteristic emission spectrum flatter than that of the synchrotron component: Thermal radiation from an ionized gas (where $S \propto \ln[\nu^{-1}]$, is the obvious candidate mechanism. This suggestion is discussed further below. Alternatively the anti-correlation can also mean that in the weak continuum regions a greater proportion of the background comes from beyond the galaxy (Bridle, 1968).

b) Emission from the Spiral Arms

One unavoidable conclusion that must be reached when thinking of our Galaxy as a spiral is that if regions of intense radio-frequency emissivity are distributed spatially in approximately the same manner as the stars and gas—a reasonable assumption in models with local pressure equilibrium between the magnetic field, relativistic particles, and thermal gas—then a substantial enhancement in the brightness distribution is expected at longitudes where the spiral arms are seen tangentially. This result which, we have seen, Reber (1944) employed very early to (erroneously) interpret the source Cyg A was not easily verifiable on the early continuum maps because of their rather poor resolution. In fact, it was not until measurements from the Sydney 3.5-m cross with a beamwidth of 50 minutes of arc were available that anything resembling an enhancement at discrete longitudes could be resolved. Even then, the very intense peaks anticipated were absent; rather only broad "steps" were seen in the brightness distribution (Mills, 1959). This behavior is illustrated in Figure 1.4, where the antenna temperature is plotted as a function of galactic longitude at two frequencies and at latitudes up to $\pm 5°$ from the plane.

Several steps are evident in this figure as abrupt increases in the measured antenna temperature distribution; it is these features that are ascribable to spiral arms seen tangentially. In particular, steps at longitudes $l = 14°, 35°, 305°, 325°,$ and $345°$ are apparent in one or more of the figures. While not all of these steps are necessarily due to spiral structure, support for the identification of several of these features with spiral arms as opposed to local irregularities comes from 21-cm neutral hydrogen observations that also show steps in the brightness at similar longitudes. The observation of such abrupt but nearly smooth increases of intensity in directions along spiral arms that are only 10 or 20% above the adjacent background rather than the very intense and narrow peaks predicted assuming the galaxy is an ordered spiral, leads to an important conclusion: If synchrotron radiation is responsible for the non-thermal emission, then the magnetic field cannot be well aligned along the spiral arms. Such a result bears critically on early theories of the galactic magnetic field in which the field lines were thought to delineate the spiral arms. The radio continuum observations suggest instead that the galactic magnetic field, even within spiral arms, is highly disordered.

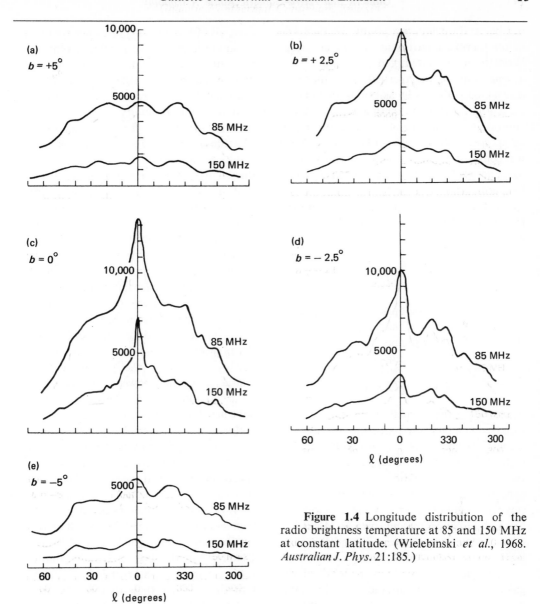

Figure 1.4 Longitude distribution of the radio brightness temperature at 85 and 150 MHz at constant latitude. (Wielebinski *et al.*, 1968. *Australian J. Phys.* 21:185.)

c) Structure in the Nonthermal Continuum

An inspection of the spectral characteristics of the emission from the galactic plane as derived from a comparison of the Mills' cross 85.5-MHz survey with the Leiden 1390-MHz survey led Westerhout (1958) to suggest that a substantial fraction of the high-frequency galactic background was in fact thermal radiation. This conclusion, which follows from the apparent flattening of the spectral index at low galactic latitudes vis-a-vis that observed at high latitudes—that is, the spectral index in the plane $|\alpha(b = 0°)| < |\alpha(b \gtrsim 10°)|$, the index at higher latitudes—seemed to indicate that the ratio of thermal to nonthermal radiation present in the galactic disk was considerably higher than the corresponding ratio out of the plane. Consequently, a reliable analysis of the disk component of the nonthermal emission and its associated magnetic fields and relativistic particles cannot be attained unless account can be taken of the thermal contribution to the emission. Westerhout first attempted to facilitate such a separation of the thermal from the nonthermal components by assuming that (1) the ratio of nonthermal to thermal emissivity was constant along the line of sight, (2) the intrinsic spectral index between 1390 and 85 MHz was constant, (3) any variation in this spectral index across the galactic plane was ascribable to thermal emission, and (4) all the thermal gas had a temperature of $10^4 °K$. Within the framework of this model the brightness temperature at a frequency ν is

$$T(\nu) = \left(\frac{T_e + T_{nt}}{\tau}\right)(1 - e^{-\tau}) \qquad (1.16)$$

where $\tau = \tau(\nu)$ is the optical depth, T_e is the electron temperature, and T_{nt} is the nonthermal brightness temperature. In the limit where the medium is optically thin, $\tau \ll 1$,

$$T(\nu) \simeq \tau T_e + T_{nt} \qquad (1.17)$$

where τT_e is the thermal brightness temperature (see Chapter 3 for a further discussion). Using these equations, together with Equation (1.13) for the free-free attenuation in a thermal plasma and Equation (1.9) to express the synchrotron brightness temperature, it is possible to compare directly the isophotes on the high (1390 MHz) and low (84 MHz) frequency maps and derive τT_e and T_{nt} independently as a function of galactic coordinates.

Through such analyses a model of the galactic disk was built up; the resulting brightness temperatures, τT_e and T_{nt}, across the sky were markedly different for the thermal and nonthermal components. It appeared that the thermal radiation was more concentrated toward the plane than was the nonthermal (the half-thicknesses of the galactic disk were ~ 200 and ~ 400 pc for the thermal and nonthermal components, respectively) and essentially all the radiation within a few degrees of the galactic center was thermal. Also it was concluded that over most of the sky the contributions of thermal and nonthermal emission were approximately equal at 1.4 GHz. How-

ever, all of these results rest on a comparison of radio surveys whose angular resolution was not better than 50 minutes of arc; before this picture can be fully substantiated one must ask what changes are manifest at still higher resolution and sensitivity.

In 1968 Altenhoff and his collaborators (Altenhoff, 1968) reported the preliminary results of a 2.7-GHz survey that was conducted with the NRAO 140-foot telescope that provides an 11-arc-minute beamwidth at this frequency; scans were analyzed within $2°$ of the galactic equator over $135°$ of longitude that included the galactic center. The purpose of this survey was to try to separate contaminating discrete sources of radio emission from the continuous background component. Altenhoff's results were particularly remarkable. More than 300 sources were detected with flux densities exceeding 2 flux units, and the distribution of these sources appeared narrowly confined to the galactic plane, the characteristic half-power width being about 24 arc minutes. After these sources had been subtracted from the total galactic 2.7-GHz emission it was found that the remaining unresolved background was mainly nonthermal, being well represented by a power law spectrum of spectral index 0.9. The so-called thermal disk component studied in earlier surveys was completely resolved into discrete sources (see however the discussion on the diffuse thermal component inferred by Gordon, Chapter 3 of this book). The distribution of the remaining nonthermal background in latitude is wider than that of the sources, $\sim 2°$, but its longitude distribution is quite similar. In particular the nonthermal brightness temperature seems to be rather uniform around the sky, although it exhibits "step" maxima at $l = 25$ and $50°$; both these longitudes coincide with HI spiral arms seen tangentially. The highly structured appearance of the galactic disk noticed in earlier surveys evidently arises from a superposition of many such discrete galactic sources rather than from spiral arms. The evident flattening of the spectral index at frequencies $\nu \lesssim 1$ GHz must be an intrinsic property of the distribu-

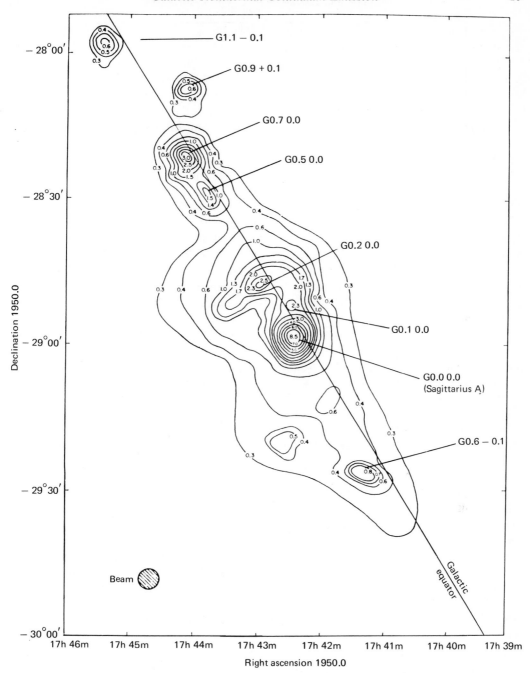

Figure 1.5 The galactic center region at 8.0 GHz. (Downes and Maxwell, 1966. *Astrophys. J.* 146:653.)

tion of relativistic electrons at energies below $\lesssim 10^3$ MeV (see Section 1.4 in this regard).

1.3.2 The Galactic Center Region

From the time of Jansky's earliest mea-

surements it was recognized that the most intense radio emission was received from the direction of the galactic center. The early low-resolution maps showed the region as a single, rather extended source, whereas with increasing resolution it became clear that the emission at the galactic center could be described as a 3.'5 component (Sgr A) that was superimposed on a larger diffuse emission region. More extensive observations, particularly those of Downes and Maxwell (1966), whose 8.0-GHz map of the galactic center region is shown on Figure 1.5, revealed that the entire region encompasses several discrete sources surrounding the very intense Sgr A source.

Sgr A itself has been the subject of considerable interest because it is thought to be coincident with the nucleus of the galaxy. Certainly this source merits special scrutiny because it is here, in the nucleus, that peculiar objects—black holes and matter-antimatter "furnaces"—have been postulated to exist. Unfortunately to study Sgr A by itself requires very high angular resolution (Figure 1.5), particularly at low frequencies; many useful measurements are consequently difficult to obtain. Sgr A appeared to be a smooth point source until recently, when interferometer data at 2.7 and 5.0 GHz indicated the existence of fine structure on a scale at least as small as 10 arc seconds (Downes and Martin, 1971). It has been possible, however, to obtain integrated radio fluxes for Sgr A at frequencies from 100 MHz to 100 GHz, and these measurements show the spectrum to be well fitted by a nonthermal power law $S \sim \nu^{\alpha}$ with $\alpha = -0.7$ at $\nu \gtrsim 1$ GHz. Below this limit the spectrum flattens considerably, to the extent that the source is very strongly suppressed below ~ 200 MHz. Indeed, the difference between the measured flux at 200 MHz and the limit obtained at 100 MHz is more than a factor of 10, *viz.*, $S(100 \text{ MHz})/S(200 \text{ MHz}) < 10^{-1}$ (Brezgunov *et al.*, 1971). The interpretation of this result is that the low-frequency flux is probably attenuated in the intervening HII gas (cf. Chapter 3). On the other hand at very short wavelengths—in the far infrared—the spectrum changes dramatically. From measurements at 100 μ Low and Aumann (1970) find that the integrated emission from Sgr A is enormous, $L \sim 10^7 L\odot$; the spectrum between 100 and 3000 GHz (100 μ) must change

abruptly as a consequence of this intense infrared component. The theoretical implications of such a large luminosity are of great interest, but a thorough discussion of these matters is beyond the scope of this chapter (see however Burbidge, 1970). Here we note only that an extrapolation of the power law radio component of Sgr A much beyond 100 GHz is not justified.

The discrete sources in the galactic center region share many characteristics in common with Sgr A with one important exception: Nearly all of the sources have thermal spectra. These galactic center HII regions are of interest in their own right, as several are also sources of intense far infrared emission (Hoffman *et al.*, 1971), or they contain interstellar molecules of current concern (especially GO.7−0.0, Sgr B2, see Chapter 9), or the hydrogen recombination lines from these sources can be used to delineate very peculiar features present in the velocity field in the general galactic center region (E. Churchwell, private communication). A detailed elucidation of the radio continuum characteristics of these sources as well as those of the galactic center region as a whole seems to await high-resolution interferometric studies from the Southern Hemisphere.

1.3.3 Emission from High Galactic Latitudes

An inspection of recent moderately high-resolution galactic continuum maps (*e.g.*, Figure 1.3) reveals that even at rather high latitudes irregularities exist on both large and small scales. Two features of this map are of fundamental interest. First, the broad prominent feature that can be seen arising near the plane at longitude 30° and that extends to nearly 80° latitude strikingly accounts for much of the marked high-latitude structure: This continuum "spur" and others of somewhat less intensity are subjects of much current controversy and are discussed in more detail in the following section. Secondly, the continuum brightness temperature over the sky but away from the galactic disk behaves in many ways as if the emission arose in an extended spherical volume centered at the galactic nucleus. Such a model receives support from the fact that the intensity in the direction of the North Galactic Pole exceeds that at moderate latitudes, $b \sim 50°$ toward the anti-center $l = 180°$, and moreover, the brightness at nearly all longitudes tends to uniformly decrease as b increases away from the plane. An analysis of the observations relating to such a hypothetical radio-emitting sphere, the so-called galactic "halo," is presented in Section 1.3.5 and is not considered further here. In this section we wish to inquire into the high-latitude brightness expected from two particular contributors: the superposition of extra-galactic sources and the emission from the local region.

An indication that a superposition of discrete galactic sources may contribute substantially to the high-latitude sky brightness comes from the Cambridge sky surveys in which the (isotropic) brightness temperatures that may be attributed to individual discrete sources with flux densities $> 0.25 \times 10^{-26}$ W m^{-2} Hz^{-1} is on the order of 15°K at 150 MHz. Accounting for all those radio galaxies that are less intense than the Cambridge detection limit requires an understanding of both the evolution of radio sources and the time-evolution of the universe, *i.e.*, the shape of the "log N–log S" curve below 1 flux unit (cf. Chapter 13). While a reliable knowledge of these functions is as yet nonexistent, nevertheless making reasonable assumptions, in particular that the flux density of all radio sources at meter wavelengths has the same spectral shape $S \propto \nu^{-0.75}$, it may be inferred from the work of Bridle (1967) that the integrated extra-galactic emission at 150 MHz is $48 \pm 11°$K. Comparing with the whole sky map, Figure 1.3, this contribution is roughly one-third to one-half the minimum integrated brightness at this frequency, so that although such an isotropic component is non-negligible, a substantial amount of radiation still remains unexplained.

Some contribution to the background emission is also to be expected from synchrotron radiation occurring in the local region.

Both studies of nearby stars and 21-cm hydrogen observations indicate that the Sun lies somewhere within a spiral arm or a very extended feature (Chapter 4), so that an estimate of the local synchrotron emission rests on a knowledge of the local spectrum of relativistic electrons and on the magnitude and orientation of the field in the local arm. While it is possible to directly obtain some idea of the distribution of energetic electrons (cf. Section 1.4), an analysis of the structure and properties of the magnetic field is considerably more difficult. The orientation of the local magnetic field has been investigated through studies of the polarization of the continuum radiation by Mathewson and Milne (1965) and others, with the result that nearly all the polarized emission arises in a band about 60° wide that contains the great circle extending through the galactic poles intercepting the plane at $l = 160$ and $340°$. This observation provides additional support for the model in which the Sun is located inside a spiral arm in which relativistic electrons gyrate around magnetic field lines that are directed along the arm toward $l = 70$ and $250°$. Such a configuration predicts that the band of intense local emission, together with the integrated radiation of the other spiral arms, should produce four broad minima in the continuum maps near $l = 70°, b = \pm 35°; l = 250°, b = \pm 35.°$ Comparing this prediction with Figure 1.3 it can be seen that three of these minima are present near their expected positions; only the one at $l = 250°$, $b = +35°$ is missing, and this particular region may be confused by the presence of a small "spur." This agreement lends strong support to arguments that much of the high-latitude brightness is in fact emission from the local arm (Hornby, 1966). Note, however, that estimates on the magnitude of this contribution depend sensitively on the value adopted for the local magnetic field. Following Yates (1968) it can be shown that essentially all the high-latitude emission can be accounted for by a combination of local-arm emission plus a superposition of extra-galactic sources if the local magnetic field is ~ 12 μgauss, a value somewhat higher than that derived from Faraday rotation observations (Chapter 8), but which may be applicable to regions of high synchrotron emissivity.

1.3.4 Galactic Spurs

Several large-scale irregularities conspicuous at high latitudes on continuum maps of the galaxy have recently been studied in considerable detail. These features—galactic "spurs"—share many properties: (1) they become individually discernible from the overall background radiation near the galactic equator because they appear to be oriented perpendicular to the plane; they subsequently can be traced to high latitudes; (2) their brightness temperature, while rather patchy, tends to decrease as b increases; and (3) their spectral and polarization characteristics are distinct from those of the adjacent background radiation. Three spurs (Table 1.3) noticed early in the Northern Hemisphere have been fitted to arcs of circles of very large radii, and these objects serve as prototypes for a class of looplike structures. Regarding for a moment these spurs as formal circles, observations indicate that along all parts of the spur the gradient of the brightness away from the outer edge of the circle is quite pronounced, whereas the brightness falls off only very slowly on the inside. This configuration suggests emission from a shell of gas expanding into a static ambient medium; the sharp outer gradient is then a shock front. Such a model seems to demand a supernova origin for these features, but as we shall see below, this interpretation is far from unique.

Under sufficient angular resolution the spectral and polarization characteristics of the spurs may be studied. Low-frequency observations of Loop I by Bridle (1967) indicated that the spectrum of the spur between 81.5 and 17.5 MHz seemed to be somewhat lower than the background, as if the radiation were being absorbed or suppressed whereas at higher frequencies ($\sim$1440 MHz,) the spectral index appeared slightly steeper than that in the surrounding regions. This suggests that the spurs are well-defined objects separable from the adjacent background. Further support for such a conclusion comes from polarization measurements of Mathewson and Milne (1965) and others, which show that the strongly polarized regions coincident with the spurs form elongated structures of rather high inferred magnetic field ($\sim$50 μgauss). Further radio measurements reveal that the magnetic field in the spur seems to be oriented parallel to its length, and this orientation agrees with that obtained from optical data on nearby stars (see however Chapter 8). The latter result, together with the observation that the polarization increases rapidly for stars at distances exceeding 100 pc, allows a distance estimate to be made for the spur Loop I; the inferred distance is 100 ± 20 pc (Bingham, 1967).

Of the many explanations advanced for the galactic spurs, only two have been considered in any detail; these are theories in which spurs are regarded either as local supernova remnants or as discrete features in the helical field of the local arm. The supernova model is supported by the morphological character of the loops, which exhibit both a shock-front appearance and a high degree of filamentary structure, as would be expected for remnants of an explosive event. Further, in

Table 1.3 Prominent Galactic Spurs

| Spur | Diameter | Center | |
		α(1950)	δ
Loop I (North Polar Spur)	113°	330°	19.5°
Loop I (Cetus Arc)	92°	100°	−30°
Loop III	71°	124°	11.5°

all these loops mentioned in Table 1.3 optical filaments rich in Hα emission have been identified on the outer edges of the radio filaments precisely in accord with models of a shock-heated gas (Elliott, 1970). Finally, evidence that the spectral index at frequencies $\gtrsim 1$ GHz seems to be steeper in the spur than that in adjacent regions suggests that the distribution of relativistic electrons in the spur is distinct from that in the more extensive galactic disk.

The supernova model is not without its difficulties, the most severe of which is that the observations demand the rather improbable occurrence of three recent supernovae very near the Sun. Instead, Mathewson (1968) and others argue that observations of polarization in the spurs, in the high-latitude radio background, and in nearby stars strongly imply that the spurs arise naturally from the helical structure of the magnetic field in the local arm. In this case it is noticed that when the distribution of spurs, ridges, and other highly polarized regions of the galactic radio emission is compared with the pattern of electric vector orientations obtained from optical polarization measurements of stars, one finds a striking similarity. It is principally this agreement that motivates the identification of the spurs as being features in the magnetic field structure of the local spiral arm. Presently the supernova interpretation is perhaps the best-supported explanation for the spurs [particularly by Elliott's (1970) recent Hα observations], but other local models certainly warrant further investigation and support.

1.3.5 Observations Relating to the Putative Galactic Radio Halo

As was mentioned previously in the introductory section, one remarkable result easily seen in the first low-resolution sky surveys was that the distribution of radio brightness, while concentrated to the galactic plane, nevertheless was still rather intense toward the poles. Westerhout and Oort (1951) recognized the implication of this result and suggested that it be interpreted as a spherical volume of radio emission that had the galactic nucleus as its apparent center; the radiation was considered to arise either in a local population of "radio stars" or in a superposition of extra-galactic radio sources. Even though Oort and Westerhout tended to support the latter conclusion, Shklovsky (1952) argued in favor of a modified version of the former. He preferred to regard the radio observations as resulting from some process occurring uniformly in a large-scale spherical halo or corona coextensive with the galaxy. In such a model the minimum brightness temperature was not expected to occur at the galactic poles; rather it should have been apparent in a direction where the path length through the halo was a minimum. The absolute minimum path length (from the position of the Sun) lies in the direction of the anti-center, but since the brightness at $l = 180°$, $b = 0°$ is confused by an additional contribution due to the galactic disk, one would anticipate instead that the observed minimum would be apparent at $l = 180°$, $b = \pm 45°$. Moreover, as the path length through a spherical halo in the direction $l = 0°$, $b = \pm 45°$ is roughly three times that toward $l = 180°$, $b = \pm 45°$, it was predicted that the intensity in the former regions would be approximately three times that in the latter. The confirmation of both predictions, the position of the minimum brightness, and that of the ratio T_b ($l = 0°$, $b = \pm 45°$)/T_b ($l = 180°$, $b = \pm 45°$) provided compelling evidence for the halo concept (Baldwin, 1955).

In the next few years similar startling results were forthcoming from the first detailed cosmic ray studies that were conducted above the atmosphere. One particular result, that the anisotropy of cosmic rays above 10^{12} eV was less than 10^{-4}, augered well for the halo concept. As Biermann and Davis (1958) realized, an inescapable conclusion associated with the halo concept is that such a large volume containing cosmic ray particles would effectively isotropize these particles, and its existence would, therefore, account for the very small cosmic ray anisotropy observed.

This important result that followed closely Baldwin's verification of the radio predictions of the halo concept led Pawsey (1961) to declare that "recognition of the corona in our Galaxy from observations of the nonthermal component of cosmic radio waves is one of the outstanding astronomical discoveries of the century."

Subsequent progress on the galactic distribution of continuous radio emission and on the propagation and confinement of cosmic rays has dampened this enthusiasm for the halo theory.

Interpretation of the high-latitude galactic radio continuum is complicated both by possible spurious instrumental effects and by the necessity to eliminate from the background those contributions resulting from the superposition of extra-galactic sources and from local features. In particular, measurements of the brightness temperature at $l = 0°$, $b = \pm 45°$ require a determination to be made of a radio brightness roughly 1% as strong as that of the nearby galactic center; to do this reliably one needs an explicit knowledge of the instrumental sidelobe response far from the main beam. Baldwin's early work that gave $R = T_B (l = 0, b = \pm 45°)/T_B (l = 180°, b = \pm 45°) = 3$, a ratio consistent with a spherical halo, may have been influenced by such spurious instrumental responses. More recently Burke (1967) repeated these measurements and found $R = 1$ as expected in the absence of the halo. From this and similar observations it is concluded that if the radio halo exists, it must be either very inhomogeneous or quite weak, $T < 30°K$ at 234 MHz.

Separation of the superposition of extra-galactic sources from the measured sky brightness is feasible if a reasonable value for the mean spectral index can be established (cf. Section 1.3.3). Estimating the influence of local features, however, is considerably more difficult. Several independent observations indicate that the distribution of radio emission at high latitudes is highly patchy; such irregular structure may be expected in directions toward the anti-center (or more explicitly

when $90° < l < 270°$) because the line of sight passes through <5 kpc of galactic medium, causing the relatively few discrete patches that contribute to the brightness to show up individually. The situation changes when the line of sight is along $l < 90°$. In these directions the path length through the emitting gas is >10 kpc, so that the number of contributing irregularities should be very large indeed. One would expect, therefore, a rather smooth distribution for these paths, but such an expectation is quite contrary to the very irregular structure observed. These observations can be reconciled only if the radiation arises comparatively nearby. Moreover, generalizing from the last section, it may be inferred that much of the high-latitude emission may be contained in the three extensive loop structures or in the scores of smaller-scale ridges identified by Merkelijn and Davis (1967) or Large *et al.* (1966), all of which are likely to be associated with features in the local arm. In order to assess the reality of a galactic halo or the magnitude of the brightness expected in this model it is necessary to subtract the effect of all the spurs and fine structure along with the extra-galactic contribution from the integrated background. While this task may be properly untenable at present, nevertheless rough attempts confirm that radiation from an extensive halo is no longer demanded by the observations (Baldwin, 1963; Yates, 1968).

Now consider further the support for the halo model provided by the strict cosmic ray anisotropy limits. Although recent cosmic ray measurements confirm the high degree of isotropy observed earlier, it is not presently believed that this observation demands a large isotropizing halo. Consider first the principal objection to the "cosmic ray" halo, which is as follows: The observed presence of spallation nuclei Li, Be, B in the cosmic ray flux implies that the average amount of matter traversed by the primary cosmic rays is $x \simeq 3.5$ g/cm^2. The diffusion time for particles out of a chaotic halo is quite long ($\tau \sim 3 \times 10^8$ yr), so that the total mass in the halo is $M = \rho V = xV/c\tau \simeq 10^{11} M\odot$. Since a halo mass much in excess of 10^9 $M\odot$ is gravitationally unstable, it must be concluded that inferences from cosmic ray confinement require either no halo or a "leaky" halo (Daniel and Stephens, 1970). It must be realized, however, that a galactic corona is not the only possible region of cosmic ray confinement. Extensive studies have been completed recently on models in which cosmic rays are produced, isotropized, and confined in the galactic disk. Here the very large energy requirements demanded by the short ($\tau \sim 3 \times 10^5$ yr) lifetime for diffusion out of the disk $I \sim 10^{42}$ erg sec^{-1} may be met by a reasonable spatial distribution of pulsars which, according to current theoretical models, are copious sources of cosmic rays. Further, the high degree of cosmic ray isotropy is to be expected, since any anisotropy will excite Alfven waves in the interstellar plasma, which in turn will interact back on the cosmic ray distribution so as to reduce that anisotropy. This disk confinement model now seems largely satisfactory.

In summary, the concept of a galactic halo or corona—an extensive spherical volume surrounding the galaxy that is laced by magnetic fields and permeated by a flux of energetic particles—is no longer demanded either by observations of the radio background or by implications drawn from the properties of cosmic rays. However, further radio observations that bear on problems associated with the separation of local irregularities and extra-galactic contributions from the integrated sky brightness are clearly desirable.

1.3.6 Very-Low-Frequency Observations of the Galactic Background

Recent advances in earth satellite technology have provided an opportunity to extend the observed galactic radio background spectrum to very low frequencies while at the same time avoiding troublesome ionospheric attenuation. Presently measurements have been obtained by the Radio Astronomy Explorer Satellite (RAE-1) at frequencies from 6 down to 0.4 MHz. Early results taken in directions toward the galactic center, the anti-center, and the North Galactic Pole all show a peak brightness of

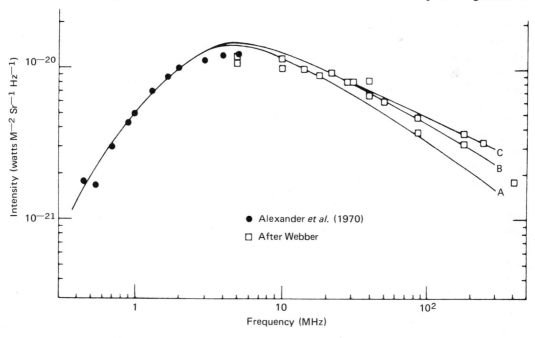

Figure 1.6 Low-frequency observations of the galactic anti-center region. (Goldstein *et al.*, 1970. *Phys. Rev. Lett.* 24:1193.)

about 10^{-19} W m^{-2} Hz^{-1} at a frequency near 3 MHz (Alexander *et al.*, 1970). At the lower frequencies (0.4 to 3.0 MHz) the spectrum exhibits a positive slope $\propto \nu^a$, where $\alpha = 1.5 \pm 0.1$, whereas at $\nu \gtrsim 3$ MHz the power level decreases as $\nu^{-0.5 \pm 0.1}$. These low-frequency data obtained when the antenna was directed toward the galactic anti-center are shown in Figure 1.6, along with a representative sample of higher-frequency data from the same region. However, the RAE-1 observations are still being analyzed in an attempt to account for subtle corrections such as impedance effects due to interactions between the booms; the spectral slope given in Figure 1.6 may therefore be subject to some change.

a) Free-Free Attenuation in the Interstellar Medium

The peak brightness found near ~ 3 MHz and the decrease at lower frequencies are spectral characteristics expected from an inverse power law nonthermal spectrum that is attenuated by free-free absorption in a medium coextensive with the emitting volume. The free-free absorption coefficient μ (Equation 1.13) is a function of the density, n_e, and temperature, T_e, of the ambient thermal gas. These quantities are intimately related to the ionization rate and thermodynamic structure of the interstellar medium. Observations of interstellar absorption lines, 21-cm studies, and much current theoretical thought support a model of the interstellar gas that has two components: (1) cool dense clouds in pressure equilibrium with (2) a tenuous hot intercloud medium. As the physical parameters in these two media are expected to be quite different, the net absorption coefficient for radio-frequency waves in traversing such a heterogeneous gas will depend on the optical depths in the cloud and intercloud regions and on the fractional volume occupied by each phase. Defining τ_c and τ_i as the optical depths corresponding, respectively, to a path length through one cloud and to the path length through the intercloud medium for a distance equal to the mean separation between clouds, the resultant radio intensity from the anti-center is given by

$$I(\nu) = \frac{\varepsilon(\nu)}{\mu(\nu)}[1 - \exp(-\tau_i)]$$
$$\times [1 - \exp(-L(\tau_c + \tau_i))][1 - \exp(\tau_c - \tau_i)]^{-1} \quad (1.18)$$

where $\varepsilon(\nu)$ is the synchrotron emissivity (Equation 1.9) and L is the size of the emitting region in kiloparsecs. Comparing such a model to the low-frequency observations, Goldstein *et al.* (1970) find that the absorption occurs primarily in the intercloud gas, not in the clouds, and that the observed radio spectrum can be reproduced if $n_e = 0.03$ and $T_e = 4000°$K are parameters characterizing the tenuous intercloud medium and if it is assumed that the emission occurs over a dimension ~ 4 kpc.

The above theoretical description, while being consistent with the observations, is by no means unique. In fact, since the RAE-1 antenna has a very wide beam (*i.e.*, very low angular resolution), $\sim 60°$, absorption occurs in regions having a wide range of physical conditions. In this case—even in the two-component interstellar model—the attenuation depends critically not only on the mean values τ_c and τ_i introduced above, but also on the dispersion in these quantities. This important point which has been emphasized by Gould (1971) has not as yet been included in an analysis of the very-low-frequency radio background.

b) Theoretical Implications

The above analysis of the low-frequency attenuation of the galactic radio emission has provided theoreticians with an important parameter, an estimate of the temperature of the intercloud medium. Using this value, it is possible to work from the equation of statistical ionization equilibrium for hydrogen,

$$\zeta_H n(H) = \alpha(T) n_e n(H^+) \quad (1.19)$$

to deduce the mean electron density in the intercloud gas. Here ζ_H, the hydrogen ionization rate, is a quantity which can be inferred either from a comparison of pulsar dispersion measures with 21-cm absorption in directions toward the pulsars or from an analysis of the 21-cm optical depth and the corresponding continuum optical depths in specific galactic sources; $\alpha(T)$ is the hydrogen recombination rate. In this way we find $n_e = 0.047$ in the intercloud medium. Alternatively the inferred value of $T_e = 4000°$K can be compared with theoretical models of the thermodynamic structure of the interstellar medium that attempt to predict this temperature. From the recent work of Habing and Goldsmith (1971) it is possible to conclude that an intercloud temperature of 4000°K is lower than that predicted by either cosmic ray or X-ray heating of the interstellar medium, but such a value is somewhat more in accord with X-ray heating. Clearly, the tentative nature of these conclusions should be emphasized, and again much work remains to be done.

Finally, the identification of the low-frequency turnover seen in Figure 1.6 as being the result of free-free attenuation in a thermal gas has itself been questioned. Lerche (1971) suggests that if the interstellar electron spectrum has a low-energy cut-off near 200 MeV ($\gamma_c \sim 400$) due presumably to ionization losses in the production sources, then the Razin-Tsytovich effect will attenuate the radio spectrum below a frequency that reflects this cut-off (compare Equation 1.14):

$$\nu_R \simeq 500 \, n_e^{3/4} \, B^{-1/2} \, \gamma_c^{1/2} \qquad (1.20)$$

This result obtains because in the presence of a thermal plasma of number density n the critical synchrotron frequency is not given by the vacuum expression, Equation (1.5); rather it becomes

$$\nu_c = \frac{\omega_o}{2\pi\gamma} \left[\gamma^{-2} + \frac{\omega_p^2}{(2\pi\nu)^2} \right]^{-3/2} \qquad (1.21)$$

where $m\omega_p^2 = 4\pi n e^2$. The medium will influence the synchrotron emission when $2\pi\nu \lesssim \omega_p\gamma_c$, and in this case the condition that $\nu \lesssim \nu_c$ (in order for significant radiation to be emitted) gives

$$2\pi\nu \gtrsim \omega_p \left(\frac{\omega_p \gamma_c}{\omega_o} \right)^{1/2} \qquad (1.22)$$

When this condition is not satisfied, *viz.*, at frequencies $\nu < \nu_R$ (Equation 1.20), the radiation will consequently be strongly suppressed. If this is the case, then an analysis of the radio turnover seen by Alexander *et al.* (1970) based on free-free absorption is misleading [because $I(\nu)$ is abruptly suppressed by the Razin-Tsytovich effect at frequencies somewhat larger than 4 MHz] and another source of radio emission at low frequencies is required. Lerche proposes that relativistically enhanced collective bremsstrahlung may provide a possible interpretation for the "missing" sub-megahertz radiation. These ideas require further consideration.

1.4 The Spectrum of Galactic Cosmic Ray Electrons

Studies of the spatial and spectral distribution of cosmic ray electrons in the galaxy are of particular interest because they provide, prospectively, the best possible approach to problems involving the propagation and confinement—and ultimately of course the production—of very energetic particles. Moreover, since electrons as well as nuclei suffer the modulation influences of the solar wind, a comparison of the data for these two species may allow some conclusions to be drawn regarding the relative contributions caused by rigidity-dependent and velocity-dependent modulation mechanisms. Two procedures exist to study these relativistic cosmic electrons: Direct (extraterrestial) measurement, and inferences drawn from the nonthermal galactic radio continuum.

1.4.1 Inferences from the Nonthermal Radio Continuum

Cosmic ray electrons, quite unlike the nucleonic components, are readily manifest by the characteristic synchrotron radiation they emit as they traverse the weak magnetic fields expected to thread the general interstellar medium. Given this particular feature it is possible to infer directly the electron

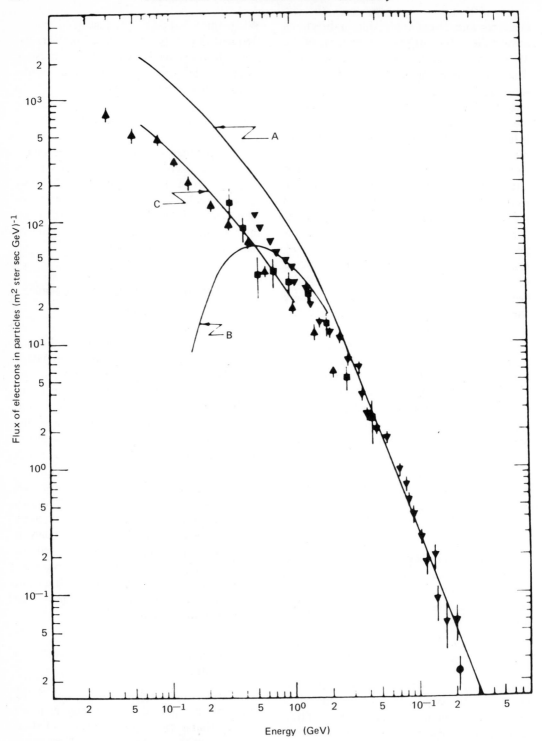

Figure 1.7 Measurements of the differential energy spectrum of cosmic ray electrons. Curve A is the electron spectrum derived from observations of the galactic radio background. Curves B and C refer to modulated spectra expected near the Earth when curve A suffers rigidity-dependent or velocity-dependent modulation, respectively. (Anand *et al.*, 1968. *Nature* 217:25.)

spectrum required to reproduce the observed background radio spectrum for a given value of the magnetic field. Following Daniel and Stephens (1970) the region of the galactic disk most likely to be similar to the interstellar environment near the Sun is in the direction toward the galactic anti-center. Assuming that all of the observed radiation arises from the synchrotron process in the galactic disk and that possible contributions from discrete galactic sources and from the integrated extra-galactic background do not supply more than a few percent of the total flux, then the radio intensity at a frequency ν is related to the electron energy spectrum by the synchrotron relation, Equation (1.9). Without introducing the *a priori* power law approximation for the radiating electrons it is possible to derive the shape $N(E)dE$ by trial and error. Doing this for the anti-center over a path length $L \sim 1.25 \times 10^{22}$ cm, the theoretical curve A in Figure 1.7 has been determined, and this procedure provides an estimate of the distribution of radio-emitting electrons. The mean value of the perpendicular component of the magnetic field used in this derivation is $\langle B_\perp \rangle = 5$ μgauss.

1.4.2 Direct Observations of High-Energy Electrons

Since it has been possible to directly infer the interstellar electron spectrum from the radio continuum fluxes, it is now feasible to compare these results with the electron distribution measured locally and obtain some information on solar modulation effects. Observations carried out from balloons and satellites have provided a measure of the differential flux of cosmic ray electrons from energies of a few MeV up to several hundred GeV. From the available data (Figure 1.7) it can be seen that at high energies ($E \gtrsim 5$ GeV), the electron distribution is well described by a power law

$$\frac{dJ}{d\varepsilon} = 126 \, E^{-26.2} \text{ electrons}$$

$$\times \text{ m}^{-2} \text{ sr}^{-1} \text{ sec}^{-1} \text{ GeV}^{-1} \quad (1.23)$$

but at lower energies the spectrum gradually flattens. The spectral index near 100 MeV is approximately -1.0. Moreover, note that the observed and inferred electron fluxes at energies exceeding ~ 5 GeV are in good agreement, but at lower energies, where, presumably, solar modulation becomes important, a real discrepancy is apparent. To emphasize the influence of solar modulation, curves B and C corresponding, respectively, to rigidity-dependent and velocity-dependent solar modulation of the radio-emitting electron spectrum, have been included on this figure. Evidently the solar modulation between 0.5 and 5.0 GeV is largely rigidity-dependent, whereas at energies less than 0.5 GeV the modulation seems to be velocity-dependent. Similar results have been obtained previously for nucleonic modulation (Daniel and Stephens, 1970).

It is precisely because of the feasibility of performing an analysis such as this, and those mentioned earlier, from which information on a wide variety of important astrophysical topics may be obtained—ionization and thermodynamic structure of the interstellar gas, the large-scale distribution of galactic magnetic fields, the spectrum and distribution of relativistic particles, and even the magnitude and shape of the solar modulation function—that a continuing study of the galactic nonthermal continuum emission remains currently rewarding.

References

Alexander, J. R., L. W. Brown, T. A. Clark, and R. A. Stone. 1970. *Astron. Astrophys.* 6:476.

Alfven, H., and N. Herlofson. 1950. *Phys. Rev.* 78:616.

Altenhoff, W. 1968. In *Interstellar Ionized Hydrogen*, Y. Terzian, ed. New York: Benjamin, p. 519.

Anand, K. C., R. R. Daniel, and S. A. Stephens. 1968. *Nature* 217:25.

Baldwin, J. E. 1955. *Monthly Notices Roy. Astron. Soc.* 115:690.

——. 1963. *Observatory* 83:153.

Biermann, L., and L. Davis. 1958. *Z. Astrophys.* 51:19.

Bingham, R. G. 1967. *Monthly Notices Roy. Astron. Soc.* 137:157.

Blumenthal, G. R., and R. J. Gould. 1970. *Rev. Mod. Phys.* 42:237.

Bolton, J. G. 1956. *Observatory* 76:62.

———, and K. C. Westfold. 1951. *Australian J. Sci. Res.* 4:476.

Brezgunov, V. N., R. D. Dagkesamansky, and V. A. Udal'tsov. 1971. *Astrophys. Lett.* 9:117.

Bridle, A. H. 1967. *Monthly Notices Roy. Astron. Soc.* 136:219.

———. 1968. Doctoral thesis. Cambridge University.

Burbidge, G. R. 1970. *Ann. Rev. Astron. Astrophys.* 8:369.

Burke, B. F. 1967. *IAU Symposium No. 31*, p. 361.

Daniel, R. R., and S. A. Stephens. 1970. *Space Sci. Rev.* 10:599.

Downes, D., and A. Maxwell. 1966. *Astrophys. J.* 146:653.

———, and A. H. M. Martin. 1971. *Nature* 223:112.

Elliott, K. H. 1970. *Nature* 226:1236.

Ginzburg, V. L., and S. I. Syrovatsky. 1964. *The Origin of Cosmic Rays*. New York: Macmillan.

———. 1965. *Ann. Rev. Astron. Astrophys.* 3:279.

Goldstein, M. L., R. Ramaty, and L. A. Fisk. 1970. *Phys. Rev. Lett.* 24:1193.

Gould, R. J. 1971. *Astrophys. Space Sci.* 10:265.

Habing, H. J., and D. W. Goldsmith. 1971. *Astrophys. J.* 166:525.

Hoffman, W. F., C. L. Frederick, and R. J. Emery. 1971. *Astrophys. J.* 170:L89.

Hornby, J. M. 1966. *Monthly Notices Roy. Astron. Soc.* 133:213.

Jackson, J. D. 1962. *Classical Electrodynamics*. New York: John Wiley.

Jansky, K. 1932. *Proc. Inst. Rad. Eng.* 20:1920.

———. 1933. *Proc. Inst. Rad. Eng.* 21:1387.

———. 1935. *Proc. Inst. Rad. Eng.* 23:1158.

Kaplan, S. A., and V. N. Tsytovich. 1969. *Soviet Phys.-Us.* 12:42.

Kiepenheuer, K. O. 1950. *Phys. Rev.* 79:738.

Landecker, T. L., and R. Wielebinski. 1970. *Australian J. Phys. Astrophys. Suppl.* 16.

Large, M. I., M. J. S. Quigley, and C. G. T. Haslam. 1966. *Monthly Notices Roy. Astron. Soc.* 131:335.

Lerche, I. 1971. *Astrophys. J.* 166:311.

Low, F. J., and H. H. Aumann. 1970. *Astrophys. J.* 162:179.

Mathewson, D. S. 1968. *Astrophys. J.* 163:147.

———, and D. K. Milne, 1965. *Australian J. Phys.* 18:635.

Merkelijn, J. K., and M. M. Davis. 1967. *Bull. Astron. Inst. Neth.* 19:246.

Mills, B. Y. 1959. *Proc. I.A.U. Symposium. No. 9*, New York: Academic Press, p. 431.

Pacholczyk, A. G. 1970. *Radio Astrophysics.* San Francisco: W. H. Freeman.

Pawsey, J. L. 1961. In *Galactic Structure*, A. Blaauw and M. Schmidt, eds. Chicago: University of Chicago Press, p. 219.

Piddington, J. H. 1951. *Monthly Notices Roy. Astron. Soc.* 111:45.

Reber, G. 1940a. *Proc. Inst. Rad. Eng.* 28:68.

———. 1940b. *Astrophys. J.* 91:621.

———. 1944. *Astrophys. J.* 100:279.

Scheuer, P. A. G., and M. Ryle. 1953. *Monthly Notices Roy. Astron. Soc.* 113:3.

Shklovsky, I. S. 1952. *Astron. Zh.* 29:418.

Westerhout, G. 1958. *Bull. Astron. Inst. Neth.* 14:215.

———, and J. H. Oort. 1951. *Bull. Astron. Inst. Neth.* 11:323.

Whipple, F. L., and J. L. Greenstein. 1937. *Proc. Nat. Acad. Sci.* 23:117.

Wielebinski, R., D. H. Smith, and X. G. Cardenas. 1968. *Australian J. Phys.* 21:185.

Yates, K. W. 1968. *Australian J. Phys.* 21:167.

General References

1. I. S. Shklovsky. *Cosmic Radio Waves.* Harvard University Press. 1960. Introduction to early observations of the galactic continuum.

2. J. L. Pawsey in *Galactic Structure*, A. Blaauw and M. Schmidt, eds. University of Chicago Press. 1960. General review of observational problems and techniques.

3. B. Y. Mills in *Annual Reviews of Astronomy and Astrophysics*, Vol. 2. 1964. Review of the observational situation and the physical interpretation.

4. S. A. Kaplan and V. N. Tsytovich. *Soviet Phys.-Us.* 12:42 (1969). Summary of plasma mechanisms in astrophysics.

5. R. R. Daniel and S. A. Stephens. *Space Sci. Rev.* 10:599 (1970). Study of cosmic ray electrons in the galaxy.

CHAPTER 2

INTERSTELLAR NEUTRAL HYDROGEN AND ITS SMALL-SCALE STRUCTURE

Gerrit L. Verschuur

2.1 General Theoretical Considerations

2.1.1 The 21-cm Emission Line of Neutral Hydrogen

Approximately 5 to 10% of the mass of the Milky Way is in the form of interstellar atomic hydrogen. The study of the physical properties of this matter is possible only because the ground state ($1^2 S_{1/2}$) can undergo a hyperfine transition, giving rise to radio emission (or absorption) at a wavelength of 21.1 cm. The transition occurs when the electron reverses its spin relative to that of the proton, the higher energy state being the one in which the two spins are parallel (the magnetic moments are parallel) and the lower state the one when they are anti-parallel. The prediction that this should be an observable transition was made by van de Hulst in 1944 well before receivers existed to make the measurements, and in 1951 two groups reported the detection of emission from interstellar neutral hydrogen (HI) in the September 1 issue of *Nature* (Ewen and Purcell, 1951; and Muller and Oort, 1951).

The frequency of the transition was accurately determined in the laboratory during the 1960's, using hydrogen masers. It is 1420.405752 MHz (Kerr, 1968).

2.1.2 The Emission Spectrum

We shall now summarize the basic equations relevant to the study of 21-cm radiation from interstellar HI. For more extensive derivations see Kerr (1968) or Spitzer (1968).

The equation of transfer can be written in the usual form

$$\frac{dI_v}{dx} = j_v - k_v I_v \qquad (2.1)$$

where I_v is the specific intensity at frequency v at a distance x from the observer. The volume coefficients of emission and absorption are j_v and k_v following the notation used by Kerr (1968).

The general solution to this equation is

$$I_v = \int_0^\infty j_v \exp\left(-\int_0^r k_v dx^1\right) dx \qquad (2.2)$$

and we may express the observed intensity as a brightness temperature, T_B, which is customary in radio astronomy.

In the case of thermodynamic equilibrium, the Rayleigh-Jeans law states that

$$T_B = \frac{I_v c^2}{2kv^2} \qquad (2.3)$$

and similarly the emission coefficient can be expressed in temperature units by writing

$$J_v = \frac{j_v c^2}{2kv^2} \qquad (2.4)$$

If a cloud is in thermodynamic equilibrium at temperature T_s, the number of atoms in the upper and lower states (n_1 and

n_0) is given by the Boltzmann distribution law,

$$\frac{n_1}{n_0} = \frac{g_1}{g_0} e^{-(h\nu_{10})/(kT_s)} \qquad (2.5)$$

where g_1 and g_0 are the statistical weights of the upper and lower levels, respectively. We find, since $g = 2F + 1$, where F is the sum of the electronic and nuclear spins and therefore takes on the values 1 or 0, that $g_1 = 3$ and $g_0 = 1$. Therefore,

$$n_1 = 3n_0\, e^{-(h\nu_{10})/(kT_s)} \qquad (2.6)$$

The important point to note is that even if the cloud is not in thermodynamic equilibrium, we can define a temperature that would account for the distribution of atoms between the two states, and it is this temperature which is called the spin or excitation temperature, T_s.

It is generally assumed that the population of the states is determined primarily by collisions between atoms, since Field (1958) has shown that other factors, such as Lyman α, light, or microwave radiation, cannot compare with collisions in importance. In this case T_s will equal the kinetic temperature. Normally the probability of a spontaneous transition occurring is given by the Einstein coefficient

$$A_{10} = 2.85 \times 10^{-15}\ \text{sec}^{-1}$$

which implies one event per atom per 11 million years. Collisions, however, increase this rate to one transition per 400 years. Now if we use Kirchhoff's law, relating the emission and absorption coefficients with a temperature we can write

$$J_\nu = K_\nu T_s \qquad (2.7)$$

Combining Equations (2.4), (2.3), and (2.2) we can get

$$T_B(\nu) = \int_0^\infty T_s K_\nu \exp\left(-\int_0^x k_\nu\, dx'\right) dx$$

$$= \int_0^{\tau'_\nu} T_s\, e^{-\tau_\nu}\, d\tau_\nu \qquad (2.8)$$

where $\tau_\nu = \int_0^x K_\nu\, dx'$ is the optical depth to a distance x at frequency ν, and τ'_ν is the optical depth at $x = \infty$.

If T_s is constant along the line of sight, we derive the basic equation for hydrogen-line work:

$$T_B(\nu) = T_s[1 - e^{-\tau(\nu)}] \qquad (2.9)$$

where $\tau(\nu)$ is the total optical depth at frequency ν, through the entire line of sight.

2.1.3 Line-of-Sight Hydrogen Content

Again following Kerr, we note that the atomic absorption coefficient, a_ν, is related to the volume coefficient of absorption by

$$K_\nu = a_\nu\left(n_0 - n_1 \frac{g_0}{g_1}\right) \qquad (2.10)$$

from Milne (1930).

In radio astronomical work $h\nu \ll kT_s$, and therefore using Equations (2.5) and (2.10) we get

$$K_\nu = a_\nu n_0 \frac{h\nu}{kT_s} \qquad (2.11)$$

Now a_ν is related to the Einstein coefficient A_{01} (absorption) by

$$a_\nu = A_{01} \frac{c^2}{8\pi\nu^2} \frac{g_1}{g_0} f(\nu) \qquad (2.12)$$

where $f(\nu)\, d\nu$ is the probability that a transition occurs between ν and $\nu + d\nu$.

From Equation (2.6) we note that, for $h\nu \ll kT_s$, the total number of atoms in the cloud (n) is given by $n \simeq 4\,n_0$.

Combining Equations (2.11), (2.12), and (2.8), and considering only a unit frequency interval of 1 kHz, the total number of hydrogen atoms in a cylinder of unit cross-section (1 cm^2) along the whole line of sight is given by

$$N_H = 3.88 \times 10^{17} \int_{-\infty}^{\infty} T_s\, \tau(\nu)d\nu \qquad (2.13)$$

or, if a 1 km sec^{-1} interval is used,

$$N_H = 1.823 \times 10^{18} \int_{-\infty}^{\infty} T_s\, \tau(\text{v})d\nu$$

A more readily useable version of Equation (2.13) is obtained by noting that Equation (2.9) reduces to

$$T_s(\nu) = \tau(\nu)\, T_s$$

for low optical depth. Provided we can assume low optical depth, and this is usually done in hydrogen-line work since any other assumption makes the job more difficult, we can rewrite Equation (2.13) as

$$N_H = 3.88 \times 10^{17} \int_{-\infty}^{\infty} T_B(\nu)\,d\nu \qquad (2.14)$$

which is a quantity that can be derived from the spectrum, since the integral is the area under the observed spectral line. Burton (1972) has suggested that within our Galaxy the data can be interpreted as if the HI is optically thin.

2.1.4 The Line Shapes

From the fact that the number of atoms at a given velocity is a Gaussian function of velocity (or frequency), we can write that the number at frequency ν, in terms of the number at some central frequency ν_0, is

$$N_\nu = N_{\nu_0}\, e^{-(\nu-\nu_0)^2/2\sigma^2} \qquad (2.15)$$

But from Equation (2.13) we see that

$$\tau_\nu = \text{constant } \frac{N_H}{T_S}$$

Therefore we can write

$$\tau_\nu = \tau_{\nu_0}\, e^{-(\nu-\nu_0)^2/2\sigma^2} \qquad (2.16)$$

This means that the hydrogen emission (or absorption) line is Gaussian on a linear optical depth scale and is Gaussian in brightness temperature only if $\tau \ll 1$, since then $T_B = \tau T_S$.

Gaussian shapes (in optical depth) are found for the narrowest absorption lines, but the narrowest emission lines observed to date do not show a Gaussian shape when plotted as T_B versus ν. This forces us to conclude that, in these cases at least, $\tau \gtrsim 1$ (Verschuur and Knapp, 1971).

The natural line width of the transition is given by A_{10}/π, which is insignificantly small,

but the Doppler broadening produced by thermal motions in the clouds gives the line a finite width. Other motions within the clouds, such as turbulence, rotation, expansion, and contraction, will also produce line broadening. In Chapter 4 the effects of streaming as a broadening mechanism are discussed.

Within each cloud the atoms have a Maxwellian velocity distribution due to their thermal motions, which means that the numbers at any velocity are proportional to a term like $\exp(-mv^2/2kT)$. This has the shape of a Gaussian function, $\exp(-v^2/2\sigma^2)$, where σ is the dispersion. For a single Gaussian profile the dispersion is therefore related to the temperature of the gas. Equating exponents in the two expressions above, we obtain

$$\sigma^2 = \frac{kT}{m}$$

The kinetic temperature, T_K, is then given by

$$T_K = \frac{m}{k}\,\sigma^2$$

$$= 121\ \sigma^2 \text{ if } \sigma \text{ is measured in km sec}^{-1} \qquad (2.17)$$

$$\text{or } = 5.4\ \sigma^2 \text{ if } \sigma \text{ is measured in kHz} \qquad (2.18)$$

2.2 Absorption Spectra and Emission Spectra Contrasted

2.2.1 Measuring Absorption Spectra

Absorption and emission-line observations reveal very different aspects of the interstellar HI. The first most obvious difference is that single-dish observations of emission or absorption spectra have very different angular resolutions. All the emission in the beam of the telescope contributes to the emission spectrum, whereas only the matter directly in front of the radio source, whose angular size may be orders of magnitude less than the telescope beamwidth, contributes to the absorption spectrum.

The second important difference is shown by a discussion of the detailed equations relating to the absorption measurements.

If a radio telescope is pointed at a radio source producing a brightness temperature, T_{BS}, which is assumed constant over the frequency range in which we are interested, then a cloud of optical depth τ will absorb this radiation to give a spectrum like

$$T_B^1(\nu) = T_{BS}\, e^{-\tau(\nu)} \qquad (2.19)$$

However, the emission in the beam of the telescope will add to this absorption spectrum and produce a spectrum given by

$$T_B(\nu) = T_{BS}\, e^{-\tau(\nu)} + T_S\,[1 - e^{-\tau(\nu)}] \quad (2.20)$$

In practice the emission spectrum [the second term in Equation (2.20)] has to be determined separately and subtracted from $T_B(\nu)$ in order to derive the true absorption spectrum $T_B^1(\nu)$. This can be done by assuming that the hydrogen structure varies slowly around the source and is uniform within the beam. These assumptions are invariably wrong, however; but if $T_{BS} \gg T_B(\nu)$ and $T_S \ll T_B(\nu_0)$, then the errors introduced by such assumptions are considerably reduced.

A practical observation of an absorption spectrum necessitates making a measurement at a frequency away from that of the spectral line so that receivers are used which switch between a measurement of T_{BS} (unabsorbed) and T_{BS} (absorbed). In other words the data are usually better expressed by

$$T_B(\nu) = T_{BS}\, e^{-\tau(\nu)} + T_S\,[1\ e^{-\tau(\nu)}] - T_{BS}$$

$$= (T_S - T_{BS})\,[1 - e^{-\tau(\nu)}] \qquad (2.21)$$

Now note that for large τ

$$T_B(\nu) = T_S - T_{BS} \qquad (2.22)$$

which is constant over that part of the spectrum where large τ holds. We then say that the line is saturated. Clearly a very cold cloud will give the largest negative going signal (T_S small), whereas a hot cloud will give a smaller signal.

For low τ, $T_B(\nu) = \tau(\nu)\,(T_S - T_{BS})$, and it is still true that the colder cloud gives a larger signal than the hot cloud.

This may be contrasted with Equation (2.9) for the emission spectrum alone. For high optical depth

$$T_B(\nu) = T_S \qquad (2.23)$$

so the hotter the cloud, the larger the signal. For low τ, $T_B(\nu) = \tau(\nu)\,T_S$, so we still observe the larger signal for the hotter cloud.

Hot clouds are more readily observed in emission spectra work, but cold clouds are selected out in absorption work. Comparison of the results of emission surveys or absorption-line surveys is therefore not valid, since they appear, to a first approximation, to deal with different clouds and also with different resolutions. However, the picture is not quite so simple, since in many cases very cold clouds can clearly be seen in emission. In fact, the coldest clouds observed to date are seen in emission, but such clouds are seen far from the galactic plane only in regions where the profiles are not confused by too much other hydrogen along the line of sight.

2.2.2 Obtaining an Accurate Absorption Line Spectrum

We note briefly that in order to derive the two unknowns T_S and $\tau(\nu)$ in Equation (2.20), interpolation between surrounding emission spectra for estimation of the second term was required; then the two unknowns could be solved. A more accurate method exists however. It would be ideal to observe the spectrum in the presence, and then in the absence, of the radio source. This is possible for pulsars, since they switch themselves on and off. It is also possible during the observation of variable radio sources which do not actually switch themselves completely off. Now there are two equations like Equation (2.20) with T_{BS}^1 and T_{BS}^2, which are measurable, and T_S and τ can be obtained unambiguously. Alternatively, a linearly polarized radio source can be observed parallel and normal to the plane of polarization, but this means that the polarization characteristics of the source and feed have to be very well understood before reliance can be placed in the difference spectrum that emerges.

Lastly, an interferometer can be used to

get rid of the unwanted emission in the primary beam. If the lobe size is comparable to the radio source size, then the surrounding emission will, it is hoped, cancel out in the other lobes of the interferometer, and a true absorption spectrum can be obtained. Such cancellation is automatically assured, of course, if a full and complete aperture synthesis is carried out.

2.2.3 Superposition of Clouds

Notice that if several clouds are lined up at the same velocity in front of a radio source, the absorption they produce is given by

$$T_{BS} e^{-\tau} = T_{BS} e^{-\tau_1} \cdot e^{-\tau_2} \cdot e^{-\tau_3}$$

$$\therefore \tau = \tau_1 + \tau_2 + \tau_3 + \cdots \qquad (2.24)$$

so that optical depths add in absorption lines. For an emission line produced by several clouds aligned in the line of sight we must write

$$T_B(\nu)_1 = T_{S1} (1 - e^{-\tau_1})$$
$$+ T_{S2} (1 - e^{-\tau_2}) e^{-\tau_1} + \cdots \qquad (2.25)$$

and provided $T_{S_1} = T_{S_2} = T_S$, we find that

$$T_B(\nu) = T_S [1 - e^{-(\tau_1 + \tau_2 + \cdots)}] \qquad (2.26)$$

Therefore optical depths add only if the spin temperatures are constant. For large τ in emission, the second and subsequent terms in Equation (2.25) are 0 so that only the nearest cloud at a particular velocity is seen.

Kahn (1955) has suggested that the maximum brightness temperature observed in any direction will be the harmonic mean of the spin temperatures in the line of sight, *i.e.*,

$$T_B(\text{max}) = \left(\overline{\frac{1}{T_S}} \right)^{-1}$$

Since the maximum temperature observed anywhere in 21-cm emission work is $135°K$ (Burton, 1970), many workers have assumed that $T_S = 135°K$ everywhere. The assumption is that the maximum T_B is seen where the optical depth is greatest and therefore the cloud responsible has $T_S = 135°K$. This temperature may equally well be pro-

duced when the hot intercloud medium of $1000°K$ has a maximum optical depth of 0.135.

2.3 Presentation of 21-cm Data

The brightness temperature, T_B, of the 21-cm emission can be measured only as a function of velocity, V, and two position coordinates, *e.g.*, right ascension and declination (α and δ). We therefore have a temperature as a function of three coordinates and can combine the various parameters in different ways in order to best illustrate the points we need to make, using the two-dimensional space available to us for display purposes. The data can be shown in the following ways.

(1) Profiles. Temperature as a function of velocity (or frequency) at a given position. $T_B(V)|_{(\alpha, \delta)}$.
(2) Contour maps. Temperature as a function of velocity and one position, the other position coordinate being kept constant. $T_B(V, \alpha)|_{\delta}$.
(3) Contour maps. Temperature at a given velocity as a function of two position coordinates. $T_B(\alpha, \delta)|_V$.
(4) Velocity contour maps. The velocity of some point in the spectrum (peak, median, etc.) as a function of position. $V(\alpha, \delta)|_{T_{B(\text{max})}}$.
(5) Drift scans. Variation of temperature at a given velocity with one position. For example, $T_B(\alpha)|_{V, \delta}$.

Other parameters derived from the above data, in particular the profiles, are,

(6) Integrated density maps. $N_H(\alpha, \delta)$.
(7) Maps of the variation of other parameters, such as the dispersion or temperature over the cloud. $\sigma(\alpha, \delta)$.

Two basic research aims can be said to exist in motivating 21-cm hydrogen-line observations. These are

(a) large-scale surveying, with the results

presented in the form of a catalog of data or a map of the galaxy

(b) small-scale surveys of specifically interesting regions, with the results presented in the form of the various physical properties of the region studied.

The former necessitates the publication of the data in a form as close to the originally obtained data as possible so they are available as reference material to other workers without having suffered anyone's personal interpretations, *e.g.*, by the original observer. It has been found that the presentation of a contour map, like (2) above, enables the maximum amount of information to be conveyed in the most efficient way. (See for example the Maryland–Green Bank Survey data, Westerhout, 1966.)

There is much information "hidden" in the data, however, which must be "extracted" and presented in a different way so as to illustrate and highlight the properties of given concentrations of neutral hydrogen. These data contain basically three parameters that describe the object studied.

(1) The total number of hydrogen atoms per column of unit cross-section in any given direction is an important quantity. It was shown above [Equation (2.14)] that the area under the spectrum observed is related to the total number of hydrogen atoms that produced the original radiation.

By integrating the profiles at each point in a grid on the sky N_H can be mapped over the region of interest. If the astronomer can determine the depth of the cloud of neutral hydrogen through which the line of sight has passed, then he can determine the density of the cloud in terms of the number of hydrogen atoms per cubic centimeter and subsequently the mass of the cloud.

(2) The velocities of the peaks in the profiles give data about the motions of the cloud as a whole with respect to some locally defined system (local standard of rest, or the Sun) and they also tell us about the motions that might vary systematically within the cloud itself.

These data are usually obtained by find-

ing the velocity of the peak of the spectral line. Some observers use a different velocity, *e.g.*, one at which the areas below the curve at higher and lower velocities are equal. In many profiles there are several peaks, and it might be more meaningful to use the equal area velocity.

(3) The width of the spectral lines indicates the range of velocities within the clouds themselves, at the positions at which the spectra are obtained. For a relatively simple cloud a study of the way the line width varies over this area might give useful information about the variations of temperature or turbulence within the cloud or whether the cloud is expanding or contracting.

(4) It is also possible at a given velocity to plot a contour map of temperature as a function of two spatial coordinates. This enables the observer to see where relative maxima and minima occur at a given velocity. These may then be compared with other properties of the area surveyed, *e.g.*, the distribution of dust clouds or HII regions. This type of data display will often highlight aspects of the region completely missed when the data were studied in different forms. (See, for example, Verschuur, 1970.)

2.4 Observations of the Small-Scale Structure in the Interstellar HI

2.4.1 Clouds or No Clouds?

First we need to clarify the use of the word "cloud" in referring to the interstellar HI. Many radio astronomers have objected to the use of the cloud concept because it may not be applicable to the HI. Clouds in the Earth's atmosphere with which we are familiar are discrete physical entities with well-defined boundaries containing matter clearly distinguishable from the immediate surroundings. Interstellar space contains widespread HI, and from this HI stars ultimately form in some way. Obviously, when the stars have formed, it is possible to clearly distinguish between them and their surroundings. Furthermore, there are discrete physical entities which are called dust clouds and which have well-defined boundaries, visible on photographs of the Milky Way. So at least we know that there are clouds in interstellar space and at some time a proto-star must also separate itself from the more general distribution of matter. The question now is, at what stage can the distribution of the more common interstellar HI best be described by invoking a cloud concept or will this really misrepresent the true situation? Alternative words, such as a concentration or a complex, have been used but these do not really make any (but a semantic) difference.

Let us suggest the following definition for a "cloud" in interstellar space. It is an entity which has clearly defined borders, indicated either by a transition between the presence and absence of the material under consideration or by a distinct transition in some other property of the medium. The border may be an "apparent" one if the transition from cloud to intercloud space involves only a change in temperature. It is conceivable that a radio astronomical observation sensitive only to cold matter would reveal such a cloud in "temperature space," *e.g.*, if there were a cold region suspended in a hot medium. The density might remain constant, but with our definition we would still be entitled to talk about a discrete entity, or "cloud." Note, however, that large changes in temperature with no opposing changes in density result in large pressure differences which will be very transitory.

Another situation which could be envisaged is that the velocity of material in a particular region of space might suffer a sudden discontinuity. An observation of the way the intensity of emission at a given velocity varies with position might indicate the disappearance of emission at that velocity at some point in space with the *appearance* of emission at some different velocity at an adjacent point. Clearly it would be difficult, from such a limited amount of information, to be sure whether the observation included two "clouds" at physically different distances or

one coherent entity with sudden velocity changes within it.

These considerations suggest a definition of a cloud as an object within which the various observable properties such as density, temperature, or velocity remain coherent. Sudden discontinuities of any of these, which cannot be accounted for by a simple physical model of a single object, would necessitate invoking more than one "cloud" to account for the data. The cloud concept is therefore used in this chapter, although there may well be many marginal cases where the boundaries between the cloud and intercloud medium appear so vague that the use of the word "cloud" may not be justified. The two-component model now being constantly referred to in discussions of interstellar HI, *i.e.*, a model in which two distinct phases of the interstellar HI exist—a cold cloud phase, and a hot intercloud phase—presents some justification for continuing to use the concept of clouds, because distinct phases are predicted theoretically (see Field *et al.*, 1969, and Hjellming *et al.*, 1969). Naturally we should expect that they may well have a range of temperatures and densities. Many examples of "real" clouds are discussed below.

2.4.2 HI in Clusters, Associations, and HII Regions

At present, stars are known to form in regions where the density of interstellar matter is high. We might still expect to see an excess of matter in these regions unless there has been very highly efficient star formation which has used up all the gas. Several observers have investigated the HI distribution in those directions where star formation has recently occurred or in which it might still be occurring. These are the young population I objects such as HII regions, stellar associations, and galactic clusters. However, many of these high-density regions are well shielded by dust; under such conditions hydrogen molecules may be formed so that 21-cm data may give incomplete information on the interstellar gas in these regions.

The results can be summed up as follows: There is strong evidence of excess HI in the immediate vicinity of several very compact HII regions, but observations in the direction of at least 28 clusters show no excess HI emission which can be unambiguously associated with those objects. For the case of some other clusters and associations, HI is found in their directions, but no obvious connection between the stellar groups and the HI "clouds" is apparent.

The search for HI associated with HII regions was motivated by the hope of finding expanding shells of neutral matter around the ionized region. Riegel (1967) examined 27 galactic HII regions and showed that in the direction of four of them neutral hydrogen components are found which are clearly related to the HII regions, although there is no evidence for expansive motions around any of them. The data he shows do reveal, rather strongly, that IC 5146 is indeed associated with an HI cloud of about 700 M$\odot$. He finds that the mass of HI associated with another HII region, NGC 281, is at least 1.6 $\times$ 10^4 M$\odot$. Kesteven and Bridle (1970) found a large HI cloud of 10^5 M$\odot$ associated with the extended thermal radio source NRAO 621, of which two compact HII regions, K3-50 and NGC 6857, form a part. They suggest that this region may be in an early stage of formation of a star cluster.

2.4.3 The Orion Region

The Orion Nebula is a well-known HII region associated with the Trapezium stars and is located in a larger area showing strong optical emission and reflection, and containing dust clouds. The latitude of the Nebula, $-20°$, makes it a favorable one for detailed study, since there is not too much foreground or background matter at such a high latitude. C. P. Gordon (1970) has reported on a comprehensive hydrogen-line survey of this general region and finds evidence for a large HI complex showing rotational motions and centered on the Nebula. Again, no evidence for expansion was found. A maximum in the

HI was found to be associated with the optical object, but no excess hydrogen emission could be associated with Barnard's ring, the giant loop of bright material in the neighborhood of the Orion Nebula. The mass of gas associated with the Nebula is found to be about 7 $\times$ 10^4 M$\odot$, and Gordon found that a distinct cloud at $-$10 km sec^{-1} with respect to the local standard at rest covered about 30% of the Orion region.

Menon (1970) has discussed evidence for the existence of excess emission associated with the dense conical absorption band seen in projection against the Nebula. This is confirmed by the data shown in Figure 2.1. These are several spectra, both on the position of the Nebula and 15 minutes and 30 minutes of arc away from the source to the north, south, east, and west. The only way to explain the variation of the brightness at $+$13 km sec^{-1} is to place a cloud, at this velocity, at the position of the dense dust cloud.

The conclusions we may draw from these discussions of the hydrogen distribution in and around these various population I objects is that sometimes we see excess HI emission and sometimes we do not. This statement holds true for any direction observed in space, of course, so the results are inconclusive. We can state only that the very young objects are located in regions generally containing much HI (*e.g.*, spiral arms), which is not surprising. Since most of the objects studied are in the plane, we cannot attach much significance to the apparent lack of excess emission at the position and velocity of these objects since there is so much HI around that it effectively "obscures" a good look at the object in question.

2.4.4 Other Interesting Low-Latitude Features

One striking HI cloud, not obviously associated with anything else, such as a dust cloud or any population I objects, is an extensive object covering at least 650 square degrees between longitude about 345 and 25°. It is seen in absorption against all the radio

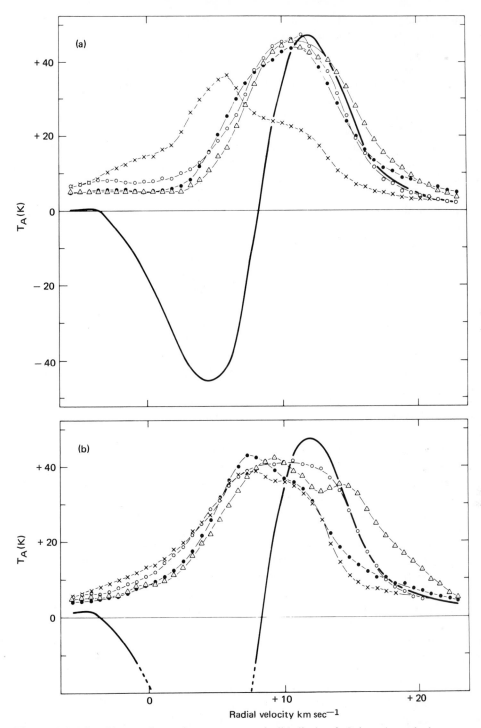

Figure 2.1 The 21-cm absorption spectrum (solid line) of Orion A and the surrounding emission spectra. (a) Spectra 15 minutes of arc away and (b) 30 minutes of arc away from Orion A. Triangles = north of source; filled circles = east of source; crosses = south of source; open circles = west of source. Absorption is still present 15 minutes of arc off the source, since the observing beamwidth was 20 minutes of arc. (*Nature* 223:140, 1968.)

sources in the plane in this part of the sky and is also visible in the surrounding emission spectra as a deep absorption line. This means that it must be very close to the Sun for it to be absorbing the more distant hydrogen emission. Riegel and Jennings (1969) have estimated that this cloud has a probable temperature of about 20°K and is located within 1 kpc of the Sun. The cloud is interesting because it is not obviously associated with anything else in this part of the sky and its velocity ($+7$ km sec^{-1}) is constant over much of its extent.

Supernovae shells must at some point in their evolution have expanded so far that they merge with the normal interstellar medium. It has been suggested that they push HI along with them in their slow expansion phases before they disappear entirely. At least two papers have presented evidence for expanding motions at low galactic latitudes. Katgert (1969) discusses a shell centered at $l = 63°$, $b = 0°3$, with the mass of hydrogen involved being 5×10^5 M$\odot$, assuming a reasonable distance of 5 kpc for the object. He admits that there is only a " . . . rather weak resemblance with a ring" in the HI he maps, which is moving 7 km sec^{-1} faster than the maximum velocity allowed for this longitude, according to the Schmidt galactic rotation law. In Chapter 4 it is noted that noncircular motions the order of 10 km sec^{-1} are often found in our Galaxy.

A much larger explosive phenomenon has been suggested by Rickard (1968) in order to explain his observations of the HI in the Perseus arm ($l = 100$ to $140°$). It is well known that optical absorption lines in this part of the sky show a double-peaked structure and that the hydrogen emission and absorption lines also show this very strongly. Rickard explains this in terms of an expanding ring or shell in the galactic plane which contains about 10^7 M$\odot$ and is expanding at about 20 to 30 km sec^{-1}, centered on $l = 120°$. Obviously no single supernova is involved in this case, but he suggests that several may have been required. However, alternative explanations have been suggested to account for the observations (*e.g.*, Roberts, 1972). The

ring of hydrogen is not unlike those rings seen in the Magellanic Cloud by Hindman (1967). We might note in passing that pulsars are supposed to be the last remains of stars which exploded to give rise to supernovae. Three are located within 5° of the center of Katgert's HI ring, and the nearest observed in the direction of Rickard's event is about 8° from the suggested center. However, it is unlikely that the pulsars are located at the distances of several kpc required for association with the hydrogen shells.

2.4.5 Dust and HI

One of the most interesting aspects of the HI distribution problem is whether or not the gas is correlated with interstellar dust. Before we discuss the present situation we should state strongly that there is a very large amount of HI in the Milky Way, and since we always have many clouds in the line of sight, especially at latitudes less than about 10°, picking out one and associating it with some other feature such as a dust cloud is usually going to be fairly difficult.

It is often stated that we are fortunate to have the 21-cm line because it enables us to see right through the obscuring dust which hampers the optical study of distant parts of the galaxy. However, although large numbers of nearly transparent HI clouds lined up one behind the other might not actually obscure the view, untangling the data becomes so confusing that the possibilities for a detailed study of interstellar matter in the plane are considerably hampered.

The study of the correlation between dust and HI has produced results ranging from good correlations to no correlations. A recent paper on the topic is aptly entitled "Correlation between Gas and Dust?" by Wesselius and Sancisi (1971). They compared the distribution of dust, as derived from galaxy counts, with the integrated hydrogen density averaged over the same areas used for the galaxy counts, and find no "general" correlation between gas and dust. However, some parts of the sky, when examined more closely, do show a positive correlation, in particular two parts of the Gould's belt system.

In the search for more detailed correlation between HI and dust clouds it is generally necessary to investigate whether there is an excess of emission at any velocity over the area covered by a dust cloud. One such association has been pointed out before, in the discussion of the Orion region.

A preliminary examination by this author of maps showing the intensity of the HI emission at various velocities as a function of position in a region of sky at about $+15°$ latitude, reveals that at *some* velocities *some* peaks coincide with *some* dust clouds, and at other velocities some minima coincide with dust clouds and some dust clouds coincide with neither peaks nor minima at any velocity. This result would probably be true for any areas chosen randomly on the maps. Such a project needs to be followed through rigorously to test the significance of any correlations found.

Various workers, notably Heiles (1967a), have noticed the apparent anti-correlation between dust and excess HI emission. His data have instead suggested that distinct minima, probably absorption features, are associated with some dust clouds. A notable example is the HI spectrum in the direction of the Taurus dust cloud, where the dip in the hydrogen spectrum is very deep and is present only at the position of the dust cloud. In general such a depression or minimum in the spectrum could be produced by three situations:

(1) a true lack of hydrogen gas
(2) an apparent lack of hydrogen, produced because it is all in the molecular form
(3) an excess of very cold HI, which produces an absorption line.

In the last case the associated HI need not even be significantly denser than the surroundings, for if its temperature is low enough and its optical depth high enough, it will absorb the background HI emission.

Heiles (1967a), in the case of the Taurus dust cloud, and Garzoli and Varsavsky (1970) for the same general region, claim that the apparent anti-correlation between dust and HI brightness is explained by the presence of much molecular hydrogen in the dust clouds concerned. Wesselius and Sancisi (1971) have argued strongly that the more plausible explanation is simply that the neutral hydrogen is cooler in these regions and in particular in the Taurus dust cloud (cloud #2 of Heiles). Spin temperatures of 30 to $50°K$ combined with optical depths of ≥ 0.5 would produce the observed effects. However, it is very nearly impossible to favor either interpretation unless strong additional arguments are made for one or the other case. Knapp (1972) has examined many more selected dust clouds and finds that the minima in the 21-cm spectra are in fact due to absorption by cold clouds, since many of the line shapes indicate saturation, *i.e.*, high optical depths. A mere lack of HI associated with an excess of H_2 seems unlikely to produce such narrow "absorption" lines.

Lastly, reference might be made to two papers discussing the dust-HI relationship in a region in Ophiuchus. Mészáros (1968) finds no marked excess of HI at the position of some of Khavtassi's well-mapped dust clouds and suggests that there is some evidence for a decrease in N_H for sufficiently high dust density, from which the presence of molecular hydrogen is again inferred. Sancisi and van Woerden (1970) examined the same region and note only the presence of a distinct elongated (although not completely mapped) feature not associated with the obscured regions. They note that their elongated feature shows "no connection with any optical features." Some people have observed that this "nonconnection" is so striking as to be an anti-correlation!

2.4.6 HI and Other Interstellar Atoms

The presence of several heavy atoms in interstellar space has been known for decades because of the absorption lines they produce in the light from distant stars. The distribution of interstellar calcium and sodium has been studied in this way. The velocities of the CaII or Na D lines can be compared with the 21-cm emission spectra velocities in the same direction to determine whether the same clouds contain both the HI and heavier atoms. Takakubo (1967) found that on the average, the velocities of the CaII lines and HI peaks within ∼2° of the position of the star observed agreed fairly well, showing an r.m.s. velocity difference of only 1 km/sec. These observations were made at intermediate latitudes, and since the hydrogen-line spectrum is typically 5 km/sec wide, this is not too surprising; otherwise the Ca could not even

be considered to be in the disk of the galaxy. Takakubo's results do not show conclusively that Ca and HI coexist in the same clouds. Habing (1969) examined this problem in more detail and noted that the near coincidence of low-velocity CaII and HI lines merely indicated that they were part of the local gas. Higher-radial-velocity HI components did not show associated CaII lines, except in the cases where the stars were more than 1 kpc from the Sun. In some cases CaII lines with no corresponding HI lines were found. Goldstein and MacDonald (1969) also noticed that there was little agreement between CaII and HI line velocities for radial velocities greater than 23 km/sec. This is strong support for the model

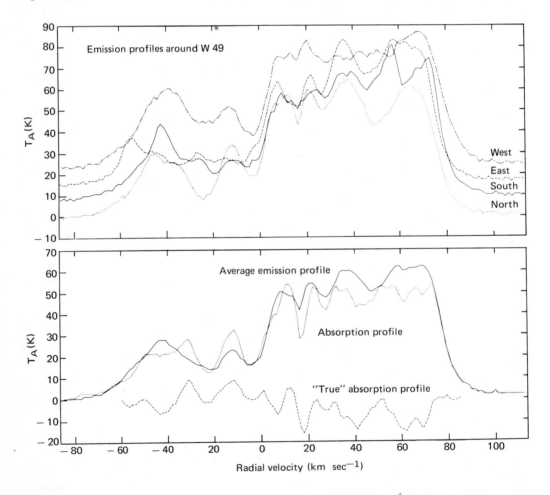

Figure 2.2 The 21-cm absorption spectrum in the direction of W 49 and emission spectra around the source, observed with the 140-foot telescope.

suggested by Habing, that the higher-velocity material is located well above the plane, out at > 1 kpc.

The picture of the coincidence (or otherwise) of Ca lines and HI emission features is far from simple, and the reader is referred to the source papers so that he might make up his own mind concerning the significance of the data. A detailed comparison of the high-resolution sodium-line data of Hobbs (1969) and HI data has not been made yet.

2.5 21-cm Absorption Experiments

2.5.1 W 49

The earliest published works on 21-cm absorption spectra dealt only with the strongest radio sources, and the spectra looked much the same no matter which telescope was used. The interpretations of the data were also basically similar. But recently, as these measurements have been carried forward to include weaker radio sources, the situation becomes more and more confused, especially when it comes to an examination of the conclusions reached by various authors. Let me stress that these remarks apply mainly to single dish measurements. Interferometer measurements, it is hoped are less subject to errors produced by the presence of small HI clouds in their beam.

We will consider the present situation by first examining the rather unusual case of W 49. Six papers totaling 24 pages of print, all concerning the absorption spectrum of W 49, have appeared in journals. Three papers are based on the same data: the Maryland–Green Bank Galactic 21-cm Line Survey! There are some startling differences between the conclusions drawn. Hughes and Routledge (1970) find a spin temperature of between 60 and 70°K for a cloud at +17 km sec^{-1}, whereas Gosachinskii and Bystrova (1968) analyzed the same data differently to obtain 22°K. The relevant cloud covers the two components of W 49 unequally, and Hughes and Routledge find optical depths of 2.5 and 1.8 in front of W 49A and B, whereas

Sato *et al.* (1967) used a 210-foot dish with a beamwidth not very different from that of the 300-foot to obtain optical depths of 1.3 and 0.6 for the same features!

The original use to which the study of the absorption spectrum of W 49 was put was to derive the distance to the HII region, since it was so highly obscured optically as to be invisible. 21-cm absorption line data are useful for this, because if absorption features are found out to a given velocity, then one can estimate from the galactic rotation law (Chapter 4) what is the lower limit at the distance to the radio source. Sato *et al.* (1967) and Sato (1968) believe that the two components of W 49, one an HII region and the other a supernova remnant, are at least 1 kpc apart in the line of sight.

Our Figure 2.2 shows the absorption spectrum of W 49 taken with the 140-foot telescope with a beamwidth of 20 minutes of arc, which includes both W 49 A and B. Also shown are four spectra taken 30 minutes of arc away from the source to north, east, west, and south. The so-called "true" absorption spectrum is also shown. This spectrum illustrates perfectly the problems that are faced in interpreting single dish 21-cm absorption-line data. It is customary to compare the average of the four spectra taken around the source with the on-source spectra to obtain the true absorption line.

There are nine "absorption" minima in Figure 2.2 and two (possibly four) excess-emission maxima. These positive-going peaks suggest that around the position of the source there is certainly excess emission at several wavelengths. Now which are "true" absorption lines? Clearly the feature at +17 km sec^{-1} is one because it is most strongly visible. If the highly asymmetric feature around +40 km sec^{-1} and the clear double line near +63 km sec^{-1} are real absorption lines, then are we not also entitled to attach significance to the two positive-going features at −30 and −12 km sec^{-1}? Do they indicate hydrogen which must be associated with W 49, or are these merely statistical fluctuations in the HI distribution? On the other hand, if we do not attach any weight to the presence of large positive-going signals of 10°K, what significance can we attach to negative-going signals of this size? If a 10°K antenna temperature indicates the random spatial fluctuations to be expected in this part of the sky, then we might "believe" only the +17 km sec^{-1} line, since it is the only one clearly visible at the position of the source.

In the absence of any other information, we would suggest that the 140-foot data (bandwidth 6.5 kHz, beamwidth 20 minutes) are too confusing to allow a rigorous analysis. This is probably also true of the other papers, which are no less subject to the confusing effects of the spatial structure of the HI in the plane.

We might suggest that unless the absorption lines are deeper than about twice the peak-to-peak variation of the surrounding emission spectra, not too much weight should be placed on single dish data.

2.5.2 Virgo A

Another good example of difficulties encountered is illustrated by the case of Virgo A. Using an 85-foot telescope (36 minutes of arc beamwidth) Williams (1965) found evidence for an absorption line of optical depth 0.17, using the "classical" method of observation, *i.e.*, comparing the on-source spectra with the average of several off-source spectra. Radhakrishnan and Murray (1969) disagree with this result and set an upper limit of 0.005 to the optical depth. They used a 210-foot dish, although Radhakrishnan *et al.* (1972c) have since repeated the experiment and found a line at -7.3 km sec^{-1} with $\tau = 0.008$. Measurements made with the 140-foot telescope (20 seconds of arc beamwidth) show a line with $\tau = 0.020$ at -6 km sec^{-1}! The optical depth here appears to depend inversely on the size of the telescope used!

Clearly there is only one reliable way to check the result and that is to observe the source with an interferometer. This has been done, and Colvin *et al.* (1970) report the presence of an absorption line at -8.5 km sec^{-1} with $\tau = 0.019$. This result is based on observations at three baselines—200, 400, and 800 feet—indicating that the measured optical depth has ceased to be a function of telescope size!

Virgo A is at a very high latitude ($+75°$) so that one would hope that the confusion due to small-scale HI structures around the source direction would be small, but clearly this is not the case—even there. Obviously one must not take the interpretation of single dish data too far unless the results are confirmed by observations with a very different resolution or unless the absorption lines are so intense as to be considerably in excess of the fluctuation in the emission around the source. This condition is satisfied for Cas A, Tau A, Cygnus A, and to a lesser extent for about only four or five other sources in the Northern Hemisphere.

2.5.3 Interpretation of Other Absorption-Line Data

Shuter and Verschuur (1964) decomposed their single dish observations of absorption lines into individual Gaussian components and estimated that the median of the temperatures of the cool clouds is 67°K. Clark (1965), using the Cal Tech interferometer, observed several more sources and obtained brightness distributions for several absorbing clouds. Clark suggested that the data could be well approximated by involving a two-component model, one a cold cloud medium at $\sim 100°$K, the other a hot intercloud medium at about 1000°K.

Hughes, Thompson, and Colvin (1971) find a mean temperature of $72 \pm 9°$K for the clouds seen in absorption against 64 sources in the Northern Hemisphere. Optical depths greater than 0.5 are found only within 20° of the galactic plane, except in the region of the Cetus arc, where it was found that $\tau = 1.4$ at a position near $b = -40°$. Hughes *et al.* (1971) used the Cal Tech interferometer to make their measurements and were able to obtain apparently reliable results on many more sources than would have been possible with a single dish.

More recently a series of papers based on 21-cm absorption-line data obtained with the Parkes interferometer have appeared. At least 60 sources are listed by Radhakrishnan *et al.* (1972a) and Goss *et al.* (1971); the equipment is described by Radhakrishnan *et al.* (1972a). Radhakrishnan and Goss (1972b) have performed a statistical analysis of the Gaussian fits to the absorption spectra and have concluded that the hydrogen seen in the galactic disk has a mean value of N_H/T_S of 1.5×10^{19} atoms cm^{-2} K^{-1} kpc^{-1}. The line of sight intersects on the average 2.5 clouds per kpc, which compares well with the estimate of Clark (1965), who found 4.1 per kpc.

2.6 "Nonplanar" Neutral Hydrogen

2.6.1 Introduction

Interstellar neutral hydrogen, being population I material, is confined mainly to the disk of the Galaxy, and this means that when viewed from the Sun, most of it appears at low galactic latitudes. We define "nonplanar" HI as the gas observed at galactic latitudes greater than 10°. This hydrogen should, to a first approximation, be relatively close to the Sun, assuming that the typical spiral arm thickness is of the order 200 to 400 pc. The study of this so-called nonplanar gas has several advantages compared with the study of low latitudes insofar as the study of detailed cloud structures and temperatures are concerned. At these higher latitudes there are smaller quantities of matter in the line of sight than nearer the plane, and therefore the observations are less subject to confusion effects introduced by the enormous quantities of foreground and background hydrogen.

Several lower-resolution surveys of nonplanar, *i.e.*, intermediate- and high-latitude, HI have been made, and the results of these surveys have usually appeared as broad generalized statements concerning the apparent cloud structures and average properties of the matter observed. We will show that much of

these data do not really help us to understand what is happening up there. Instead, the complete sky surveys recently performed at Berkeley and Groningen and the high-resolution surveys of small sections of sky performed at the NRAO show that the interstellar medium is very complex indeed and cannot be adequately studied by making incomplete samples of isolated regions or of the whole sky.

At this stage we might be advised to consider just how close high-latitude (low-velocity) material is likely to be. Until relatively recently the average spiral arm thickness was reckoned to be 200 pc, but Kepner (1970) has found evidence for matter in some of the outermost spiral arms at distances of up to 3.5 kpc from the plane. If these so-called high *z*-extensions are found to be common and to exist above and below the plane, our picture of spiral arm shapes and formation will be severely affected. There is a suggestion that the *z*-extension increases with distance from the galactic center, but even so the *z*-extensions might be at least 1 kpc in the solar neighborhood. Therefore, high-latitude clouds with little motion relative to the local standard of rest may not necessarily be matter within 100 pc of the Sun.

2.6.2 Distribution of the Low-Velocity Gas

If all the neutral hydrogen in the so-called local spiral arm were to share uniform galactic rotation, then most of the matter at latitudes above 20° should move at velocities very close to the local standard of rest. Random motions of about 10 km sec^{-1} are thought to be common in the plane, and at high latitudes we find that most of the gas is observed within the range $+10$ to -10 km sec^{-1} with respect to the local standard of rest. Not all these motions are random, however, as is discussed below.

The maps in Figure 2.3 show the longitude velocity plots for the neutral hydrogen emission at latitude $+17.9°$ from the Berkeley survey (Heiles and Habing, 1972). The effect of galactic rotation can still weakly be seen in

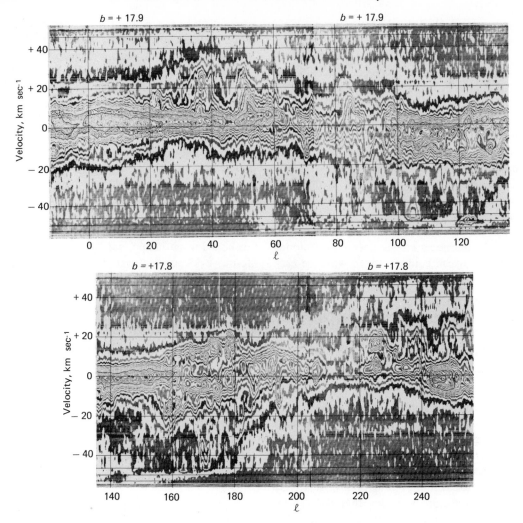

Figure 2.3 An example of the T_A $(l, v|b)$ maps from the Berkeley high-latitude 21-cm line survey (by permission of C. Heiles). The contour interval between dark lines is 2°K antenna temperature.

the distribution of velocities, and in addition it is quite apparent that in some longitude ranges the emission is very "disturbed" and in others it is relatively uniform. We might paraphrase Newton's first law of motion and say that interstellar HI will continue in its apparent state of rest or uniform motion (relative to the Sun) unless it is acted upon by a disturbance of some sort! This would suggest that the so-called random motions of clouds with respect to one another should have an explanation which we might try to seek. If this were not so, we would be very hard-pressed to explain why in some parts of

the sky the random motions are apparently absent. The point here is that several papers have appeared discussing the value of the r.m.s. velocity, but none have noted that there are parts of the sky where no such random motions are present. We will therefore tend to seek explanations for particularly striking velocity patterns in maps such as shown in Figure 2.3 rather than saying that they are merely turbulent phenomena. Of course this suggests that the concept of turbulence should not be used to describe the velocity patterns one sees in plots like Figure 2.3. At low latitudes these "disturbances" appear to be of a type that are well ordered over large regions, but at high latitudes this is not immediately obvious.

An exception is the very-large-scale disturbance of low-velocity material noted by Blaauw *et al.* (1967) several years ago in a part of the sky in which large quantities of matter were moving at 30 to 70 km sec^{-1} toward the Sun. In the region of the occurrence of this intermediate-velocity matter (all negative velocities) there is a relative lack of zero- or near-zero-velocity material. This, Blaauw *et al.* (1967) suggested, was strongly indicative of a large disturbance, perhaps of an explosive nature, which has accelerated the otherwise local material to the observed negative velocities. Figure 2.4 shows the integrated hydrogen density in various velocity intervals for the northern sky from Tolbert (1971); the lack of low-velocity matter around $l = 140°$ and $b = +60°$ is very striking. This is roughly the same area which shows the negative-velocity material excess.

It is surprising that relatively few high-resolution observations (10 seconds of arc) have been made of this matter, and Heiles (1967b) and Verschuur (various papers) appear to be the only researchers who have looked in any detail at this material. Other larger-scale surveys made with smaller telescopes, using broad bandwidths and a very open grid of observing points, are available, but few detailed discussions of the low-velocity features exist in the literature. Recently the University of California 85-foot telescope has been used to map the whole sky at latitudes above 10°, and this will give useful information about where the most interesting objects, suitable for further detailed study, are located.

2.6.3 Heiles' Survey

Heiles (1967b) mapped a strip $l = 100$ to 140°, $b = 18$ to 17° with the 300-foot telescope. He assumed a common distance of 500 pc for all the hydrogen in his survey and found three distinct physical components; a diffuse smooth background, two large sheets of gas moving in slightly different ways, and velocities in which the densities are about 2 cm^{-3}. These sheets show much small-scale structure. He found evidence that little hydrogen exists near the Sun in these directions, based on the way the number of hydrogen atoms/cm^2 falls off with latitude. We note that the apparent decrease with latitude is offset by the increase in the velocity of the material at higher b, supporting the model for a disturbance in the velocity pattern of local matter. This does not necessarily mean the latitude dependence of N_H found by Heiles should be interpreted in this way. Tolbert (1971) shows that when N_H is taken over a broader velocity range, the latitude dependence is not particularly unusual.

Heiles suggested that the small-scale structure of the HI distribution could best be described in terms of a hierarchy of clouds, from large condensations to cloudlets, which were usually found in groups in a larger cloud concentration. A cloudlet in the data of Heiles appears as a set of closed contours in the RA versus velocity maps he studied. At higher latitudes or at velocities well removed from the main band of emission, a cloudlet would always be clearly distinguishable, but then it might be called a cloud! There is a question as to whether any additional concept such as a cloudlet is needed to explain the data. Clearly, models on the nature of the interstellar medium depend on how the data are presented and studied. This is further illustrated below.

2.6.4 High-Resolution Studies of Low-Velocity Clouds

I have used the 300-foot telescope of the NRAO in Green Bank, West Virginia, for studying nonplanar neutral hydrogen. This study has been most informative, since it has revealed the existence of many isolated HI structures which appear to represent very good examples of specific phenomena that might be expected in interstellar space. These have included possible examples of shells of HI, rotating clouds, colliding clouds, and very

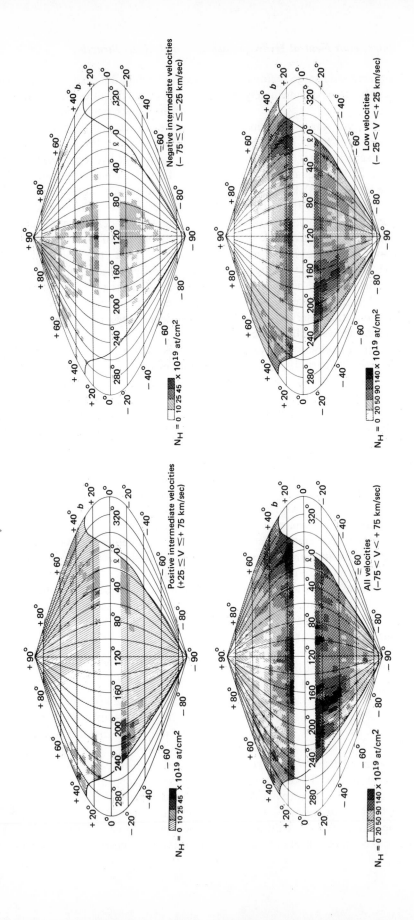

Figure 2.4 Maps of the distribution of N_H over the sky in various velocity intervals. (From Tolbert, 1971. *Astron. Astrophys. Suppl.* 3:349.)

cold clouds of HI (see references to Verschuur).

When we refer to certain objects on the maps as rotating clouds, we are interpreting the change in velocity with position, *i.e.*, the velocity gradients within particularly discrete clouds, as rotation effects. This need not be so of course. Shearing motions might lead to the same observed picture. Rotation is expected to exist in stable clouds to a certain degree; we therefore favor this interpretation. Whatever the cause for the strong velocity gradient in many clouds, they are clearly important enough to produce line broadening within the beam of the radio-telescopes commonly used in the study of 21-cm emission lines, *i.e.*, 85-foot-diameter telescopes with 40 minutes of arc beamwidths. Interpretation of line widths observed with small telescopes is therefore severely effected by nonthermal motions such as rotation. Narrow lines due to obviously very cold clouds are one particular type of emission feature that merits more discussion at this point.

2.6.5 The Temperature of the Gas

Common features occur in all these clouds. One is that very distinct boundaries exist for many of these clouds, suggesting a scale length of less than the beamwidth (10 seconds of arc) in many cases. Very considerable velocity changes with change in position exist in many clouds. Those in which this velocity gradient is zero are the ones which appear the "coldest," probably because the line widths observed have not been broadened by velocity gradients. Further observation should show how common narrow emission features, *i.e.*, cold clouds, are. Until recently it was thought that the average HI cloud had a temperature of 100°K. However, high-resolution observations made with the 300-foot telescope have revealed the presence of very cold clouds ($T_S < 40°K$) in surveys of emission spectra at high galactic latitudes. These low temperatures were inferred from the narrow line widths in the first instance. More recently Verschuur and Knapp (1971) analyzed the line shapes (using Equations 2.9

and 2.16) in two clouds to show unambiguously that the clouds in question had values of T_S of 20 and 12°K. Such clouds are probably very common, since the cold HI features associated with dust clouds (Knapp, 1972) and the large cold cloud in the galactic center direction (Riegel and Jennings, 1969) all indicate temperatures of this order.

It will be interesting to examine how the observed peak, T_B, varies with latitude in the full-sky survey of Heiles and Habing (1972). It has often been argued that the peak T_B in the plane of 135°K (Burton, 1970) indicates the existence of a common temperature in the clouds of 135°K and that there the optical depth is infinite. We know that at higher latitudes, maximum T_B values of only 20°K are seen, but mostly T_B is less. This would not be inconsistent with the idea that spin temperatures of 20 to 30°K were common and the optical depths in the high-latitude clouds were large. 21-cm absorption lines observed at high northern latitudes do not show many large optical depth clouds above 20° latitude, however. On the other hand, Verschuur and Knapp (1971) show that two intermediate-latitude ($+45°$) clouds have $\tau = 1$ and 2, which contrasts with the absorption-line data. The way out of this apparent discrepancy is to consider what the chances are that a line of sight to a radio source, several seconds of arc in diameter, manages to avoid a small high-optical-depth HI cloud, if these are indeed very common. If one examines the maps of Tolbert and Fejes (1966) and assumes that each discrete cloud seen on their maps, *i.e.*, objects with closed contours (preferably more than one) are large-optical-depth objects, then one can examine the list of Hughes, Thompson, and Colvin (1971) to see how many of the radio sources they studied are in the direction of these presumed high-optical-depth clouds.

The results are startling. Twenty-four sources from Tables 1 and 2 of Hughes *et al.* (1971) are located in directions covered by Tolbert and Fejes (at $b > +25°$). For the sources whose latitudes lie between two maps of Tolbert and Fejes we interpolate to the source position. This is probably not valid, but nevertheless, of the 24 sources, only 4 lie in the direction of discrete clouds. Two of those directions showed absorption lines of $\tau \approx 0.5$. The others showed no absorption lines, ($\tau < 0.2$ and < 0.08). The line of sight to the other 20 sources missed any significant HI concentration according to the HI emission maps. Two sources in Table 2 of Hughes *et al.* (1971) lie just outside the boundaries of clouds A and B studied by Verschuur and Knapp (1971), which have $\tau = 1$ and 2, respectively; and Hughes *et al.* were able to set limits of ~ 0.1 to the optical depths in those directions. A better comparison can be made by examining some of the data already circulated by Heiles and Habing. Emission-line data in the direction of 11 of the Hughes *et al.* sources exist, and this time we place a constraint that the data within about $1/2°$ of the source position should be used. Six sources are located in the direction of distinct peaks (three weak ones) in the emission. These reveal absorption features with $\tau = \sim 0.6$, 1.0, 0.8, 0.4, 0.3, 0.2, and 0.1. The other five are complete misses and no absorption lines were detected. The most interesting case is that of 3C 48 whose latitude is $-28°.7$. There is virtually no HI emission seen in its immediate vicinity in the data of Heiles and Habing, and naturally no absorption line was found. This contrasts with the general presence of large optical depths at high southern latitudes as noted by Hughes *et al.* (1971).

None of these considerations contradict a model in which there are many high-optical-depth cold clouds at latitudes $> +25°$, and hence throughout the Galaxy. Absorption-line data have not revealed these clouds, since due to the distribution of material and sources, the lines of sight have simply not intersected more than a few of them.

2.6.6 The North-South Asymmetry in τ

Hughes *et al.* (1971) note that there are many sources at relatively high southern latitudes with large-optical-depth absorption lines in their direction. They suggest a coincidence with the presence of the Cetus arc, the continuum ridge of emission seen in the

southern skies, and invoke the presence of an excess of gas in the south to explain this phenomenon. This conclusion is arrived at after comparing the results with the Northern Hemisphere picture, which is presupposed to be "usual."

I want to suggest that it is the Northern Hemisphere HI which is unusual and that the apparent correlation with the Cetus arc is not real. Let us consider the thesis that the Southern Hemisphere HI distribution is the one that is "normal" and then see whether we can explain why the Northern Hemisphere is "unusual." It is rather easy to do this. Above we pointed out that very few sources in the north happen to lie behind discrete HI clouds, and the reason for this is that there are relatively fewer low-velocity clouds at northern latitudes than at southern latitudes. This apparent lack of low-velocity material in the north was noted by Blaauw *et al.* (1967) and is clearly evidenced in the new data of Tolbert (1971). This lack of high-latitude low-velocity matter, with a particularly dramatic lack around zero velocity, is associated with the presence of an excess of intermediate negative-velocity material, which has no southern counterpart.

Blaauw *et al.* (1967) suggested that a disturbance had removed the low-velocity matter and imparted a negative velocity to it. If this disturbance also heated much of the matter, the lack of 21-cm absorption lines showing large τ clouds at northern latitudes is not unexpected.

Our thesis is that the southern material is normal, but that the northern sky has had, over at least half of its extent, the normal HI disturbed, resulting in the asymmetry now noted in the observations. Let us now discuss the properties of the peculiar velocity gas in more detail.

2.7 Anomalous Velocity Neutral Hydrogen

2.7.1 General Comments

Surveys of the distribution of 21-cm emission at latitudes greater than 10°, where we expect the gas to have velocities close to

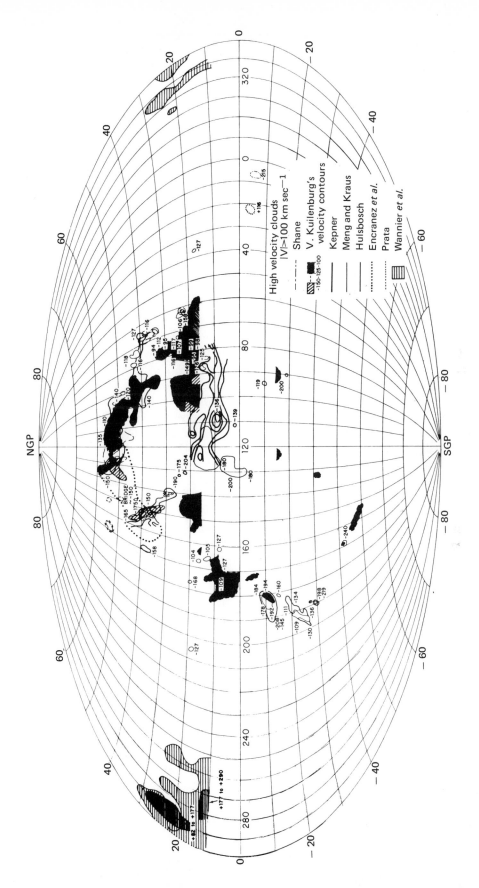

Figure 2.5 The distribution of high-velocity clouds $|v| > 100$ km sec^{-1} in the sky, plotted in new galactic co-ordinates. (See Verschuur, 1973a. *Astron. Astrophys.* 22:139.)

zero in the conventional model of the Galaxy (*i.e.*, spiral arms ~200 pc thick), have revealed an overwhelming preponderance of material with large negative velocities ranging from -30 to -200 km sec^{-1} (with respect to the l.s.r.). This apparent excess of negative-velocity matter now appears to be a possible selection effect, because equally large positive velocities are observed in high-latitude clouds in the southern skies. In the discussion below we shall, as a first approximation, make the distinction between intermediate-velocity (-30 to -100 km sec^{-1}) clouds (IVC's) and high-velocity (< -100 km sec^{-1}) clouds (HVC's). This is probably an unrealistic division in velocity because it is not yet proven whether there are any important differences between them. The boundary is often drawn at -70 km sec^{-1}, but we will show that at least some clouds moving at -100 km sec^{-1} belong to the former class and a number of HVC's are very clearly different from the IVC's. Figure 2.5 shows the distribution in the sky of the HVC's, taken from a summary prepared by Verschuur (1973a).

2.7.2 The Highest-Velocity Clouds

The originally discovered group of clouds around $l = 120°$, $b = 40°$ (complex C) of Hulsbosch and Raimond (1966) and Hulsbosch (1968) consists of two HVC's and one IVC, according to our definition, and these clouds lie in the general part of the sky where the lack of zero-velocity matter is most evident and in which the great excess of intermediate-velocity matter is found. This led to Oort's (1967) suggestion that matter was here seen to be falling into our Galaxy, and we observe the lack of low-velocity matter because it had been swept to the observed negative velocities. Tolbert (1971) shows that the total number of HI atoms in the line of sight at all velocities in this part of the sky is more or less consistent for its latitude and that much of it has a large velocity (mostly around -50 km sec^{-1}, as is evident in Meng and Kraus, 1970). Oort (1970) has suggested that the infall of matter is associated with the more general processes of galaxy formation

and that this infall is due to matter still being accreted by the galaxy. We might ask why we do not see this 21-cm emission from similar matter falling into other more distant spiral arms or why the anti-center group of HVC's should have the highest negative velocities and yet the lowest latitudes. If those clouds are nearby, their orbits must have been most unusual for them to "appear" to be coming toward the Sun at 200 km sec^{-1} at longitude 190°, latitude $-10°$.

Dieter (1971) has recently been able to trace low-latitude ($|b| < 10°$) high-velocity matter at all longitudes, and he invoked a model in which the galaxy has a ring of this material at its boundaries which is falling inward.

We should mention that from a gravitational stability consideration alone, the HVC's must be at about a 500-kpc distance to be coherent, gravitationally stable entities, and the negative velocities observed may be apparent ones due to the Sun's motion about the galactic center (Verschuur, 1969b). For example, M31 and M33 have negative velocities with respect to the l.s.r. of the same order as the HVC's, and they lie in the same general longitude range as the clouds. Kerr and Sullivan (1969) considered the longitude dependence of the galacto-centric velocity of the HVC's and showed that this was not inconsistent with the clouds being in a nearby (50 kpc) orbit of our galaxy.

An observation that might have some bearing on the HVC problem is the fact that the outermost spiral arms are seen to extend to very large distances from the plane (hi-z). Kepner (1970) discussed this in some detail. If there were to have been an explosive event in the outer arm around $l = 120°$, we might find a natural explanation for the Northern Hemisphere HVC's, since they form an obvious loop in the sky and their negative velocities would merely indicate that they were located in the outer arm. This does not explain the South Pole complex or the anti-center group, however, and this problem always occurs. One model explains one group of clouds, but not the other.

2.7.3 The Intermediate-Velocity Clouds.

The distinction between the intermediate-velocity hydrogen and the high-velocity hydrogen is somewhat vague. Separating them presupposes that they are different phenomena.

Stability arguments, similar to those applied to the HVC's, suggested that the IVC's need only be located on the order of thousands of parsecs away to be gravitationally bound, assuming that most of their mass is neutral hydrogen. In this respect they are very different from the HVC's which needed to be hundreds of kiloparsecs away. There also exists a suggestion that the IVC emission-line widths are typically of the order of 5 to 8 km sec^{-1}, whereas the HVC's have lines 20 km sec^{-1} wide. However, high-resolution observations of HVC's have shown that they too may have line widths around 6 km sec^{-1} (Verschuur, Cram, and Giovanelli, 1972; Giovanelli *et al.*, 1973).

2.7.4 High-Velocity Matter as Part of Spiral Structure

This author has recently suggested that the high-velocity material, both HVC's and IVC's, are parts of distant spiral arms beyond the solar circle (Verschuur, 1973a, b). This conclusion is based on an examination of constant latitude scans made with the 300-foot telescope using a narrow-bandwidth multi-channel receiver at 21 cm. These data show that the HVC's previously cataloged lie, in many cases, in strings of clouds which have a pattern in velocity-longitude space which is consistent with differential galactic rotation effects expected for distant matter in the galaxy. Basically negative-velocity clouds (HV and IV) lie between $l = 0$ and 180° and positive-velocity clouds lie between $l = 180$ and 360°. In addition there appear to be serious noncircular motions acting in the most distant spiral arms (10 kpc from the Sun, 20 kpc from the galactic center), and a detailed picture of the spiral structure based on the HV data is difficult to construct.

The distant spiral arms do appear to extend to large distances from the galactic plane (8 kpc) and are also not centered on the galactic plane. These arms are probably extensions to higher z-distances of the well-known "hat brim" effect in the Galaxy, in which the plane is tilted upward around $l = 90°$ and downward around $l = 270°$. This in turn might be the result of interaction with the Magellanic Clouds in some way. On this model the intermediate-velocity clouds are parts of high z-extensions in nearby spiral arms, whereas the high-velocity clouds are parts of high z-extensions in more distant arms. The region of sky near the North Galactic Pole containing intermediate-velocity matter apparently associated with the local lack of HI is probably a different phenomenon.

References

Blaauw, A., I. Fejes, C. R. Tolbert, A. N. M. Hulsbosch, and E. Raimond. 1967. *Proc. I.A.U. Symp. No. 31.* H. van Woerden, ed. New York: Academic Press, p. 265.

Burton, W. B. 1970. *Astron. Astrophys.* 10:76.

———. 1972. *Astron. Astrophys.* 16:158.

Clark, B. G. 1965. *Astrophys. J.* 142:1398.

Colvin, R. S., M. P. Hughes, A. R. Thompson, and G. L. Verschuur. 1970. *Astrophys. Lett.* 6:211.

Dieter, N. H. 1971. *Astron. Astrophys.* 12:59.

Ewen, H. I., and E. M. Purcell. 1951. *Nature* 168:356.

Field, G. B. 1958. *Proc. I.R.E.* 46:240.

———, D. W. Goldsmith, and H. J. Habing. 1969. *Astrophys. J.* 155:L149.

Garzoli, S. L., and C. M. Varsavsky. 1970. *Astrophys. J.* 160:75.

Giovanelli, R., G. L. Verschuur, and T. Cram. 1973. *Astron. Astrophys. Suppl.* In press.

Goldstein, S. J., and P. D. MacDonald. 1969. *Astrophys. J.* 157:1101.

Gordon, C. P. 1970. *Astron. J.* 75:914.

Gosachinskii, I. V., and N. V. Bystrova. 1968. *Soviet Astron.—AJ.* 12:548.

Goss, W. M., V. Radhakrishnan, J. W. Brooks, and J. D. Murray. 1971. *Astrophys. J. Suppl.* 24:123.

Habing, H. J. 1969. *Bull. Astron. Inst. Neth.* 20:120.

Heiles, C. 1967a. *Astrophys. J.* 156:493.

———. 1967b. *Astrophys. J. Suppl.* 15:97.

———, and H. J. Habing. 1972. Berkeley Hi-Lat HI Survey, in preparation.

Hindman, J. V. 1967. *Australian J. Phys.* 20:147.

Hjellming, R. M., C. P. Gordon, and K. J. Gordon. 1969. *Astron. Astrophys.* 2:202.

Hobbs, L. M. 1969. *Astrophys. J.* 157:135.

Hughes, M. P., A. R. Thompson, and R. S. Colvin. 1971. *Astrophys. J. Suppl.* 23:323.

Hughes, V. A., and D. Routledge. 1970. *Astron. J.* 75:1148.

Hulsbosch, A. N. M. 1968. *Bull. Astron. Inst. Neth.* 20:33.

———, and E. Raimond, 1966. *Bull. Astron. Inst. Neth.* 18:413.

Kahn, F. D. 1955. *Proc. I.A.U. Symp. No. 2.* Amsterdam: North Holland Publishing Co. p. 60.

Katgert, P. 1969. *Astron. Astrophys.* 1:54.

Kepner, M. 1970. *Astron. Astrophys.* 5:444.

Kerr, F. J. 1968. *Star and Stellar Systems.* Vol. VII. "Nebulae and Interstellar Matter." B. M. Middlehurst and L. H. Aller, eds. Chicago: University of Chicago Press, p. 575.

———, and W. J. Sullivan. 1969. *Astrophys. J.* 158:115.

Kesteven, M. J. L., and A. H. Bridle. 1970. *Astron. J.* 75:902.

Knapp, G. R. 1972. Doctoral thesis. University of Maryland.

Meng, S. Y., and J. D. Kraus. 1970. *Astron. J.* 75:535.

Menon, T. K. 1970. *Astron. Astrophys.* 5:240.

Mészáros, P. 1968. *Astrophys. Space Sci.* 2:510.

Milne, F. A. 1930. Handbuch d. Ap., Vol. 3, p. 159.

Muller, C. A., and J. H. Oort. 1951. *Nature* 168:357.

Oort, J. H. 1967. *Proc. I.A.U. Symp. No. 1*, p. 279.

———. 1970. *Astron. Astrophys.* 7:381.

Radhakrishnan, V., and J. D. Murray. 1969. *Proc. Astron. Soc. Australia* 1:215.

———, J. W. Brooks, W. M. Goss, J. D. Murray, and W. Schwarz. 1972a. *Astrophys. J. Suppl.* 24:1.

———, and W. M. Goss. 1972b. *Astrophys. J. Suppl.* 24:161.

———, J. D. Murray, P. Lockhart, and R. P. J. Whittle. 1972c. *Astrophys. J. Suppl.* 24:15.

Rickard, J. J. 1968. *Astrophys. J.* 152:1019

Riegel, K. W. 1967. *Astrophys. J.* 148:87.

———, and M. C. Jennings. 1969. *Astrophys. J.* 157:563.

Roberts, W. W. 1972. *Astrophys. J.* 173:259.

Sancisi, R., and H. van Woerden. 1970. *Astron. Astrophys.* 5:135.

Sato, F. 1968. *Publ. Astron. Soc. Japan.* 20:303.

———, K. Akabane, and F. J. Kerr. 1967. *Australian J. Phys.* 20:197.

Shuter, W. L. H., and G. L. Verschuur. 1964. *Monthly Notices Roy. Astron. Soc.* 127:387.

Spitzer, L. 1968. *Diffuse Matter in Space.* New York: Interscience.

Takakubo, K. 1967. *Bull. Astron. Inst. Neth.* 19:125.

Tolbert, C. R. 1971. *Astron. Astrophys. Suppl.* 3:349.

———, and I. Fejes. 1966. Preprint.

Verschuur, G. L. 1969a. *Astron. Astrophys.* 1:473.

———. 1969b. *Astrophys. J.* 156:771.

———. 1970. *Astrophys. Lett.* 6:215.

———. 1973a. *Astron. Astrophys.* 22:139.

———. 1973b. *Astron. Astrophys.* In press.

———, and G. R. Knapp. 1971. *Astron. J.* 76:403.

———, T. Cram, and R. Giovanelli. 1972. *Astrophys. Lett.* 11:57.

Wesselius, P. R., and R. Sancisi. 1971. *Astron. Astrophys.* 11:246.

Westerhout, G. 1966. *Maryland–Green Bank Galactic 21-cm Line Survey.* 1st ed. University of Maryland.

Williams, D. R. 1965. *Quasi-stellar Radio Source and Gravitational Collapse.* I. Robinson, A. Schild, and E. Shucking, eds. Chicago: University of Chicago Press p. 213.

CHAPTER 3

THE RADIO CHARACTERISTICS OF HII REGIONS AND THE DIFFUSE THERMAL BACKGROUND

M. A. Gordon

3.1 Ionized Interstellar Hydrogen

3.1.1 Emission Nebulae

The astronomical sky contains a large number of optically luminous objects which could not be resolved into stars by the small-aperture telescopes used before the 19th century. Historically, objects of this type were called *nebulae* (Latin for "clouds") and were the subject of considerable speculation then as they are today. Careful observation with improved telescopes separated these objects into six distinct types: *external galaxies*, the nearest of which can frequently be resolved into stars; *globular clusters* of stars in our own galaxy; *supernova remnants* radiating by means of electrons moving with speeds close to that of light but trapped in magnetic fields; *reflection nebulae*, which are concentrations of cold gas and dust reflecting the light of nearby stars; *planetary nebulae*, which are comparatively small, shell-like structures of hot gas expanding away from a central star; and *bright diffuse nebulae*, which are concentrations of hot gas, often asymmetrical, surrounding early-type stars. Because it is hotter than the surrounding medium, the gas of planetary and bright diffuse nebulae as well as supernova remnants emits spectral lines which appear in emission. These nebulae are categorically known as *emission nebulae*.

In the late 18th century many astronomers were interested in the detection and observation of comets. Frequently nebulae were inadvertently identified as comets. To prevent such mistakes, Charles Messier compiled a list of objects which might be mistaken for comets. Thus a list of the brightest "nebulae" has been available for 200 years, and their catalog number is still used to identify these objects today.

Although it is often difficult to classify emission nebulae as planetary or diffuse nebulae on the basis of their optical appearance, certain guidelines appear to hold. Planetary nebulae are distributed in the galaxy as population II stars; *i.e.*, they have little concentration toward the galactic plane and may well be connected with advanced stages of the stellar life cycle. Bright diffuse nebulae are strongly concentrated toward the galactic plane, are asymmetrical in shape to some degree, and are suspected to be connected with the birth of stars. Because of their propensity to lie in the galactic plane, diffuse nebulae are often obscured by intervening dust and absorbing matter.

For both planetary and diffuse emission nebulae, the basic energy source is the ultraviolet radiation emitted by the central star. Ultraviolet photons with wavelengths shorter than 912 Å (the longest wavelength which will ionize hydrogen) travel outward from the star until they encounter and ionize a hydrogen atom. The newly created free proton and electron can now either recombine and emit a photon, or collide with other atoms and thereby heat the gas. Unlike the original ionizing photon, the photon emitted in the process of radiative recombination can be emitted in any direction. This scattering process dilutes the original radiation field emitted by the star. Furthermore, there may or may not be a degradation of the original photon in terms of energy. In the recombination process, the probability is greater that an electron-proton pair will recombine into an excited state rather than into the ground state; thus the original photon energy is likely to split into at least two parts, each of which is unlikely to ionize additional atoms. Such a process is called *incoherent* scattering because the original photon energy is reradiated at other wavelengths.

The effects of the interaction of UV radiation and gaseous nebulae have been well studied at optical wavelengths. Optical spectra show the Balmer lines of hydrogen, principally Hα (6562 Å), Hβ (4861 Å), and Hγ (4340 Å). In addition, there are the famous lines of "nebulium," so named because of early problems of identifying singly ionized oxygen (OII) and nitrogen (NII) and doubly ionized oxygen (OIII) which emitted the lines. The most prominent of these lines are the N_1 (5007 Å), N_2 (4959 Å), and 4363 Å lines of OIII; the corresponding 6584 Å, 6548 Å, and 5755 Å lines of NII; and the 3726 Å and 3729 Å lines of OII. These lines are forbidden by the simplest of the selection rules and, because of long lifetimes, their upper states are easily depopulated by collisions, which explains why they appear in the tenuous environment of the interstellar medium and not in the higher-density environment of terrestrial light sources. The ratio of the 3729 Å and 3726 Å lines of OII is nearly independent of temperature and gives density information, while the ratio of the 4363 Å and 5007 Å lines of OIII can be used to obtain temperature information. The Balmer lines, together with the nebular lines,

can also be used to determine physical conditions of emission nebulae. Bound levels of hydrogen involve energies less than 13.6 ev, compared to the 25 to 55 ev of the nebular lines (transition energies are smaller, however). Thus the region of oxygen and nitrogen emission may not always coincide with regions of Balmer emission, and physical characteristics of emission nebulae determined from these two sets of lines may not agree.

The effects of the UV radiation in emission nebulae also include substantial continuum and line radiation in the radio wavelength range. Even at high temperatures the majority of neutral hydrogen atoms are in the ground state simply because, after recombination into an excited state, the newly bound electrons are most likely to jump quickly to the ground state. But a few electrons will jump to adjacent lower levels and emit photons in the radio spectrum which, although extremely weak, can be detected as line radiation. During the short time they are free, the electrons and protons absorb and radiate continuum radiation detectable by radio-telescopes. This radiation not only gives additional information about nearby nebulae, but permits the study of distant nebulae obscured by dust and thereby inaccessible to optical astronomy. Because the radio emission is due principally to hydrogen, we should not necessarily expect physical characteristics determined from radio information to agree with those deduced from the nebular lines.

Most of this chapter describes the radio characteristics of emission nebulae exclusive of supernova remnants. In dense regions having kinetic temperatures greater than 4000°K, virtually all hydrogen is ionized, and thus we deal with the transfer of radiation through a fully ionized plasma. Such regions are also called HII regions, a term (originally suggested by Sharpless) used to denote that, in these regions, very nearly all hydrogen (including that in the stars heating this gas) is ionized. In recent years the term HII region has come to be used to describe the ionized gas exclusive of the stars.

3.1.2 Diffuse Thermal Background

In addition to discrete concentrations of ionized gas, there are spatially extended thermal sources of interest to radio astronomers. First of all, there is the 21-cm emission of neutral hydrogen discussed in Chapter 2. There is also free-free continuum and line emission throughout the galaxy, which can be detected by dispersion of pulsar bursts, by absorption of radiation from discrete nonthermal sources, and by line emission from highly excited states of hydrogen. The characteristics of the extended emission vary considerably over the galaxy, and at this writing, are not yet well understood.

3.2 Radiation Transfer

To interpret radio observations quantitatively, it is necessary to understand just how radiation travels through a medium. As a first step, we consider loss and gain mechanisms in the direction of the observer without regard to the details of atomic processes making up these mechanisms. We also assume all processes to be stationary, *i.e.*, all parameters are time-independent over the time scale of our observations. Consider the situation sketched in Figure 3.1. The differential intensity dI contributed from an elemental volume of path length dx is

$$dI = \underbrace{-I\kappa dx}_{\text{absorption}} + \underbrace{jdx}_{\text{emission}} \quad (3.1)$$

where κ is the linear absorption coefficient, and j is the linear emission coefficient in the direction of the observer. Integrating from the far side of the nebula toward the radio-telescope, we find the observed specific intensity to be

$$I = \underbrace{I(0)e^{-\langle\kappa\rangle L}}_{\substack{\text{attenuated} \\ \text{background}}} + \underbrace{\int_0^L \frac{j}{\kappa} e^{\kappa(x-L)}\kappa dx}_{\substack{\text{contribution} \\ \text{from the} \\ \text{nebula}}} + \underbrace{I(x > L)}_{\text{foreground}} \quad (3.2)$$

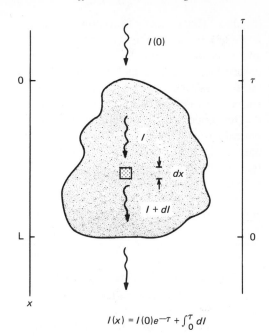

$$I(x) = I(0)e^{-\tau} + \int_0^\tau dI$$

Figure 3.1 Background radiation $I(0)$ travels through the gas cloud, where it is either strengthened or weakened, toward the radio-telescope.

where L is the near edge of the nebula. The parameter $\langle\kappa\rangle$ is the mean of the absorption coefficient over the physical length of the nebula. The radiation received by the radio-telescope is simply the sum of (1) the attenuated background radiation, (2) the contribution from the nebula, and (3) the contribution of foreground sources (if any). Equation (3.2) is perhaps the most basic equation in astrophysics because it relates the physical processes in a medium, as described by κ and j, to the radiation intensity measured by the telescope. The remainder of this chapter is concerned with the evaluation of j and κ and their effects upon the radiation field.

In practice, it is convenient to lump κ and L together by defining a parameter τ called *optical depth*.

$$\tau \equiv \int_{x_1}^{x_2} \kappa \, dx \approx \langle\kappa\rangle (x_2 - x_1) \quad (3.3)$$

In a sense the optical depth may be thought to be the probability of a photon at point x_1, reaching point x_2. Its particular appeal is that it reduces the number of unknowns in Equation (3.3) by combining $\langle \kappa \rangle$ and $(x_2 - x_1)$, and it is much more likely to be a direct observable. Note that, as Figure 3.1 shows, τ increases as x decreases. For our purpose, we will assume the near limit of the path to be the telescope itself, $x_2 = 0$, and designate optical path by the most distant point in the path, *i.e.*, $\tau = \tau(x_1)$. In this notation the transfer equation given by Equation (3.2) becomes

$$I(x) = I(0)\,e^{-\tau(0)} + \int_0^{\tau(0)} \frac{j}{\kappa}\,e^{-t}\,dt + I(x > L)$$

$$(3.4)$$

3.3 Continuum Thermal Emission (Bremsstrahlung)

The most intense radio emission from HII regions is that caused by interactions between unbound charged particles, sometimes called free-free radiation, or bremsstrahlung (braking radiation). Because the particles are free, their energy states are not quantized, and the radiation resulting from changes in their kinetic energy is continuous over the radio spectrum.

3.3.1 Absorption and Emission Coefficients

The determination of the continuous absorption coefficient, κ_c, and the corresponding emission coefficient, j_c, is not a trivial matter. The calculations involve modeling not only the interaction between two charged particles, not always of the same charge polarity, but also the distribution of the particles as a function of velocity. Classically, the encounter of two moving charged particles involves changes in their directions— either toward each other for unlike-charge encounters, or away from each other in the case of like-charge encounters. These direction changes require accelerations and, from the work of Hertz, we know that accelerating charged particles radiate. Close encounters involve substantial Coulomb forces, accelerations are large, and the radiation is apt to be

in the X-ray range. Radiation in the radio range is produced by the more distant encounters where Coulomb forces are small, and the particles can be considered to continue traveling in a straight line. In any case, the emission coefficient is determined by integrating the emission produced during each encounter, over a velocity distribution of particles (usually Maxwellian).

Some approximations are required to make this integration. The free electrons tend to shield the electric field of the ions, and thus the force field is effective only over some finite distance. Furthermore, it is usual to assume (1) that the energy radiated is small compared to the kinetic energy of the electron moving past the ion, and (2) that the reciprocal of the radiated frequency is small compared to the time for the electron to undergo a 90° deflection.

Using these assumptions, Oster (1961) derived an expression for the free-free absorption coefficient in the radio domain:

$$\kappa_v = \frac{N_e N_i}{v^2} \cdot \frac{8\,Z^2 e^6}{3\sqrt{3}\,m^3\,c} \cdot \left(\frac{\pi}{2}\right)^{1/2} \cdot \left(\frac{m}{kT}\right)^{3/2} \langle g \rangle$$

$$(3.5)$$

where N_e and N_i are the number densities of electrons and ions, v the wave frequency, Z the ion charge, e the electronic charge, m the electronic mass, c the speed of light, k Boltzmann's constant, T the kinetic temperature, and $\langle g \rangle$ the Gaunt factor averaged over a Maxwellian velocity distribution. All units are cgs. For temperatures less than 892,000°K,

$$\langle g \rangle = \frac{\sqrt{3}}{\pi} \ln \left[\left(\frac{2kT}{\gamma m}\right)^{3/2} \cdot \frac{m}{\pi\gamma Z e^2 v} \right] \quad (3.6)$$

and for temperatures greater than 892,000°K,

$$\langle g \rangle = \frac{\sqrt{3}}{\pi} \ln \left[\frac{2kT}{\pi v \hbar \gamma} \right] \quad (3.7)$$

where γ is Euler's constant.

Under some conditions of low frequencies and low temperatures, Equation (3.5) no longer holds, because the basic assumptions (1) and (2) above are violated. In this case the Gaunt factors must be evaluated for each particular case. Oster (1970) has done this for a large range of low temperatures and wave frequencies.

There is an approximation (Altenhoff *et al.*, 1960) to Equation (3.5) which is often used in the analyses of radio observations of HII regions because of its simplicity.

$$\kappa_c \approx \frac{0.08235 \, N_e N_i}{v^{2.1} T_e^{1.35}} \qquad (3.8)$$

where v is in units of GHz. For densities and temperatures encountered in HII regions this approximation is accurate to within 5%. Mezger and Henderson (1967) tabulate a correction which improves the approximation. Here κ_c is in pc^{-1}.

Having the free-free absorption coefficient, we may calculate the emission coefficient by

$$j_v = \kappa_v \, B_v \, (T_e) \qquad (3.9)$$

where $B_v(T_e)$ is the Planck function which, if $hv \ll kT_e$, can be approximated by $2kT_e v^2/c^2$, the well-known Rayleigh-Jeans approximation.

3.3.2 Transfer Equation for Continuum Radiation

The substitution of the Rayleigh-Jeans approximation for the ratio j/κ and of the continuum optical depth τ_c are all that is needed to prepare Equation (3.4) to describe the observed free-free emission of HII regions in the radio range, as a function of frequency

$$I_c(x) = I(0) \, e^{-\tau_c(0)} + \int_0^{\tau_c(0)} \frac{2kT_e v^2}{c^2} \, e^{-t} \, dt$$
$$+ \, I(x > L) \qquad (3.10)$$

where $\tau_c = \int \kappa_c dx$.

For simplicity, let us assume the nebula to be homogeneous in density and temperature and the foreground emission to be zero. Thus Equation (3.10) can be evaluated by moving the integrand out from under the integral, and

$$I_c(x) = I(0) \, e^{-\tau_c(0)} + \frac{2kT_e v^2}{c^2} \, (1 - e^{-\tau_c(0)}) \qquad (3.11)$$

should describe the intensity of free-free emission of the nebulae observed at the radio-telescope.

In practice, the radio astronomer uses units of temperature to measure the intensity of radiation. The telescope and receiver can be considered to be a resistor generating noise in proportion to its effective temperature. As a radio source moves in and out of the telescope's beam, the temperature (and thus the noise output) of this resistor rises and falls in response to changes in the radiant energy collected by the telescope. If the angular extent of the source is larger than the beam, it is often convenient to rewrite Equation (3.11) in terms of *brightness temperature*, T, by substituting the Rayleigh-Jeans approximation to the Planck function for the specific intensity. Because the telescope is never free of losses, the actual *antenna temperature*, T_A, is less than the brightness temperature by an efficiency factor, ε, and

$$I = \frac{2kT v^2}{c^2} = \frac{2kT_A v^2}{c^2} \Big/ \varepsilon \qquad (3.12)$$

Equation (3.11) becomes

$$T(x) = T(0) \, e^{-\tau_c(0)} + T_e(1 - e^{-\tau_c(0)}) \qquad (3.13)$$

which can be compared directly with observations.

3.3.3 Low-Frequency Observations of HII Regions

At low frequencies the frequency dependence of the free-free absorption coefficient given by Equation (3.8) causes τ_c to become much greater than 1 for most HII regions. Here Equation (3.13) becomes

$$T(x) \approx T(0) \, e^{-\tau_c} + T_e \qquad (3.14)$$

In the direction of the galactic plane the background emission is mainly nonthermal and, as such, is very intense at low frequencies. Moving the beam of the radio-telescope across an HII region lying in this direction causes a decrease in the observed brightness temperature $T(x)$ because of absorption by the HII region. If the optical depth is large enough, the background contribution to $T(x)$ is negligible, and one is able to measure the electron temperature of the HII region directly.

3.3.4 High-Frequency Observations of HII Regions

At high frequencies the nonthermal radiation of the galactic background becomes negligible compared to the thermal radiation of the HII region. The v^{-2} dependence makes $\tau_c \ll 1$ at high frequencies, and Equation (3.13) becomes

$$T(x) \approx T(0) + T_e \tau_c \qquad (3.15)$$

which, of course, predicts an increase in antenna temperature as the beam is scanned over the source.

3.3.5 Integrated Observations of HII Regions

It is often difficult to identify HII regions as such at large distances from the Sun. The optical obscuration caused by gas and dust lying in the plane of the galaxy can make it difficult to use the standard techniques of examining the optical emission lines of OII, OIII, and NII. To classify distant galactic radio sources as being thermal or non-thermal, it is useful to develop some classification technique which does not depend upon optical measurements or on angular dimensions of the source relative to the beam of the radio-telescope.

For the radio astronomer many sources have angular dimensions much less than those of the antenna beam, and it is usual to consider measurements of the *spectral flux density* of a source simply because, in this case, it is independent of the source size. By definition,

$$S = \int_{\text{source}} I_v \, d\Omega \approx \frac{2kv^2}{c^2} \int T_{\text{source}} \, d\Omega \quad (3.16)$$

where Ω is the solid angle subtended by the source. At frequencies where the beam is less than the source, the spectral flux density is determined by mapping the source, thereby performing the integration in Equation (3.16).

The basic transfer equation, Equation (3.11), can be rewritten in terms of S:

$$S(x) = S(0) \, e^{-\tau_c(0)} + S(1 - e^{-\tau_c(0)}) \quad (3.17)$$

Neglecting the background $S(0)$, we can investigate the spectral flux density of an HII region as a function of frequency in the radio domain. At high frequencies where the gas is optically thin,

$$S(x) \approx S\tau_c \qquad \tau_c \ll 1 \qquad (3.18)$$

But $S \propto v^2$ and $\tau_c \propto v^{-2}$, and thus at high frequencies $S(x)$ is independent of frequency. [The Gaunt factor of Equation (3.6) makes $\tau_c \propto v^{-2.1}$, and thus $S(x)$ is really a weak function of frequency.]

Figure 3.2 shows a plot of S versus v for the Orion Nebula. For frequencies less than 200 MHz, S varies as v^2, as we should expect from Equation (3.19) and the Rayleigh-Jeans law. For frequencies exceeding 2000 MHz, S varies as $v^{0.1}$, as we would expect from Equation (3.18).

The intermediate-frequency range is often known as the "turnover" range. In practice it is usual to describe a thermal spectrum in terms of its *turnover frequency*, this being the frequency at which $\tau_c = 1$. (Some authors use $\tau_c = 1.5$.) Neglecting the background contribution, we use Equation (3.17) to calculate the flux decrease at the turnover frequency to be

$$S(x)_T = S(1 - e^{-1}) = 0.632 \, S \quad (3.19)$$

3.3.6 The "Homogeneous Nebula" Approximation

The foregoing analysis of continuum radiation of an HII region assumes the ionized gas to be homogeneous in density and temperature, although not marked explicitly. The major justification for this procedure is that it is usually impossible to determine the detailed structure of an HII region. For most

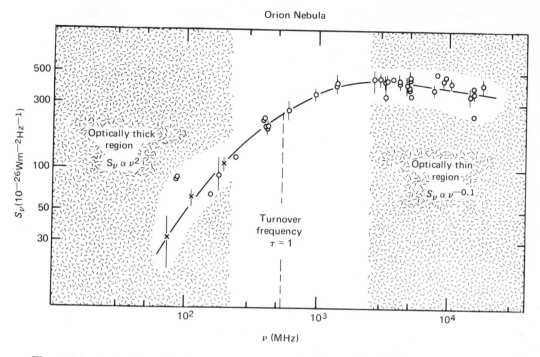

Figure 3.2 Spectral flux density of the Orion Nebula plotted against frequency. The shaded regions mark the optically thick and thin regions of the spectrum. (Terzian and Parrish, 1970. *Astrophys. Lett.* 5:261.)

radio observations a substantial fraction of the HII region is contained in the antenna beam. Even if the beam is small compared to the angular extent of the nebula, the beam includes a long column of gas through the nebula if the gas is optically thin. The physical characteristics of the gas may vary considerably (and undetectably) along this column. To interpret radio observations in a general way, it is usual to assume the nebula is homogeneous in density and temperature.

The homogeneity assumption has physical justification in some instances. If the cold hydrogen surrounding the evolving star is initially quite uniform in density, then the ionization front will expand outward in a spherical fashion until the total number of recombinations within the sphere (Strömgren sphere) bears a special relationship to the number of photons, having $\lambda < 912$ Å, leaving the star. The ionized gas within the sphere will be reasonably uniform in density and temperature in this idealized situation.

However, it seems more likely that the initial distribution of HI has large density fluctuations. For how else could the star initially condense? In this case the velocity with which the ionization front moves through the gas will vary as a function of position. The accompanying expansion of the heated newly ionized gas will be asymmetrical, and we should expect variations in density and perhaps in temperature of the HII gas over the nebula.

Invoking the homogeneity assumption, we can use the observations of the spectral flux density to determine the electron temperature and electron density of the HII regions. In essence, we are fitting a model—an isothermal HII region of constant density—to the observations. The basic technique is to measure the flux density at two different frequencies. Thus, using the theoretical form of the spectrum given by Equation (3.17), we form a system of two equations and the two unknowns T_e and N_e—providing we know the physical size of the source (Wade, 1958). Variations of this method involve procedures such as fitting the entire flux density spectrum by a least-squares fit of Equation (3.17) (cf. Terzian *et al.*, 1968) or the specification of the turnover frequency and the peak flux density. As we shall see, the "homogeneous nebula" approximation usually leads to underestimations of the electron temperature and of density.

3.4 Radio Recombination Lines

After ionization, an ion and an electron may find themselves in such close proximity that, after a suitable release of energy, they become bound together with the electron in an energy level having a large principal quantum number. Although most of the newly bound electrons immediately jump directly to the ground state, some of the electrons can cascade downward to adjacent levels, emitting a series of spectral lines known as *recombination lines*. At any instant the number of electrons in these highly excited levels is approximately 10^{-5} of those in the ground state. Levels having large principal quantum numbers differ little in energy, and line emission resulting from transitions between these levels appears in the radio spectrum. Transitions between different angular momentum levels are not important observationally because the observed lines are very broad, and thus we need only consider transitions between spectroscopic *terms*, which are completely designated by principal quantum number.

3.4.1 Line Frequencies

Because the Coulomb fields are hydrogenic and because of the large line widths, the rest frequencies of the radio recombination lines can be adequately calculated by considering changes only in principal quantum number. Therefore, the Rydberg formula is used to calculate frequencies of the lines:

$$\nu = Rc\,Z^2 \left[\frac{1}{n^2} - \frac{1}{(n + \Delta n)^2} \right] \quad (3.20)$$

For detectable recombination lines in the radio wavelength domain, $\Delta n \ll n$, and

$$\nu \approx 2\,Rc\,Z^2\,\frac{\Delta n}{n^3}\left(1 - \frac{3}{2}\frac{\Delta n}{n}\right)$$

$$\approx 2\,Rc\,Z^2\,\frac{\Delta n}{n^3} \quad (3.21)$$

where R is the Rydberg constant of the element giving rise to the line, c the speed of light, Z the electric charge of the nucleus, and Δn the change in principal quantum number n. The Rydberg constant is a function of mass.

$$R = R_\infty \left(1 - \frac{m}{M}\right) \quad (3.22)$$

where R_∞ is the constant for infinite mass, m the electronic mass, and M the total mass (including the electrons) of the atomic species involved.

As can be seen from Equation (3.20), recombination lines appear throughout the spectrum. In the radio domain where $\Delta n \ll n$, we can use Equation (3.21) to calculate the separation $\Delta \nu$ between adjacent frequencies of transitions of the same type (Δn):

$$\Delta \nu = 2Rc\,Z^2\,\Delta n \left[\frac{1}{n^3} - \frac{1}{(n + \Delta n)^3}\right]$$

$$\approx 6\,Rc\,Z^2\,\frac{\Delta n}{n^4} = \frac{3\nu}{n} \quad (3.23)$$

Because of the enormous number of detectable transitions, a convention is used to identify each transition. For example, if $\Delta n = 1$, then the transition is referred to as an α-type; $\Delta n = 2$, a β-type; etc. The transition is further identified by the atomic species and the principal quantum number of the lower level, *viz.*, H157α, He109β. The most intense transitions are those of the H$n\alpha$ type because of the large abundance of hydrogen and the large oscillator strengths of α transitions.

Table 3.1 lists the frequencies and frequency separations for α transitions. As can be seen in Equation (3.23), the relative frequency separation decreases as the fourth power of quantum number.

Table 3.1* Frequencies and Frequency Separations of α Transitions

Principal quantum number of lower term n	Line frequency (MHz)	Line wavelength	Frequency separation (MHz)
13	3×10^6	100μ	690,000
28	3×10^5	1 mm	32,000
60	3×10^4	1 cm	1,500
130	3×10^3	10 cm	69
280	3×10^2	1 m	3.2
600	3×10	10 m	0.15

* After Kardashev, 1959.

3.4.2 Shape Function

Observations of HII regions involve measurements of gas having a Maxwell-Boltzmann velocity distribution (we exclude large-scale turbulence in this discussion). In the absence of magnetic fields, an HII region having an electron temperature of $10^{4\circ}$K and an electron density of 10^2/cc will thermalize in minutes following any perturbation in the velocity distribution of the charged particles. For larger electron densities, thermalization occurs in even shorter times.

Because the gas cloud can be considered to be composed of particles having a Maxwellian velocity distribution, the number of atoms with velocity components between v_x and $v_x + dv_x$ along the line of sight is

$$dN(v_x) = N \sqrt{\frac{M}{2\pi kT}} e^{-mv_z^2/2kT} \, dv_x \quad (3.24)$$

where N is the total number of particles and M is the mass of particles of that species. Using the classical Doppler formula* for the observed frequency

$$\nu = \nu_o \left(1 - \frac{v_x}{c} \right) \quad (3.25)$$

and differentiating,

$$dv_x = -c \cdot \frac{d\nu}{\nu_o} \quad (3.26)$$

* Radio Astronomical convention only; the rigorous formula is

$$\nu = \nu_o \left[\left(1 - \frac{v}{c} \right) \Big/ \left(1 + \frac{v}{c} \right) \right]^{1/2}$$

we convert Equation (3.24) into line intensity on the assumption that the gas is optically thin—in other words, that the observed line intensity I is proportional to the number, N, of emitters and that $dI(\nu) \propto dN(v_x)$,

$$dI(\nu) = I \sqrt{\frac{4ln2}{\pi}} \cdot \frac{1}{\Delta\nu} \exp\left[-4ln2 \left(\frac{\nu_o - \nu}{\Delta\nu} \right)^2 \right] d\nu \quad (3.27)$$

where the Doppler width $\Delta\nu$ is defined to be the full width of the line at half-intensity,

$$\Delta\nu \equiv \left(4ln2 \cdot \frac{2kT}{mc^2} \right)^{1/2} \nu_o \quad (3.28)$$

Equation (3.27) tells us that the emitted line will have a Gaussian shape when the gas is optically thin. It is convenient to define a line shape function $f(\nu)$ such that

$$dI \equiv I f(\nu) \, d\nu \quad (3.29)$$

By comparison with Equation (3.27), the shape function here is

$$f(\nu) = \sqrt{\frac{4ln2}{\pi}} \cdot \frac{1}{\Delta\nu} \cdot \exp\left[-4ln2 \left(\frac{\nu_o - \nu}{\Delta\nu} \right)^2 \right] \quad (3.30)$$

Because of its definition in Equation (3.29), $\int f(\nu) \, d\nu = 1$.

In deriving the Gaussian shape function above, we have neglected two important mechanisms which can and do affect the shapes of radio recombination lines emitted from HII regions. First, HII regions are known to have large asymmetrical velocity fields, presumably arising from differential expansion of newly ionized hot gas against remnants of the original neutral cold gas. Thus small-scale (micro) turbulence—having physical dimensions considerably less than the beamwidth of the radio-telescope—convolves with the Gaussian line shapes to broaden them such that the observed Doppler width is characteristic of a kinetic temperature several times larger than the actual temperature. Typically, a Doppler width of a recombination line emitted by an HII region may be as large as 30,000°K, compared to the actual kinetic temperature of 10,000°K. Large-scale (macro) turbulence—having physical dimensions larger than the beam)—is apt to cause asymmetries in the line profiles such that they are no longer Gaussian and cannot be described by Equation (3.27).

The other mechanism affecting the line shape is that of *pressure broadening*—in this case, linear Stark effect. Such line profiles are a blend of a Gaussian core and Lorentzian wings if the line is optically thin. (Velocity and Stark broadening are discussed more fully in Sections 3.5.3 and 3.5.4.) In any case, the total power emitted in the line is an invariant; the shape function describes only its distribution in frequency-space.

3.4.3 Absorption and Emission Coefficients

As in the case of the free-free emission, the calculation of absorption and emission coefficients is the major step necessary for the prediction of line intensities as a function of physical conditions in the gas. Linear absorption coefficients are additive, and since we have calculated κ_c in Equation (3.5), we need only calculate κ_L to determine the total absorption coefficient in the frequency range containing the recombination line.

We consider two energy levels designated by principal quantum numbers n' and n. Radiation interacts with such levels in two basic ways. First, an electron in the upper state n' may spontaneously radiate and jump to the lower state n. This type of process is called *spontaneous emission*, and its probability (sec^{-1}) is designated $A_{n'n}$. Second, the electron may be induced to radiate (or to absorb) by the ambient radiation field. This type of process is called *stimulated emission* (or absorption), and its probability is designated $I B_{n'n}$ or $(I B_{nn'})$. We will consider the stimulated emission term to be negative absorption, and thus the linear absorption coefficient is

$$\kappa_L = \frac{h\nu}{4\pi} f(\nu) \left(N_n B_{nn'} - N_{n'} B_{n'n} \right) \quad (3.31)$$

Here we assume the emission and absorption line profiles to be identical. The parameter N_i is the number of atoms having electrons in level i, which are given by the Boltzmann formula

$$\frac{N_{n'}}{N_n} = \frac{\varpi_{n'}}{\varpi_n} e^{(-h\nu/kT_{ex})} \quad (3.32)$$

where T_{ex} is the excitation temperature of the levels n' and n, or in terms of the electron temperature, T_e,

$$\frac{N_{n'}}{N_n} = \frac{b_{n'}}{b_n} \frac{\varpi_{n'}}{\varpi_n} e^{(-h\nu/kT_e)} \quad (3.33)$$

Here, the depletion factor b_i is the ratio of the actual number of atoms having electrons in level i to the number which would be there if the populations were in thermodynamic equilibrium at the electron temperature T_e. The ϖ_i is the statistical weight of level i. Using the known relationships between the transition probabilities and the above equations, we transform Equation (3.31) into

$$\kappa_L = \frac{h\nu}{4\pi} \cdot f(\nu) \cdot N_n \cdot B_{nn'} \left[1 - \frac{b_{n'}}{b_n} e^{(-h\nu/kT_e)} \right]$$

$$(3.34)$$

For our applications, $h\nu \ll kT_e$, and we simplify Equation (3.34) to

$$\kappa_L \approx \kappa_L^* b_{n'} \left(1 - \frac{d\ln b_{n'}}{dn} \cdot \Delta n \frac{kT_e}{h\nu} \right) \quad (3.35)$$

where Δn is the change in principal quantum number, $d\ln b_{n'}/dn$ is the slope of the function b_n at the point $b_{n'}$, and κ_L^* is the linear absorption coefficient of the line under conditions of thermodynamic equilibrium,

$$\kappa_L{}^* = \frac{h^2\nu^2}{4\pi k T_e} f(\nu)\cdot N_n{}^* B_{nn'} \qquad h\nu \ll kT_e$$
$$\text{(3.36)}$$

where $N_n{}^*$ is the *TE* population of level n. It is convenient to rewrite the Einstein coefficient $B_{nn'}$ in terms of the more classical oscillator strength $f_{nn'}$, which is tabulated by Goldwire (1969) and by Menzel (1970),

$$\kappa_L{}^* = 1.070 \times 10^7 \, \Delta n \cdot \frac{f_{nn'}}{n} N_e N_i T_e{}^{-2\cdot 5}$$

$$\times \exp\left(\frac{E}{kT_e}\right)\cdot f(\nu) \qquad h\nu \ll kT_e \quad \text{(3.37)}$$

where we have also related the population N_n to those of the unbound states by means of the Saha-Boltzmann equation. Equation (3.37) has the population densities measured in cm^{-3}, T_e in $^\circ K$, and $\kappa_L{}^*$ in pc^{-1}, and E is the energy of the upper quantum level. For hydrogen, $E/kT_e = (1.579 \times 10^5)/n^2/T_e$.

The line-emission coefficient can be related to κ_L simply by changing the definition

$$j_L \equiv N_{n'} A_{n'n} \frac{h\nu}{4\pi}\cdot f(\nu) \qquad \text{(3.38)}$$

to

$$j_L = \kappa_L{}^* b_{n'} B_\nu (T_e) \qquad \text{(3.39)}$$

by using the physical relationship between the Einstein transition probabilities, Equation (3.33), and the relationship between the Planck function, $B_\nu (T_e)$, and the linear emission and absorption coefficients as given by Equation (3.9).

3.4.4 Transfer Equation for Line Radiation

Following procedures used in Section 3.2, we can write that the intensity at some frequency within the recombination line is the sum of that in the underlying continuum (I_c) and that from the line (I_L).

$$I = I_L + I_c$$
$$= S\left[1 - e^{-(\tau_c + \tau_L)}\right] \qquad \text{(3.40)}$$

where the source function S is given by

$$S \equiv \frac{j_c + j_L}{\kappa_c + \kappa_L} \qquad \text{(3.41)}$$

and by substitution for the line-associated coefficients,

$$S = \frac{\kappa_c + \kappa_L{}^* b_{n'}}{\kappa_c + \kappa_L{}^* b_{n'}\, \gamma} B_\nu(T_e) \qquad \text{(3.42)}$$

where

$$\gamma = 1 - \frac{d\ln b_{n'}}{dn}\cdot \Delta n\cdot \frac{kT_e}{h\nu} \qquad \text{(3.43)}$$

The line optical depth $\tau_L = \tau_L{}^* b_{n'}\, \gamma$, where the asterisk marks the value of the optical depth of the line under *TE* conditions.

Equation (3.40) provides a general description of the intensity of radiation emitted by a (homogeneous) thermal gas. At frequencies where no recombination lines can be detected, $\tau_L + \tau_c \approx \tau_c$, $S \approx B_\nu (T_e)$, and the equation describes the intensity of the free-free emission alone. If lines are emitted under *TE* conditions, γ and $b_{n'}$ are unity, $\tau_L = \tau_L{}^*$, and $S = B_\nu (T_e)$. On the other hand, if the gas is not in *TE*, then the b_n function must be determined from other calculations in order to predict the correct intensity of the radiation in the line. Such calculations are usually done assuming that any given level n is populated in statistical equilibrium, *i.e.*, the sum of all processes populating a given level is equal to the sum of all processes depopulating that level. These calculations are a function of T_e and N_e, and have been performed for a wide range of astrophysical conditions by Sejnowski and Hjellming (1969), Brocklehurst (1970), and Dupree (1972).

3.5 Radio Recombination Lines and the Physical Characteristics of HII Regions

3.5.1 Local Thermodynamic Equilibrium (LTE)

As a first approximation, radio recombination lines initially observed from HII regions were assumed to be emitted in thermodynamic equilibrium. At frequencies in the line, the intensity is

$$I = I_L + I_c = B_\nu(T_e)[1 - e^{-(\tau_c + \tau_l^*)}] \quad (3.44)$$

and the intensity contributed by the line alone is

$$I_L = I - I_c = B_\nu(T_e)e^{-\tau_c}(1 - e^{-\tau_l^*}) \quad (3.45)$$

$$\approx B(T_e)\tau_L^* \qquad \tau_L^*, \tau_c \ll 1 \quad (3.46)$$

Under the same conditions, $I_c \approx B_\nu(T_e)\tau_c$, and the ratio of the total energy emitted in the line to that emitted in the underlying continuum is

$$\int_{\text{line}} \frac{I_L d\nu}{I_c} = \int \frac{\tau_L^* d\nu}{\tau_c} \approx \int \frac{\kappa_L^* d\nu}{\kappa_c} \quad (3.47)$$

The observed parameter on the left has the advantage that beam efficiencies and dilution factors apply more or less equally to numerator and denominator and hence more or less cancel. Thus the parameter is easy to measure.

Combining Equations (3.8) and (3.37) we find an expression relating to observations to physical conditions in the gas.

$$\int \frac{I_L d\nu}{I_c} = 1.299 \times 10^5 \, \Delta n \cdot \frac{f_{nn'}}{n} \, \nu^{2\cdot1} \, T_e^{-1\cdot15} \, F^{-1}$$

$$\times \exp\left[\frac{1.579 \times 10^5}{n^2 \, T_e}\right] \quad (3.48)$$

where $d\nu$ must be measured in kHz. The factor F accounts for the fraction of the free-free emission due to interactions of He^+ with electrons:

$$F = \left(1 - \frac{N_{He}}{N_H}\right) \quad (3.49)$$

where N_{He} and N_H are the relative number densities of helium and hydrogen ions, respectively. Observations of radio recombination lines in HII regions have established that N_{He}/N_H is approximately 0.08, thereby giving a value for F of 0.92.

In Equation (3.48) the exponential term is nearly unity for lines within the radio range for all but very low temperatures. The line-to-continuum ratio varies nearly inversely with

temperature but increases with frequency; the reason is that the free-free radiation has a temperature and frequency dependence quite different from that of the line radiation.

Equation (3.48) provides an important tool in astronomy; the electron temperature of an optically obscured HII region can be calculated from simple observations—if the level populations are in thermodynamic equilibrium. All early observations of α-lines ($\Delta n = 1$) from HII regions tended to give the same values for T_e, approximately 6000°K. Therefore many astronomers concluded that 10,000°K, which had previously been assumed to be the electron temperature of HII regions, was much too large. Other astronomers—Goldberg (1966) in particular—felt that the levels involved in the observed radio lines were out of *TE* in such a way that the line intensities were enhanced over their *TE* values. Therefore, Equation (3.48) would give values of electron temperatures which were too small. Quantitatively, the observed lines were enhanced by an average of 80% owing to departures of the level populations from thermodynamic equilibrium values.

3.5.2 Departures from Local Thermodynamic Equilibrium

The concept of departures from thermodynamic equilibrium is easy to understand in the case of radio recombination lines by using a classical picture of the atom. We visualize the electron orbiting the nucleus at a radius of $r = 0.529 \, (n^2/Z)$ Å, an equation derived simply from Bohr's theory of atoms. The parameter Z is the atomic number and n is the principal quantum number. Thus a hydrogen atom having its electron in principal quantum level 110, as for the 109α line, has a radius 12,000 times larger than that of the ground state. Its "target area" would be approximately 1.4×10^8 times larger than in the ground state, from a classical point of view. Therefore the populations of these quantum levels are much more apt to be influenced by collisions than those of levels with smaller principal quantum numbers. In short, we expect the collision cross-sections to go something like n^4. Thus the populations of very large quantum levels should have excitation temperatures close to the electron temperature of the gas. Collisions should have decreasingly smaller effects as the atom "shrinks," and the b_n should become less than unity as n decreases. At some point the radiative rates should wholly control the population of the levels. Figure 3.3 shows the kinds of population (b_n) curves we should expect. If the electron density increases, the effects of collisions become more important and cause the transition region between the collision and the radiative asymptotes to shift toward lower principal quantum numbers.

Population curves are calculated from a system of equations in which all the ways out of a level n are equated with all the ways into that level:

$$N_n \sum_{n \neq m} P_{nm} = \sum_{n \neq m} N_m P_{mn} \qquad (3.50)$$

Where P_{ij} is the transition rate from i to j and N_i is the population of level i. The system of equations described by Equation (3.50) is not closed; there is always one more unknown than equations. The necessary extra equation is provided by normalizing the populations N_i to the N_i^* expected in thermodynamic equilibrium; in other words, by creating the dimensionless variable $b_i \equiv N_i/N_i^*$. The solution of Equation (3.50) is then possible and will lead to curves similar to that shown in Figure 3.3 as a function of N_e and T_e. For most situations encountered in radio recombination lines, the stimulated rate IB_{ij} is always less than the spontaneous rate A_{ij}. Also, collisions of the atoms with neutral particles are less important than those with electrons.

Knowing the general shape of the population curves, we can predict their effects upon the observed intensities of the radio recombination lines. Lines formed by transitions between large quantum numbers will lie in the collision region, and their intensities will be the LTE values. On the other extreme, lines involving small quantum numbers will lie in the radiative region, and their intensities will be weaker than the LTE values, simply because there are fewer atoms in those levels, i.e., $b_n < 1$. In the transition region the situation is more complicated because of the substantial slope of the b_n curve as shown in the lower half of Figure 3.3. Here, although both upper and lower levels are underpopulated with respect to LTE, the upper level is slightly overpopulated with respect to the lower level. Thus we should expect an enhancement of the line intensity. And yet the b_n of the upper level is less than 1, which should weaken the line intensity. The point here is that in the transition, these are two competing effects: line enhancement because of the slope of the b_n curve, and line weakening because $b_n < 1$. Only a quantitative analysis can predict whether a line formed in this region will be enhanced or weakened.

The correct form of the transfer equation for lines out of LTE is found by substituting the non-LTE source function (Equation 3.41) and line absorption coefficients (Equation 3.35) into the transfer equation (Equation 3.40). If τ_c and $\tau_L \ll 1$, as is the general case for HII regions emitting radio recombination lines in the centimeter range, the exponentials can be expanded to second-order to give the line intensity as

$$I_L \simeq I_L^* \, b_{n'} \left(1 - \frac{\tau_c}{2} \gamma \right) \qquad |\gamma| \gg 1 \quad (3.51)$$

where γ is defined by Equation (3.43). Equation (3.51) shows clearly the effects of competing line-weakening and line-enhancing processes. If the term $\tau_c \gamma$ containing the slope of the b_n curve is very much less than 0, then $I_L > I_L^*$, and the line is enhanced. On the other hand, if $\tau_c \gamma$ is nearly 0, then $I_L < I_L^*$ and the line is weakened. In this case, the free-free optical depth τ_c can be considered to be an index to the number of amplifiers along the line of sight, each having a gain of $-\gamma^{1/N}$. Substitution of numbers appropriate to HII regions into the correct transfer equation—the essence of which is given by Equation (3.51)—predicts that the lines will be enhanced over LTE values, and thus analysis by the LTE

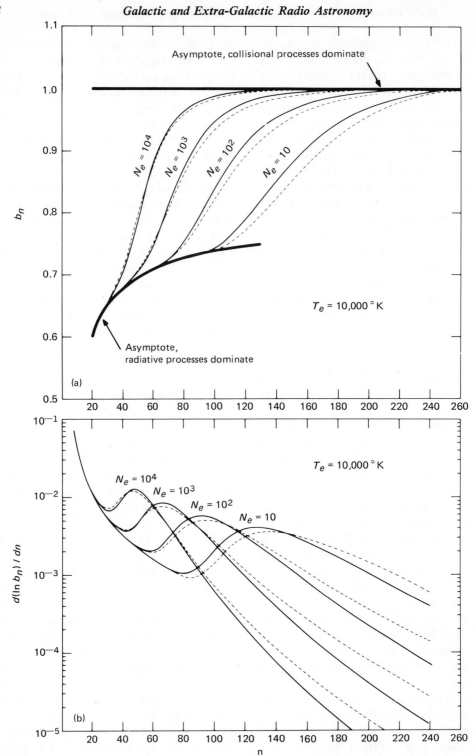

Figure 3.3 Top: The population factor b_n is plotted against principal quantum number n for hydrogen gas having electron densities N_e. Bottom: The slope function is plotted against principal quantum number, for the above curves. Dashed lines mark the results obtained with other types of collision cross sections. (Sejnowski and Hjellming, 1969. *Astrophys. J.* 156:915.)

equation (Equation 3.48) will indeed lead to an underestimate of the electron temperature.

The effects of departures from LTE can be checked by another method. Within any given frequency range of a radio receiver, there are apt to be transitions of higher-order lines: β, γ, etc. Because the beamwidth is the same for each of these lines, these lines are produced by the same volume of gas. As Equation (3.21) shows, their frequencies must be such that $\Delta n/n^3$ must be nearly a constant for that frequency range. In the $\lambda 6$-cm range we might expect to find 110α, 138β, 158γ, 173δ, 186ϵ, etc. As the order of each transition increases, so does the principal quantum number of the state. Thus the highest-order transitions are formed increasingly toward (or in) the collision region of the b_n curves where the line intensities are in LTE. From the b_n curve and Equation (3.51) we, should expect T_e to increase systematically as a function of transition order toward the actual electron temperature. Figure 3.4 illustrates expected results, and Table 3.2 shows that this is exactly what is observed.

The degree of line enhancement should vary as the beam is moved across the HII region. The underlying assumption that the gas is optically thin in both line and continuum requires the amount of gas contained within the beam to decrease considerably from the center of the HII region toward the edge because of the geometry of the nebula. The mean temperature of this gas column should vary only a little, because of the nature of the cooling mechanisms in the HII region. Because the optical depth is a function of the emission measure, τ_c should also decrease

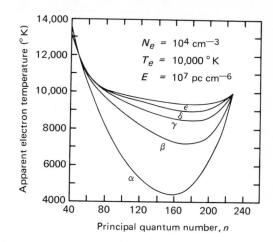

Figure 3.4 Apparent electron temperature plotted against principal quantum number for $n\alpha$ through $n\epsilon$ transitions. (Goldberg and Cesarsky, 1970. *Astrophys. Lett.* 6:93.)

from center to edge, thereby reducing the departures from LTE. Therefore, Equation (3.51) predicts the line enhancement to decrease toward the edge of an HII region.

Figure 3.5 shows a map of the parameter Q over the HII region W49A. The parameter Q is the ratio of the electron temperature calculated from the H109α line to the value calculated from the H137β line, assuming the gas to be in LTE. This ratio should be 1 if the gas is in LTE, values of Q less than unity indicating departures from LTE. Note that Q is smallest in the center and increases toward the edge, as we would expect.

Another method of determining the electron temperature is independent of the line intensities. The method, originally devised by Baade *et al.* (1933), makes use of spectral lines emitted by atoms of two different

Table 3.2 Apparent Electron Temperatures for the Orion Nebula*

Transition	Frequency (MHz)	T_e (°K)
H110α	4874.157	6990 $\pm$ 130
H138β	4897.779	9150 $\pm$ 400
H158γ	4862.780	9130 $\pm$ 900
H173δ	4909.394	10,260 $\pm$ 1000
H186ϵ	4910.858	10,430 $\pm$ 1600

* From Davies, 1971.

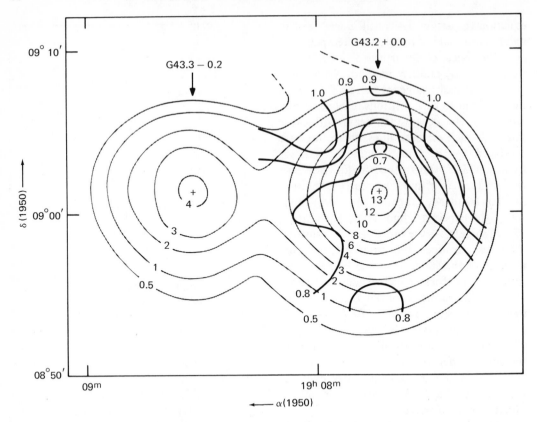

Figure 3.5 Smoothed contours of the parameter Q over the HII region W 49A. Q is the ratio of electron temperature calculated from the H109α line to that calculated from the H137β line, thereby constituting a measure of departures from LTE. (Gordon and Wallace, 1970. *Astrophys. J.* 167:235.)

masses. Since microturbulence also gives a Gaussian line shape, each line will have a width $\Delta\nu$ due to microturbulence v_t which *does not* scale with mass and to thermal broadening which *does* scale with mass.

$$\Delta\nu = \frac{2\nu}{c}\left[ln2\left(\frac{2kT}{M} + \frac{2}{3}v_t^2\right)\right]^{1/2} \quad (3.52)$$

Comparing line widths of the 85α recombination lines of hydrogen and helium, Gordon and Churchwell (1970) derived a temperature of $12,000 \pm 500°$K for the Orion Nebula. This temperature compares favorably with those calculated from the higher-order lines listed in Table 3.2, and we can be certain that non-LTE analyses give reasonably correct electron temperatures for HII regions.

Having determined the correct electron temperature for any given HII region, we can

deduce mean values of the emission measure $\int N_e^2 dx$ (from τ_c) and the electron density (from b_n) for the line-emitting regions. For the Orion Nebula these values are approximately 10^7 pc cm^{-6} and 10^4 cm^{-3}, respectively (Hjellming and Gordon, 1971). The insertion of these values into Equation (3.47) leads to an optical depth of $\tau_c = 3.28/\nu^{2.1}$, or a turnover frequency of 1.76 GHz—far larger than the 0.5-GHz value actually seen from the measurements of the spectral flux density of the nebula (Figure 3.2). The disagreement can be understood if the nebula contains substantial variations in density. The lines are weighted heavily by the high-density and hence high-emission measure regions, which become optically thick at $\nu < 1.76$ GHz. At lower frequencies the continuum emission is mainly from the lower-density gas of the nebula. (As we shall see, there is another interpretation which argues that the lines come principally from the low-density gas.)

3.5.3 Stark Broadening

The presence of charged particles in HII regions in substantial densities means that recombination lines may be broadened by means of the linear Stark effect. This interaction between the free, charged particles and the emitted wave train is usually calculated by means of one of two limiting approximations: *quasi-static* (when the wave disruption is so rapid that the particles can be assumed to be at rest during the disruption of the wave train), or *impact* (when the disruption occurs so slowly that the particle(s) move substantial distances during the disruption). For recombination lines emitted from HII regions and from the interstellar medium, only impact-broadening by free electrons is important.

Until recently, theoretical descriptions of Stark broadening always assumed adiabatic interactions between the perturbers and the emitted wave train. For recombination lines in the radio region of the spectrum this assumption is not valid, and the classical formulae cannot be used. Recombination lines involve transitions between levels degenerate in angular momentum. The particle-wave interaction can easily cause changes in the quantum numbers of the bound electron, thereby constituting nonadiabatic encounters. This kind of interaction results in cancellation of upper- and lower-state perturbations, causing the actual line-broadening to be substantially less than would be predicted by the classical formulae.

Nonadiabatic encounters not only reduce Stark effects but also make them more difficult to calculate. Nonadiabatic effects are particularly important for radio recombination lines, where the levels are separated by small amounts in energy, because the Stark effects arise in the impact regime. Here the perturber moves a long way before the emitted wave train is perturbed substantially, and the aspect angle between atom and perturber

changes considerably during the phase change in the emitted wave train. Thus there can be considerable coupling between the angular momentum of the atom and of the perturber which causes transitions between the degenerate energy levels.

The result of Stark effects is to redistribute the energy in the line over a larger frequency interval than would be expected from thermal and turbulent broadening alone. In particular, the energy is apt to be distributed into a line shape function known as a Voight profile—a blend of a Gaussian core and nearly Lorentz wings. One method of characterizing the amount of Stark broadening in a spectral line is to find the ratio of the full widths at half-maximum of the Lorentz portion of the profile $\Delta \nu^J$ to that of the Gaussian portion $\Delta \nu^D$ for α-type lines (Brocklehurst and Seaton, 1972).

$$\frac{\Delta \nu^J}{\Delta \nu^D} = 0.14 \left(\frac{n}{100}\right)^{7.4} \left(\frac{10^4}{T_e}\right)^{0.1}$$

$$\times \left(\frac{N_e}{10^4}\right) \left(\frac{M}{M_H} \times \frac{2 \times 10^4}{T_D}\right)^{1/2} \quad (3.53)$$

where n is the principal quantum number, T_e is the electron temperature in $°K$, T_D is the kinetic temperature in $°K$ required to account for the width of the Gaussian component (including the effects of electron temperature and microturbulence), M is the mass of the species, M_H is the mass of hydrogen, and N_e is the electron density in cm^{-3}. Using typical values found for the Orion Nebula of $T_e = 10^4 °K$, $N_e = 10^4$, and $T_D = 2 \times 10^4 °K$, we find that the ratio $\Delta \nu^J / \Delta \nu^D \approx 4$ for the H 157α line (λ18 cm). Such broadening is not observed!

The disagreement between the observed and the predicted Voight profiles for HII regions suggests that either the Stark broadening theory or the non-LTE analysis of HII regions is incorrect. However, Brocklehurst and Seaton (1972) pointed out that if the electron density of an HII region decreased outward, a recombination line seen by an observer would have a line profile heavily weighted by that of the tenuous regions closest to the observer. In essence, they suggest that the dense high-emission measure regions predicted by the non-LTE analysis and subsequently observed by ratio interferometers must be concentrated toward the core of the HII regions. Subsequent to the approach of Hjellming and Gordon (1971), Brocklehurst and Seaton show that amplification is greatest in the tenuous gas of the HII regions. Since line enhancement in HII regions is something like a maser process, the energy input at the core will last undergo amplification in the least dense regions of the HII region, and the

shapes of the observed lines will be weighted heavily by the line absorption coefficients in the outer regions of the HII regions, which are Gaussian. Thus, while Lorentz wings will be present, they will be much less intense than if the line radiation was generated uniformly from only the dense clumps of ionized gas, and might well escape detection.

3.5.4 Velocity Fields

Perhaps the most direct result of observations of radio recombination lines is the radial

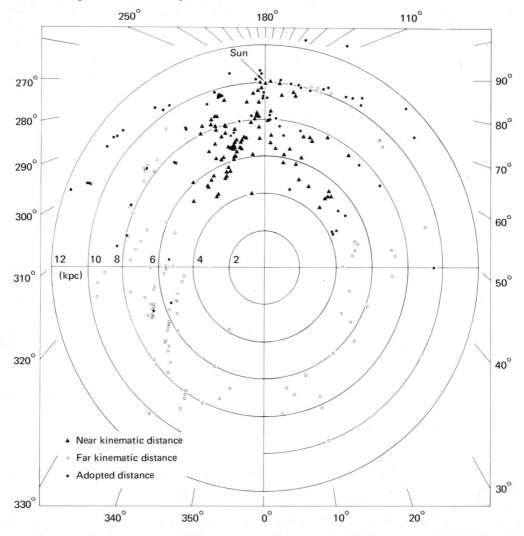

Figure 3.6 The spatial distribution of HII regions as determined by the velocities of radio recombination lines and the Schmidt model of galactic rotation. (Wilson *et al.*, 1970. *Astron. Astrophys.* 6:364.)

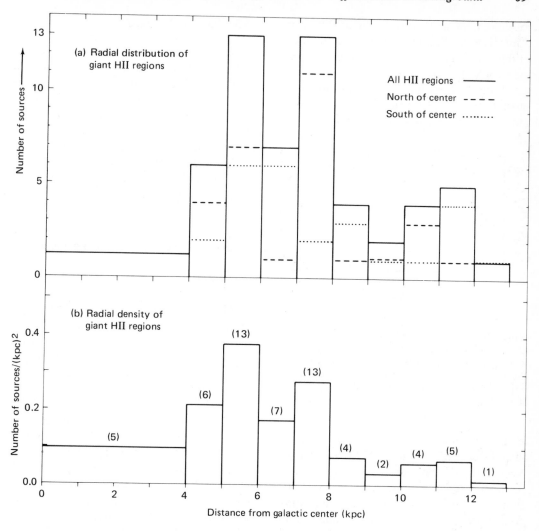

Figure 3.7 The radial distribution of giant HII regions in the galaxy. (Mezger, 1970. *I.A.U. Symposium* 38:107.)

velocity indicated by the line centers. Analyses of the radial velocities of all HII regions identifiable on existing maps of the galactic plane seem to show that velocities of HII regions agree well with velocities of the most intense neutral hydrogen radio emission, with some unexpected exceptions (Reifenstein *et al.*, 1970; Wilson *et al.*, 1970). First of all, as can be seen in Figure 3.6, there are few HII regions within 4 kpc of the galactic center, if the Schmidt rotation model is used to translate velocity into galactic radius. Outside 6 kpc the number of sources drops consider-

ably. If one selects only the giant HII regions, one finds them to be even more markedly concentrated within this zone, as shown in Figure 3.7 (Mezger, 1970). The HII regions do not appear to form distinct spiral patterns, but such spiral patterns are not usually observed for HII regions in external galaxies either. Figure 3.6 also suggests a population asymmetry between northern and southern hemispheres of the galaxy, but this may also be an instrumental effect.

The microturbulence contributing to line broadening seems to be a function of angular resolution, as can be seen in Figure 3.8, from Sorochenko and Berulis (1969). Extrapolation of this curve agrees well with optical measurements of higher resolution.

With high-resolution instruments it is possible to investigate the variation of peak radial velocity across HII regions of large

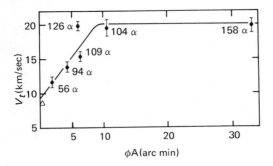

Figure 3.8 The turbulence velocity determined from radio recombination lines of the Orion Nebula plotted against the beamwidth used for the observations (Sorochenko and Berulis, 1969, *Astrophys. Lett.* 4:173). The optical point is taken from Smith and Weedman (1970), *Astrophys. J.* 160:65.

angular extent. Figure 3.9 shows the results of such investigations for IC1795, the Orion Nebula, the thermal part of W49, and Orion B. In all cases there appears to be a velocity gradient across the nebulae, suggesting rotation. However, this is more likely due to differential expansion of the hot gas against the inhomogeneous, surrounding HI gas. The virtual "rotation periods" are in the order of 10^6 years.

Having determined the mean electron temperature for a particular line of sight through an HII region, one can separate the thermal and velocity components of line broadening in the predominantly Gaussian lines. Figure 3.10 shows the distribution of velocity broadening over the HII region Orion B (Gordon, 1969). The decrease in the velocity broadening from center to edge can be explained by a radial expansion or contraction of the HII region at a velocity of 6.5 km/sec. Alternatively, the distribution suggests a decrease in turbulence from center to edge.

3.5.5 Planetary Nebulae

Radio recombination lines have recently been detected from planetary nebulae (Terzian and Balick, 1969). Electron temperatures computed under the assumption of LTE range from 6600 to 17,300°K. The lines are weak because of the small angular sizes of planetary nebulae and the correspondingly large beam-dilution factors, and it will be difficult to detect the higher-order transitions necessary to investigate departures from LTE.

3.6 Diffuse Thermal Background

3.6.1 The Theoretical Picture

Our galaxy contains a great deal of gas not in the form of stars or other discrete objects. The presence of this "interstellar medium" (ISM) was first suggested by optical absorption lines of interstellar calcium and sodium detected in stellar spectra in the early part of this century. Later, the 21-cm radio line of neutral hydrogen made it possible to explore the characteristics of the hydrogen component of the ISM, as described in Chapters 2 and 4. More recently, the absorption of nonthermal radio sources, the frequency dispersion of pulsar signals, and the detection of radio recombination lines have added greatly to our knowledge of the ISM.

Hot stars and supernovae can radiate energetic photons in the UV and X-ray regions of the spectrum, and low-energy cosmic rays which heat the ISM and maintain some degree of ionization. If the heating mechanism is uniformly distributed (Field *et al.*, 1969), the interstellar gas might have a temperature-versus-density curve like that shown in Figure 3.11. At low densities the gas remains hot because there are few mechanisms available to radiate energy away. At high densities, however, collisions are much more frequent and make it possible to transmit kinetic energy into the excitation of levels of C^+, Si^+, F_e^+, and O^+, from which energy is lost via infrared radiation. Thus the high-

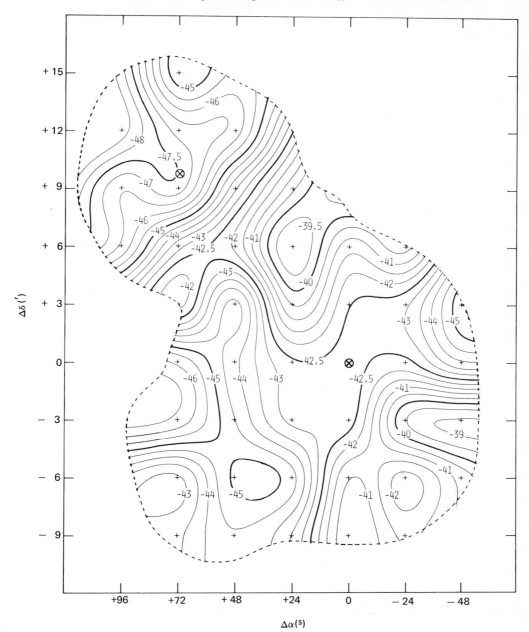

Figure 3.9 (A) Radial velocities (in km/sec) of hydrogen gas observed in IC1795 (W3). (Rubin and Mezger, 1970, *Astron. Astrophys.* 5:407.)

density gas tends to cool quickly and will have a low temperature.

The temperature-density curve can be transformed into the plot of pressure-density seen in Figure 3.12. The shape of the curve indicates that there are, in general, three values of density which have the same pres-

sure, and thus gas clouds having any of these densities could exist in pressure equilibrium with one another. Gas in region F is tenuous and has a kinetic temperature of $10^{4\circ}$K; gas at point G is less tenuous and 5000°K; gas at point H is dense and $10^{2\circ}$K. The point G region is thermally unstable. If a parcel of gas

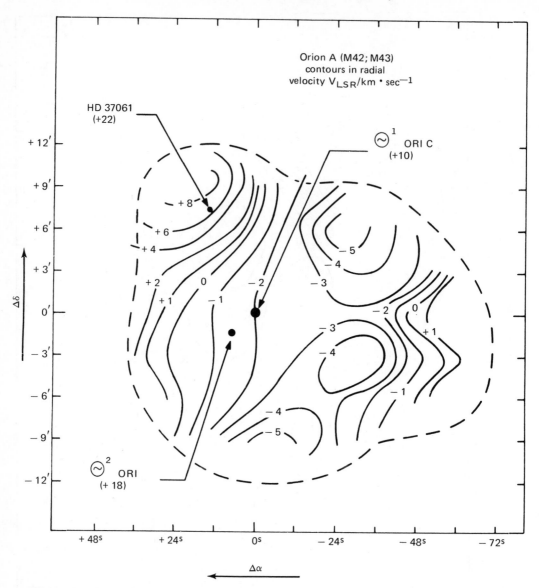

Figure 3.9 (B) Radial velocities (in km/sec) of hydrogen gas observed in the Orion Nebula. (Mezger and Ellis, 1968. *Astrophys Lett.* 1:159.)

is compressed slightly, its pressure will decrease because of the temperature drop caused by an increase in cooling efficiency, and it will move along the curve toward point H. Time scales for this are estimated to be ~10⁶ years. Consequently, after a long time we would expect the ISM to consist of cold dense clouds (point H) in pressure equilibrium with hot tenuous gas (point F).

Observations of 21-cm radiation of

hydrogen with high-angular resolution do indeed suggest such a two-phase medium. However, no one has yet detected temperatures greater than 6000°K for the neutral hydrogen. If the time scale required to bring gas of intermediate densities (point G) into the stable regions is a great deal longer than 10⁶ years, then we must expect a large amount of gas in the process of moving toward a stable point and to have temperatures

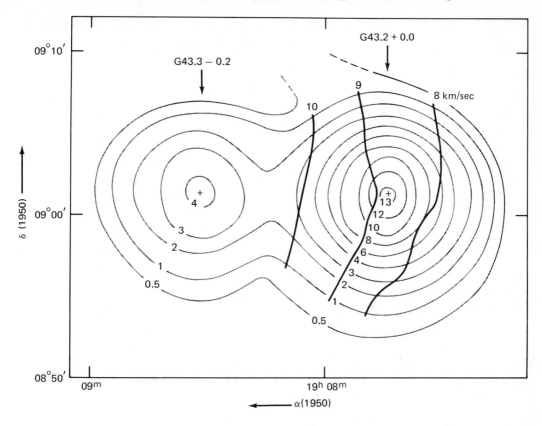

Figure 3.9 (C) Radial velocities (in km/sec) of hydrogen gas observed in W49. (Gordon and Wallace, 1970. *Astrophys. J.* 167:235.)

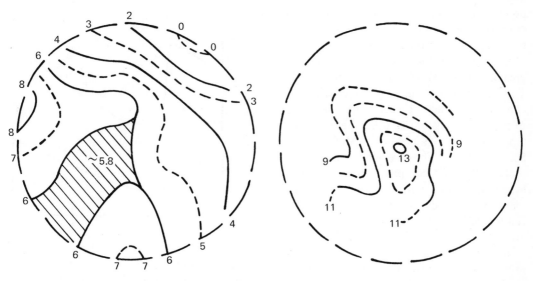

Figure 3.9 (D) Radial velocities (in km/sec) of hydrogen gas observed in Orion B. (Gordon, 1969. *Astrophys J.* 158: 479.)

Figure 3.10 A map of velocity broadening in the H94α lines from Orion B. Contour intervals in km/sec. (Gordon, 1969. *Astrophys. J.* 158:479.)

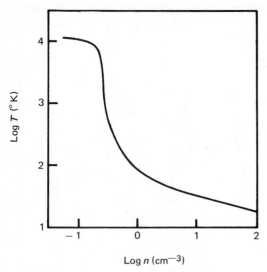

Figure 3.11 Kinetic temperature as a function of number density for neutral regions heated by 100-MeV cosmic rays. (Field *et al.*, 1969. *Astrophys. J.* 155:L144.)

between those of the stable regions, 10^2 and $10^4°$K.

It is possible that the heating mechanism is not uniformly distributed over the ISM. Because those agents which are most effective in ionizing the ISM necessarily have a limited travel range, a number of people feel that the ionization is apt to be more of a stochastic process, both in time and position. For example, the UV and X-ray emission flash from supernovae could create dense regions of ionization having temperatures initially as great as $10^5°$K (Morrison and Sartori, 1969). The newly ionized gas would expand and compress surrounding neutral gas, then cool to approximately 100°K while maintaining a large fractional ionization (McCray and Schwarz, 1971). Thus we would also expect more or less of a two-phase ISM from these events but with a larger range of permitted temperatures and fractional ionization. Recently, Hills (1972) suggested that stars evolving from red giants to white dwarfs would emit substantial UV radiation. This radiation could constitute another important stochastic source of energy for the interstellar medium.

Whatever the details, theory predicts a wide range of temperature and density for the interstellar medium—which in fact is observed.

3.6.2 Pulsar Dispersion

An important characteristic of pulsars (see Chapter 6) is the variation of pulse arrival time as a function of frequency. This behavior results from refraction in the ISM between the pulsar and Earth. A measurement of this dispersion determines the integral of electron density over the path, $\int N_e \, dl$. By observing the 21-cm line absorption of the pulse, we can determine the velocities of the intervening absorption clouds, thereby the position of the pulsar in the galaxy from the rotation law, and hence its distance L from Earth. The depth of the absorption line determines the column density of neutral hydrogen along the same path $\int N_H \, dl$. If the gas ionization is in statistical equilibrium, then the total ionization rate ζN_H is equal to the total recombination rate $\alpha N_e N_i$, and the effective ionization coefficient is

$$\zeta = \frac{\alpha N_e N_i}{N_H} \qquad (3.54)$$

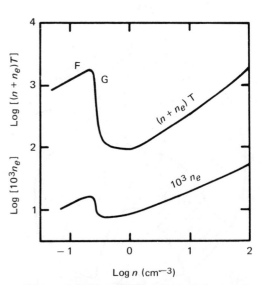

Figure 3.12 Pressure (without Boltzmann's constant) plotted against number density of neutrals. Upper curve is total pressure, bottom curve is partial pressure of electrons. (Field *et al.*, 1969. *Astrophys. J.* 155:L144.)

Assuming that the ion (proton) density $N_i = N_e$, we can use the pulsar observations to solve for ζ as a function of temperature because α is an atomic parameter dependent upon temperature. Calculations show that $\zeta \approx 10^{-15}$ sec^{-1} for pulsars observed within 2 kpc of the Sun (Hjellming *et al.*, 1969). Such an ionization rate is about what would be expected from extrapolation of the observed cosmic ray spectrum down to energies where they could cause substantial ionization of the interstellar medium.

3.6.3 Turnover Frequencies of Supernova Remnants

The effective ionization rate for the ISM can be obtained in another way. The spectral flux density of supernova remnants (SNR) is usually an uncomplicated function of frequency (see Chapter 5). Toward low frequencies, S increases logarithmically with frequency and then suddenly decreases. This turnover frequency can be due to suppression of synchrotron radiation (Razin effect), or to synchrotron self-absorption within the SNR itself, or to free-free absorption between source and observer. If it is free-free absorption, then the measurement of the turnover frequency determines the integral

$$\int N_e N_i T_e^{-1.35} \, dl$$

over the path (see Equation 3.8). Corresponding measurements of the absorption line at 21-cm by means of a spectral-line interferometer determine $\int N_H \, dl$ and L. These measurements can be used in Equation (3.54) to determine the effective ionization coefficient ζ and, in fact, lead to a value of 10^{-15} sec^{-1} in excellent agreement with that obtained from pulsar observations. The fact that the two values of ζ agree implies that the ionization of the medium is fairly homogeneous because the value of N_e determined from the first moment of electron density over the path, $\int N_e \, dl$, apparently agrees with that obtained from the second moment, $\int N_e N_i \, dl$, over the same path—at least in these directions which

more or less avoid the central regions of the galactic plane.

3.6.4 Radio Recombination Lines

It is possible to detect recombination line emission of hydrogen from the diffuse interstellar medium by two different techniques. The first makes use of departures from LTE to amplify the line brightness. In theory, line emission from the ISM in the directions of bright background sources should result in line enhancement, owing to emission stimulated by the sources; thus, the increased intensity should make the lines easier to detect. Indeed, such lines have been found against bright background sources which have velocities similar to those of HI, and hence are characteristic of the ISM rather than of the background source (Ball *et al.*, 1970). However, to date, such lines have been seen only against HII regions, and then, only as asymmetries in the line profiles emitted by the HII region. No lines have been detected against nonthermal objects such as SNR. It may well be that these lines are emitted by HI shells immediately surrounding HII regions rather than from the distributed ISM between source and observer. At the moment, more research is clearly needed to establish the origin of these lines.

A second way to detect hydrogen recombination lines from the ISM is to search in directions which avoid discrete sources (Gottesman and Gordon, 1970). Such lines cannot avail themselves of enhancement effects, and their intensities are weak and therefore difficult to observe. The integral of brightness temperature over the line is $\int T_L dl \approx \kappa_L T_e dl$ if the line is optically thin and may be calculated from Equation (3.37) to be

$$\int T_L dl = 1.070 \times 10^4 \cdot \Delta n \frac{f_{nn'}}{n}$$

$$\times \int N_{HII} N_e \cdot b_n \cdot T_e^{-1.5}$$

$$\times \exp\left(\frac{1.579 \times 10^5}{n^2 T_e}\right) dl \qquad \tau_L, \tau_c \ll 1$$

$$(3.55)$$

where the brightness temperature T_L of the line excess to the continuum background is in °K, ν is in kHz, the densities N_i are in cm^{-3}, and the path length l is in pc.

In Equation (3.55) the basic information contained in these recombination lines lies within the integral. For most circumstances in the ISM, the exponential is close to unity. The path length l is determined by the velocity extent of the line profile and the rotation curve of the galaxy. (Inside the solar circle there is a distance ambiguity; see Chapter 4.) Thus the measurements establish the quantity $\langle N_{\mathrm{HII}} N_e \cdot b_n \cdot T_e^{-1.5} \rangle$ averaged over the path.

The $\langle b_n \rangle$ will range between 10^{-2} and 0.5, depending upon the temperature and density of regions along the path.

Unlike the case of HII regions, it is not possible to uniquely separate density from temperature. In the galactic plane the background continuum is a mixture of nonthermal and thermal emission in uncertain proportions; one cannot use the free-free continuum emission to separate $\langle N_e \rangle$ and $\langle T_e \rangle$. Even if one could determine the free-free emission accurately, the difference in temperature dependence (see Equations 3.8 and 3.37) weights the free-free emission along the path

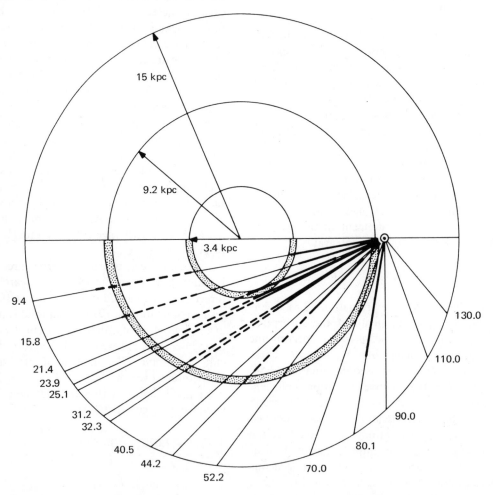

Figure 3.13 Lines of sight in the galactic plane which have been searched for recombination lines from diffuse material. Accented parts (solid and dashed) mark possible regions where the observed lines could be generated; the dashed part marks regions of position ambiguity. (Gordon and Cato, 1972. *Astrophys. J.* 176:587.)

differently than the line emission. Because the ISM has two components, direct comparison of observed free-free and line emission may involve comparison of slightly different regions and accordingly give erroneous results.

A survey (Gordon and Cato, 1972) of recombination lines of hydrogen from regions of the galactic plane believed to be free of discrete radio sources reveals the lines to come from a zone bounded by galactic radii 3 and 9 kpc—more or less the zone containing the giant HII regions, but not the Sun. The distribution of this material may be patchy (Davies *et al.*, 1972). Figure 3.13 shows lines of sight observed and the inferred position of the line-emitting zone. Within the zone the effective ionization rates ζ (Equation 3.54) may be as large as 10^{-14} to 10^{-13} sec if the gas is in statistical equilibrium. That these rates are 1 to 2 orders of magnitude larger than those measured from pulsars and from some SNR shows that conditions in this zone are considerably different from those outside the zone.

Because of similarity between integrals in expressions for $\kappa_L{}^*$ and κ_c, we can use the line observations (Gordon, 1972) to predict turnover frequencies ν_T for SNR lying on the far side of the zone,

$$\nu_T{}^{2.1} \approx 7.969 \times 10^{-6} \frac{n}{\Delta n \cdot f_{nn'}} \langle T_e \rangle^{0.15} \int T_L d\nu$$

$$(3.56)$$

where ν_T is measured in GHz when T_e is in °K, and $\int T_L d\nu$ is in °K·kHz. Here we assume that b_n is unity. The factor $\langle T_e \rangle^{0.15}$ accounts for the small difference in temperature dependence in expressions for κ_L and κ_c. Substitution of appropriate parameters into Equation (3.56) gives values of ν_T ranging from 30 to 100 MHz, depending upon whether T_e is taken to be 10^2 or 10^4°K, respectively. Departures from LTE would be greatest at 10^2°K and, if considered, would raise the 30-MHz value toward the 100-MHz value. Recent observations (Dulk and Slee, 1972) of distant SNR remnants lying in, or at the far edge of, the zone suggest values of $\nu_T \approx 80 \rightarrow$

90 MHz, in excellent agreement with the frequencies predicted from the line observations. Unfortunately, the good agreement between the SNR spectra and the recombination lines does not help to pin down the effective temperature and density of the intervening gas.

3.7 The Anomalous Recombination Lines

3.7.1 Carbon

Shortly after the discovery of radio recombination lines from HII regions, observations showed that the lines from helium often were asymmetrical, with a narrow feature visible on the high-frequency side of the line profile. Figure 3.14 shows observations (Churchwell, 1970) of the feature in the helium lines of the Orion Nebula at λ3, 6, and 18 cm. On the basis of abundance and line frequencies, carbon is the most likely candidate for this feature. Its higher mass would cause its recombination spectra to shift to shorter wavelengths, as shown by Equations (3.21) and (3.22).

Subsequent observations of this line show it to be narrow and to have radial velocities unlike those of the HII regions. The effective Doppler temperature of the line width is less than 10^3°K, suggesting that the lines are formed in the ISM rather than in the 10^4°K kinetic field of the HII region.

Dupree and Goldberg (1969) have suggested that the line is enhanced over its LTE level by emission stimulated by the intense background radiation of the HII regions. If so, then these lines should also be seen in the direction of other intense sources such as SNR. They have not been detected under such circumstances, the negative results implying that "carbon" is associated peculiarly with HII regions. However the available data are scanty, and additional observations are required.

The intensity of the line varies with frequency in a different manner from those of the hydrogen and helium lines (Churchwell, 1970). Toward low frequencies (larger princi-

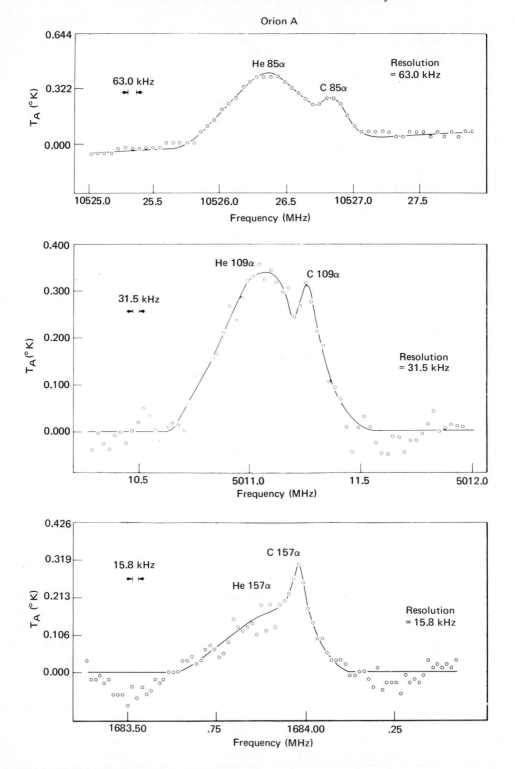

Figure 3.14 Blended helium-carbon line profiles at λ3, 6, and 18-cm in the Orion Nebula. (Churchwell, 1970.)

pal quantum numbers) the intensity of the carbon line increases relative to that of the helium line, which is known to be emitted by the HII region. The same behavior occurs with increasing transition order; that is, the ratio of the intensities of α to β lines at the same frequency is smaller for "carbon" lines than for hydrogen or helium lines (see Dupree, 1971), again showing that the carbon line varies with increasing quantum number in a different manner than helium or hydrogen lines. At this writing it is not clear whether such behavior results from departures from LTE or from optical depth effects due to actual variations in rest frequencies. Additional observations will be required to explore whether these elements have actually been detected.

For the Orion Nebula and IC 1795 (W 3) the angular extent of the carbon-line emission has been explored (Churchwell, 1970) relative to the emission of the background HII regions. In Orion the carbon emission appears to be most intense in the vicinity of the dust absorption covering the eastern side of the nebula. In IC 1795 the maximum of the carbon emission is in the center of the HII region.

3.7.2 Other Lines

As sensitivity continues to improve, recombination lines of elements of atomic numbers greater than carbon will be detected. As of this writing, some workers (Chaisson *et al.*, 1972) have taken spectra at $\lambda 18$ cm of Orion B and IC 1795, which show a feature to the high-frequency side of the "carbon" line. If the gas emitting this feature has the same radial velocity as the carbon line, then it could be due to ^{24}Mg, ^{28}Si, ^{32}S, or ^{56}Fe, or some combination of these. These atoms have ionization potentials less than that of carbon, and thus they could be expected to give rise to recombination lines in regions where the carbon line appears. It is always difficult to separate frequency shifts due to velocities from those due to mass, but when the masses are large, the lines will tend to blend together.

3.8 Measurements of the Relative Abundance of Helium and Hydrogen

3.8.1 Cosmological Significance

Observations of the ratio of helium to hydrogen might provide a method for investigating the ways and places in which the hierarchy of elements was formed. If the major production of helium results from hydrogen burning in the core of stars, it is difficult to see how the helium can be released into the ISM. Although supermassive stars can explode and thereby release elements to the ISM, most main sequence stars will continue into higher stages of fusion in which the newly formed helium will be converted into other elements. Diffusion of large amounts of core material into stellar atmospheres seems unlikely. Also spallation processes in the stellar atmospheres seem unlikely to result in substantial abundances of He/H in the interstellar medium. If helium is formed by nucleosynthesis in stars, then one would expect the ratio of helium to hydrogen to be no greater than 0.01 by number (Hoyle and Tayler, 1964).

On the other hand, the concept of helium formation in supermassive objects or in the initial stages of the formation of the universe seems to permit a wider range of N_{He}/N_H. Gravitational collapse of supermassive objects $(M < 10^3 \, M_\odot)$, which would later explode and disperse, would lead to core temperatures of $10^{10}\,°$K—two orders of magnitude greater than those permitted in normal stars. Such temperatures also occur in initial stages of the "big bang" formation of the universe. At these temperatures nuclear reactions proceed quickly, and the elemental hierarchy is produced on time scales of 10 to 10^3 sec. The principal advantage of high-temperature synthesis is the speed with which nucleosynthesis proceeds; the elemental abundances are more or less already formed before stellar evolution begins, and more or less uniform relative abundances should be observed throughout the universe.

Thus the measurement of helium to hydrogen abundances may restrict theories of the formation of the elements. For example, if the ratio is found to be substantially greater than 0.01 by number, it is unlikely that helium is formed by nucleosynthesis in the cases of main sequence stars. Furthermore, if the ratio is observed to vary widely, then formation processes must be localized and the "big bang" mechanism may be ruled out.

3.8.2 Observations

Radio recombination lines of helium have been detected in the radiation from HII regions. Comparison of the ratios of α to β lines shows helium to have the same departures from LTE as hydrogen. For nebulae excited by stars of spectral type 07 or earlier, the Strömgren spheres of HII and HeII should very nearly coincide. Therefore, the comparison of the relative intensities of helium and hydrogen radio recombination lines from emission nebulae should give estimates of N_{He}^{+}/N_{H}^{+} unaffected by dust or other obscuration. Because $N_{He}^{++} \ll N_{He}^{+}$, the observed ratio should yield accurate values of their relative abundances by number, because in most HII regions, hydrogen and helium are ionized in proportion to their abundance. Thus a measurement of N_{He}^{+}/N_{H}^{+} is that of N_{He}/N_{H}.

The most recent theory concerning the radiation transfer of radio recombination lines in HII regions shows line enhancement to be greatest in the outer (low-density) regions of the nebulae. Batchelor and Brocklehurst (1972) note that most HII regions have considerable small-scale structure and, because of the enhancement effect, that helium and hydrogen lines can be enhanced by different amounts depending upon the angular size of the radio beam and the relative sizes of and density variations within the hydrogen and helium Strömgren spheres. They predict variations in N_{He}/N_{H} deduced from observations of radio recombination lines to deviate as much as $\pm 10\%$ from the actual value. Therefore, the observer must investigate the departures from LTE for both helium and hydrogen lines when using them to determine

their relative abundances in HII regions.

There are some observational problems. Frequently, the helium recombination line is blended with that of "carbon" such that high-precision measurements are required to separate the blends. Furthermore, there are apparent emission nebulae, like NGC 2024 (Orion B), which are excited by stars of spectral type later than 07. Nevertheless, careful measurements in selected nebulae should give good results.

The most recent observations of helium and hydrogen recombination lines suggest N_{He}/N_{H} to be approximately 0.07 (Gordon and Churchwell, 1970). This value is somewhat less than previous radio determinations (Palmer *et al.*, 1967) because improved measurements show the helium line to be blended with lines from other elements. Unfortunately, this value is consistent with all theories of helium formation described in the previous section!

References

Altenhoff, W., P. G. Mezger, H. Wendker, and G. Westerhout. 1960. *Veröff. Sternwärte*, Bonn, No. 59, p. 48.

Baade, W., F. Goos, P. P. Koch, and R. Minkowski. 1933. *Z. Astrophys.* 6:355.

Ball, J. A., D. Cesarsky, A. K. Dupree, L. Goldberg, and A. E. Lilley. 1970. *Astrophys. J.* 162:L25.

Batchelor, A. S. J., and M. Brocklehurst. 1972. *Astrophys. Lett.* 11:129.

Brocklehurst, M. 1970. *Monthly Notices Roy. Astron. Soc.* 148:417.

———, and M. J. Seaton. 1972. *Monthly Notices Roy. Astron. Soc.* 157:179.

Chaisson, E. J., J. H. Black, A. K. Dupree, and D. A. Cesarsky. 1972. *Astrophys. J.* 173: L131.

Churchwell, E. 1970. Unpublished doctoral dissertation. Indiana University.

Davies, R. D. 1971. *Astrophys. J.* 163:479.

———, H. E. Matthews, and A. Pedlar. 1972. *Nature Phys. Sci.* 238:101.

Dulk, G. A., and O. B. Slee. 1972. *Australian J. Phys.* 25:429.

Dupree, A. K. 1971. *Astrophys. J.* 170:L119.

———. 1972. *Astrophys. J.* 173:293.

————, and L. Goldberg. 1969. *Astrophys. J.* 158:L49.

Field, G. B., D. W. Goldsmith, and H. J. Habing. 1969. *Astrophys. J.* 155:L144.

Goldberg, L. 1966. *Astrophys. J.* 144:1225.

————, and D. A. Cesarsky. 1970. *Astrophys. Lett.* 6:93.

Goldwire, H. C., Jr. 1969. *Astrophys. J. Suppl.*, No. 152, 17:445.

Gordon, M. A. 1969. *Astrophys. J.* 158:479.

————. 1972. *Astrophys. J.* 174:361.

————, and E. Churchwell. 1970. *Astron. Astrophys.* 9:307.

————, and D. C. Wallace. 1970. *Astrophys. J.* 167:235.

————, and T. Cato. 1972. *Astrophys. J.* 176:587.

Gottesman, S. T., and M. A. Gordon. 1970. *Astrophys. J.* 162:L93.

Hills, J. G. 1972. *Astron. Astrophys.* 17:155.

Hjellming, R. M., C. P. Gordon, and K. J. Gordon. 1969. *Astron. Astrophys.* 2:202.

————, and M. A. Gordon. 1971. *Astrophys. J.* 164:47.

Hoyle, F., and R. J. Tayler. 1964. *Nature* 203:1108.

Kardashev, N. S. 1959. *Soviet Astron.—AJ* 3:813.

McCray, R. A., and J. Schwarz. 1971. In *Gum Nebula Symposium* (Greenbelt, Maryland).

Menzel, D. H. 1970. *Astrophys. J. Suppl.*, No. 161, 18:221.

Mezger, P. G. 1970. "The Spiral Structure of Our Galaxy," *I.A.U. Symposium No. 38.* Dordrecht:Reidel, p. 107.

————, and A. P. Henderson. 1967. *Astrophys. J.* 147:471.

————, and S. A. Ellis. 1968. *Astrophys. Lett.* 1:159.

Morrison, P., and L. Sartori. 1969. *Astrophys. J.* 158:541.

Oster, L. 1961. *Rev. Mod. Phys.* 33:525.

————. 1970. *Astron. Astrophys.* 9:318.

Palmer, P. 1968. Unpublished doctoral dissertation. Harvard University.

————, B. M. Zuckerman, H. Penfield, A. E. Lilley, and P. G. Mezger. 1967. *Nature* 215:40.

Reifenstein, E. C., III, T. L. Wilson, B. F. Burke, P. G. Mezger, and W. J. Altenhoff. 1970. *Astron. Astrophys.* 4:357.

Rubin, R. H., and P. G. Mezger. 1970. *Astron. Astrophys.* 5:407.

Sejnowski, T. J., and R. M. Hjellming. 1969. *Astrophys. J.* 156:915.

Smith, M. G., and D. W. Weedman. 1970. *Astrophys. J.* 160:65.

Sorochenko, R. L., and J. J. Berulis. 1969. *Astrophys. Lett.* 4:173.

Terzian, Y., P. G. Mezger, and J. Schraml. 1968. *Astrophys. Lett.* 1:153.

————, and B. Balick. 1969. *Astrophys. Lett.* 4:195.

————, and A. Parrish. 1970. *Astrophys. Lett.* 5:261.

Wade, C. M. 1958. *Australian J. Phys.* 11:388.

Wilson, T. L., P. G. Mezger, F. F. Gardner, and D. K. Milne. 1970. *Astron. Astrophys.* 6:364.

Recommended General References

1. Comprehensive Review of Interstellar Matter
   L. Spitzer, Jr. *Diffuse Matter in Space.* New York: John Wiley. 1968.

2. Emission Nebulae
   L. H. Aller. *Gaseous Nebulae.* New York: John Wiley. 1956.
   M. J. Seaton. "Planetary Nebulae," *Rep. Progr. Phys.* 23:313 (1960).

3. Line Formation
   J. T. Jefferies. *Spectral Line Formation.* Waltham: Blaisdell. 1968.

4. Radio Recombination Lines
   L. Goldberg, in *Interstellar Ionized Hydrogen.* Y. Terzian, ed. New York: Benjamin. 1968, p. 373.
   A. Dupree and L. Goldberg. "Radio Frequency Recombination Lines," *Ann. Rev. Astron. Astrophys.* 8:231 (1970).
   R. M. Hjellming and M. A. Gordon. "Radio Recombination Lines and Non-LTE Theory: A Reanalysis," *Astrophys. J.* 164:47 (1971).
   M. Brocklehurst and M. J. Seaton. "On the Interpretation of Radio Recombination Line Observations," *Monthly Notices Roy. Astron. Soc.* 157:179 (1972).

5. Thermodynamics of Interstellar Matter
   G. Field. "Thermodynamic Structure," in *Molecules in the Galactic Environment.* M. A. Gordon and L. E. Snyder, eds. New York: John Wiley. 1973, p. 19.

6. Table of Radio Recombination Line Frequencies
   A. E. Lilley and P. Palmer. "Tables of Radio Frequency Recombination Lines," *Astrophys. J. Suppl., No. 144,* 16:143 (1968).

CHAPTER 4

THE LARGE-SCALE DISTRIBUTION OF NEUTRAL HYDROGEN IN THE GALAXY

W. Butler Burton

4.1 Observations of Neutral Hydrogen

Although insight into the structure of our Galaxy can be sought in a number of different ways, the study of the neutral hydrogen component is particularly suitable. Neutral hydrogen is the main observed constituent of the interstellar medium. The physical characteristics of the hydrogen are closely related to the characteristics of other galactic constituents, both stellar and interstellar. The interstellar medium is transparent enough to hydrogen radio emission that, with the exception of a few directions along the galactic equator, investigation of the entire Galaxy is possible. This transparency allows investigation of regions of the Galaxy which are too distant to be studied optically. Interstellar neutral hydrogen is so abundant and is distributed in such a general fashion throughout the Galaxy that the 21-cm hyperfine transition line has been detected in emission for every direction in the sky at which a suitably equipped radio-telescope has been pointed. No time variation of a neutral hydrogen line has been found. What we detect is a line profile giving intensity, usually expressed as a brightness temperature, as a function of frequency. The frequency measures a Doppler shift from the natural frequency of 1420.406 MHz. In practice the measured frequency shifts are converted to radial velocities (1 km sec^{-1} = 4.74 kHz), and in Milky Way studies, they are corrected to the local standard of rest. The local standard of rest is conventionally

defined by the standard solar motion of 20 km sec^{-1} toward α, δ = 18^h, + 30° (1900).

Table 4.1 lists the main surveys which have been made for galactic-structure studies and which have become available since 1966. Kerr (1968) has given a similar table for earlier surveys. These tables summarize the equipment parameters, the region surveyed, and the form in which the results are published. By way of example, reference will be made in this chapter to the observations shown in Figure 4.1 as a velocity-longitude map displaying neutral hydrogen brightness temperature contours. These observations were made along the galactic equator at half-degree longitude intervals between l = 354° and l = 120°.

The natural width of the neutral hydrogen line is 10^{-16} km sec^{-1} and is thus infinitesimally small compared with what can be measured by radio-astronomical methods. However, profiles observed near the galactic plane typically extend over about 100 km sec^{-1}. This broadening occurs through several mechanisms; consequently, the observing bandwidth is chosen according to the needs of the investigation, taking into account also that the sensitivity varies as (bandwidth)$^{-1/2}$. The broadening corresponding to the thermal velocities of atoms within a single concentration of gas will produce a line with a Gaussian shape characterized by a dispersion $\sigma \approx 0.09 \sqrt{T}$ km sec^{-1}. For a kinetic temperature of 100°K, σ is 0.9 km sec^{-1}, which corresponds to a full width between half-intensity points

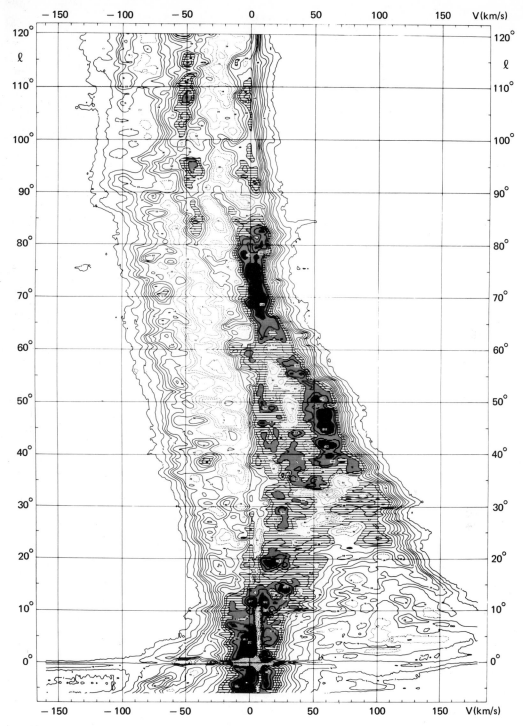

Figure 4.1 Contours of neutral hydrogen brightness temperatures in the galactic plane. Broken-line contours enclose regions of relatively low brightness temperatures. The velocity is with respect to the local standard of rest. The observations were made with a bandwidth of 1.7 km sec^{-1} and a half-power beamwidth of 0°.6. The observations were spaced at half-degree intervals of longitude. Temperatures in the ranges $70°K < T_b < 90°K$, $90°K < T_b < 100°K$, and $T_b > 110°K$ are indicated by successive degrees of shading. The dotted portion of the map near $l = 0°$, $V = 0$ km sec^{-1} is a region of absorption where contour lines would be overcrowded due to steep temperature gradients. (Burton, 1970a. *Astron. Astrophys. Suppl.* 2:261.)

Table 4.1 Surveys of Hydrogen Emission from near the Galactic Equator

Authors	Publication date	Beam	Band-width (kHz)	Approximate region	Approximate interval	Form of publication
P. O. Lindblad	1966	0°.6	10	l = 173 to 243° $b \approx$ −15 to 2°	Irregular Irregular	Profiles*
A. P. Henderson	1967	10′	8	l = 16 to 230° b = −10 to 10°	Δl = 5° Scanned	$(b, V)\|_l$ maps
F. J. Kerr and R. Vallak	1967	14′.5	38	l = −1 to 1° b = −1 to 1°	Δl = 0°.1 Δb = 0°.1	$(b, V)\|_l$ maps
F. J. Kerr	1969b	14′.5	38	l = −296 to 63°.5 b = 0° l = 300 to 60° b = −2 to 2°	Δl = 0°.1 Δl = 1 or 5° Δb = 0°.1	$(l, V)\|_b$ maps $(b, V)\|_l$ maps
G. Westerhout	1969	10′	8	l = 11 to 235° b = −1 to +1°	$\Delta \delta$ = 5′	$(\alpha, V)\|_\delta$ maps
W. B. Burton	1970a	0°.6	8	l = 354 to 120° b = 0°	Δl = 0°.5	Profiles and $(l, V)\|_b$ maps
W. B. Burton	1970b	0°.6	10	l = 43 to 56° b = −4°.5 to 4°.5	Δl = 0°.5 Δb = 0°.5	Profiles*, $(l, V)\|_b$ and $(b, V)\|_l$ maps
P. C. van der Kruit	1970	0°.6	50	l = 8 to 10° b = −5 to 5°	Δl = 1° Δb = 0°.5	$(b, V)\|_l$ maps
J. V. Hindman and F. J. Kerr	1970	14′.5	38	l = 190 to 299° $b \approx$ −5 to 5°	Δl = 5° Δb = 0°.1	$(b, V)\|_l$ maps
F. J. Kerr and J. V. Hindman	1970	14′.5	38	l = 185 to 63° b = 0°	Δl = 1°	$(l, V)\|_b$ maps
L. Velden	1970	0°.6	12	l = 120 to 240° b = −30 to 30°	Δl = 10° Δb = 0°.5	Profiles*
V. R. Venugopal and W. L. H. Shuter	1970	0°.6	10	δ > −29°	Δl = 5° Δb = 5°	Profiles
W. W. Shane	1971a, 1971b	0°.6	20	l = 22°.3 to 42°.3 b = −6 to 6°	Δl = 1° Δb = 0°.5	Profiles*, $(b, V)\|_l$ maps
N. H. Dieter	1972	0°.6	10	l = 10 to 250° b = −15 to 15°	Δl = 2° Δb = 2°	$(b, V)\|_l$ maps
W. B. Burton and G. L. Verschuur	1973	30′	10	l = 20 to 230° b = −20 to 20°	Scanned Δb = 2°	$(l, V)\|_b$ maps
		10′	10	l = 15 to 140° b = −30 to 30°	Δl = 5° Scanned	$(b, V)\|_l$ maps
P. O. Lindblad	1973	12′.5	10	l = 12 to 72° b = 1 to 10°	Δl = 3° Scanned	$(b, V)\|_l$ maps
		26′	5	l = 339 to 12° b = −15 to 15°	Δl = 3° Scanned	$(b, V)\|_l$ maps
S. C. Simonson and R. Sancisi	1973	0°.6	16	l = 354 to 24° b = −5 to 5°	Δl = 0°.5 Δb = 0°.2	Profiles, $(l, V)\|_b$ and $(b, V)\|_l$ maps
M. A. Tuve and S. Lundsager	1973	0°.9	10	l = 336 to 270° b = −16 to 16°	Δl = 4° Δb = 2°	Profiles*
H. Weaver and D. R. W. Williams	1973	36′	10	l = 10 to 250° b = −10 to 10°	Δl = 0°.5 Δb = 0°.25	Profiles
G. Westerhout	1973	12′.5	10	l = 13 to 235° b = −2° to 2°	Scanned Δb = 0°.2	$(l, V)\|_b$ maps

* Gaussian decomposition of line profiles available.

of 2.1 km sec^{-1}. Turbulent motions within a concentration of neutral hydrogen gas will also produce profile broadening of this order of magnitude. Large-scale streaming motions with amplitudes of the order of 10 km sec^{-1} have been observed in a number of regions of the Galaxy, and these motions influence the measured width of features in a profile. But none of these broadening mechanisms is sufficient to account for the observed characteristic widths of 100 km sec^{-1}, although these

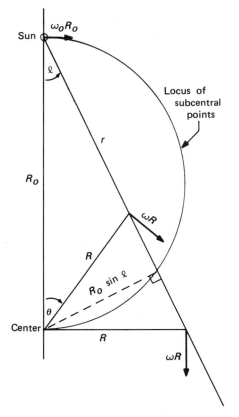

Figure 4.2 Diagram illustrating the construction used in the derivation of Equation (4.2).

mechanisms will account for much of the structure *within* a profile and will prevent isolated peaks from having sharp edges. Most of the total broadening comes from differential galactic rotation. This is of particular importance because it means that 21-cm profiles can be interpreted to give information about differential galactic rotation.

4.2 Kinematics of Galactic Neutral Hydrogen

4.2.1 Velocities Due to Differential Galactic Rotation

Although it is clear that reality is more complicated, in order to interpret the observations in terms of large-scale structure, it has been common to assume that the motions of the gas about the center of the Galaxy are everywhere circular and that the angular velocity, $\omega(R)$, is a decreasing function only of distance, R, from the center of the system. We write l for galactic longitude and θ for galactocentric azimuth, both angles being measured as in Figure 4.2. The distance from the observer to the emitting concentration of hydrogen is r. The observer is of course located within the system, at a distance R_o from its center, and is rotating about this center with angular velocity ω_o. The velocity, V, which is measured (by convention positive in sign if the emitting gas is moving away from the observer) is the Doppler-shift velocity of the gas along the line of sight:

V = component of ωR along line of sight minus the component of $\omega_o R_o$ along line of sight

$\quad = \omega R \cos(90° - l - \theta)$
$\qquad\qquad - \omega_o R_o \cos(90° - l)$

$\quad = \omega R (\sin\theta\cos l - \cos\theta\sin l)$
$\qquad\qquad - \omega_o R_o \sin l \quad (4.1)$

This, since $r\sin l = R\sin\theta$ and $R\cos\theta = R_o - r\cos l$, becomes

$$V = R_o [\omega(R) - \omega_o]\sin l \quad (4.2)$$

This is the fundamental equation of 21-cm galactic structure analysis. If the function $R_o[\omega(R) - \omega_o]$ is known, then in principle distances along the line of sight can be attributed to each measured V. It is worth emphasizing that distances cannot be determined directly, but require, in advance, accurate knowledge of the velocity field throughout the Galaxy.

Practical application of Equation (4.2)

raises a number of problems, some of which are discussed in this chapter. The first of these problems involves the accuracy with which the angular-velocity rotation curve $\omega(R)$ can be determined. Within the framework of the assumptions of the preceding paragraph this can be determined as follows from 21-cm measurements for regions along the line of sight where $R < R_o$. For any reasonable rotation law, V will vary as shown schematically in Figure 4.3. Consider a particular line of sight with l in the range $0° < l < 90°$. As the distance from the Sun, r, increases along this line of sight, the distance to the galactic center at first decreases. This corresponds to increasing $\omega(R)$ and thus to increasing V. However, as r increases further, a point on the line of sight is reached which is closest to the galactic center. At this "subcentral point," $R = R_{\min} = R_o \sin l$, and $\omega(R)$ and thus V reach maximum values. For still larger r, R increases so that V decreases from the value it reached at the subcentral point. By measuring the cut-off value of V on a profile observed

at each longitude and attributing this "terminal velocity" to the distance from the center, $R_o \sin l$, corresponding to that longitude's subcentral point, one obtains $\omega(R)$, provided R_o and ω_o are known from other methods. The terminal velocity is, in general, $V_T = R_o [\omega(R_o|\sin l|) - \omega_o] \sin l$. In determining this velocity from the observations, it must be taken as a suitably defined point on the high-velocity edge of each profile, since the edges are in practice not sharp but are blurred by various broadening mechanisms. It is also necessary, of course, to assume that hydrogen is actually present at the subcentral point. The linear-velocity rotation curve, giving the circular velocity $\Theta(R) \equiv \omega R$ as a function of V_T, is then obtained using

$$\Theta(R_o|\sin l|) = |V_T| + \omega_o R_o|\sin l| \quad (4.3)$$

Note that this terminal velocity method cannot be used for $R > R_o$. As we penetrate along any line of sight in the longitude range $90° < l < 270°$, R (which in these directions will always be greater than R_o) becomes larger and larger, so that $\omega(R) - \omega_o$ becomes more and more negative and V increases or decreases smoothly, depending on the sign of $\sin l$. There is no unique velocity corresponding to a known unique distance, as there is for the directions $0° < l < 90°$ and $270° < l < 360°$. Because of this the rotation curve for $R > R_o$ is not derived directly from the 21-cm observations. From the 21-cm rotation curve at $R < R_o$, which is assumed to represent only gravitational forces, a model of the mass distribution of the galaxy is constructed (see Schmidt, 1965), and the rotation at $R > R_o$ is in turn derived from this dynamical model.

In practice the terminal velocity procedure fails at directions within about 20° of the direction of the galactic center, where the assumption of circular rotation is clearly violated. The procedure is also weak at longitudes $75° < l < 90°$ and $270° < l < 295°$; because of the geometry in these directions, R and thus V change very slowly with increasing distance along the line of sight. This makes it difficult to assign a unique distance to the terminal velocity. Consequently the

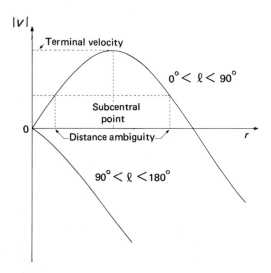

Figure 4.3 Schematic run of velocities with respect to the local standard of rest as a function of distance from the Sun. For the longitude range $0° < l < 90°$, the diagram illustrates the distance ambiguity, the terminal velocity, and the subcentral point distance. The schematic situation in the longitude quadrants $270° < l < 360°$ and $180° < l < 270°$ is similar to that illustrated here, except for a reversal of the sign of the velocity.

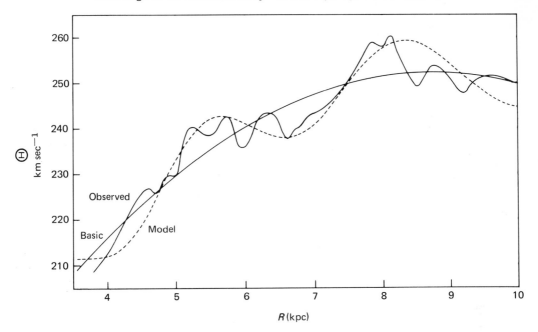

Figure 4.4 Apparent galactic rotation curves. The irregular curve is the one derived by Shane and Bieger-Smith (1966) from observations in the longitude range $22° < l < 70°$. The smooth curve is the basic rotation curve which represents the observed one freed of the perturbations of streaming motions. The dashed curve shows the basic rotation curve perturbed by the density-wave-theory streamings discussed in Section 4.2.3. In the presence of deviations from circular motion the maximum velocity might not come from the subcentral point; nevertheless the figure is drawn assuming that it does.

rotation curve of our Galaxy is best determined from 21-cm observations over the range $4 < R < 9$ kpc. Optical observations, especially of OB stars, provide kinematic information for the region where $R \approx R_o$. However, the comparison of spectroscopically determined optical distances with kinematically determined radio distances is difficult because optical distances are most accurate close to the Sun, where the accuracy of the kinematic distances is poor.

The rotation curve derived by Shane and Bieger-Smith (1966) from 21-cm observations in the first quadrant of galactic longitude is shown in Figure 4.4. We shall see in the following section that the irregularities in this rotation curve can be attributed to large-scale streaming motions. Over much of the Galaxy the deviations from circular velocity are of the order of 3% of the rotation velocities. In the central region where $R < 4$ kpc the noncircular motions are on the same order as the rotation velocities.

It is also clear from Figure 4.4 that over much of the Galaxy Θ changes much more slowly with respect to R than would be the case for solid-body rotation. This strong differential rotation indicates a strong increase in total mass toward the center of the galactic system.

It is sometimes useful to have available a simple analytic expression for the rotation curve freed of the perturbations of streaming motions. The expression (Burton, 1971)

$$\Theta(R) = R\,\omega(R) = 250.0 + 4.05\,(10 - R)$$
$$- 1.62\,(10 - R)^2 \quad (4.4a)$$

is a satisfactory fit to the apparent curve in Figure 4.4. It is valid for the region $4 < R < 10$ kpc in the first quadrant of galactic longitude. For the extrapolated region $10 < R < 14$ kpc, a fit to the rotation required by the Schmidt (1965) mass model is given by

$$\Theta(R) = R\,\omega(R) = 885.44\,R^{-1/2} - 30{,}000\,R^{-3}$$
$$(4.4b)$$

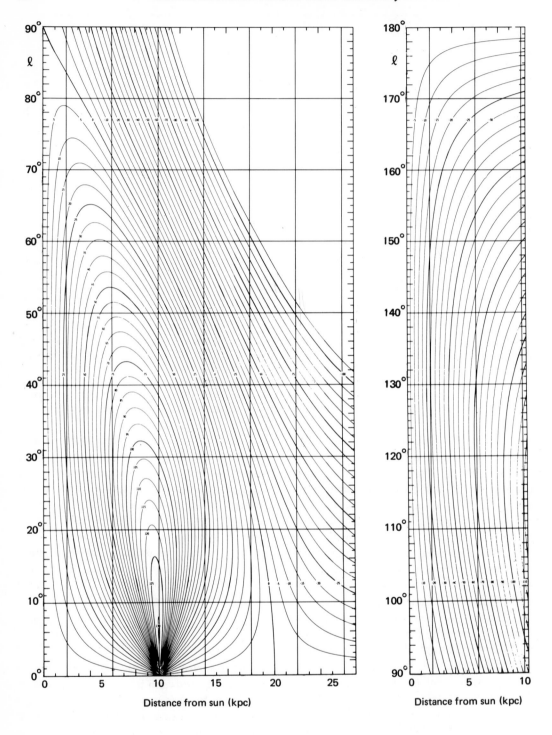

Figure 4.5 Contours of radial velocity with respect to the local standard of rest, expected for the circular rotation described by Equation (4.4), plotted as a function of longitude and distance from the Sun.

The Keplerian term dominates in this expression; the effect of the non-Keplerian term decreases as R increases.

Velocities with respect to the local standard of rest, calculated from Equation (4.2) using the velocity field described by Equation (4.4), are plotted as a function of distance from the Sun in Figure 4.5. This figure shows in more detail the features illustrated schematically by Figure 4.3. It shows the "distance ambiguity" at positive velocities, the unique distance corresponding to the maximum velocity at each longitude in the first quadrant, and the particularly slow variation of velocity with increasing distance at longitudes $75° < l < 90°$. If one trusts the rotation curve given by Equation (4.4) and if the radial velocity of a feature is known, then this figure can be used to estimate its distance. For example, at $l = 30°$ a radial velocity of $V = +75$ km sec^{-1} places the feature at a distance of either 5.3 or 12.0 kpc from the Sun.

Since the observer is located inside the Galaxy and is rotating with it, we must have information on his position in the Galaxy, R_o, and on his rotational velocity, $\Theta_o = \omega_o R_o$, in order to fix the scale and zero-point of the rotation curve. The determination of these basic quantities lies primarily in the optical domain and has been discussed by Schmidt (1965). The distance to the center of the Galaxy can be measured directly by finding the distance to the density maximum of some type of object, such as RR Lyrae variable stars, observed in directions near that of the galactic center. A distance $R_o = 10$ kpc has been adopted as standard, although the determination is sensitive to the correction for interstellar absorption and to the adopted absolute magnitude of the RR Lyrae stars. The rotation velocity at the Sun, Θ_o, is best determined not directly, but rather in terms of the Oort constants of galactic rotation A and B. These constants are defined as

$$A \equiv \frac{1}{2}\left[\frac{\Theta_o}{R_o} - \left(\frac{d\Theta}{dR}\right)_{R_o}\right] = -\frac{1}{2}R_o\left(\frac{d\omega}{dR}\right)_{R_o}$$
(4.5a)

and

$$B \equiv -\frac{1}{2}\left[\frac{\Theta_o}{R_o} + \left(\frac{d\Theta}{dR}\right)_{R_o}\right]$$
(4.5b)

so that $\Theta_o = (A - B)R_o$. The values of the Oort constants which have been adopted as standard by the I.A.U. (1966) are

$$A = 15 \text{ km sec}^{-1} \text{ kpc}^{-1}$$
(4.6a)

and

$$B = -10 \text{ km sec}^{-1} \text{ kpc}^{-1}$$
(4.6b)

so that $\Theta_o = 250$ km sec^{-1}, with the uncertainty in the value of Θ_o coming primarily from the uncertainty in the value of B. It is not easy to measure the error in the quantities Θ_o and R_o; a 20% error cannot be ruled out. A subsequent revision of these two quantities may change the scale of the rotation curve, but the shape of it and the general kinematic conclusions drawn from it will not be affected.

The value of A has been determined from radial velocities and distances of galactic clusters and cepheids. Direct information on the value of B comes from proper motion studies, although the available observational data are so weak that B is more reliably determined indirectly. Thus the standard value of B is based on studies of the velocity ellipsoid (see Schmidt, 1965), since the ratio of A and B is related to the ratio of the axes of the velocity ellipsoid in the galactic plane.

The 21-cm observations cannot provide A or R_o separately, but they can provide a determination of the product AR_o. Measurement of this product gives a valuable independent check of the values of A and R_o determined separately by other methods. Radio determinations of AR_o lie in the range 135 to 150 km sec^{-1}. The 21-cm observations provide AR_o as follows. The angular velocity $\omega(R)$ is expanded in a Taylor approximation:

$$\omega(R) = \omega_o + (R - R_o)\left(\frac{d\omega}{dR}\right)_{R_o}$$
(4.7)

This first-order approximation is applicable where $R - R_o$ is small. Using the definition of the Oort constant A,

$$\omega(R) = \omega_o - (R - R_o)\frac{2A}{R_o}$$
(4.8)

so that substitution into Equation (4.2) gives

$$V = -2A(R - R_o)\sin l \qquad (4.9)$$

Still assuming circular rotation, there is no distance ambiguity for the terminal velocity, V_T, since this velocity corresponds to $R = R_{min} = R_o|\sin l|$. Consequently measurements of the terminal velocities give

$$AR_o = V_T[2\sin l(1 - |\sin l|)]^{-1} \qquad (4.10)$$

The product should be determined from observations at longitudes not too much less than $90°$ in order for the assumption $R \approx R_o$ to remain valid. The variations observed in V_T, which are of the order of 5 km sec^{-1} and are attributed to systematic streaming motions, will introduce large errors in AR_o, since the denominator in Equation (4.10) is less than 1. In principle the product AR_o could be determined in a similar fashion using stellar objects. However, it is difficult to be sure that the stellar objects really are at R_{min}.

Using the above values of R_o and Θ_o, we find that the period of revolution of the local standard of rest around the galactic center is $2\pi R_o/\Theta_o = 2.5 \times 10^8$ years. This is only 1 or 2 % of the age of the galactic disk. One of the classic questions of galactic structure research has asked how structure in the disk can be maintained against the shearing forces of the strong differential rotation for times longer than one galactic revolution.

4.2.2 Deviations from Circular Symmetry and Circular Motions

One important factor in the rotation curve derivation is the assumption that there is enough hydrogen present near the subcentral point to actually determine the cut-off of the profile. Suppose that there is not enough hydrogen; then the measured terminal velocities would be contributed by gas in a region with $R > R_o|\sin l|$ and thus $\omega < \omega(R_o|\sin l|)$, so that the derived $\Theta(R)$ would be less than the true velocity. In fact, the observed rotation curve plotted in Figure 4.4 shows irregularities in the velocities plotted against distance from the center. In the origi-

nal determination of the rotation curve by Kwee, Muller, and Westerhout (1954) this sort of irregularity was attributed to extended regions that did not contain enough gas near the subcentral point to determine the profile cut-off. The idea was that if there was not much gas at the subcentral point, the observed cut-off velocity would be due to lower velocity gas at $R > R_o|\sin l|$. If this is the case, then the rotation curve should be drawn using the upper envelope of the observed terminal velocities. Circular motion could in this way be retained. However, Shane and Bieger-Smith (1966) showed that the irregularities could better be attributed to large-scale streaming motions, *i.e.*, to systematic deviations from circular rotation. They rejected the possibility of large empty regions because low-intensity extensions on the ends of profiles are not observed at longitudes corresponding to the dips in the run of observed terminal velocities. Indeed, one of the most pronounced characteristics of the observations in Figure 4.1 is the persistence of high intensities near the maximum velocity. A model based on *circular* rotation and reproducing most characteristics of the terminal velocities could be constructed only by using an unacceptably artificial density distribution. In order for the whole maximum-velocity part of the profile to shift to lower velocities there would have to be essentially no hydrogen (about a factor 100 less than the average hydrogen density) along the whole region of the line of sight with velocities within approximately 10 km sec^{-1} of the velocity corresponding to the subcentral point. This means (as is clear from inspecting Figure 4.5), that regions of 4 or 5 kpc extent would have to be essentially empty of gas and that these regions would have to have a preferential orientation with respect to the observer. This is implausible. The conclusion that the observed irregularities in the run of the terminal velocities reflect corresponding irregularities in the galactic velocity field is an important one.

There are other indications that the assumption of circular rotation is less than

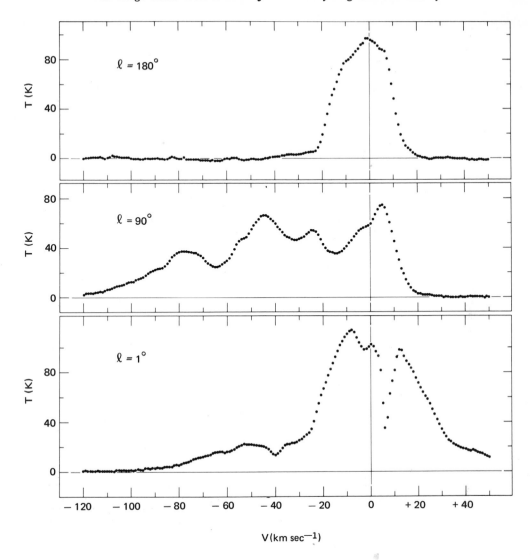

Figure 4.6 Line profiles observed in cardinal directions at $b = 0°$, unambiguously showing systematic deviations from circular motion. Because of very strong absorption at $l = 0°$, the profile at $l = 1°$ is substituted for it. The profile at $l = 180°$ was observed by Velden (1970); the other two profiles were observed by Burton (1970a).

satisfactory. Unambiguous proof of non-circular motions in the Galaxy is given by profiles observed in the galactic plane in the cardinal directions $l = 0°$, $90°$, and $180°$. The velocities expected in terms of circular rotation at these longitudes are illustrated in Figure 4.5, while Figure 4.6 shows the profiles observed. At longitudes $0°$ and $180°$ all circular motions should be perpendicular to the line of sight, and in the presence of only such

motions the profiles should have one peak symmetric with respect to zero velocity. Actually, strong radial motions are observed. For $l = 90°$, circular motion would imply no positive velocity peak in the profile. Actually, there is a peak at $V \approx +6$ km sec^{-1}.

Since there is no reason to expect these irregularities to be axially symmetric or even to have characteristic length scales of more than a kiloparsec or so, the rotation curve illustrated by the wavy line in Figure 4.4 is an *apparent* one only, since it describes motions along one particular locus, and not over the entire galactic plane. This locus is not necessarily the locus of subcentral points, since in the presence of deviations from circular motion, the terminal velocity does not necessarily originate at the subcentral point. Thus it is not surprising that the apparent rotation

curve derived from observations of the terminal velocities in the longitude quadrant $270° < l < 360°$ has irregularities somewhat differently placed, although of about the same amplitude. What is more disturbing is that the *unperturbed* or *basic* rotation curve, obtained by drawing a smooth curve through the irregular apparent one, is different when determined for the fourth quandrant than

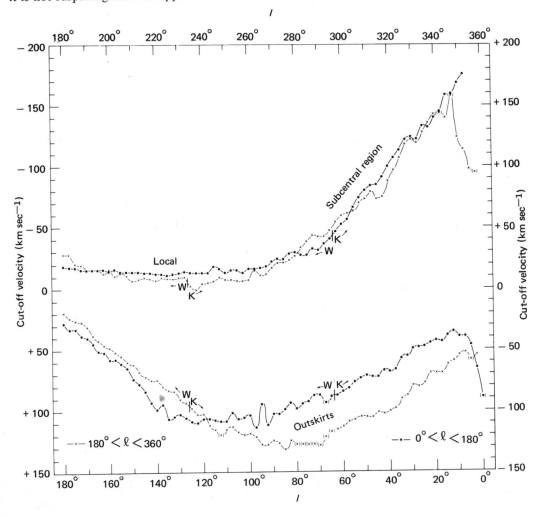

Figure 4.7 Comparison of cut-offs on positive- and negative-velocity wings of profiles observed in the longitude range $0° < l < 180°$ (dots, right-hand scale) with those in the corresponding longitudes in the range $180° < l < 360°$ (crosses, left hand scale). The systematic differences in the cut-off velocities imply large-scale deviations from axial symmetry. The data are taken from the surveys of Westerhout (1969) and Kerr and Hindman (1970), as indicated. For the purposes of this figure, the cut-off was defined as the velocity at a particular antenna temperature: this was $T_A = 5°K$ for the "W" points and $T_A = 6°K$ for the "K" points.

when determined for the first quandrant (see Kerr, 1969a). The systematic difference of about 10 km sec^{-1} between the two curves over the region between $R = 5$ and 8 kpc is further evidence requiring us to accept kinematic asymmetries on a very large scale.

Deviations from circular symmetry in our Galaxy can be demonstrated in more detail by comparing the cut-offs on both the positive-velocity and negative-velocity wings of profiles observed at corresponding longitudes. Such a comparison is shown in Figure 4.7. The differences in the cut-offs at corresponding longitudes are large and systematic. The cut-offs plotted in Figure 4.7 provide information from different parts of the Galaxy, as suggested by the labels "local," "subcentral region," and "outskirts" in the figure. Although different explanations may have to be found for each region, explanations in terms of kinematics seem more plausible than explanations in terms of the density distribution.

The general phenomenon of deviations from circular symmetry has important consequences for the derivation of the galactic distribution of the neutral hydrogen. These consequences are the subject of Section 4.3.2.

4.2.3 Noncircular Motions Predicted by the Linear Density-Wave Theory

The density-wave theory of galactic rotation has had some success in explaining the observed velocity characteristics of the Galaxy. It also provides a plausible mechanism for maintaining spiral structure against the strong shearing forces of differential rotation. The theory is discussed in detail in a monograph by Lin and Shu (in preparation), but it is useful to briefly describe here the observable consequences of the theory.

Spiral arms are considered as *waves* in the theory (see Lindblad, 1963). The rigorous development of the theory is due to C. C. Lin and collaborators; see, *e.g.*, Lin *et al.* (1969). Stars and gas move through the spiral arms, but since they stay longer in the arms, the arms are density maxima. The theory provides for the maintenance of such density waves, although it does not provide for their origin. There is still little agreement on the origin of the spiral perturbation in the first place. A spiral-like density wave is assumed to be superimposed as a perturbation on an axisymmetric background mass distribution. The superimposed pattern has the form of a two-arm trailing spiral rotating with constant angular velocity, ω_p, with respect to an inertial system. Thus the wave pattern moves relative to the material and does not follow differential rotation. The resultant spiral gravitational field produces streaming motions and a subsequent redistribution of densities which can maintain the imposed perturbation. Although the imposed perturbation is only a few percent of the total mass density, the interstellar gas responds quite strongly to it. In the linear theory the effect of the wave on the gas and on the stars is in phase; the effect differs only in the amplitude of the perturbation. According to the theory the response of a particular population of the galactic mass is determined by the population's turbulent velocity dispersion; populations with small velocity dispersion have a large response. The interstellar hydrogen is characterized by velocity dispersions of about 5 km sec^{-1}, whereas most of the galactic mass is contributed by older stars with dispersions characteristically of the order of 30 km sec^{-1} (see Delhaye, 1965). The density-wave theory predicts observable streaming motions also for the youngest stars, characterized by velocity dispersions of about 10 km sec^{-1}. For the gas component the theory predicts both large streaming motions, typically 7 km sec^{-1}, and a large ratio between the density in the arm and in the interarm regions of about 3:1. Such streaming motions or such density variations would, by themselves, have easily observable consequences in the profiles. Since in fact such effects are superimposed on top of each other, separating the kinematic from the density characteristics represented in the profiles is a challenging problem. In principle, the density variations and the streaming motions are related to each other through Equations (4.12). However, the pattern speed enters the relationship. Since the pattern speed is not known directly from the theory or from observations, the separation of density and velocity characteristics is not easy.

The density-wave theory relates the streaming motions and the densities as follows. The deviations from circular motion are expressed in terms of the peculiar motions, V_R and V_θ, taken, respectively, to be positive in the directions of increasing radius and azimuth. These peculiar motions are assumed to vary periodically:

$$V_R = -a_R \cos(\chi(R, \theta))$$
$$V_\theta = a_\theta \sin(\chi(R, \theta)) \quad (4.11)$$

where $\chi(R, \theta) = 2\theta - \phi(R)$ is the phase of the superimposed two-arm spiral gravitational potential, and $\phi(R)$ is the radial phase function. The amplitude functions of the peculiar motions are given by Lin *et al.* (1969) in terms of the density contrast between the gas surface density in a unit column perpendicular to the galactic plane through the center of a spiral arm, σ_{max}, and the gas surface density between arms, σ_{min}:

$$a_R = \frac{\sigma_{max} - \sigma_{min}}{\sigma_{max} + \sigma_{min}} [\omega(R) - \omega_p] \frac{2}{-k(R)}$$
$$a_\theta = \frac{\kappa^2}{4} \frac{1}{(\omega(R))^2 - \omega(R)\omega_p} a_R \quad (4.12)$$

Here $\omega(R)$ is the basic unperturbed rotation of the background mass, such as given by Equation (4.4). The epicyclic frequency, κ, is defined by (see Oort, 1965)

$$\kappa^2 = [2\,\omega(R)]^2 \left[1 + \frac{R}{2\omega(R)} \frac{d\omega}{dR}\right] \quad (4.13)$$

The radial wave number of the spiral pattern is $|k(R)|$, where $k(R) = d\phi/dR$. The spacing between adjacent arms is $\lambda = 2\pi/|k(R)|$.

The geometry of the spiral pattern, $R(\theta) = R_o \exp(-t(\theta - \theta_o))$, is characterized by a tilt angle, t, defined as the acute angle between

the spiral and a galactocentric circle (positive in sign for trailing arms). The tilt angle is related to the wave number by

$$k(R) = \frac{\partial \chi}{\partial R}\bigg|_\theta = \frac{d\phi}{dR} = 2 \frac{\partial \theta}{\partial R}\bigg|_\chi = -\frac{2}{R \tan t} \quad (4.14)$$

These equations determine the streaming motions predicted by the first-order density-wave theory for certain assumed parameters, which with certain restraints may be adjusted in accordance with the observations. The sense of the streaming is apparent from Equations (4.12), since both a_R and a_θ are always positive for $k(R) < 0$ (trailing arms) and for $\omega_p < \omega(R)$. For our own Galaxy the predicted radial velocity with respect to the local standard of rest becomes, in the presence of the density-wave peculiar motions,

$$V(R,l,\theta) = V(R,l) - V_R \cos(l + \theta) + V_\theta \cos(90° - l - \theta) \quad (4.15)$$

Here $V(R,l)$ is the unperturbed basic rotation described by Equation (4.2). The density-wave streaming motions (for a particular set of parameter values) relative to the local standard of rest and given by the last two terms on the right in Equation (4.15) are plotted in Figure 4.8. Indeed it is qualitatively understandable that a stretched-out structural feature will, by its own gravitational forces, induce streaming motions in the sense indicated. Material on the outer side of the feature will be pulled toward it, increasing the material's angular momentum and thus increasing its velocity in the direction of rotation. The situation will be the other way around for material on the inner side of such a feature.

The density-wave theory is applicable over the range of the galactic disk for which the conditions

$$\omega(R) - \frac{\kappa}{2} < \omega_p < \omega(R) + \frac{\kappa}{2} \quad (4.16)$$

are satisfied. The pattern speed ω_p has been chosen by Lin *et al.* (1969) so that this range is as large as possible, as well as to satisfy the

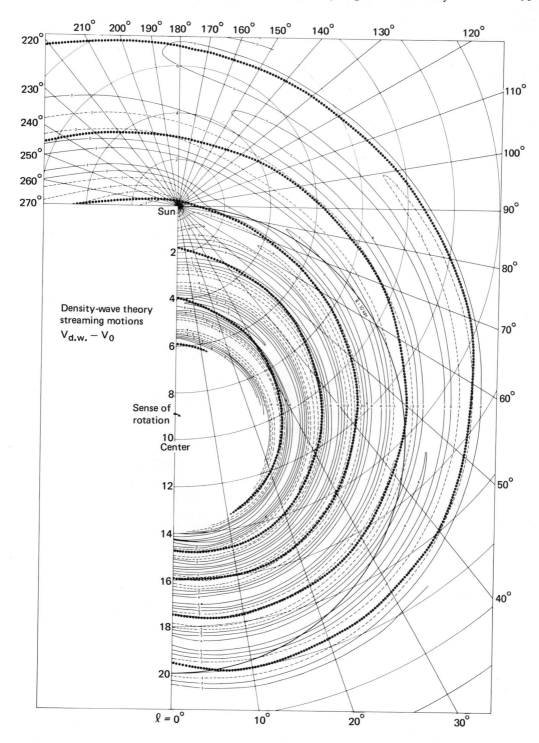

Density-wave theory
streaming motions
$V_{d.w.} - V_0$

Sense of
rotation

Center

Sun

Figure 4.8 Streaming motions relative to the local standard of rest predicted for the gas by the density-wave theory, for a particular set of parameters. The contours represent the last two terms on the right in Equation (4.15). The heavy dots show the location of the spiral-arm potential minima.

presumed galactic spiral pattern, in particular the observed location of the two bumps in the rotation curve (see Figure 4.4), contributed by two major spiral arms. At the borders of this range are the locations of the so-called Lindblad resonances. According to the theory the spacing between arms approaches zero at the resonances and the pattern ends as a ring. Outside this range the theory is no longer applicable. The observational consequences of these resonances are presently under investigation. For our own Galaxy and a pattern speed of 13 km sec^{-1} kpc^{-1}, the inner Lindblad resonance occurs at $R \approx 4$ kpc. It has been suggested that resonance phenomena might account for the large radial motions observed in the 21-cm line near the "3-kpc arm" (Shane, 1972; Simonson and Mader, 1973).

The density-wave theory has been applied

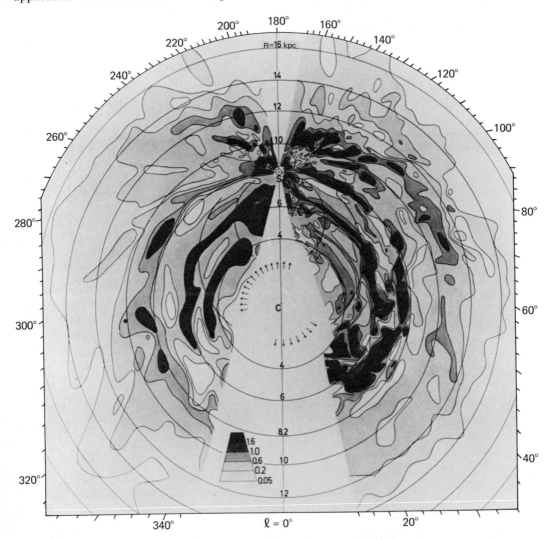

Figure 4.9 Distribution of neutral hydrogen densities in the galactic plane as determined from the Dutch and Australian surveys. This map was drawn using Equation (4.2) (with $R_o = 8.2$ kpc) and depends for its validity on the assumptions inherent in that equation. Because of this, the map is difficult to interpret and may contain serious errors. Nevertheless, it seems that it would be difficult to produce an improved version valid over the whole galactic plane. (Oort *et al.*, 1958. *Monthly Notices Roy. Astron. Soc.* 118:379.)

in some detail to the interpretation of the velocity and density patterns observed in the Milky Way. Streaming motions of the sort predicted by the theory and illustrated by Figure 4.6 have been observed in the gas (Burton, 1966, 1971, 1972; Burton and Shane, 1970; Roberts, 1972; Shane, 1972; Tuve and Lundsager, 1972, 1973) and in the youngest stars (*e.g.*, Feast, 1967; Humphreys, 1970, 1971, 1972). The derivation of an apparent rotation curve, by using Equation (4.15) and best-fit density-wave parameters, results in the dashed curve labeled "model" in Figure 4.4. This calculated rotation curve is certainly a better fit to the one derived from the first-quadrant observations than the one derived using the circular rotation equation, Equation (4.2). The theory has also been used to interpret observations of a few external Galaxies (M101, Rogstad, 1971; M51, Mathewson *et al.*, 1972, and Tully, 1972).

4.3 Determination of Galactic Structure from Hydrogen Observations

The standard overall picture of the neutral hydrogen distribution in the Galaxy is still the one based on early Dutch and Australian observations (Schmidt, 1957; Westerhout, 1957; Kerr, 1962). This map, shown in Figure 4.9, was derived using the basic equation, (4.2). We have seen that the assumptions of axial symmetry and circular rotation inherent in this equation are not generally valid. It now remains to be seen how this should influence the interpretation of the standard map in Figure 4.9, and to discuss some of the problems involved in the derivation of such a map.

4.3.1 Line Profile Characteristics Caused by Geometrical Effects

It is instructive to look in more detail at the change of V with distance along the line of sight calculated for the simple circular rotation described by Equation (4.4). This velocity with respect to the local standard of rest is plotted (using a full-drawn curve)

against distance from the Sun for two typical longitudes, $l = 50°$ and $l = 75°$, in Figure 4.10(a).

Two things are immediately evident from this figure. In the first place, there are *two* regions on the line of sight which contribute to each positive velocity, whereas only *one* region contributes to each negative velocity. This distance ambiguity is expected in all directions where $0° < l < 90°$ or $270° < l < 360°$. If the hydrogen gas is optically thin and generally distributed throughout the Galaxy, the effect of this double-valuedness at positive velocities (for $0° < l < 90°$) should show up in the observations, since the intensities at positive velocities should then be typically about twice what they are at negative velocities. The actual existence of the *intensity cut-off near zero velocity* in the range $20° < l < 70°$, evident in the reference map in Figure 4.1, indicates that on the largest scale the hydrogen gas at these longitudes and located at small values of $|V|$ is indeed optically thin. In the second place, Figure 4.10(a) shows that the velocity observed from regions near the subcentral point changes relatively slowly along the line of sight. Consequently the profiles contain, near the terminal velocity, a contribution from an especially long path length. This crowding in velocity results in the *high-velocity ridge pattern*, which is a striking characteristic of the observations in the reference map. Velocity crowding can cause the profiles to approach saturation in certain directions and at the velocities in question. Probable examples of optically thick gas occur at small values of $|V|$ in the directions $l \approx 0°$, $180°$, and $75°$. On a smaller scale there are of course cold hydrogen clouds which are optically thick.

The density of hydrogen atoms in a column of $1\,cm^2$ cross-section per unit interval of velocity is $n_H\,(|dV/dr|)^{-1}$, where n_H is the number of hydrogen atoms per cubic centimeter. Assuming that the hydrogen is more or less evenly distributed, the relative contribution to the profiles at each velocity is determined by the rate of change of the velocity with distance. This is illustrated by Figure

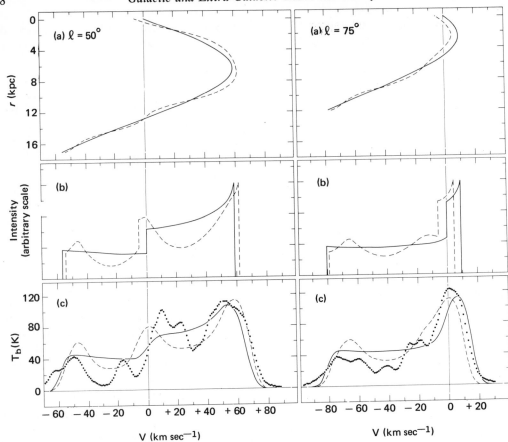

Figure 4.10 Diagram illustrating for two typical longitudes the importance of the geometrical velocity-crowding effects for the case of purely circular rotation (full-drawn lines) and for the density-wave velocity field (dashed lines), which incorporates noncircular motions of about 5 km sec^{-1}. (a) Velocity with respect to the l.s.r. as a function of distance from the Sun. (b) Schematic line profiles each considered as the sum, for the near and far side of the subcentral point, of the slopes $(dV/dr)^{-1}$ of the corresponding curve in (a). These profiles show the relative geometrical enhancement at each velocity. (c) Theoretical line profiles calculated with the velocity fields illustrated in (a) and a completely uniform distribution of hydrogen density, temperature, and dispersion. The structure in the profiles is attributed to regions along the line of sight where the velocity changes slowly with distance. The heavy dots represent the observed profiles.

4.10(b). Here the reciprocal of the change of velocity with distance from the Sun, $|dV/dr|^{-1}$, summed at positive velocities where there is the distance ambiguity between the near and the far side of the subcentral point, is plotted against velocity. The vertical scale is $|dV/dr|^{-1}$, but since intensities at velocities at which $|dV/dr|$ is small will be proportionately enhanced on the profiles, the vertical scale can be interpreted as an intensity scale. Thus the plots in Figure 4.10(b) can be considered schematic line profiles.

The full-drawn curves in Figure 4.10 are calculated for simple circular rotation. The dashed curves in the figure, however, are calculated for a rotation law in which large-scale streaming motions are present. At this stage it is not important that the rotation law used is one derived using density-wave-theory kinematics and illustrated in Figure 4.4. What is important is that in the presence of deviations from circular motion, the variation of V along the line of sight will not be as regular as in the circular rotation case, but will show the sort of structure illustrated in Figure 4.10(a). Irregularities in the plot of V against r will show up as structure in the schematic profile constructed from the $|dV/dr|^{-1}$ against r relation.

Theoretical line profiles which represent the geometrical effects in a more realistic way can be calculated, assuming a rotation law and a *completely uniform hydrogen distribution*. Such profiles are shown in Figure 4.10(c). The full-drawn profiles are calculated for the circular rotation described by Equation (4.4), for $l = 50°$ and $l = 75°$. These calculated profiles illustrate that structure in the observed profiles is to be expected even for a structureless distribution of hydrogen throughout the Galaxy. The cut-off near zero velocity, which is due to the fact that two regions contribute to positive velocities, and the enhanced intensities near the maximum velocity, which are due to the crowding in velocity near the subcentral point, are geometrical effects which are *model-independent* in the sense that structure of this sort would be present in the profiles for any reasonable rotation law and density distribution. Obviously, this sort of profile structure must be satisfactorily accounted for in the subsequent analysis of profiles and not attributed to spurious characteristics of the hydrogen distribution. Although there are numerous cases in the literature where this has not been done, it should not be too difficult to take the model-independent effects into account.

Making allowance for the effects of systematic streaming motions is a different matter, however. The dashed-line profiles in Figure 4.10(c) were calculated using the density-wave-theory velocity field illustrated by Figure 4.10(a) and, again, a completely uniform hydrogen distribution. The profiles show the model-independent effects, but these effects are now superimposed on the geometrical effects attributable to deviations from circular rotation. The dashed-line profiles illustrate the efficiency with which systematic streaming motions of about 5 km sec^{-1} amplitude will distort the observations; it would require large density differ-

ences to achieve intensity differences equivalent to those obtained by systematic streaming motions of only a few km sec^{-1}. The observed profiles are included in Figure 4.10(c) as dots, in order to show that the structure in the observed profiles is of the same magnitude as the structure in the calculated profiles.

4.3.2 The Kinematic Distribution of the Neutral Hydrogen

It is important to emphasize that profiles are more sensitive to small variations in the streaming motions than to even substantial variations in the hydrogen density (Burton, 1971, 1972; Tuve and Lundsager, 1972, 1973). Line profiles are sensitive to the velocity field because what is observed is the amount of hydrogen *per unit velocity*. In the presence of streaming motions there will be regions on the line of sight where the radial velocity changes relatively slowly with increasing distance from the Sun. The relative contribution to the observed profiles from these regions where $|dV/dr|$ is small will be enhanced. In the absence of saturation, irregularities in the velocities are themselves sufficient to cause the appearance in the profiles of structure that has generally been interpreted in terms of density concentrations on galactic structure maps derived assuming purely circular rotation.

In order to exploit the fact that line profiles are very sensitive to velocity variations, we can assume for the sake of the argument that *all* structure in the profiles has a kinematic origin. By adjusting the line-of-sight streaming motions in a model, but maintaining a uniform density and temperature distribution, *any* profile can be reproduced to within a scale factor. This method of perturbing the line-of-sight velocity field, illustrated in Figure 4.11, involves fitting model profiles to observed profiles by perturbing the line-of-sight velocity field from the basic circular-rotation velocity field described by Equation (4.4). The difference between the perturbed and the basic line-of-sight velocity fields, $V_p - V_o$, gives informa-

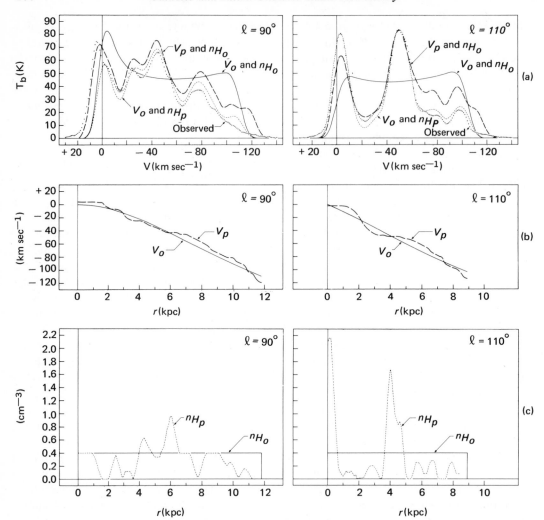

Figure 4.11 Diagram illustrating the kinematic profile-fitting approach for two typical longitudes. (a) Comparison of the observed profiles (dots) with model profiles. The zero-order profile (full-drawn line) was calculated with the basic circular-rotation velocity field $V_o(r)$ and a completely uniform hydrogen distribution, $n_{Ho} = 0.4$ cm^{-3}. The heavy-dashed-line profile was calculated by perturbing only the velocity field, giving $V_p(r)$, leaving the density uniform. The profile illustrated by the light dashed line was fit to the observations by varying the density, giving $n_{Hp}(r)$, while retaining the circular velocity field $V_o(r)$. (b) Perturbed line-of-sight velocity field, $V_p(r)$, which together with a uniform hydrogen distribution results in the heavy-dashed-line profile in (a). Note that the kinematic approach can account for the "forbidden-velocity" peak at $l = 90°$ in a natural manner and that the required streaming amplitudes appear reasonable. (c) Perturbed density distribution, $n_{Hp}(r)$, which together with the circular-rotation velocity field results in the light-dashed-line profile in (a). Note that the density approach fails to account for the "forbidden-velocity" hydrogen and that extended, essentially empty, interarm regions are necessary. (Burton, 1972. *Astron. Astrophys.* 19:51.)

tion on the spatial distribution of the streaming motions. Application of the method to profiles observed in the galactic plane results in the distribution of the streaming parameter, $V_p - V_o$, shown in Figure 4.12. It is

clear from this figure that the observations are satisfactorily reproduced with line-of-sight streaming amplitudes of the order of 7 km sec^{-1}. This amplitude is about 3% of the velocity of rotation about the galactic

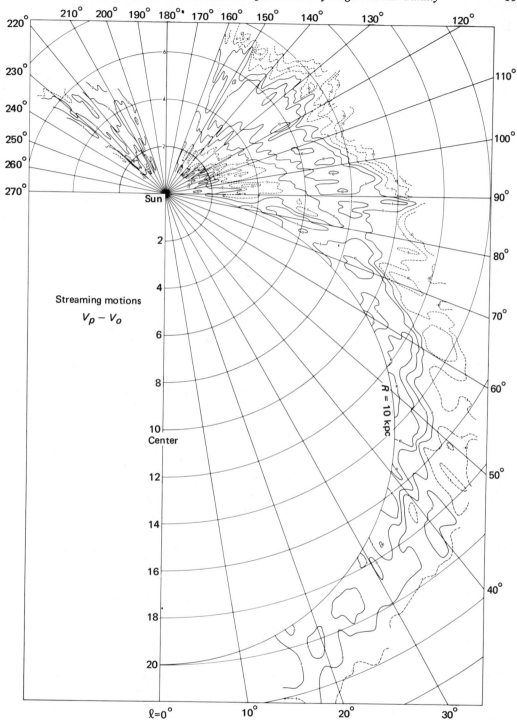

Figure 4.12 Line-of-sight streaming motions, for $R > R_o$, derived from observations in the galactic plane using the kinematic procedure illustrated in Figure 4.11. The contour values express in km sec^{-1} the difference between the perturbed velocity field and the circular-rotation velocity field. These streaming motions alone (thus retaining constant gas density) are sufficient to account for the structure in the observed profiles. This figure was derived from observations made by Westerhout (1973).

center. The sense and magnitude of the kinematic variations are consistent with gas
motions observed in a number of regions and
with motions predicted by the density-wave
theory. The spatial distribution of $V_p - V_o$,
provided by the procedure, is a measure
which can be compared with optically derived
velocity residuals, using optical distances.

The kinematic approach outlined here is
discussed in more detail by Burton (1972) and
Burton and Bania (in preparation). The
detailed interpretation of a map such as the
one in Figure 4.12 requires the adoption of a
theory relating the velocity and density fields,
since it is clear that fluctuations in velocity
and density will indeed accompany each other.
In any case, the approach serves to emphasize
the extreme sensitivity of the observations to
streaming motions.

4.3.3 The Model-Making Approach to the Derivation of the Large-Scale Structure

Because streaming motions and density
variations will generally be associated, attempts to derive the spatial distribution of the
gas should be based on simultaneously
derived solutions for the galactic velocity
field. We have seen that even slight variations
in local velocity conditions can lead to
substantial variations in the observed profiles. It is necessary to account for the consequences of both the adopted velocity field
and the density distribution when deriving
the large-scale structure. A good way to do
this is to calculate model line profiles whenever one produces a map of the galactic
structure.* These line profiles would of course
contain the effects discussed in Section 4.3.1
inherent in the overall geometry and in the
particular velocity field adopted. It would be
very difficult to account for these effects
otherwise. Once it is established that the
structure map is reasonable in the sense of the
comparison of the model profiles with the
observations, the map can further be judged
in terms of the reasonableness of the adopted

*A procedure for calculating model line profiles has
been outlined by Burton (1971).

velocity field and resultant density distribution.

Figures 4.13 and 4.14 show velocity-
longitude contour maps constructed from
model line profiles. These maps can be compared with the observed contour map in
Figure 4.1. The model in Figure 4.13 was
constructed using a velocity field of the type
predicted by the density-wave theory and
described by Equation 4.15. The parameters
in the density-wave-theory formulation were
adjusted so that the model contour map
agrees as well as possible with the map in
Figure 4.1. The choice of the parameters
was guided primarily by the location of the
bumps in the distribution of terminal
velocities and by the relative contrast between
the regions of low and high intensities in the
observations. The resultant streaming amplitudes in this model vary between 3 and 8
km sec^{-1}, the arm-interarm density contrast
is typically 3:1, and the tilt of the spiral
pattern is about 7°.

The map in Figure 4.14 illustrates some
of the failures of kinematic models based on
circular rotation. This map was constructed
using the same density distribution and the
same spiral pattern as in the Figure 4.13
model, but using only the basic rotation of
Equations (4.2) and (4.4), without the
density-wave-theory perturbation. The difference in appearance between the models in
Figures 4.13 and 4.14 is due entirely to the
difference in the adopted velocity fields. It is
clear that the second model is a poorer fit to
the observations.

Figure 4.15 illustrates this model-making
approach in more detail for the interior
portion of the galaxy between $l = 40°$ and
$l = 90°$. The figure is a composite consisting
of the observed velocity-longitude contour
map [Figure 4.15(a)], the model contour map
which is a best fit to the observations [Figure
4.15(b)], and the spatial map [Figure 4.15(c)],
illustrating the density distribution used in
deriving the model.

The model used for Figure 4.15, which
is based on density-wave kinematics, is
similar to the one represented in Figure 4.13

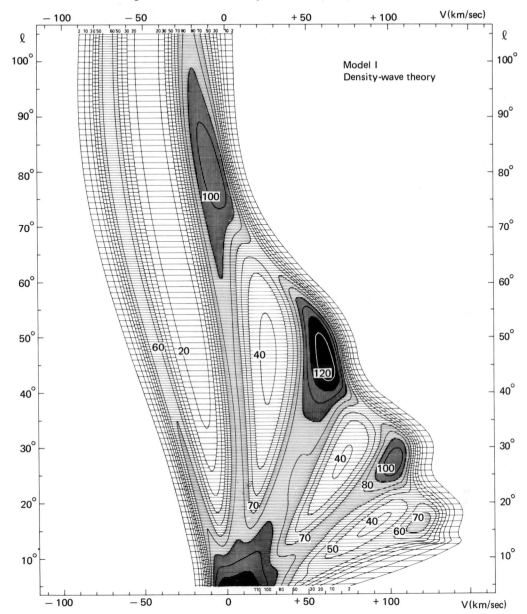

Figure 4.13 Model velocity-longitude contour map for the galactic plane based on the density-wave theory. (Burton, 1971. *Astron. Astrophys.* 10:76.)

in most respects. The fit of the model to the observations is judged by comparing the maps in Figures 4.15(a) and 4.15(b). The high intensities near the terminal velocities, the cut-off in intensities near zero velocity, and the enhanced intensities between $l \approx 65°$ and $85°$ are interpreted in terms of the model-independent effects discussed in Section 4.3.1.

The density-wave theory has been able to account for the run of terminal velocities with longitude, as well as the general appearance of the observations associated with the Sagittarius spiral arm, which is seen tangentially at $l \approx 50°$. The shaded part of Figure 4.15(c) is the region where the density is above average; as has been stressed, the exact density

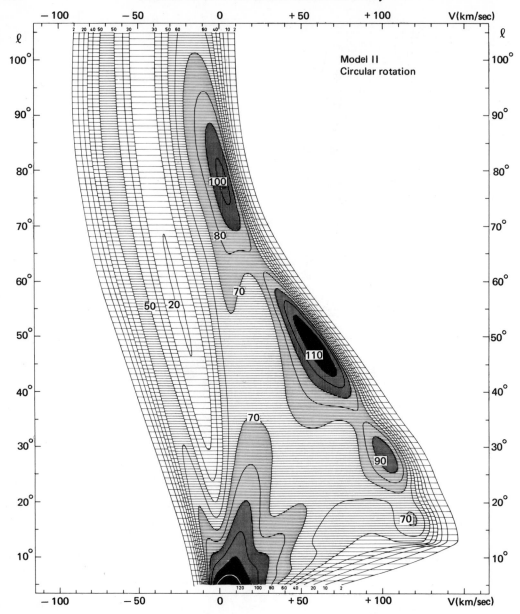

Figure 4.14 Model velocity-longitude contour map based on circular galactic rotation and the same hydrogen distribution as derived for the model in Figure 4.13. A comparison of Figures 4.13 and 4.14 shows that the velocity characteristics are responsible for the predominant contribution to the structure in Figure 4.13 and that the density effects illustrated in Figure 4.14 are secondary. Model-independent features are, of course, present in both figures. (Burton, 1971. *Astron. Astrophys.* 10:76.)

distribution is important in the model only because it determines the streaming amplitudes through the equations of Section 4.2.3. The borders of the shaded regions in Figure 4.15(c) are the loci of maximum $|V_\theta|$ and the loci of $V_R = 0$. In spite of the very simple spatial distribution input, the resulting agree-

ment with the observations is rather detailed when viewed in the observed velocity space.

4.3.4 Some Remarks on the Spiral Structure of the Galaxy

It is clear from Figure 4.10(c) that any deviations from the adopted rotation law of

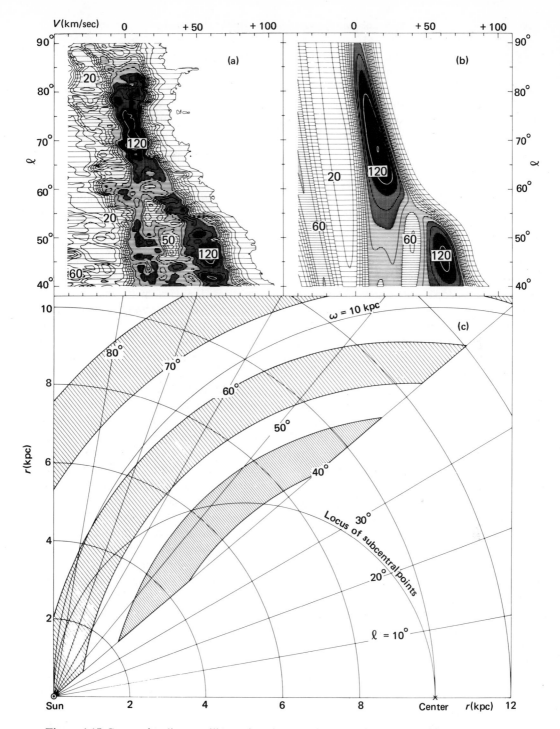

Figure 4.15 Composite diagram illustrating the model line-profile approach for the hydrogen distribution near the galactic plane in the region $40° < l < 90°$. (a) Observed l, V contour map. (b) Model contour map based on self-consistent density-wave streaming and hydrogen distribution, derived as a best fit to the observations. Although there is no detailed structure in the input, a certain amount of detailed agreement with the observations is evident after transformation to the observed l, V space. (c) Map of the hydrogen distribution corresponding to the density distribution inherent in part (b) of the figure. The shaded regions are regions of above-average hydrogen density. The emphasis in deriving the distribution is on the agreement of Figure 4.15(b) with Figure 4.15(a), since both of these figures represent the same coordinate-velocity space. (Burton, 1971. *Astron. Astrophys.* 10:76.)

only 1 or 2 km sec^{-1}, which are systematic over a few degrees on the sky, will probably cause serious errors in the interpretation of the profiles. Since the kinematics dominate the appearance of the profiles, interpretation of the structure observed in the profiles in terms of a map of the density distribution requires, in advance, very accurate knowledge of the velocity field throughout the Galaxy. However, the velocity field can be directly measured only along a restricted region near the locus of subcentral points. At the same time it is also obvious that structure in the *temperature* distribution will also result in structure in the observed profiles. Consequently it is necessary to know the temperature distribution on a large scale. There is abundant evidence that the temperature varies on a small scale, but the observational evidence for a large-scale variation in the harmonic mean temperature of the neutral hydrogen is rather shaky so far. Maps of the hydrogen distribution have generally been based on the assumptions that the temperature of the gas is constant and that the optical depth is low. Although these assumptions are not generally valid, they have been necessary in order for the analysis to proceed.

We saw in the preceding section that streaming motions of the sort known to exist are themselves capable of producing structure in the profiles of the same sort that is observed. This implies that the interpretation of 21-cm profiles in terms of a map of the density distribution on a galactic scale is, at best, difficult. Even if the peaks in the profiles are correlated with density concentrations, it will be difficult to account for the profile distortions caused by streaming motions and temperature variations in terms of the hydrogen density distribution. Only in the special case of gas concentrations (1) with no systematic streaming, (2) free from the shearing effects of differential rotation, (3) with constant temperature throughout, and (4) isolated from the model-independent geometrical effects, can the relative intensities of peaks and valleys in the profiles be interpreted directly in terms of a density contrast

between "cloud" and "intercloud" regions. Similarly only in such a situation will the measured dispersion be a direct indication of the random cloud velocities. Although such a situation may pertain locally at large distances from the galactic plane, it certainly does not pertain to the plane itself.

In fact, it is not clear to what extent a particular peak in a line profile owes its characteristics to streaming motions, to a density concentration, or to a variation in temperature. Although it is certain that fluctuations in the velocity, density, and temperature will accompany one another, the interstellar hydrogen is embedded in such a complicated environment that it does not seem possible to determine, directly from the observations, the relative importance of the variables for each peak or valley in a profile. Near the galactic plane this very complicated environment includes HII regions, supernovae and their expanding shells, regions of star formation, gravitational effects from mass concentrations, and the largely unknown effects of magnetic fields, which are probably coupled to the neutral gas by collisions between the plasma component and the neutral one. Each of these mechanisms can produce differences in the motions, densities, and temperatures on a scale large enough to affect the appearance of the profiles and thus of the subsequent map. This means that hydrogen density variations in the galactic plane cannot be determined with any accuracy directly from observations. Similarly it is not possible to determine the true velocity dispersion from observations in the plane.

The standard map presenting the distribution of neutral hydrogen densities in the galactic plane is shown in Figure 4.9 (Oort *et al.*, 1958). It must be interpreted in terms of the above remarks. The region in this map near the locus of subcentral points deserves special attention for additional reasons. We saw that irregularities in the velocity field are observed along the locus of subcentral points. If distances are derived using a mean rotation curve such as the curve described by Equation (4.4), then there will be some hydrogen emission observed at velocities higher than the maximum velocity predicted by the rotation curve. The rotation curve cannot be used to assign a distance to this gas. In preparing the standard map of the Galaxy, hydrogen contributing such intensities was distributed uniformly over 2 kpc (2.4 kpc if corrected from the old value $R_o = 8.2$ kpc to $R_o = 10$ kpc) of line of sight, situated symmetrically with respect to the subcentral point distance (Schmidt, 1957). The region near the subcentral points in the classic map gets relatively low weight also for the following reason. In preparing the map, the distance ambiguity problem for $R < R_o$ was approached by assuming that the hydrogen layer had a constant thickness. By measuring the distribution in latitude at each velocity an attempt was made to separate the material on the near side of the subcentral point, at a distance r_1 from the Sun, from the material on the far side of the subcentral point, at a distance r_2 where it would subtend a smaller angle. For the region $(r_2/r_1) < 1.8$, which is defined on the map in Figure 4.9 by heavy lines, the method was not considered accurate enough to allow a separation. For lack of any other information, the contributions at r_1 and r_2 were assumed to be equal. The resulting regularity in the region where $(r_2/r_1) < 1.8$ is therefore probably not realistic.

Hydrogen associated with small-scale regions with peculiar kinematics, such as expanding associations, could also have influenced the appearance of Figure 4.9. Clube (1967) suggested that the unique kinematics of the local system of luminous B stars within 400 pc of the Sun, known as Gould's Belt, is sufficient to cause the appearance of a spurious structure on a map of the large-scale structure drawn using a velocity field described by simple circular rotation. (Clube's suggestion illustrates an important point, although it seems that in this particular case the latitude distribution of Gould's Belt is much larger than that of a large-scale spiral feature.)

The map of the large-scale hydrogen distribution in Figure 4.9 thus has to be considered with reserve. The apparent regularity near the locus of subcentral points has low weight. Some other apparently regular features of the map may be spurious. Most of the assumptions upon which the validity of the fundamental equation depends are not strictly correct; even small deviations from the assumed situation can result in substantial errors in the map. However, it should be added that in many cases, even if a peak in a profile owes its prominence to streaming motions, such streaming motions can be seen as perturbations on the basic rotation described by Equation (4.4). Consequently the distance assigned to peaks using the basic rotation may be correct to the first order.

Even with these reservations, it seems safe to conclude that 21-cm observations do indicate structure on a large scale in the galactic plane, although it is not immediately clear whether this structure owes its prominence to velocity, density, temperature, or even perhaps magnetic fluctuations. In some cases the structure defines extended regions of some kiloparsecs length. These extended regions are situated, roughly, along galactocentric arcs. We call this structure *spiral* structure, primarily in view of our knowledge of other galaxies. This is not to imply that spiral structure does not exist in our Galaxy. After all, we do have knowledge of other galaxies, and spiral structure is present to some degree in almost all galaxies with sufficient gas in a disk distribution. Hence it is reasonable to expect it in our own. Although an examination of a number of other galaxies in photographs such as those in the *Hubble Atlas of Galaxies* shows that the structure is more often than not very irregular, in a number of galaxies a general spiral "grand design" emerges from a more or less tangled background structure. Galaxies with such a grand design can be expected to provide much of the information necessary for a confrontation with theories of spiral structure. But it does not seem that the 21-cm observations have demonstrated that our Galaxy does, or does not, belong to the galaxies which display such a grand design.

The problem of the origin and maintenance of spiral structure in galaxies is obviously of fundamental importance. To confront theories of spiral structure, observations are necessary which will provide answers to questions which are so far not satisfactorily answered for our own Galaxy. These questions include the following: Does our Galaxy exhibit a "grand design" of spiral structure? If our Galaxy does have more or less regular arms, are these arms trailing or winding with respect to galactic rotation? What pitch angle and what spacing between arms characterize the structure? What is the density distribution across an arm? What are the motions of the gas between arms? What are the motions and distribution of the gas relative to the stars? Solutions to these spiral structure problems require that distances be determined with an accuracy substantially better than the characteristic width of a spiral feature.

The weakness of our answers to these questions is due, more than anything else, to our vantage point within the Galaxy. Our vantage point for observing other galaxies is better. It appears that many of the problems which, according to the plans of a decade or so ago, were to be solved for the Milky Way system using 21-cm methods, can now better be approached through investigations of other galaxies. This requires line receivers of great sensitivity and telescopes of high resolution.

At the same time the Milky Way problems should be investigated using all the material which can be made available. The picture of the Galaxy which will emerge will be a synthesis of information from a variety of studies (see Bok, 1971). Insight into the large-scale structure of our Galaxy will come from studies which might include: O- and B-type stars and their associations, M-type supergiants, long-period Cepheid variables, the distribution of optical and radio polarization vectors, the distribution of the continuum background radiation, the integrated properties of 21-cm profiles, the information contained in the latitude variations of 21-cm

parameters, optical and radio studies of HII regions, recombination-line studies, optical and radio studies of absorption lines, and molecular-line surveys (especially of OH, H_2CO, and CO).

There is also much which can be, and has been, learned about the neutral hydrogen distribution using less accurate distances. This is discussed in the following sections.

4.4 The Neutral Hydrogen Layer

Observations of the hydrogen distribution in the z-direction (perpendicular to the galactic plane) have shown that the gas is confined to a thin and rather flat layer at distances $R < R_o$. In the outer parts of the Galaxy the layer is thicker than in the inner parts and is systematically distorted from the plane defined by $b = 0°$.

This layer can be studied at distances $R < R_o$ by measuring at each longitude the distribution in latitude of the intensities observed near the terminal velocity. At the terminal velocity there is no distance ambiguity, so the linear thickness of the layer can be measured. The deviation of the center of the hydrogen distribution from the galactic plane can also be measured. Studies of this sort have shown that, although latitude variations in various parameters are present, the main structure is confined to a remarkably thin and flat layer at distances from the center less than $R \approx R_o$. The average full thickness of the layer to half-density points is about 220 pc at $R \lesssim 9$ kpc. This thickness depends on the subcentral point distances, which in turn depend geometrically on the distance scale of the whole galaxy. In the original thickness determination by Schmidt (1957), $R_o = 8.2$ kpc was used. Although subsequent determinations of the layer thickness using telescopes of higher resolution have shown the angular thickness to be somewhat smaller than in the original determination, the revision of the distance scale to $R_o = 10$ kpc allows the characteristic thickness of about 220 pc to be retained. This value is an upper limit, since it is obtained from observations which contain at the subcentral point velocity a contribution from an extended line-of-sight region, along which the mean z may fluctuate, thereby increasing the apparent thickness. In the region where $R \lesssim 100$ pc (and in the inner parts of the galaxy generally) the thickness is quite a bit less than 220 pc (*e.g.*, Rougoor and Oort, 1960). Over much of the region within $R \approx R_o$ the deviation of the center of mass of this layer from the galactic equator defined by $b = 0°$ is less than 30 pc (Gum *et al.*, 1960). There are also regions within this distance where the deviation from the central layer is larger; Miller (1971) found that the HI gas in the range $280° < l < 295°$ lies about 150 pc below the galactic plane at 4 or 5 kpc from the Sun. These quantities can be compared with the diameter of the layer, which is about 30 kpc.

From the general flatness of the central layer inside the solar distance it seems plausible to expect that systematic motions in the z-direction are smaller than a few kilometers per second. The random gas velocities in the z-direction must increase toward the galactic center in order to dynamically maintain the constant layer thickness against the total mass density, which increases strongly toward the center of the Galaxy. In order to maintain constant thickness the velocity dispersion in the z-direction should increase proportionally to $(\partial K_z / \partial z)$, the derivative near the galactic plane of the gravitational force in the z-direction (Schmidt, 1957). This derivative increases by a factor of about 3 between $R = 10$ and 4 kpc. The average random motion in the z-direction at $R = 100$ pc might be as high as 50 km sec^{-1}. We saw in the preceding sections that the actual random gas velocities are difficult to measure near the galactic plane. Furthermore, it is possible to get information only on the θ- and R-components of the dispersion, except in the immediate vicinity of the Sun. The question naturally arises, although it is not satisfactorily answered so far, whether the θ-, R-, and z-components are the same. There is some evidence that the dispersion of

the neutral gas increases with decreasing R (Burton, 1971). Observations of the dispersion will be easier to make in other galaxies when adequate equipment is available.

The density of the neutral hydrogen gas remains roughly constant over the major part of the galactic disk, again in contrast to the stellar density, which increases strongly toward the center (Westerhout, 1957; see Figure 17 in Kerr and Westerhout, 1965).

Although the average thickness of the layer does not vary much in the part of the Galaxy with $R < R_o$, this is not to imply that the layer is symmetric with respect to $b = 0°$, either in temperature or velocity characteristics. In particular, velocity-shearing motions are commonly observed but not completely understood (see Harten, 1971; Yuan and Wallace, 1972).

The characteristics of the mean layer of the disk in the transition region near $R = R_o$ are difficult to measure because of the superposition of local gas on the parts of all profiles near zero velocity and because of very poor distance determinations near $R = R_o$.

At distances $R > R_o$ the mean gas layer is not centered at $b = 0°$. In several cases features extend to a height of several kiloparsecs above the plane (Habing, 1966; Kepner, 1970; Verschuur, 1973). These extensions seem to be associated with the spiral structure in the plane because the extensions occur at the same velocity interval as the structure near $b = 0°$. There is presumably no distance ambiguity in this velocity association because of the one-to-one correspondence of velocity to distance in the outer regions. (Primarily because of this one-to-one correspondence the parts of the profiles contributed by the outer regions of the Galaxy are simpler in appearance than the parts contributed by the interior regions, although this of course does not imply that the physical structure is simpler.) The example of a high-z extension in Figure 4.16 is at a height of $z = 3.6$ kpc at $b = 10°$. The derivation of this height requires that the distance to the arm in the plane be known. The velocity of the arm projected onto the

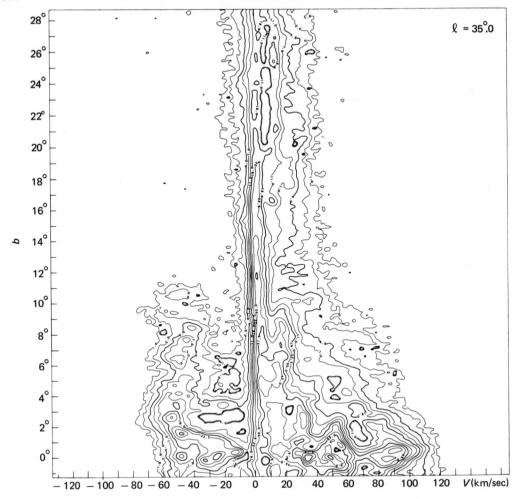

Figure 4.16 Contours of antenna temperature in the velocity-latitude plane illustrating for $l = 35°$ the deviation of the outer part (at this longitude: negative velocities) of the hydrogen layer from the galactic equator $b = 0°$. (Burton and Verschuur. 1973. *Astron. Astrophys. Suppl.*, 12: 145.)

plane is about -50 km sec^{-1}. The distance corresponding to this velocity, 20.9 kpc, can be read from Figure 4.5. However, even an error of 3 kpc in this distance will result in an error of only 0.5 kpc in the z-height.

The high-z extensions remain poorly understood. It is natural to ask how the material reached such large distances into the halo, and once there how it is maintained in the rather narrow velocity slot corresponding to the spiral feature in the plane. It is also difficult to understand how these extensions have preserved their small internal velocity dispersions and have not evaporated into the halo. This is all the more enigmatic since it is

conceivable that a violent event or events shot the gas to the large z-distances in the first place. Oort (1970) has suggested that if the extensions are expelled from the disk layer, explosions such as those observed by Hindman (1967) in the Small Magellanic Cloud could be a possible source of energy. It seems clear that the material would return to the central layer under gravitational attraction on a short time scale of about 10^7 years. The inference is that some replenishment or maintenance mechanism is necessary, since it does not seem reasonable to assume that this rather general phenomenon occurred so recently.

Although the observations of these extensions are still scanty, it is known that the extensions show strong asymmetries with respect to the central layer. In the first quadrant the extensions are primarily to positive latitudes, with no corresponding feature on the negative latitude side of the galactic equator. This is consistent with a general warping of the central layer. This warping is such that the centroid of maximum neutral hydrogen density is located at positive latitudes in the first longitude quadrant and at negative latitudes in the fourth quadrant. Figure 4.16 serves as a reminder that the structural properties derived from observations at one latitude, such as the kinematic distribution shown in Figure 4.12, need not be representative of a particular longitude's properties at an adjacent latitude.

It is not known to what extent the total galactic mass participates in this warping. However, Graham (1970) showed that the layer of OB-type stars in the Carina section of the Milky Way follows the galactic plane to distances of about 3 kpc, but at distances greater than 4 kpc the mean layer is distorted by 2 or $3°$ to negative latitudes. This is consistent with the 21-cm data in this direction. Gum and Pawsey (1960) showed that the continuum radiation is concentrated to the same plane of symmetry as the neutral hydrogen gas.

There have been several theoretical attempts to explain the curious warping of the hydrogen layer. These attempts are reviewed by Hunter and Toomre (1969). Realizing that the Large Magellanic Cloud is at the galactocentric longitude corresponding to the maximum downward bending, Burke (1957) and Kerr (1957) suggested that the bending might be a tidal distortion by the Magellanic Clouds. Although these authors doubted that the Clouds could be massive enough to account for the distortion, others recently re-examined the suggestion and concluded that resonance could build up adequate effects if (and this is a stringent requirement) the Large Magellanic Cloud has remained in a closed orbit around the galaxy

for about 15 revolutions (Elwert and Hablick, 1965; Avner and King, 1967). Habing and Visser (1967) and Hunter and Toomre (1969) also favor a tidal distortion, but one which arose from a single close transit of the Large Cloud. Hunter and Toomre's model requires a passage of the Large Cloud at a minimum distance of about 20 kpc from the center of the galaxy. It has also been suggested that close passages of the Clouds might furnish the initial perturbation for a spiral density wave in the Galaxy (Toomre, 1969). Alternative interpretations of the observed warping have been given by Kahn and Woltjer (1959), who suggested that the distortion might be due to a flow of intergalactic gas past the Galaxy, and by Lynden-Bell (1965), who considered a free oscillation mode of the spinning galactic disk.

Although most of the hydrogen is confined to the central layer, it appears that there is a diffuse and structureless component of hydrogen extending at least several hundred parsecs above the main concentrations. The distribution of densities in the central layer in the direction perpendicular to the galactic plane is approximately Gaussian. However, the distribution does deviate from Gaussian in the form of low-intensity wings extending to higher z-distances (Shane: see Oort 1962; Burke *et al.*, 1964; Mebold, 1972). A decomposition of individual line profiles into Gaussian components by Shane (1971b) has shown evidence for a background envelope of hydrogen with a larger characteristic velocity dispersion (≈ 11 km sec^{-1}) and a larger thickness in latitude to half-density points (≈ 700 pc) than exhibited by the more intense structure in the central layer. Emission from this diffuse background shows a much smoother distribution than that shown by the main concentrations of hydrogen nearer the galactic equator.

4.5 Neutral Hydrogen in the Galactic Nucleus

The nuclei of large galaxies commonly show signs of eruptive activity. A growing body of evidence suggests that an understanding of the phenomena occurring in galactic nuclei is necessary for an understanding of phenomena observed throughout galaxies. The neutral hydrogen in the central region of our own Galaxy has been studied in detail, and although no complete dynamical explanation of the observations has been given so far, a rather clear picture of the kinematics of the central region has emerged.

The reference map in Figure 4.17 shows 21-cm observations made in the galactic plane in the directions between $l = 345$ and $16°$. What has been called the "nuclear disk" appears in this map as the narrow ridge of intensities between $l = 0°$ and $l = 358°5$, extending through negative velocities to about -210 km sec^{-1}. There is a similar ridge on the other side of the galactic center between $l = 0°$ and $l = +1°5$, at positive velocities, which although somewhat confused with other material, is clearly symmetric with the ridge at negative velocities. This symmetry, both with respect to the direction of the center and with respect to zero velocity, strongly suggests that the radiation originates in the central region. The angular extent of the disk implies, adopting $R_o = 10$ kpc, a radius of about 260 pc.

That the velocities in the disk are rotational velocities seems certain in view of the fact that no sign of noncircular motions attributable to the disk is evident at $l = 0°$, either in emission or in absorption. The rotational velocity increases very rapidly going out from the center.

The outer boundary of the disk is quite sharp. The map in Figure 4.17 illustrates the structureless appearance of the nuclear disk, especially for the negative velocities which are uncontaminated by foreground or background emission. This smooth appearance, together with the lack of radial motions, implies that the disk has not been disrupted by violent events in the nucleus.

The nuclear disk extends to $l \approx \pm 1°5$. Between $l = 356°5$ and $355°5$, centered at $V \approx -240$ km sec^{-1}, there is another concentration which also has a symmetric

counterpart at positive velocities at $l \approx 3°$. This "nuclear ring" also has very sharp boundaries. In particular, the intensities at the uncontaminated negative velocities show that the region between the nuclear disk and the ring, at $l \approx 357°$, contains relatively little neutral hydrogen at $b = 0°$. At adjacent latitudes, however, emission at negative velocities is observed (Simonson and Mader, 1973). As is the case with the disk, the asymmetry with respect to both $l = 0°$ and $V = 0$ km sec^{-1} implies that the ring is concentric with the Galaxy. The radius of the ring is about 750 pc, and there appear to be only rotational motions within it.

Rougoor and Oort (1960) (see also Oort, 1971) have derived the mass density in the nuclear region of the Galaxy, assuming that the motions in the disk and ring are governed by gravitation only. They showed that the velocities observed are approximately the same as the velocities which one would expect if the mass distribution in the nuclear region of our Galaxy were similar to that inferred for the Andromeda galaxy from the distribution of light. Thus 21-cm observations have given information on the galactic rotation curve, not only for the region $R > 4$ kpc but also for the region $R < 750$ pc.

From the angle it subtends in latitude, the linear thickness of the disk between half-density surfaces is estimated to be less than 100 pc at distances closer to the center than about 300 pc, but about 230 pc in the outer parts of the disk. The density of observed neutral hydrogen averages about 0.3 atom cm^{-3} and is remarkably constant over the entire disk. Although the total amount of neutral hydrogen observed within 750 pc is roughly 4×10^6 M$_\odot$, the total amount of hydrogen must be much larger, since much of it must be in the form of unobserved hydrogen molecules. The total mass within 750 pc, presumably due for the most part to old and well-mixed population II objects, was estimated by Rougoor and Oort to be 2×10^{10} M$_\odot$. The ratio of neutral hydrogen mass to total mass is thus very much less in the nuclear region than elsewhere in the Galaxy.

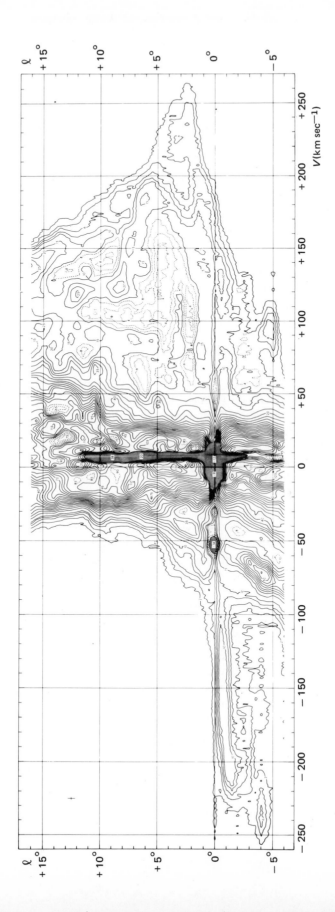

Figure 4.17 Contours of neutral hydrogen brightness temperatures in the plane in directions near that of the galactic center. Broken-line contours enclose regions of relatively low brightness temperatures. The bandwidth (1.7 km sec^{-1}) and the half-power beamwidth (0°.6) are indicated by a cross in the upper left-hand corner. The shaded portions of the map are regions of absorption where contour lines would be overcrowded due to steep temperature gradients. Observations are spaced at half-degree intervals of longitude. (Burton, 1970a. *Astron. Astrophys. Suppl.* 2:261.)

4.6 Radial Motions in the Central Region

So far we have discussed regions where the deviations from circular motion are less than a few percent of the rotational velocities. The situation is fundamentally different in the region between the nuclear disk and $R \approx \frac{1}{3}R_o$, where noncircular motions of the same order as the rotational velocities are observed.

One of the most regular 21-cm features observed is the expanding "3-kpc arm" (van Woerden *et al.*, 1957; Rougoor and Oort, 1960; Rougoor, 1964). This feature is evident in the Figure 4.17 reference map as the ridge of intensities extending from $l = 354°$ at $V \approx -80$ km sec^{-1}, across the Sun-center line $l = 0°$ at $V = -53$ km sec^{-1}, and blending into other hydrogen at $l \approx 5°$. There is strong absorption at $l = 0°$, $V = -53$ km sec^{-1}, indicating that this branch of the arm is located between the Sun and the central continuum source Sagittarius A. The negative radial velocity of the absorption dip indicates that the feature has a net expansion away from the center. Because of this expansion motion, the distance scale of the arm cannot be measured by the kinematic procedure of Section 4.2.1. Instead, the distance scale of the feature has been determined geometrically. The feature can be traced to $l \approx 338°$, where it is seen tangentially. Here $R = R_o \sin 22° = 3.0$ kpc if $R_o = 8.2$ kpc (hence the feature's name) or 3.7 kpc if $R_o = 10$ kpc. Admittedly, the longitude at which the feature becomes tangential to the line of sight is not very easy to determine. This is nevertheless the only point on the arm at which a distance can be estimated. Its distance from the center at $l = 0°$ is therefore uncertain, except that it must pass between the Sun and the center.

Considering now the positive velocities, we see at $l = 0°$ emission at velocities up to about $+190$ km sec^{-1}. There is no appreciable absorption at the high velocities, implying that this gas is located on the far side of the central source and expanding away from it. There are a number of different features evident at positive velocities. The distances to these features, and their interconnections, are highly hypothetical because of the lack of distance criteria.

Besides the gas in the plane, there also appears to be considerable neutral hydrogen, both above and below the galactic plane, which is also moving in a manner forbidden in terms of circular motion (Shane: see Oort, 1966; van der Kruit, 1970; Sanders and Wrixon, 1972a, 1972b). Although there is no satisfactory way to measure the distance of this gas, the observations suggest that it is moving away from the galactic center in two roughly opposite directions. The mass of the neutral hydrogen involved in these motions is of the order of 10^6 M$_\odot$. One of the most prominent concentrations has a velocity of -130 km sec^{-1} at $l = 0°$ and a mean latitude of $-2°5$. One plausible interpretation of these observations, suggested by Oort (1966, 1971) and worked out by van der Kruit (1971), involves gas ejected from the galactic nucleus with a high velocity and at an angle with respect to the galactic plane. The angle under which the ejection took place, 25° to 30°, would have allowed the nuclear disk to survive. According to this model, the material, which has been ejected with velocities of about 600 km sec^{-1}, would return to the galactic plane between 3 and 5 kpc from the center, where the expansion would be braked by the mean gas layer. In this way one could account for the velocity structure of the 3-kpc arm in terms of the low angular momentum of this ejected gas. The energy involved in the ejection would be about 10^{56} ergs. It is not known what mechanism could produce such a tremendous amount of activity in the nucleus. According to the Oort–van der Kruit model the active stage of the nucleus lasted about a million years and took place about 13 million years ago. During this stage the nucleus of our Galaxy was similar to a Seyfert nucleus.

Other models for the motions in the inner regions of the Galaxy are presently being considered. It has been suggested by Shane (1972) and by Simonson and Mader (1973) that gravitational resonance mechanisms, of the sort predicted by the density-wave theory, could produce effects similar to those observed. If these results are correct, then it is unnecessary to invoke explosions or other activity in the galactic nucleus to account for the "3-kpc arm."

It is, however, not clear that only one type of mechanism is responsible for all the motions in the central region. It is also not clear if the motions we observe are transient and perhaps rare phenomena, or if on the other hand we are observing a permanent flow of gas. The flux involved in the gas flow is of the order of 1 or 2 solar masses per year. If the flow is a steady phenomenon, a mechanism for a circulation of gas through the center is necessary. Furthermore, the roles of hydrodynamic and magnetic forces in the nuclear region are not understood. What is clear, in view of rapidly accumulating evidence, is that an understanding of the activity in the nuclei of galaxies is of utmost importance. This evidence includes the recent observations of infrared emission in the nuclei of galaxies, including our own, which indicates extraordinarily energetic phenomena (see, *e.g.*, Low, 1971; Becklin and Neugebauer, 1968; Hoffmann and Frederick, 1969). Ambartsumian (1961) and Arp (1969), especially, have argued that explosive phenomena in the nuclei of galaxies may provide the spiral structure observed in the outer regions.

In addition to the 21-cm data, interpretations of the central region will also have to account for the abundance of molecules, such as OH, NH_3, and H_2CO, which is high relative to other regions of the Galaxy. Sandqvist (1970) studied the absorption spectra of HI, OH, and H_2CO in the direction of the galactic center using observations made with very-high-frequency resolution. He found broad, smooth absorption features showing very little structure. It would be interesting to know how molecule formation can take place so efficiently and why the absorption features can remain so smooth in the presence of the violent motions which apparently exist in the nucleus of our Galaxy.

References

Ambartsumian, V. A. 1961. *Trans. I.A.U. XI-B*: 145–160.

Arp, H. 1969. *Astron. Astrophys.* 3:418.

Avner, E. S., and I. R. King. 1967. *Astron. J.* 72:650.

Becklin, E. E., and G. Neugebauer. 1968. *Astrophys. J.* 151:145.

Bok, B. J. 1971. *Highlights of Astronomy.* Vol. 2. C. de Jager, ed. Dordrecht: Reidel.

Burke, B. F. 1957. *Astron. J.* 62:90.

———, K. C. Turner, and M. A. Tuve. 1964. *The Galaxy and the Magellanic Clouds.* F. J. Kerr and A. W. Rodgers, eds. Canberra: Australian Academy of Science, pp. 131–134.

Burton, W. B. 1966. *Bull. Astron. Inst. Neth.* 18: 247.

———. 1970a. *Astron. Astrophys. Suppl.* 2:261.

———. 1970b. *Astron. Astrophys. Suppl.* 2:291.

———. 1971. *Astron. Astrophys.* 10:76.

———. 1972. *Astron. Astrophys.* 19:51.

———, and W. W. Shane. 1970. *The Spiral Structure of our Galaxy.* W. Becker and G. Contopoulos, eds. Dordrecht: Reidel, pp. 397–414.

———, and G. L. Verschuur. 1973. *Astron. Astrophys. Suppl.* 12:145.

Clube, S. V. M. 1967. *Observatory* 87:140.

Delhaye, J. 1965. *Stars and Stellar Systems.* Vol. 5. "Galactic Structure." A. Blaauw and M. Schmidt, eds. Chicago: University of Chicago Press. pp. 61–84.

Dieter, N. H. 1972. *Astron. Astrophys. Suppl.* 5:21.

Elwert, G., and D. Hablick, D. 1965. *Z. Astrophys.* 61:273.

Feast, M. W. 1967. *Monthly Notices Roy. Astron. Soc.* 136:141.

Graham, J. A. 1970. *The Spiral Structure of Our Galaxy.* W. Becker and G. Contopoulos, eds. Dordrecht: Reidel. pp. 262–264.

Gum, C. S., F. J. Kerr, and G. Westerhout. 1960. *Monthly Notices Roy. Astron. Soc.* 121:132.

———, and J. L. Pawsey. 1960. *Monthly Notices Roy. Astron. Soc.* 121:150.

Habing, H. J. 1966. *Bull. Astron. Inst. Neth.* 18: 323.

———, and H. C. D. Visser. 1967. *Radio Astronomy and the Galactic System.* H. van Woerden, ed. London: Academic Press. pp. 159–160.

Harten, R. H. 1971. Doctoral thesis. University of Maryland.

Henderson, A. P. 1967. Doctoral thesis. University of Maryland.

Hindman, J. V. 1967. *Australian J. Phys.* 20:147.

———, and F. J. Kerr. 1970. *Australian J. Phys. Astrophys. Suppl.* 18:43.

Hoffmann, W. F., and C. L. Frederick. 1969. *Astrophys. J.* 155:L9.

Humphreys, R. M. 1970. *Astron. J.* 75:602.

———. 1971. *Astrophys. J.* 163:L111.

———. 1972. *Astron. Astrophys.* 20:29.

Hunter, C., and A. Toomre. 1969. *Astrophys. J.* 155:747.

International Astronomical Union. 1966. *Trans. I.A.U.* XII-B: 314–316.

Kahn, F. D., and L. Woltjer. 1959. *Astrophys. J.* 130:705.

Kepner, M. 1970. *Astron. Astrophys.* 5:444.

Kerr, F. J. 1957. *Astron. J.* 62:93.

———. 1962. *Monthly Notices Roy. Astron. Soc.* 123:327.

———. 1968. *Stars and Stellar Systems.* Vol. 7. "Nebulae and Interstellar Matter." B. M. Middlehurst and L. H. Aller, eds. Chicago: University of Chicago Press, pp. 575–622.

———. 1969a. *Ann. Rev. Astron. Astrophys.* 7:39.

———. 1969b. *Australian J. Phys. Astrophys. Suppl.* 9:1.

———, and J. V. Hindman. 1970. *Australian J. Phys. Astrophys. Suppl.* 18:1.

———, and R. Vallak. 1967. *Australian J. Phys. Astrophys. Suppl.* 3:1.

———, and G. Westerhout. 1965. *Stars and Stellar Systems.* Vol. 5. "Galactic Structure." A. Blaauw and M. Schmidt, eds. Chicago: University of Chicago Press. pp. 167–202.

Kruit, P. C., van der. 1970. *Astron. Astrophys.* 4:462.

———. 1971. *Astron. Astrophys.* 13:405.

Kwee, K. K., C. A. Muller, and G. Westerhout. 1954. *Bull. Astron. Inst. Neth.* 12:117.

Lin, C. C., C. Yuan, and F. H. Shu. 1969. *Astrophys. J.* 155:721.

Lindblad, B. 1963. *Stockholm Obs. Ann.* 22: No. 5.

Lindblad, P. O. 1966. *Bull. Astron. Inst. Neth. Suppl.* 1:177.

———. 1973. *Astron. Astrophys. Suppl.,* in preparation.

Low, F. J. 1971. *Nuclei of Galaxies.* D. J. K. O'Connell, ed. Amsterdam: North Holland Publishing Co. pp. 195–208.

Lynden-Bell, D. 1965. *Monthly Notices Roy. Astron. Soc.* 129:299.

Mathewson, D. S., P. C. van der Kruit, and W. N. Brouw. 1972. *Astron. Astrophys.* 17:468.

Mebold, U. 1972. *Astron. Astrophys.* 19:13.

Miller, E. W. 1971. Doctoral thesis. University of Arizona.

Oort, J. H. 1962. *Interstellar Matter in Galaxies.* L. Woltjer, ed. New York: Benjamin, pp. 71–77.

———. 1965. *Stars and Stellar Systems.* Vol. 5. A. Blaauw and M. Schmidt, eds. Chicago: University of Chicago Press pp. 455–511.

———. 1966. Nonstable Phenomena in "Galaxies." *Proc. IAU Symp. No. 29.* Yerevan: Izd-vo Akademiia Nauk Armianskoi SSR, pp. 41–45.

———. 1970. *Astron. Astrophys.* 7:381.

———. 1971. *Nuclei of Galaxies.* D. J. K. O'Connell, ed. Amsterdam: North Holland Publishing Co. pp. 321–344.

———, F. J. Kerr, and G. Westerhout. 1958. *Monthly Notices Roy. Astron. Soc.* 118:379.

Roberts, W. W. 1972. *Astrophys. J.* 173:259.

Rogstad, D. S. 1971. *Astron. Astrophys.* 13:108.

Rougoor, G. W. 1964. *Bull. Astron. Inst. Neth.* 17:381.

———, and J. H. Oort. 1960. *Proc. Nat. Acad. Sci.* 46:1.

Sanders, R. H., and G. T. Wrixon. 1972a. *Astron. Astrophys.* 18:92.

———, and G. T. Wrixon. 1972b. *Astron. Astrophys.* 18:467.

Sandqvist, Aa. 1970. *Astron. J.* 75:135.

Schmidt, M. 1957. *Bull. Astron. Inst. Neth.* 13:247.

———. 1965. *Stars and Stellar Systems.* Vol. 5. A. Blaauw and M. Schmidt, eds. Chicago: University of Chicago Press. pp. 513–530.

Shane, W. W. 1971a. *Astron. Astrophys. Suppl.* 4:1.

———. 1971b. *Astron. Astrophys. Suppl.* 4:315.

———. 1972. *Astron. Astrophys.* 16:118.

———, and G. P. Bieger-Smith. 1966. *Bull. Astron. Inst. Neth.* 18:263.

Simonson, S. C., and G. L. Mader. 1973. *Astron. Astrophys.* 27:337.

———, and R. Sancisi. 1973. *Astron. Astrophys. Suppl.* 10:283.

Toomre, A. 1969. *Astrophys. J.* 158:899.

Tully, R. B. 1972. Doctoral thesis. University of Maryland.

Tuve, M. A., and S. Lundsager. 1972. *Astron. J.* 77: 652.

———, and S. Lundsager. 1973. Monograph,

Department of Terrestial Magnetism, Carnegie Institute of Washington, No. 630.

Velden, L. 1970. *Beitrage zur Radioastronomie.* Bonn: Max-Planck Institut fur Radioastronomie. Vol. 1, Part 7.

Venugopal, V. R., and W. L. H. Shuter. 1970. *Mem. Roy. Astron. Soc.* 74:1.

Verschuur, G. L. 1973. *Astron. Astrophys.* 22:139.

Weaver, H. F., and D. R. W. Williams. 1973. *Astron. Astrophys. Suppl.* 8.

Westerhout, G. 1957. *Bull. Astron. Inst. Neth.* 13:201.

———. 1969. *Maryland–Green Bank Galactic 21-cm Line Survey*, 2nd ed. University of Maryland.

———. 1973. *Maryland–Green Bank Galactic 21-cm Line Survey*, 3rd ed. University of Maryland.

Woerden, H., van, G. W. Rougoor, and J. H. Oort. 1957. *Compt. Rend. Acad. Sci. Paris* 244:1961.

Yuan, C., and L. Wallace. 1972. *Bull. Am. Astron. Soc.* 4, 4:316.

General References

1. A. Blaauw and M. Schmidt, eds. "Galactic Structure." *Stars and Stellar Systems.* Vol 5. Chicago: University of Chicago Press. 1965.

2. H. van Woerden, ed. "Radio Astronomy and the Galactic System." *Proc. I.A.U. Symposium No. 31.* London: Academic Press. 1967.

3. W. Becker and G. Contopoulos, eds. "The Spiral Structure of Our Galaxy," *Proc. I.A.U. Symposium No. 38.* Dordrecht: Reidel. 1970.

4. Hong-Yee Chiu and A. Muriel, eds. *Galactic Astronomy.* New York: Gordon and Breach. 1970.

CHAPTER 5

SUPERNOVA REMNANTS

David E. Hogg

5.1 Introduction

About 10 years ago much work was done on the problem of supernova remnants in the galaxy, and many remnants were identified, measurements of the integrated radio flux density were made, and theories of both the supernova event and the expansion of the remnant were postulated. Following this period, interest in supernova remnants waned, and the progress in the field was slow. Now, in response to a number of exciting new measurements, the field has again become the focus of much effort by astronomers. Among the new measurements the most important are the association of pulsars with super-novae; the detection of X-ray emission from a number of remnants; and the availability of high-resolution radio observations of both the total intensity and the polarized intensity from a large number of remnants.

The next section describes the optical properties of supernovae, determined primarily from observations of extra-galactic objects. Subsequent sections describe (1) the two best-studied remnants—the Crab Nebula and Cas A; (2) the radio properties of remnants in the galaxy; and (3) the relationship between supernova remnants and cosmic rays.

Besides the more detailed references which will be cited throughout this chapter, there are a number of excellent surveys of the field, the most important of which are Minkowski (1968), Shklovsky (1968), the Crab Nebula Symposium at Flagstaff (sum-

marized in *Publications of the Astronomical Society of the Pacific* 82, 1970), the I.A.U. Symposium No. 46 on the Crab Nebula (1971), Milne (1970), Downes (1971), Ilovaisky and Lequeux (1972a, 1972b), and Woltjer (1972).

5.2 Optical Properties of Supernovae

5.2.1 Identification of Supernova Remnants in the Galaxy

As the early surveys of radio emission were compiled, it became clear that the radiation from the galaxy was dominated by intense emission from a narrow region about the galactic plane, upon which were super-imposed numerous discrete radio sources having angular diameters of one degree or less. Some of these had thermal spectra and were identified with prominent HII regions (Chapter 3), while others with nonthermal spectra were ultimately identified as remnants of galactic supernovae.

The association between supernova remnants and radio sources was initially suggested from the identification by Bolton, Stanley, and Slee (1949) of the radio source Tau A with the Crab Nebula, the remnant of a supernova observed by the Chinese and Japanese in A.D. 1054. Subsequent work strengthened the association by showing that the strong source Cas A was situated in a filamentary nebula having properties like those of super-nova remnants, and by the successful search for radio emission in the regions of the supernovae of Tycho (A.D. 1572) and of Kepler (A.D. 1604). It is now reasonable to assume that all nonthermal galactic radio sources having diameters of a minute of arc or greater are remnants of supernovae.

The successful identifications of radio sources with SN 1054, SN 1572, and SN 1604 encouraged an examination of the ancient astronomical records of the Chinese, Koreans, and Japanese, in hopes of finding other supernovae. Such records are very difficult to use because they are contaminated by observations of comets and novae which frequently cannot be distinguished because the necessary data on changes of position with time, or of magnitude and duration of variability, are simply not available. Minkowski (1971) concludes that two other objects have been observed optically within the last 3000 years—the supernovae of A.D. 185 and of A.D. 1006. A number of other objects from the ancient catalogs have been suggested as supernovae, and a few of these might indeed gain general acceptance, but it is unlikely that the total number will exceed 10.

5.2.2 Types of Supernovae

Because there have been few outbursts observed in the galaxy, the information about types of supernovae must come from the study of extra-galactic objects. The survey of galaxies for the purpose of finding supernovae has for the past 30 thirty years been led by Zwicky, although a few very interesting objects have been found by chance by other observers. Photometric and spectroscopic observations of some of the 250 supernovae discovered show that there are two principal types of supernovae.

Type I. This type is identified by its light curve, in which the time near maximum is about 50 days, and the subsequent decay is exponential, with the brightness decreasing by $1/e$ in 50 to 70 days. The photographic magnitude at maximum, determined from the work of Kowal (1968), is given by $M_{pg} = 18.6 + 5 \log (H/100)$, where H is the Hubble constant. Also characteristic of this type is

its color (relatively red, B-V between 0.5 and 0.9) and its spectrum, which in the initial stages shows broad overlapping emission bands. The initial velocity of expansion of the ejected material is in the range 15,000 to 20,000 km sec^{-1}. Theoretical studies [Gordon (Pecker-Wimel), 1972] suggest that about 0.5 M$\odot$ is ejected and that the envelope is deficient in hydrogen.

Since all of the identified supernovae that have occurred in E or SO galaxies belong to Type I, it is natural to assume that such supernovae originate in stars of population II, of mass about 1.5 M$\odot$. However, Type I supernovae also occur in the disks of Sb and Sc galaxies, in regions thought to be predominantly of population I. One solution to this problem, proposed by Tammann (1970), is that these supernovae result from collapsing white dwarfs of an intermediate population.

Type II. The light curve shows great variations and cannot be used alone for identification of the type. If photometry is available, these objects can be distinguished by their ultraviolet excess. The best distinction is by means of the spectrum, which at maximum is featureless, with a strong blue continuum. The photographic magnitude at maximum, as determined by Kowal (1968), is $M_{pg} = -16.5 + 5 \log (H/100)$. The ejected material shows velocities of about 6000 km sec^{-1}. By comparison with novae spectra, Shklovsky (1968) estimates that as much as 1 M$\odot$ is ejected in the shell.

Type II supernovae apparently result from extreme population I objects having masses greater than 10 M$\odot$, since they appear only in spiral or irregular galaxies, often actually within spiral arms.

Zwicky suggests that, in addition, there may be three other less common types, Types III, IV, and V. These objects may simply be extreme variations of the other types, but are so rare that their properties cannot clearly be established. Thus Type III supernovae may be similar in nature to Type II, except that a larger mass is ejected with a higher (12,000 km sec^{-1}) velocity. Type V supernovae could be either dwarf supernovae or massive novae, with velocities of only 2000 km sec^{-1}.

5.3 The Crab Nebula

5.3.1 Historical Summary

Throughout the study of supernova remnants the Crab Nebula has occupied a central position. The relationship between an observed supernova and visible nebulosity was established by the work of Duyvendak, Mayall, and Oort (1942), which conclusively identified the peculiar nebula M1, the Crab Nebula, with the supernova of A.D. 1054. Subsequently the Nebula was identified as a radio source. Shklovsky (1954) proposed that the anomalous optical continuum from the Crab was synchrotron radiation, a proposal that was confirmed when the predicted optical polarization was discovered by Vashakidze (1954) and Dombrovsky (1954). Finally, the Crab Nebula was one of the first X-ray sources to be identified, and it is still the only supernova remnant showing an optical pulsar.

Because of its importance in the general study of supernova remnants, the Crab Nebula is discussed in detail in this section. However, it must be emphasized that this nebula has unique properties. The majority of the well-studied remnants are of a quite different nature, and are more nearly like Cas A, as will be shown in subsequent sections.

5.3.2 Optical Properties

The optical appearance of the Crab Nebula is dominated by an intricate network of sharp well-defined filaments which have given it its name. Although the distribution of filaments in three dimensions is difficult to reconstruct, the filaments clearly are not confined to a thin shell at the periphery. The brightest filaments are distributed irregularly over the face of the object, and, in a few cases, extend radially outward. The fainter filaments are found in almost all parts of a well-defined

elliptical region of dimension 3 minutes of arc by 2 minutes of arc.

Extensive studies of the radial velocities and proper motions of the filaments have been made by Trimble (1968). Measurements of both radial velocity and proper motion are available for 125 filaments, and these can be used to determine the distance of the nebula if the geometry of the filaments is known. Limits to the distance are obtained by assuming that the volume containing the filaments is either an oblate or a prolate spheroid. In the center the radial velocity is observed to be $v_r = 1450$ km sec^{-1}. Along the major axis the largest proper motions are 0.22 arc sec yr^{-1}, and are fairly well-behaved, but along the minor axis there are large dispersions, with values up to 0.17 arc sec yr^{-1}. The distance in pc is then

$$D = \frac{v_r}{4.74\,\mu}, \quad v_r \text{ in km sec}^{-1}, \quad \mu \text{ in sec yr}^{-1}$$

(5.1)

from which for an oblate spheroid having $\mu = 0.22$, $D = 1.4$ kpc, while for a prolate spheroid having $\mu = 0.15$, $D = 2.0$ kpc.

Alternatively, instead of assuming a model for the geometry, Woltjer (1970a) has found the maxima in the distribution of both the proper motions and radial velocities, and concludes that the best value for the distance is 1.5 kpc. This is probably a lower limit, since the measurements of radial velocities are biased toward low values. In the following, a distance of 2.0 kpc will be used.

The proper motion studies of the filaments provide two other important facts about the nebula. First, if the motions are assumed constant and extrapolated back in time, the filaments converge in A.D. 1140, with an uncertainty of 15 years. Since this is significantly later than the outburst of A.D. 1054, the expansion must be accelerating. Second, the convergent point of the nebulosity differs from the position of the pulsar, which in turn is generally accepted as being the stellar remnant of the supernova. The proper motion of the star is difficult to measure, apparently because many of the observations have been made with low angular resolution and are thus affected by the variable features in the nebulosity. However, the best estimate of the stellar proper motion is consistent with it being at the convergent point of the nebulosity in A.D. 1054.

The spectrum of the filaments is moderately rich in emission lines, with a number of hydrogen recombination lines as well as the lines of [OII], [OIII], [NII], [NeIII], HeI, and HeII. With these lines it is possible in principle to determine the physical conditions in the nebulosity. The density is simply obtained from the ratio of the lines in the [OII] doublet at $\lambda 3729/3726$, and is about 1000 cm^{-3}. The temperature, total mass, and element abundance are much more difficult problems, since they depend upon both the differential absorption across the spectrum and the excitation mechanism. Davidson and Tucker (1970) have shown that a plausible model is obtained if it is assumed that collisional ionization is unimportant, and that the ionizing radiation originates in an ultraviolet continuum which smoothly joins the optical and X-ray data. In this case the density of helium (including all levels of ionization) must be approximately equal to that of hydrogen to explain the strong helium recombination line. That helium is seven times more abundant than in the interstellar medium in general suggests that the filaments were formed from enriched material ejected at the time of the supernova outburst. The temperature of the nebula drops from 15,000°K at the center to 10,000°K at the edge, and the total mass of ionized gas is 1.5 M⊙. It is also likely that some of the filaments have cores in which the hydrogen is neutral; the mass of the neutral gas is not known.

Within the region outlined by the filaments is a bright continuum source, erroneously referred to as the amorphous component. Although generally elliptical in shape, the most intense emission comes from an "S"-shaped ridge along the major axis, and there are bays to both the east and west side where the emission is weak. It is also

weak in the center, near the pulsar. Under best seeing, the emission appears to come from a complex pattern of very fine filaments, which are of a different nature than the line-emitting filaments previously described.

The continuum emission was originally thought to originate in free-free and bound-free transitions in a highly ionized gas. However, the great difficulties with this mechanism—the large mass of ionized material required, the absence of emission lines, the strong radio emission—led Shklovsky (1954) to propose that the mechanism was synchrotron radiation. With the detection of polarization in the optical emission the synchrotron mechanism gained general acceptance, not only for the Crab Nebula but for the other remnants as well.

The best summary of the optical polarization is still that of Woltjer (1958). In the central regions the polarization is 40%, rising to 60% in the outer regions. The direction of the polarized vector is more or less uniform over the central part of the nebula, and implies a magnetic field perpendicular to the bright ridge previously mentioned. In the outer parts the directions are more scattered, but there are a number of well-defined fans. The electric vectors in the fans are radial, suggesting a magnetic field around a current flow that is perpendicular to the plane of the sky.

The "amorphous" component has one other important characteristic. It has been known for 50 years that the details seen optically in the center of the nebula change with time. Recently Scargle (1969) has given a much more detailed description of these changes (Figure 5.1). The south-preceding star S1 is now identified as the pulsar. To the northeast of the pulsar are four well-defined filaments. The thin wisp nearest the star S1 moves in a quasi-periodic fashion, away from and toward the star, with an apparent velocity of about 6×10^4 km sec^{-1} and a time scale of two years. The motion of the other filaments is less regular, but it could represent either actual mass motion, in which the gas and field move together, or com-

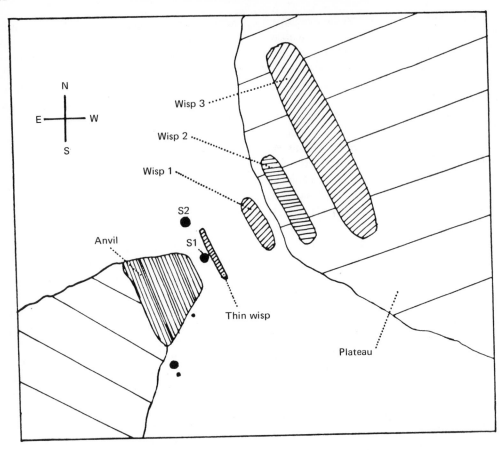

Figure 5.1 Schematic representation of the nebulosity and stars in the central region of the Crab Nebula. (J. D. Scargle, 1969. *Astrophys. J.* 156:401.)

pressional waves, which are generated near the star and move outward.

5.3.3 The Spectrum of the Continuous Emission

In the frequency range 25 to 100 MHz most of the radio emission comes from an elliptical region centered on the optical nebula but larger, of size 5.5 × 3.5 minutes of arc. Within this region, in fact at a position (as determined by long-baseline interferometry) coinciding with that of the south-preceding star, is a strong point source with a peculiar spectrum. The source accounts for 20% of the total flux density at 38 MHz and 10% at 81.5 MHz. Between these two frequencies the spectral index is −1.2; at higher fre-

quencies it must steepen to −2. Its size is uncertain, but scintillation measurements give a value of 0.2 ± 0.1 seconds of arc, implying that if the source is optically thick at 38 MHz and is radiating by the synchrotron mechanism, then the magnetic field (from Equation 12.22) is

$$B \sim 2 \times 10^{-5} \, \theta^4 \, \nu^5 \, S_m^{-2} \sim 10^{-10} \quad (5.2)$$

and the energy in particles alone would be 10^{50} ergs. Both of these values seem unacceptable. An alternative explanation—that the source radiates by plasma oscillations—is unattractive, since the predicted high degree of circular polarization has not been found. A more plausible suggestion, by Lang (1971) and Drake (1970), is that the pulsar radio emission has been scattered by the interstellar medium. The predicted values of source size and spectrum, as well as the absence of pulses, are in general agreement with the observations.

Between 0.1 and 10 GHz high-resolution observations are now becoming available. Figure 5.2 shows a map of the total intensity at 2695 MHz, made with the NRAO synthesis interferometer (Hogg, Macdonald, Conway, and Wade, 1969) having angular resolution of 10 seconds of arc. The nebula is still elliptical, but smaller than at the lower frequencies. There is good agreement between the optical and radio features; for example, the prominent optical bays have radio counterparts, as does the ridge along the major axis of the ellipse. Correspondingly

detailed maps of the polarization at 2695 MHz also show features similar to those at optical wavelengths. The degree of polarization is about 10 % near the center, and the polarization vectors have a uniform direction in the region of the central ridge. There are in the outer parts a number of distinctive features which clearly correspond to the optical fans, but the orientations of the electric vectors are much different, presumably because of rotation measures in excess of 300 rad m^{-2}.

For many years it was thought that the flux density increased by an order of magnitude at millimeter wavelengths and fell again in the infrared. Often, however, these early observations were not corrected properly for the size of the source. Where only the best

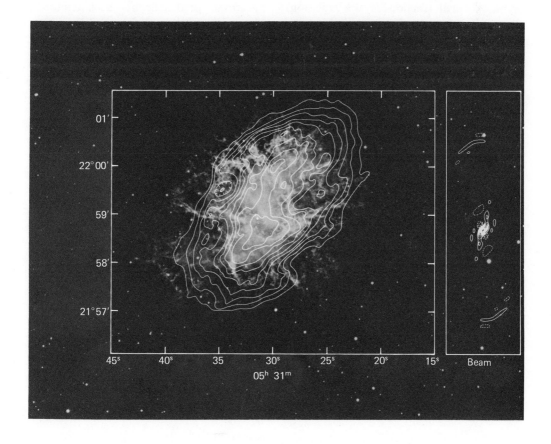

Figure 5.2 The brightness distribution over the Crab Nebula at a wavelength of 11.1 cm. The coordinates are for 1950.0. The contour interval is 620°K and the outermost contour level is 2200°K. (D. E. Hogg *et al.*, 1969. *Astron. J.* 74:1206.)

data are included, the spectrum continues smoothly from decimeter wavelengths to optical wavelengths with a spectral index of −0.26.

The characteristics of the optical emission have already been reviewed. The apparent spectral index of the optical emission is strongly influenced by the amount of extinction, which is not known accurately. If, for example, the visual absorption is 1 magnitude, then the intrinsic optical spectrum would fit smoothly to the infrared data, but the index would have steepened to −0.9.

The Crab Nebula is a strong X-ray source, which has been observed throughout the range 1 to 100 keV. The spectrum of the radiation is hard, although the spectral index of −2 is much steeper than at longer wavelengths. A small fraction of the emission, ranging from 2% at 1 keV to 15% at 100 keV, is pulsed, indicating an origin in the pulsar.

The remainder of the X-ray emission comes from an extended region of diameter 100 seconds of arc, the centroid of which lies near, but perhaps not coincident with, the south-preceding star. The X-ray emission is probably Compton-synchrotron emission from the Nebula itself; in order that it be radiation scattered from the pulsar, there must be an unacceptably large amount of dust as well as a large X-ray flux from the pulsar. That the average X-ray polarization in the range 5 to 20 keV shows polarization of 15% (Novick *et al.*, 1972) is additional support for the synchrotron model.

The Crab Nebula itself has not yet been detected at energies above 1 MeV, although pulsed γ-ray emission has been observed. The upper limits are consistent with the emission spectrum predicted by a Compton-synchrotron model, in which optical and radio photons generated by the synchrotron process

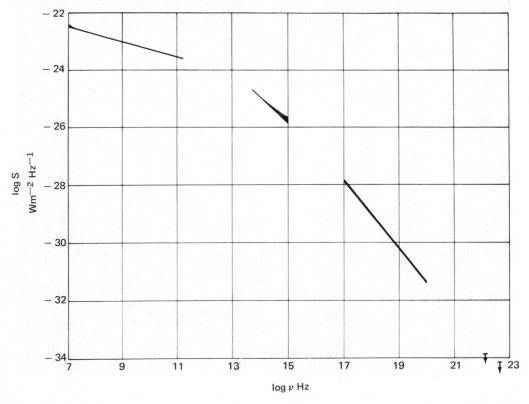

Figure 5.3 The electromagnetic spectrum of the total radiation from the Crab Nebula. The thickness of the lines indicates the uncertainty at a given frequency. (J. E. Baldwin, 1971. *I.A.U. Symposium* 46:22.)

are Compton-scattered into the γ-ray region. The limits to the γ-ray flux rule out the possibility that the high-energy electrons in the nebula are secondaries produced by the decay of π-mesons.

A summary of the spectrum of the Crab Nebula is given in Figure 5.3, taken from the work of Baldwin (1971). The synchrotron mechanism is the only radiation mechanism which provides reasonable agreement with the observations over this frequency range. The total power radiated is 1×10^{38} ergs sec^{-1}, for an assumed distance of 2 kpc. The UV and X-ray region ($\lambda < 3000$ Å) accounts for 63% of the radiated power, while 14% appears in the optical region (3000 to 10,000 Å), 23% appears in the infrared region (1 μ to 1 mm), and only about 0.5% is emitted at radio wavelengths ($\lambda > 1$ mm).

5.3.4 The Crab Pulsar

The discovery of a pulsar in the Crab Nebula is one of the most important events in the study of supernova remnants, since the pulsar holds the key to the understanding of such diverse problems as the origin of the magnetic field in the nebula and the source of the relativistic particles which produce the observed emission. It was first observed as a radio pulsar by Staelin and Reifenstein (1968). They found, in the course of a survey for pulsars, an intense but highly variable pulsar (occasional peaks of 20,000 flux units have been observed at 430 MHz) with a pulse period of 33 msec. The period is not constant, but shows an increase of 36 nsec per day. There have as well been at least two instances where the steady increase in period has been interrupted by an abrupt decrease, of about 100 nsec.

Shortly after the discovery of the radio pulsar, Cocke, Disney, and Taylor (1969) determined that the south-preceding star was optically pulsing in synchronism with the radio pulses. Figure 5.4 is the dramatic photograph by Miller and Wampler (1969) showing the absence of optical emission from the star between pulses. This star had long

been known for its peculiar spectrum, which is continuous without recognizable absorption lines, and had been suggested by Scargle (1969) as the origin of the motions of the wisps.

Subsequent work has led to the detection of the pulsar in the infrared, X-ray, and γ-ray wavelengths, and has given much detail about the shapes of the pulses, the dispersion measure, and the amplitude variations. A more complete discussion is given in Chapter 6.

It should be noted that a second pulsar, NP 0525, was found near the Crab, but it appears that it is not associated with SN 1054, since the observed decrease in period is inconsistent with the large transverse velocity required.

The decrease in the period of the Crab pulsar inspired Gold (1969) to suggest that the loss in energy from the pulsar could supply the required energy for the Nebula itself. From Equation (6.3) it can be shown that the rate of loss of energy is

$$\frac{dE}{dt} = -I \frac{4\pi^2}{P^3} \frac{dP}{dt} \qquad (5.3)$$

with I being the moment of inertia and P, the period.

For a star of one solar mass and diameter 10 km, values thought to be typical of a neutron star, Equation (5.3) predicts a loss of 4×10^{38} ergs sec^{-1}, to be compared with the radiative loss of 1×10^{38} ergs sec^{-1} from the Nebula.

At this time there is no agreement as to how the energy released by the star may be coupled to the Nebula. It is generally accepted that the pulsar is a rotating magnetic neutron star with a corotating magnetosphere. The star may be surrounded by a relatively dense plasma. Particles escaping from the magnetosphere could be accelerated to high energy by magnetic dipole radiation, as has been suggested by Ostriker and Gunn (1969), or they may be accelerated electrostatically by a component of the electric field parallel to the magnetic lines (Goldreich and Julian, 1969). Another interesting suggestion concerning

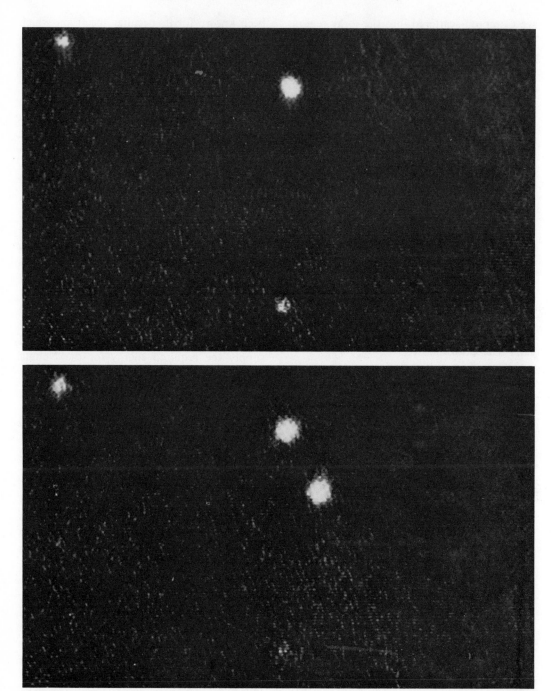

Figure 5.4 The central region of the Crab Nebula, with north at the top and east to the left. The upper picture is taken when the pulsar is near the minimum light, and shows the north-preceding star S2. The lower picture shows the pulsar near maximum light. (J. S. Miller and E. J. Wampler, 1969. *Nature* 221: 1037.)

the period decrease, or "spinup," has been made by Pacini (1971). Shortly after the event in 1969, there was some evidence that the structure of the wisps changed. The change in rotational energy of the star is insignificant, but perhaps there was a major temporary change in the structure of the magnetosphere, permitting the release of up to 10^{43} ergs of plasma energy. Pacini suggests that such an event might occur every year or two, in agreement with the time scale observed for the motion of the thin wisp.

Clearly much more work is required on the various theories, but already the close relationship between the pulsar and the physical conditions in the Nebula has been established.

5.4 The Remnant Cassiopeia A

5.4.1 Optical Properties

The optical object associated with the strong radio source Cas A (3C 461) was first identified by Baade and Minkowski in 1954. The best existing plate shows that the optical nebulosity forms an incomplete shell of diameter 4 minutes of arc, with a large number of small knots located to the north and northeast of the apparent center of expansion. Fainter and more diffuse filaments are located in the southern part of the shell. In the east the shell is broken, and a flare extends outward to a distance of about 4 minutes of arc.

The filaments and knots are of two distinct types: About 30 are very red and show radial velocities of only 30 km sec^{-1}, while the other hundred are much bluer, and have either large radial velocities of up to 6000 km sec^{-1} or large proper motions. In a recent study of the high-velocity knots van den Bergh and Dodd (1970) have shown that the motions are consistent with the theory that the knots originated at the same time, in A.D. 1667, and have not suffered significant deceleration. There is no stellar remnant brighter than $m \sim 23.5$ at the position of the center of expansion. The distance is uncertain

by about 10%; the observed proper motions of 0.5 yr^{-1}, combined with an expansion velocity of 7400 km sec^{-1}, lead to a distance of 3 kpc, which is consistent with the hydrogen 21-cm absorption profiles that have been observed.

The origin of the flare at the east is unknown. If it is as old as the filaments and knots in the rest of the shell, then the shell itself must be severely retarded, which is inconsistent with the analysis of proper motions. Moreover, the initial velocity must have been in excess of 30,000 km sec^{-1}, larger than that observed for any other supernova. Alternatively, Minkowski (1968) suggests that two shells have been ejected, at much different velocities, and that the flare is the only surviving part of the high-velocity shell, while the system of knots shows the location of the slower shell. Such an expansion has been observed in the Type III supernova in NGC 4303.

It is an interesting question as to why this supernova was not observed visually by the European astronomers, since to them the object is circumpolar. A typical supernova of Type II has a visual magnitude at maximum of -18, for a Hubble constant of 50 km sec^{-1} Mpc^{-1}. In the absence of absorption the supernova would have attained an apparent magnitude of -6, too bright to have been overlooked. However, if the visual absorption is as much as 7 magnitudes, as measurements by van den Bergh suggest, then the maximum visual magnitude would have been less than $+1$, sufficiently faint that it could have been missed.

The lifetimes of the moving knots are short, on the order of 10 years, with the smallest structures changing even within one year. New knots are continually being formed in the regions in which knots are already present, so that the general shell-like appearance of the source is maintained.

Spectra of the stationary filaments show lines of Hα and [NII], with very weak lines of [OII]. According to Peimbert and van den Bergh (1971), this requires that nitrogen be overabundant relative to oxygen if the filaments are ionized by radiation. Thus, these filaments might be formed by compression of enriched circumstellar material which was present before the supernova event.

The fast-moving knots have a much richer spectrum, characterized by the forbidden lines of oxygen, sulfur, and argon. Lines of hydrogen and nitrogen are absent. Assuming that the total visual absorption is 6 magnitudes, Peimbert and van den Bergh conclude that the abundance of oxygen is anomalously high, by a factor of up to 70, relative to nitrogen and hydrogen. This is the best direct observational evidence for the hypothesis that heavy elements are synthesized in supernova outbursts.

The line ratios also reflect the physical conditions within the knots. The electron temperature is 15,000°K, the electron density is 1×10^4 cm^{-3}, and the mass of the moving knots visible at the present time is $\sim 0.25 M\odot$. It is not known how much of the shell is neutral; if the shell were complete, the total mass could be as high as 2 M$\odot$.

5.4.2 Radio Properties

As the brightest source in the radio sky, with the exception of the Sun, Cas A has been studied extensively at a large number of wavelengths. Between 22.5 MHz and 14 GHz the flux density decreases with frequency, with a spectral index of -0.77. Below 18 MHz the flux density decreases with decreasing frequency, with an index of $+1.6$. The most likely explanation of the turnover at low frequency is absorption by free-free transitions either in the interstellar medium or within the source. A consistent picture for Cas A and a number of other sources showing a similar turnover requires that the absorption be in the interstellar medium. The low-frequency spectral index is too steep to be explained by a cut-off in the energy spectrum of the electrons which produce the synchrotron radiation; too much of the radiation comes from regions of angular size 40 seconds of arc or greater, ruling out the possibility that there is significant self-absorption (cf. Chapter 12); and the Tsytovich-Razin effect could be

important in the filaments where the density is high, but requires too small a magnetic field to be important for the bulk of the source.

The distribution of total intensity of radiation from the source has been mapped at frequencies between 1400 and 5000 MHz, with resolutions between 24 and 7 seconds of arc. The observed distribution is consistent with an emitting shell of thickness 0.4 pc and outer radius 1.9 pc. Figure 5.5 from Hogg *et al.* (1969) shows a map of the source at 11 cm, and the relationship to the optical nebulosity. The shell contains many discrete regions of angular scale 10 seconds or less which lie toward the edge of the source and show little connection with the optical filaments. At 5000 MHz about 9% of the total radiation comes from these small features (Rosenberg, 1970).

In all of these maps the region of the flare containing the high-velocity filaments is characterized by a break in the radio shell. The work of Rosenberg suggests that the spectral index, uniform over the bulk of the source with value -0.75, may be flatter in this region, with value -0.6. There is also a suggestion of an extension in the contours of low-intensity emission. It is interesting to note that at meter wavelengths Jennison (1965) has found a spur whose spectrum might be very steep, thereby favoring low-frequency emission. It will be important to settle the question of the spectral index of the spur, and to measure it at low frequency with higher sensitivity and angular resolution.

The total power radiated by Cas A in the frequency range up to 100 GHz is 4×10^{35} ergs/sec, assuming a distance of 3 kpc. The total energy radiated over the lifetime of this object, if the rate had been unchanged, is 4×10^{45} ergs. This is a small fraction of the total energy in relativistic electrons (10^{49} ergs as calculated by Rosenberg). An estimate of the magnetic field (assumed to be uniform throughout the shell) is obtained by assuming that equipartition between particle energy and magnetic field energy obtains; the field is thus 5×10^{-4} Gauss. In the small more-intense regions of radio emission, the

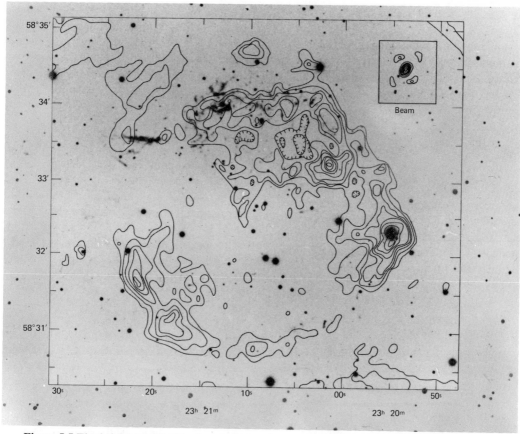

Figure 5.5 The brightness distribution over Cas A at a wavelength of 11.1 cm. The coordinates are for 1950.0. The contour interval is 1750°K and the outermost contour level is 5200°K. (D. E. Hogg *et al.*, 1969. *Astron. J.* 74: 1206.)

emissivity is approximately 100 times greater than in the shell, and the field is twice as strong.

Mayer and Hollinger (1968) discovered several years ago that the radio emission from Cas A was significantly polarized, but that it was so symmetric that there is substantial cancellation of the integrated polarization unless the observations are made with high resolution. Two such high-resolution maps— by Rosenberg (1970) at 6 cm and by Downs and Thompson (1972) at 11.1 cm—are now available. The polarized radiation is mainly concentrated in the bright ring of the source and amounts to about 5% of the unpolarized flux density. Although the data are as yet sparse, Downs and Thompson show that the complex patterns of depolarization and Faraday rotation could be explained by a

magnetic field consisting of a uniform radial component of strength 1.6×10^{-4} gauss and a random component of rms strength 5×10^{-4} gauss. If the depolarization occurs in the shell, then the required electron density is 2 cm^{-3}, and the total mass of the shell is 1.2 M$\odot$. This mass is in addition to the mass contained in the optical filaments, since the filaments occur predominantly in the northwest, while the depolarization is important in the other quadrants.

5.4.3 Secular Changes in Intensity and Structure

Over a decade ago Shklovsky (1960) showed that the flux density of a homogeneous expanding nebula should decrease with time, and that for an especially young object, such as Cas A, the change should be readily observable.

If a source is radiating by the synchrotron mechanism, the flux density, S_ν, will be related to the radius r_o of the radiating volume and the magnetic field strength $H_{o\perp}$ by

$$S_\nu \propto r_o{}^3 \, K_o \, H_{o\perp}{}^{(\gamma+1)/2} \, \nu^{(1-\gamma)/2} \quad (5.4)$$

where the energy spectrum of the radiating electrons is

$$dN(E) = K_o E^{-\gamma} dE$$

As the source expands to a radius r, and assuming both that the magnetic flux remains constant and that the particle energy is limited by statistical acceleration, the following relations obtain:

$$H = H_o \left(\frac{r_o}{r}\right)^2$$

$$E = E_o \left(\frac{r_o}{r}\right) \qquad (5.5)$$

$$\gamma = \gamma_o$$

$$K = K_o \left(\frac{r_o}{r}\right)^{\gamma-1} \left(\frac{r_o}{r}\right)^3$$

Thus the flux density at any later time is simply

$$S_\nu \propto r^{-2\gamma} \qquad (5.6)$$

Actually Cas A is not a uniform sphere, but rather a shell source. For an expanding source in which the shell thickness remains constant, Kesteven (1968) finds that the equation is modified slightly, to become

$$S_\nu \propto r^{-\frac{1}{4}(3\gamma-1)} \qquad (5.7)$$

The annual decrease in flux density is then

$$\frac{\Delta S_\nu}{S_\nu} = -\frac{(3\gamma-1)/2}{T} \qquad (5.8)$$

For Cas A, $\alpha = (1 - \gamma/2) = -0.77$; $\gamma = 2.54$; and $T = 305$ years. The annual decrease predicted by Equation (5.8) is 1.1%—in good agreement with the observed value of $1.3 \pm 0.1\%$ (Scott, Shakeshaft, and Smith, 1969).

If the radio shell is expanding at the rate given by the proper motions of the optical

filaments, it should be possible to see changes in the source structure. Rosenberg (1970) has compared maps made at Cambridge over a time interval of three years, with inconclusive results. In the near future, as the time interval approaches 10 years, it seems likely that significant structural changes will be observed.

5.4.4 X-Ray Emission from Cas A

This remnant has been detected as a source of X-rays in the energy range $1 \le E \le 10$ keV. As yet high angular resolutions are not possible, so that the structure of the X-ray source is not known. It has a hard spectrum, with a spectral index equivalent to -3.3. The flux density ranges from approximately 2×10^{-29} Wm^{-2} Hz^{-1} at 1 keV to 2×10^{-31} Wm^{-2} Hz^{-1} at 10 keV, or about a factor of 10 less than the X-ray flux from the Crab Nebula. The radio spectrum extrapolates to meet the X-ray flux at 1 keV.

It is not yet possible to determine the origin of the X-ray emission. That the radio flux extrapolates to meet the X-ray flux suggests that the X-ray emission could simply be high-energy synchrotron emission. Alternatively, the expanding shell source might have a sufficiently high temperature to produce thermal X-rays. The question of the origin of the radiation is quite critical, because of the short lifetimes involved. For example, Equation (12.6) shows that the power radiated by a relativistic electron is

$$\frac{dE_{Gev}}{dt} = -AB_\perp{}^2 E_{Gev}{}^2 \ \text{Gev/sec} \quad (5.9)$$

from which the half-life is

$$t_{1/2} = \frac{1}{AB_\perp{}^2 E_{Gev}} \ \text{sec} \qquad (5.10)$$

with $A = 3.80 \times 10^{-6}$. For a field of 10^{-4} gauss and X-radiation at 10^{18} Hz, the particle energy from Equation (12.5), assuming radiation at $0.28 \, \nu_c$, is 4.6×10^4 Gev. Equation (5.10) then predicts for such a particle a lifetime of 20 years, requiring that injection of such particles must still be occurring.

5.5 Supernova Remnants in the Galaxy

5.5.1 Some Well-Studied Remnants

There are now more than 90 objects which have been identified as supernova remnants or as possible remnants. Catalogs of these objects have been compiled most recently by Milne (1970), by Downes (1971), and by Ilovaisky and Lequeux (1972a). Attempts to detect hydrogen recombination-line emission from many of these objects have been made, and a number of HII regions which were misidentified have been found (cf. Dickel and Milne, 1972), but the majority are nonthermal galactic sources. There are as yet no data as complete as those available for the Crab Nebula and Cas A. It is hoped that in the near future better maps of radio polarization, more detections of X-ray emission, and more spectroscopic studies of visible nebulosity will be obtained for a large number of remnants. Even now, however, there are sufficient data to reveal some general properties of supernova remnants.

The basis of the discussion of the properties of supernova remnants is the group of 16 objects for which a distance has been estimated. The accuracy of the distance estimate varies greatly from source to source. For example, the three objects for which optical proper motions are available—the Crab Nebula, Cas A, and the Cygnus Loop—are at distances known to better than 20%, while only a lower limit, based on neutral hydrogen absorption, is available for Tycho's supernova. Other techniques used are the association of the radio source with a star cluster of known distance, the estimation of the distance modulus from the observed optical maximum, and the amount of inter-stellar absorption at low frequencies. The properties of these objects, adapted from the work of Ilovaisky and Lequeux (1972a), are given in Table 5.1. A number of these objects are deserving of further comment.

(1) *SN 1572* (Tycho's supernova, 3C 10). This object was observed by Tycho Brahe in 1572, when it attained a maximum apparent magnitude of −4. The observations of the light curve and color suggest that it was a supernova of Type I, and in fact Minkowski (1968) considers it to be a prototype of this class.

Table 5.1 Radio Properties of Remnants of Known Distance

Galactic source number	Source	Flux at 1 GHz flux units	Spectral index	Mean diameter arc minutes	Surface brightness at 1 GHz (MKS)	Distance (kpc)	Diameter (pc)
G 111.7−2.1	Cas A	3100	−0.77	4.3	2.00E−17	3.0	3.8
G 184.6−5.8	Tau A	1000	−0.25	3.6	9.18E−18	2.0	2.1
G 4.5−6.8	Kepler	20	−0.58	3.0	2.64E−19	6–10	5–9
G 43.3−0.2	W 49B	39	−0.33	4.8	2.02E−19	10	14
G 326.2−1.7	MHR 44	145	−0.24	9.8	1.80E−19	4	11
G 130.7+3.1	3C 58	33	−0.10	6.3	9.89E−20	8	15
G 120.1+1.4	Tycho	52	−0.74	8.1	9.43E−20	5	12
G 41.1−0.3	3C 397	35	−0.3	9.0	5.14E−20	7	18
G 332.4−0.4	RCW 103	28	−0.34	9.0	4.11E−20	4	10
G 34.6−0.5	W 44	190	−0.40	31	2.29E−20	3	27
G 189.1+2.9	IC 443	180	−0.45	40	1.34E−20	1–2	12–23
G 263.4−3.0	Vela XYZ	1800	−0.30	200	5.35E−21	0.5	29
G 327.6+14.5	SN 1006	25	−0.63	26	4.40E−21	1.3	10
G 315.4−2.3	RC W86	33	−0.5	40	2.45E−21	2.5	29
G 132.4+2.2	HB 3	36	−0.7	80	6.70E−22	2.0	46
G 74.0−8.6	Cygnus Loop	160	−0.45	180	5.94E−22	0.8	42

The remnant is seen as two filaments and an arc which are symmetric enough to allow determination of the center of expansion. The observed radial velocities are very low, presumably because the filaments are near the edge. Van den Bergh (1971) has found that the proper motions are 0.2 second of arc yr^{-1}, corresponding to a velocity of 4700 km/sec at a distance of 5 kpc. Since the observed proper motion is less than one-half that required if the nebula were expanding uniformly, the shell must be strongly decelerated. If the deceleration has been caused by interstellar matter of density 1×10^{-24} g cm^{-3}, then the mass of material already swept up is 12 M$\odot$, and the initial velocity must have been $\sim$20,000 km sec^{-1}.

The radio source has a shell structure, with a very sharply defined outer edge. The optical filaments lie close to the edge of the radio source. The thickness of the radio shell is about one-quarter of the outer radius. Observations of the polarization at 1420 and 2880 MHz by Weiler and Seielstad (1971) and at 2695 MHz by Hermann (1971) reveal that the field is primarily radial in direction, and is highly ordered, with the polarization generally about 10%, but rising to 20% in some regions. Figure 5.6, from the work of Weiler and Seielstad (1971), shows the distribution of the intrinsic position angle of the electric vector for this source.

Recently this remnant has been detected as an X-ray source. It has a relatively hard spectrum, comparable with that of Cas A.

(2) *SN 1604* (Kepler's supernova, 3C 358). In 1604 Johannes Kepler noted the appearance of this object, with a maximum apparent magnitude of -2. Although the data are not as good as for SN 1572, the supernova was probably of Type I. There are several filaments showing Hα, [NII], and [OI] with normal intensity ratios. The observed radial velocities are in the range 200 to 300 km sec^{-1}. These low velocities are very puzzling, in view of the short time available for deceleration. They imply either that the density of interstellar matter is very

much greater near SN 1604 than it is near Cas A, for example, or that the filaments are analogous to the low-velocity features in Cas A, and may therefore be compressed circumstellar material, as has been suggested by van den Bergh.

Little is known about the radio structure of this source. The recent work by Hermann at 2695 MHz shows that it is probably a shell

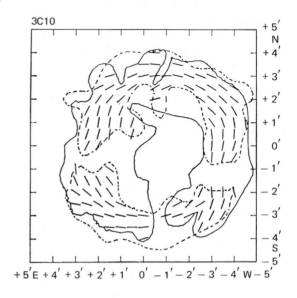

Figure 5.6 Intrinsic position angle of the electric vector for Tycho's supernova. The solid line shows the distribution of intensity of the 21.1-cm polarized radiation, at a level of 0.012×10^{-26} Wm^{-2} Hz^{-1} per square minute of arc, while the dashed line shows the 10.4-cm contour, at a level of 0.029×10^{-26} Wm^{-2} Hz^{-1} per square minute of arc. (K. W. Weiler and G. A. Seielstad, 1971. *Astrophys. J.* 163:455.)

source, with the fractional polarization reaching a maximum of 10% near the center. The limited polarization data are consistent with a radial magnetic field.

(3) *The Cygnus Loop*. This is a famous object showing well-developed filamentary structure. The northeast region is so bright that it is identified separately in the NGC catalog as NGC 6992/95.

Studies of the spectrum by Parker (1964) have led to the conclusion that the filaments are actually thin sheets of nebulosity seen edge-on. There is evidence for temperature stratification behind a shock front. The presence of [OIII] lines is best explained by a region with temperature greater than 50,000°K, while lines arising from hydrogen, nitrogen, and sulfur require temperatures of only 20,000°K. The abundance of these atoms is normal if stratification is assumed. The total mass of the visible nebulosity is 2 M⊙, a small fraction of the 100 M⊙ which might have occupied the volume that has now been swept out. Parker concludes that the distribution of visible filaments is closely related to the distribution of interstellar material with which the expanding shell can interact. The density of the interstellar clouds will govern the level of ionization and the temperature behind the shock front.

From the proper motions and radial velocity of the filaments the distance is 770 pc and the age, allowing for deceleration, is 70,000 years.

The radio emission from the Cygnus Loop shows good correspondence with the optical nebula, especially in the region of NGC 6992/95; there may be significant thermal radiation from the filaments there. There is a prominent source outside the shell at the southwest. In this region the shell is broken, and the source there might represent the loss of energetic particles from the nebula. This source, and the neighboring region in the shell, are the only places where Kundu (1969) found significant polarization at 11 cm. The polarization ranges between 15 and 25%, and if the Faraday rotation is small, shows that the magnetic field is aligned along the filaments.

The Cygnus Loop has been identified as an X-ray source having a spectrum that is much softer than that of Cas A or SN 1572. The X-ray structure measured by Gorenstein *et al.* (1971) has the same angular size as the outermost boundaries of the optical filaments, and is more or less constant across the circular region defined by the filaments. This is in contrast to the radio emission.

5.5.2 The Radio Properties of Supernova Remnants

In recent years—due in large part to the work of Milne, Dickel, Kundu, and Downes —the structure in both total intensity and in polarization has been measured for a large number of remnants. In total intensity 80% of the objects which have been studied with high resolution show definite shell structure, or at least enhanced brightness at the edge of the sources. The remainder could be similar to the Crab Nebula, although with very much lower surface brightness. Amongst the shell sources the thickness of the shell ranges from 10 to 30% of the diameter; Milne (1970) concludes that as the remnants expand, the ratio of the shell thickness to diameter remains constant.

A large number of remnants have polarized emission in which the direction of the electric vector is quite confused, so that there appears to be no systematic magnetic field. Many objects, amongst them the youngest remnants (Cas A, SN 1604, SN 1572, and SN 1006) have a highly ordered radial magnetic field. Seven others, generally with diameters greater than 15 pc, clearly have the magnetic field directed around the periphery of the shell. The picture that emerges is that the field in the younger remnants is radial, presumably a result of the general outflow of material in the expansion. As the expansion proceeds, regions in which the field is tangential are formed by the interaction of the shell with the interstellar medium. Since this latter process is dependent not only on the energy and mass distribution of the shell but also on the strength and orientation of the interstellar field, it is to be expected that remnants in this stage will show a confused field pattern, with both tangential and radial components present.

The spectral indices of the emission from remnants range between -0.1 and -0.8, with a mean value of -0.45. Most remnants have straight spectra over the observable range, except for the absorption at low frequencies by interstellar matter. Suitable observations are not generally available to determine if there are spectral index changes within a source, but it is clear, for example, that in the Cygnus Loop the spectrum of the emission from NGC 6992/95 is curved and is different from that of other parts of the source. Contrary to early suggestions, there is no change in radio spectral index as the remnant expands.

5.5.3 Evolution of Supernova Remnants

It is to be expected from Equation (5.6) or (5.7) that the flux and surface brightness of a supernova remnant will decrease as it expands into the interstellar medium. Since for most objects the decrease in flux will be undetectable, even over a decade, a number of authors have instead attempted to measure the change of flux with size. Some have plotted the distance-independent quantity surface brightness as a function of angular diameter; there will, however, be some scatter introduced by the implicit assumption that all remnants are at approximately the same distance, whereas in fact they range over distances differing by a factor of 10. The most useful plot is that of surface brightness against linear diameter for the sources of known distance (Table 5.1) because if a relationship can be established, then it can be used to find the distances of those objects for which surface brightnesses and angular diameters are known.

Figure 5.7 shows the surface brightness–linear diameter relationship for the objects of Table 5.1. Although there is considerable scatter, there is a definite decrease of surface brightness Σ with increasing diameter D. The line in Figure 5.7 is the least-squares solution, given by

$$\Sigma \propto D^{-3.7 \pm 0.4} \qquad (5.11)$$

In this solution, the Crab Nebula has been omitted, since there is clear evidence for continuing injection of particles. The sources Kepler, MHR 44, 3C 397, W 44, IC 443, SN 1006, and HB 3 have been given half weight,

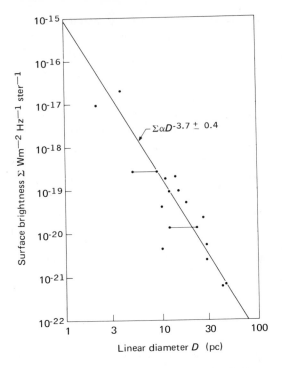

Figure 5.7 The surface brightness Σ as a function of linear diameter D for the 16 supernova remnants from Table 5.1. The least-squares solution $\Sigma \propto D^{-3.7}$ is also shown.

because of the large uncertainties in their distances.

A part of the scatter in Figure 5.6 is due to the uncertainties in the observations. Most serious are the uncertainties in the distances, where the most accurate (Cas A, Tau A, Cygnus Loop) are no better than 10%, while in the worst cases the possible error approaches 50%. The errors in surface brightness are less than 10% for the sources of high surface brightness, where the angular diameter can be accurately defined, but rise to perhaps 30% for a source like HB 3.

The scatter in Figure 5.7 is greater than that expected from these errors, however, showing that there is an intrinsic dispersion amongst the sources themselves. Certainly the extra-galactic supernovae show a range of velocity of the ejecta, and it is to be expected therefore that the galactic supernovae will have a similar dispersion in velocity, and perhaps in initial energy as well. In addition, the latter stages of evolution, even for objects which were initially identical, might be different because of differences in density of the ambient material into which they expand. Finally, it is known that in the Crab Nebula relativistic particles are still being injected 900 years after the supernova event. If continuing injection occurs in other cases as well, it must have a profound effect on the surface brightness–linear diameter relationship, leading for example to a parallel sequence of objects like the Crab Nebula.

From the Shklovsky relation (Equation 5.6) the surface brightness will be related to the linear diameter by

$$\Sigma \propto D^{-2\gamma-2} = D^{4\alpha-4} \qquad (5.12)$$

which for a typical spectral index of -0.5 gives $\Sigma \propto D^{-6}$, much steeper than is possible from Equation (5.11). Similarly, an expanding shell can also be ruled out. A remnant expanding with a shell of constant thickness requires $\Sigma \propto D^{-4.5}$ from Equation (5.7), which is included within the uncertainty of Equation (5.11); such objects are probably excluded since the remnants of low surface brightness have relative shell thicknesses that are comparable to or larger than those of the remnants with high surface brightness. Thus the surface brightness–diameter relationship (Equation 5.11) leads to the conclusion that the majority of the remnants observed are in a phase where the interaction with the surrounding medium has become important (van der Laan, 1962; Poveda and Woltjer, 1968).

Several distinct phases in the evolution of a supernova remnant can now be identified, and are described by Woltjer (1972) and by Ilovaisky and Lequeux (1972a). Initially the remnant expands freely, as a result of the outburst, in the manner described by Shklovsky (1960). This stage is quite short-lived (Woltjer estimates perhaps 100 years) because quite rapidly the expanding remnant sweeps up interstellar matter of total mass much greater than the mass initially ejected. During the second phase the expansion is adiabatic, and could lead to the development

of shell sources of radio emission as envisioned by van der Laan. Most remnants observed are in this phase. For example, Tycho's supernova has entered this phase after only 400 years, while the X-ray emission from large remnants with low surface brightness such as the Cygnus Loop shows that the radiative cooling which terminates this phase has not yet become important. Thus the second stage may last as long as 7×10^4 years. The final stages of evolution are defined only theoretically, since no known remnants are definitely identified with them. The shell might continue expansion at constant radial momentum, or it might be forced to expand by the pressure of cosmic rays trapped within the remnant. In either case the source will ultimately become indistinguishable from the galactic radio background.

A completely different mode of evolution in which the decrease of radio emission is ascribed to the decrease in particle injection by a central pulsar has been proposed by Pacini (1971). In this theory all supernova remnants have pulsars which are a source of continuous injection of relativistic particles. In a remnant such as Cas A the pulsar period is predicted to be long, and the pulsar accordingly would be difficult to detect. The magnetic field near the postulated pulsar would be weak, and the ejected particles would emit significant synchrotron radiation only in an outer shell where the interstellar field has been compressed during the expansion of the remnant. Thus the difference in structure between the shell sources and the more uniform ellipsoids like the Crab Nebula is attributed to the property of the central star.

This theory is difficult to assess, since repeated searches of supernova remnants have failed to find associated pulsars, except in the three instances discussed below. The surface brightness–diameter relationship in Equation (5.11) is of little value, since the relationship predicted in Pacini's theory depends critically on certain properties of pulsars which are not yet known. However, even if there is continuing injection of particles, the remnants must pass through the three expansion stages described above, although the time spent at each stage might be increased over the case where there is no injection after the initial event.

A potentially more sensitive method of studying the final stages of evolution is by determination of the luminosity function. In the most complete discussion at this time, Ilovaisky and Lequeux (1972a) find that for remnants having diameters between 10 and 25 pc, within 7.6 kpc of the Sun, the number having a diameter less than a given value is

$$n(<D) \propto D^{3.2 \pm 0.7} \qquad (5.13)$$

The luminosity function for remnants having diameters greater than 30 pc is dominated by observational selection effects, since such objects cannot easily be identified against the galactic background. On the other hand, there are too few objects with diameters less than 30 pc to permit an accurate determination of the number-diameter relationship; the observed exponent, 3.2 ± 0.7, cannot distinguish whether the remnants are in the deceleration phase (exponent value is 2.5), the constant radial momentum phase (exponent is 4), or the cosmic ray pressure phase (exponent is 3). Since there is little hope of increasing the number of identified remnants significantly, Ilovaisky and Lequeux conclude that the luminosity function has little value in the study of the evolution of supernova remnants.

5.5.4 Distribution of Remnants in the Galaxy

The location of all of the presently identified remnants can be determined by using Equation (5.11) to estimate the distances. The remnants are concentrated toward the Sun because of observational selection effects that are especially serious for objects of low surface brightness. An estimate of the completeness of the sample can be made by assuming that the luminosity function equation, Equation (5.13), applies not only near the Sun but throughout the galaxy as well. In this way Ilovaisky and Lequeux (1972a) find that the surface density of remnants with diameters less than 30 pc is approximately constant as a function of distance from the galactic center out to 8 kpc, with a value of $0.5 \, \text{kpc}^{-2}$, after which the surface density drops rapidly, with no remnants lying beyond 16 kpc. The total number expected in the galaxy is only 200, so that nearly half of the entire population has now been identified.

The remnants are strongly concentrated toward the galactic plane, although the scale height of the distribution increases with increasing distance from the galactic center; the scale height near the Sun is 90 pc. Milne (1970) has found that within the plane the remnants are found preferentially in the neutral hydrogen concentrations of the spiral arms.

It is clear therefore that the distribution of remnants within the galaxy is similar to the distribution of extreme population I objects. They must result from supernovae of Type II (or possibly of Type III). It is tempting to postulate that since the majority of remnants are shell sources like Cas A and the Cygnus Loop, the shell structure is a characteristic of the Type II supernova. Sources like the Crab Nebula and 3C 58 could be remnants of a fundamentally different type of event. However, both Tycho's supernova and Kepler's supernova are believed to have been of Type I, and the remnants of these also have a shell structure. Apparently present data are not sufficient to uniquely classify the type of supernova that led to a given remnant. Some other criterion, such as the spectroscopic characteristics of the filaments in the remnant, must be developed.

There is considerable interest in the comparison of the galactic distribution of supernova remnants and of pulsars, in order to test the hypothesis that pulsars are the stellar remnants of supernovae. The best evidence of an association would be to find pulsars actually in supernova remnants, but after many searches there is still only one well-established association (the Crab pulsar) and two possible associations (pulsar $0833 - 45$ with Vela X and pulsar $1154 - 62$ with G $296.8 - 0.3$ in Crux). This lack of detailed correlation might arise because the surveys for pulsars, especially of short period, are not complete; because the radiation from pulsars is sharply beamed; or because there is no real association. To evaluate this last possibility, Ilovaisky and Lequeux (1972b) have studied the galactic distribution and density of pulsars. They confirm the earlier results that the scale height of the pulsar distribution near the Sun is 120 pc, comparable with that of the supernova remnants. The surface density of pulsars of age less than 5×10^6 yr is about 100 kpc^{-2}, or 200 times greater than the surface density of remnants having diameters less than 30 pc and ages less than 1×10^4 yr. Since the surface densities are approximately in the same ratio as the lifetimes of the two classes of objects, it is concluded that the density and distribution of pulsars is consistent with their origin being in the same type of event that produced the supernova remnants.

5.5.5 Input of Energy to the Galaxy

It has long been felt that supernovae could be a significant source of energy for the galaxy, with the energy transmitted both in the form of cosmic rays and in kinetic energy of the expanding remnant. Now that the properties of remnants are better understood, it is useful to estimate the amount of energy potentially available from these objects.

The initial energy of the supernova outburst must be greater than $\sim 10^{49}$ ergs, the total amount of light emitted near maximum (Minkowski, 1968). Another lower limit is obtained from the kinetic energy of the expanding shell. For a remnant like the Cygnus Loop, which has entered the deceleration phase, Woltjer (1970b) gives the diameter D as a function of time t, energy E of the outburst, and density ρ of the surrounding medium:

$$D = 2.34 \frac{E^{1/5}}{\rho} t^{2/5} \qquad (5.14)$$

For the Cygnus Loop, $D = 42$ pc and $t \sim 5 \times 10^4$ years. Assuming $\rho \sim 1 \times 10^{-24}$ $g\, cm^{-3}$, $E = 2 \times 10^{50}$ ergs. This value is

critically dependent on the density of the interstellar matter, and could be as low as 5×10^{49} ergs. Other remnants also appear to require initial kinetic energies in the range 10^{49} to 10^{50} ergs.

These values are lower limits, because it is not known at what efficiency the initial energy of the outburst can be converted into either optical energy or kinetic energy. Presumably the supernova outburst occurs as a result of evolution of the core of a star to a density of $\sim 10^{11}$ $g\, cm^{-3}$, at which point a dynamic implosion occurs, in the manner described, for example, by Colgate and White (1966). The implosion ultimately leads to an outgoing shock wave which can eject the envelope of the star with a kinetic energy of at least 10^{50} ergs, the amount required by Equation (5.14). The explanation of the optical flash is more controversial. On the one hand Morrison and Sartori (1969) propose that the optical emission is neither from the supernova itself nor from the envelope, but rather is from a large HII region excited by a pulse of UV radiation. Such a model would require that the total energy in the event be $\sim 10^{52}$ ergs for a Type I supernova, and 10^{50} ergs for a Type II supernova. Colgate (1972) on the other hand proposes that the excitation of the HII region results from the conversion of the kinetic energy of the ejected shell; this also requires $\sim 10^{52}$ ergs. Thus, although the nature of the outburst is not yet understood, the total energy involved is 10^{51} to 10^{52} ergs, of which 1 to 10% appears as kinetic energy in the remnant.

Equation (5.14) can also be used to estimate the frequency of occurrence of supernovae in the galaxy. There are 170 remnants in the galaxy having diameters less than 25 pc (Ilovaisky and Lequeux, 1972a). These must have ages less than 1.4×10^4 years from Equation (5.14), assuming that $E/\rho = 2 \times 10^{74}$, the value for the Cygnus Loop. Thus the mean interval between supernovae is 80 years. The value obtained here is slightly higher than the 60 years obtained in a number of other studies; the difference perhaps lies in the value of E/ρ adopted.

With these numbers it is now possible to estimate the amount of energy available from supernovae. The energy appearing as kinetic energy in the expanding remnant is 8×10^{40} ergs sec^{-1}, while the energy available from the initial UV flash amounts to $\sim 10^{42}$ ergs sec^{-1}. This latter amount is large enough to be of critical importance to the thermal and ionization balance of the interstellar gas (Jura and Dalgarno, 1972).

Supernovae and their remnants have also been postulated as the primary source of cosmic rays in the galaxy. To maintain the present energy density of 1×10^{-12} erg cm^{-3} requires a continual injection of $\sim 6 \times 10^{40}$ ergs sec^{-1} (Woltjer, 1970b) for all cosmic rays, and about one-hundredth of that for the electron component. The observed remnants have an energy in relativistic electrons of $\sim 5 \times 10^{48}$ ergs, sufficient to produce an input of $\sim 10^{39}$ ergs sec^{-1} if the particles can escape. However, the mean spectral index of the remnants, -0.45, would produce a cosmic ray electron spectrum of 1.9, significantly different from the observed value of 2.5. Therefore, although sufficient energy for the electron component of cosmic rays is in principle available from supernovae, a mechanism must be found to explain the difference in spectra. To supply the cosmic ray protons would require an injection rate of 1×10^{50} ergs per event. Such energy is available at the time of the outburst, at least in the Colgate-White model, where the outer envelope is ejected at relativistic speeds. However, recent observations of the chemical composition and anisotropy of cosmic rays lead to the suggestion that the protons at least may be of extra-galactic origin (Brecher and Burbidge, 1972).

5.6 Summary

Much progress has been made in the identification of supernova remnants and in the study of their radio properties. The number of such objects known will probably not increase significantly in the next several years, although it would be important to find a few large remnants that are in the third and final stage of evolution. Perhaps one such object is the Gum Nebula. Future observations should concentrate on:

(1) the measurement of the polarization structure at radio wavelengths, in order to determine more clearly the interaction between the remnant and the interstellar gas;

(2) optical studies of filaments in remnants, in order to obtain distances and, perhaps, to distinguish the type of supernova and the stage of evolution of the remnant;

(3) X-ray studies of the remnants, in order to determine if particle injection is still continuing and to distinguish the stage of evolution of the remnant by measuring the temperature of the material inside the shell.

From the standpoint of theory, the stages of hydrodynamic evolution are generally understood, but a much more detailed model of the second stage, where the shell is decelerated by the ambient interstellar matter, would be most useful. In addition, the work by Pacini on the relationship between pulsars and remnants, especially relating to continuing injection of particles, should be followed up.

References

Baade, W., and R. Minkowski. 1954. *Astrophys. J.* 119:206.

Baldwin, J. E. 1971. *I.A.U. Symp. No. 46: The Crab Nebula.* Dordrecht: Reidel, p. 22.

Bergh, S. van den. 1971. *Astrophys. J.* 168:37.

———, and W. W. Dodd. 1970. *Astrophys. J.* 162:485.

Bolton, J. G., G. J. Stanley, and O. B. Slee. 1949. *Nature* 164:101.

Brecher, K., and G. R. Burbidge. 1972. *Astrophys. J.* 174:253.

Cocke, W. J., M. J. Disney, and D. J. Taylor. 1969. *Nature* 221:525.

Colgate, S. A. 1972. *Astrophys. J.* 174:377.

———, and R. H. White. 1966. *Astrophys J..* 143:62.

Davidson, K., and W. Tucker. 1970. *Astrophys. J.* 161:437.

Dickel, J. R., and D. K. Milne. 1972. *Australian J. Phys.* 25:539.

Dombrovsky, V. A. 1954. *Dokl. Akad. Nauk.* 94:1021.

Downes, D. 1971. *Astron. J.* 76:305.

Downs, G. S., and A. R. Thompson. 1972. *Astron. J.* 77:120.

Drake, F. D. 1970. *Publ. Astron. Soc. Pacific* 82:395.

Duyvendak, J. J., N. U. Mayall, and J. H. Oort. 1942. *Publ. Astron. Soc. Pacific* 54:91.

Gold, T. 1969. *Nature* 221:25.

Goldreich, P., and W. H. Julian. 1969. *Astrophys. J.* 157:869.

Gordon (Pecker-Wimel), Ch. 1972. *Astron. Astrophys.* 20:87.

Gorenstein, P., B. Harris, H. Gursky, R. Giacconi, R. Novick, and P. Vanden Bout. 1971. *Science* 172:369.

Hermann, B. R. 1971. Doctoral thesis, University of Illinois.

Hogg, D. E., G. H. Macdonald, R. G. Conway, and C. M. Wade. 1969. *Astron. J.* 74:1206.

Ilovaisky, S. A., and J. Lequeux. 1972a. *Astron. Astrophys.* 18:169.

———, and J. Lequeux. 1972b. *Astron. Astrophys.* 20:347.

Jennison, R. C. 1965. *Nature* 207:740.

Jura, M., and A. Dalgarno. 1972. *Astrophys. J.* 174:365.

Kesteven, M. J. L. 1968. *Australian J. Phys.* 21:739.

Kowal, C. T. 1968. *Astron. J.* 73:1021.

Kundu, M. R. 1969. *Astrophys. J.* 158:L103.

van der Laan, H. 1962. *Monthly Notices Roy. Astron. Soc.* 124:125.

Lang, K. R. 1971. *I.A.U. Symp. No. 46: The Crab Nebula.* Dordrecht: Reidel, p. 91.

Mayer, C. H., and J. P. Hollinger. 1968. *Astrophys. J.* 151:53.

Miller, J. S., and E. J. Wampler. 1969. *Nature* 221:1037.

Milne, D. K. 1970. *Australian J. Phys.* 23:425.

Minkowski, R. 1968. *Nebulae and Interstellar Matter.* B. M. Middlehurst and L. H. Aller, eds. Chicago: University of Chicago Press. p. 623.

———. 1971. *I.A.U. Symp. No. 46: The Crab Nebula.* Dordrecht: Reidel. p. 241.

Morrison, P., and L. Sartori. 1969. *Astrophys. J.* 158:541.

Novick, R., M. C. Weisskopf, R. Berthelsdorf, R. Linke, and R. S. Wolff. 1972. *Astrophys. J.* 174:L1.

Ostriker, J. P., and J. E. Gunn. 1969. *Astrophys. J.* 157:1395.

Pacini, F. 1971. *I.A.U. Symp. No. 46: The Crab Nebula.* Dordrecht: Reidel, p. 394.

Parker, R. A. R. 1964. *Astrophys. J.* 139:493.

Peimbert, M., and S. van den Bergh. 1971. *Astrophys. J.* 167:223.

Poveda, A., and L. Woltjer. 1968. *Astron. J.* 73:65.

Rosenberg, I. 1970. *Monthly Notices Roy. Astron. Soc.* 151:109.

Scargle, J. D. 1969. *Astrophys. J.* 156:401.

Scott, P. F., J. R. Shakeshaft, and M. A. Smith. 1969. *Nature* 223:1139.

Shklovsky, I. S. 1954. *Dokl. Akad. Nauk.* 98:353.

———. 1960. *Soviet Astron.* 4:243.

———. 1968. *Supernovae.* New York: John Wiley.

Staelin, D. H., and E. C. Reifenstein. 1968. *Science* 162:1481.

Tammann, G. A. 1970. *Astron. Astrophys.* 8:458.

Trimble, V. 1968. *Astron. J.* 73:535.

Vashakidze, M. A. 1954. *Astron. Tsirk.* No. 147:11.

Weiler, K. W., and G. A. Seielstad. 1971. *Astrophys. J.* 163:455.

Woltjer, L. 1958. *Bull. Astron. Inst. Neth.* 14:39.

———. 1970a. *Publ. Astron. Soc. Pacific* 82:479.

———. 1970b. *I.A.U. Symp. No. 39: Interstellar Gas Dynamics.* Dordrecht: Reidel, p. 229.

———. 1972. *Ann. Rev. Astron. Astrophys.* 10:129.

CHAPTER 6

PULSARS

Richard N. Manchester

6.1 Introduction

6.1.1 Discovery of Pulsars

The discovery of pulsars by astronomers at the University of Cambridge, England, during 1967 (Hewish *et al.*, 1968) was one of the more exciting events in astronomy of recent years. The very unusual properties of these objects generated a great deal of effort on the part of both observational and theoretical astronomers. This resulted in the production of large amounts of observational data and numerous theories attempting to explain the observed properties. Searches for pulsars were instituted at many observatories around the world, with the result that the number of known pulsars grew rapidly from the original four discovered at Cambridge to over 100 known now. Many of these pulsars have been found by searching for bursts of strong pulses on records from a large radio-telescope equipped with a receiver sensitive to rapid input fluctuations. This technique has been particularly successful at the Molonglo Radio Observatory in Australia, where about 30 of the presently known pulsars have been discovered. Pulsars were not discovered earlier because in most radio astronomical observations, receivers have relatively long time constants to smooth out random noise fluctuations. Consequently these receivers do not respond to pulsar pulses.

6.1.2 Location of Pulsars

When pulsar positions are plotted in galactic coordinates, as in Figure 6.1, there is clearly a clustering along the galactic equator. This shows that pulsars are located within our galaxy and also that they are frequently at distances large compared with the thickness of the galactic disk. Pulsars at high galactic latitudes are probably within a few hundred parsecs of the Sun, whereas those at low latitudes may be more distant. Approximate distances may be obtained from measurements of pulse dispersion (discussed in Section 6.2), and for some pulsars lower limits can be estimated from HI absorption measurements, but no accurate method of determining distances is known at present. However, it is probable that most of the observed pulsars lie within 3 kpc of the Sun.

As observations become more complete, the properties of pulsars seem to be falling into recognizable patterns, and some understanding of the mechanisms involved in pulsar emission is being obtained. Crucial to this understanding was the discovery of the fast pulsars associated with the Crab and Vela supernova remnants. The existence of pulsars with periods as short as tens of milliseconds gave strong support to the rotating neutron star model as discussed by Gold (1968). It was soon discovered that the periods of pulsars were not exactly constant but were, in general, slowly increasing. This observation, together with predictions of strong magnetic fields on pulsars, led to the oblique-rotator model (Section 6.7) for pulsars. This model is now generally accepted, and most explanations for the various pulsar phenomena are based on it.

In the following sections the observed

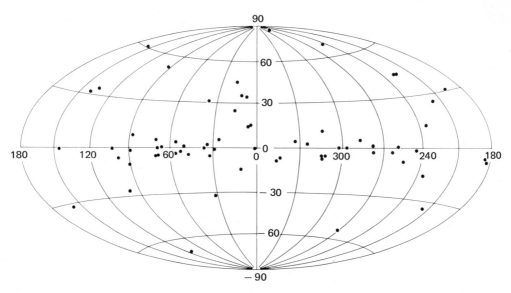

Figure 6.1 Positions of 65 pulsars plotted in galactic coordinates.

properties of pulsars are discussed. These properties fall into two categories: (1) those intrinsic to the pulsar, and (2) those resulting from propagation effects in the interstellar medium. Pulsars are unique probes of the interstellar medium, and a great deal of information can be deduced from analysis of properties falling in the second category. However, properties in the first category are emphasized in this chapter.

Pulsars are named according to their position in equatorial coordinates. When a pulsar is first discovered, it has been common practice to name it with a two-letter prefix, indicating the name of the observatory, followed by four digits giving the approximate right ascension, *e.g.*, NP 0532. Then when an accurate position is obtained, this name is replaced by one giving the (truncated) right ascension and declination (epoch 1950), with the prefix PSR indicating that the object is a pulsar, *e.g.*, PSR 0531+21.

6.2 Basic Properties

6.2.1 Pulse Period

The observed emission from pulsars consists of a series of narrow pulses of variable amplitude recurring at regular intervals. The mean time between successive pulses is known as the basic period, P_1. Observed periods range from 33 msec for the Crab Nebula pulsar, PSR 0531+21, to 3.75 sec for PSR 0525+21, the slowest observed so far. The duty cycle (pulse width/period) for most pulsars lies between 1 and 10%, with an average value of about 3%, and so the pulses are relatively narrow.

In a few pulsars a second pulse is observed between the main pulses. This pulse, known as an interpulse, often occurs almost (but not exactly) halfway between the main pulses and is generally much weaker than them. The Crab pulsar has an exceptionally strong interpulse situated 13.37 msec, or approximately 40% of the period, after the main pulse. In at least two pulsars (PSR 0531+21 and PSR 0950+08) there is significant emission in the shorter of the two intervals between the main pulse and the interpulse.

6.2.2 Dispersion

Dispersion is an important effect of the interstellar medium on pulsar signals. Because of the presence of thermal electrons in the path to the pulsar, the group velocity for radio-wave propagation is slightly less than c, the velocity of light, and is smaller for lower frequencies. Consequently if a pulse is emitted simultaneously from the pulsar at all frequencies, this pulse will arrive at the observer first at high frequencies and at later times at lower frequencies. The delay time for this effect is inversely proportional to frequency squared, so we can define a dispersion constant, DC, by

$$\Delta t = DC \cdot \Delta\left(\frac{1}{f^2}\right) \qquad (6.1)$$

This dispersion constant (in units of Hz) is directly related to the column density of electrons in the path to the pulsar. Consequently the dispersion measure, DM, which is defined by

$$DM = \int n_e dl = 2.410 \times 10^{-16} DC \quad (6.2)$$

[where n_e is the electron density (cm^{-3}), dl is an element of path (parsecs), and the integral is over the path to the pulsar] is commonly used as a dispersion parameter.

Observed dispersion measures range between 3 and 400 cm^{-3} pc for different pulsars. For the Crab pulsar the dispersion measure has been observed to vary by about 1 part in 3000 over time scales of several weeks. No significant variations in dispersion measure have so far been observed for other pulsars, the best upper limits being about 1 part in 1000.

To the extent that the electron density in the interstellar medium may be considered constant, the dispersion measure can be used as an indicator of pulsar distance. Most high-dispersion pulsars are very close to the galactic plane and are probably at relatively large distances. However, at these low latitudes the line of sight may pass through an HII region, which would make a large contribution to the observed dispersion measure. If not allowed for, this leads to an overestimation of the pulsar distance. Observations of interstellar scintillation (Section 6.4) give evidence for small-scale ($\sim 10^{11}$ cm)

fluctuations in the electron density. These fluctuations are weak (less than 1% of the mean density) and have little effect on distance estimates.

6.2.3 Pulse Shapes

Individual pulses from pulsars vary greatly in shape, intensity, and polarization. Several of these effects will be discussed in more detail in subsequent sections. It is remarkable that, despite the variation in shape of single pulses, when the average or integrated pulse shape is obtained by adding together many (500 or more) pulses synchronously with the period, it almost always has a stable shape. Observations made weeks or months apart give essentially the same integrated pulse shape for a given pulsar. The shape of the integrated profile is, however, quite different for different pulsars. Profiles for several pulsars are given in Figure 6.2, showing the wide variation in observed shapes. Many pulsars have two identifiable peaks or components in the profile, as for PSR 1919+21 in Figure 6.2, but the number of observed components varies between one and five for different pulsars. For pulsars with more than one identifiable peak in the profile, it is found that the separation of the peaks is a slow function of frequency, usually with wider separations at lower frequencies.

In two pulsars, PSR 0329+54 and PSR 1237+25, significant changes in the integrated pulse shape have been observed. Most of the time the profile has the shape illustrated in Figure 6.2, but occasionally it abruptly changes to a second stable form which persists, typically for several minutes, sometimes longer, and then abruptly reverts to the standard form. Because of these abrupt changes, this behavior is sometimes called mode changing.

6.2.4 Period Stability and Changes

The outstanding characteristic of pulsars is their extremely stable pulse repetition rate. Accurate measurements show that, over long time intervals, the period stability is often

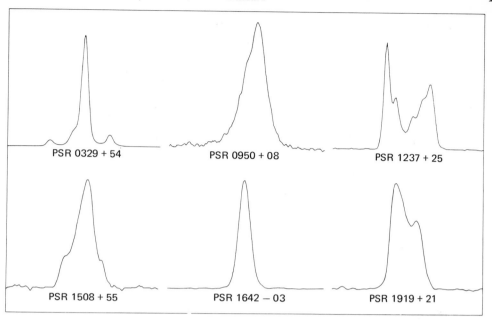

Figure 6.2 Integrated pulse profiles for six pulsars. Only a portion of the period is shown in each case.

as great as 1 part in 10^{10}. To measure these very accurate periods, pulse arrival times are measured at intervals over a long period—weeks or months. If integrated profiles are used, arrival times can often be obtained with an expected error smaller than 100 μsec. The number of pulse periods between two measured arrival times for a given pulsar is then determined using the best previously known period. If the period was known sufficiently accurately, this number will be close to an integer. It is then rounded to this integer and a more accurate period calculated by inverting the equation. Because of the Earth's motion around the Sun, and other motions, the period determined in this way will be Doppler-shifted. Therefore, to calculate accurate pulsar periods, arrival times must be referred to a frame which is not accelerated with respect to the pulsar. The frame normally chosen is that of the barycenter (center of mass) of the solar system. As the pulsar will, in general, have some velocity with respect to the barycenter, the observed period will still be Doppler-shifted from the true value. However, this

Doppler factor is constant and cannot (at present) be separately determined, and so it is normally ignored.

When accurate pulsar periods are calculated in this way, it is found that the observed arrival times gradually begin to depart from arrival times calculated by extrapolation from previous data using a constant period. The curve of residuals or differences between the observed and predicted arrival times usually has a parabolic form. This shows that the period of the pulsar was changing at a uniform rate during the time of the observations. In all cases where accurate measurements are available, the period is found to be increasing with time. To calculate barycentric arrival times, the pulsar position must be known. If this position is not correct, the observed curve of residuals will contain sinusoidal components with a period of one year. Corrections to the pulsar coordinates can be derived from the amplitude and phase of these components. Consequently, the period, period first derivative, and pulsar position are normally obtained simultaneously from a least-squares fit of the observed residuals. In some cases higher-order period derivatives are also derived.

Period derivatives determined in this way are important pulsar parameters. Short-period pulsars normally have large-period derivatives, *i.e.*, they are slowing down quickly. However, there is a considerable scatter in period derivatives for long-period pulsars, some having quite large values. For example, PSR 0525+21, the pulsar with the largest presently known period, also has the third largest known period derivative. Observed period derivatives cover a much wider range than do the periods. This is discussed further in Section 6.6.

The period increase in most pulsars is very regular, and pulse arrival times can generally be predicted far in advance. However, this is not always the case. In March, 1969, the period of the Vela pulsar, the second shortest known, at 89 msec, was observed to decrease discontinuously by about 200 nsec and then resume its regular increase of about 11 nsec/day. The actual period jump was not observed, but it took less than one week to occur. In September, 1971, a second period decrease, this time of 179 nsec, was observed for this pulsar. Similar (but much smaller) jumps have also been observed in the Crab pulsar. Timing data obtained after these jumps show that the period is not quite as predictable as usual; it appears that the pulsar requires some time to settle down. In addition to these discontinuous changes, the period of the Crab pulsar has been observed to fluctuate by a small amount (usually less than 1 part in 10^9) in an apparently random way over time scales of days and weeks. Some other pulsars also show evidence for small fluctuations superimposed on the regular increase in period.

6.2.5 Radio, Optical, and X-ray Spectrum

Radio-frequency observations show that, in most cases, the pulse energy is much less at higher frequencies—that is, most pulsars have a steep radio-frequency spectrum. In Figure 6.3 spectra are given for several

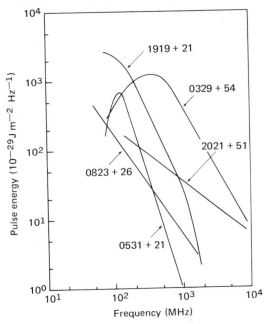

Figure 6.3 Radio-frequency spectra for several pulsars showing that most pulsars are weak at high frequencies. In some cases there is also a low-frequency turnover.

pulsars, showing that the spectral index varies with frequency for a given pulsar and is quite different for different pulsars at a given frequency. In most pulsars the spectrum is convex in form, with a steeper fall-off at high frequencies. In several pulsars a low-frequency cut-off has also been observed. Pulsar spectra are extremely difficult to determine reliably because of the large variations in intensity observed over various time scales (see Sections 6.3 and 6.4). High time resolution observations show that the different pulse components in pulsars with double, or more complex, pulse shapes generally have different spectral indices. Therefore pulse shapes for these pulsars vary with frequency.

The Crab pulsar PSR 0531+21 is so far unique in having been observed outside radio wavelengths. It emits radiation pulsed at the same rate as the radio emission at infrared, optical, X-ray, and γ-ray wavelengths. Cocke, Disney, and Taylor (1969) were first to observe the optical pulses which come from the south-

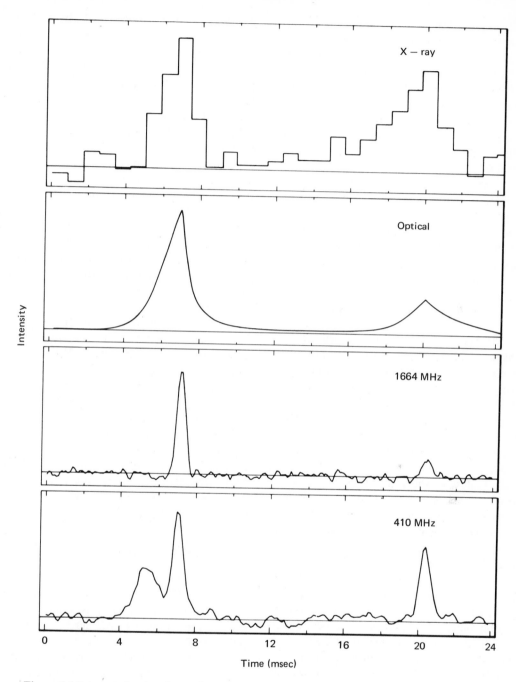

Figure 6.4 Integrated pulse shapes for the Crab pulsar PSR 0531+21 at radio, optical, and X-ray frequencies.

preceding of the two stars near the center of the Crab Nebula. Observations with high time resolution have shown that the emission from this star, which was identified as the remnant of the supernova explosion by Baade and Minkowski as early as 1942, is entirely in the form of pulses. Pulse shapes at several frequencies are given in Figure 6.4, showing that the optical and X-ray pulse shapes resemble the radio shape with a main pulse and interpulse. The pulse peaks in the optical profile are very sharp, being unresolved in observations with a sampling interval of 32 μsec. Pulse arrival time measurements show that, within the observational uncertainty of about 200 μsec, the radio and optical pulses are emitted simultaneously from the pulsar.

The optical and X-ray spectrum is not continuous with the radio spectrum, but has a peak near optical wavelengths. This, and the different pulse shapes at radio and optical frequencies, indicates that the optical emission mechanism is different from that at radio frequencies.

6.3 Amplitude Fluctuations

In Section 6.2 it was mentioned that the shape of individual pulses varies greatly from one pulse to the next. Generally individual pulses are narrower than the mean pulse profile and appear to have a rather simple shape, although high time resolution observations of several pulsars have shown that complex structure is often present within a single pulse. PSR 0950+08 has structure on time scales as short as 10 μsec (Hankins, 1971). The intensity of individual pulses is variable, but is not strongly dependent on where within the mean pulse profile the pulse falls. Strong components in the mean pulse profile usually result from a greater frequency of pulses at the phase of the strong component rather than pulses being stronger at this point. Therefore, the mean pulse profile represents a probability distribution for the occurrence of individual pulses. This is illustrated by the phase-time diagram in Figure 6.5(a) (Taylor and Huguenin, 1971), where each horizontal line on the diagram represents a portion of a period and successive periods aligned in the vertical direction. The random location of individual pulses can be seen.

6.3.1 Drifting Subpulses

In most pulsars the phase of single pulses is random, as described above; however, there exists a class of pulsars in which the phase variation is not random, but highly systematic. In these pulsars the individual pulses move regularly from the trailing to the leading edge of the mean profile in successive pulses. This effect is known as drifting subpulses. A phase-time diagram for PSR 0809+74 is given in Figure 6.5(b) showing the regularly drifting subpulses. Frequently two subpulses are present on one line of this diagram; the spacing between these subpulses is often called P_2 to distinguish it from the basic period P_1. A further period, P_3, the interval between successive crossings by subpulses of a line of constant phase, can also be defined. P_3 is called the pattern repetition period, and the ratio P_2/P_3 is known as the phase drift rate.

At the present time, drifting subpulses have been observed in the phase-time diagrams of six pulsars out of 21 for which suitable observations have been made. For most of these the drifting pattern is not as regular as that shown in Figure 6.5(b), and it is sometimes difficult to see. The phase drift rate may vary in successive bands of subpulses although it is usually constant within a given band. In one case (PSR 0031−07) the drift rate appears to be quantized and to take only one of several harmonically related values.

6.3.2 Periodic Fluctuations

In a pulsar such as PSR 0809+74 [Figure 6.5(b)] the energy of successive individual pulses exhibits a periodic fluctuation as the subpulses drift across the mean pulse profile. This periodic component is shown clearly in spectral analysis of a series of pulse energy values. Power spectral analysis of this type shows that periodic variations in pulse intensity are present in many pulsars— only in some cases are the periodic features related to drifting subpulses. For example PSR 0834+06 has a very strong periodic pulse intensity variation, with a period of slightly more than two pulse periods. Con-

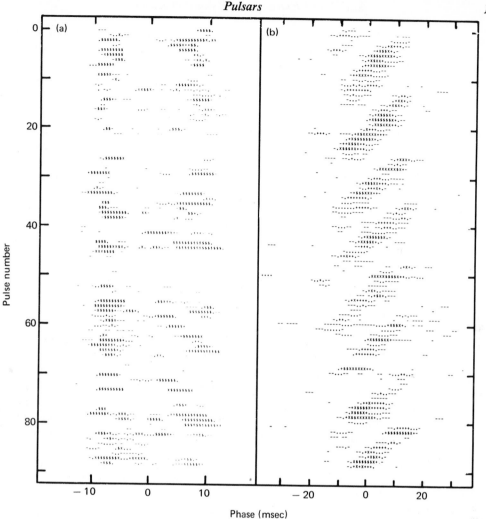

Figure 6.5 (a) Phase-time diagram for PSR 1133 + 16 showing the narrow subpulses occurring at various phases. In this type of diagram each horizontal line represents the portion of the period containing the pulse, and successive lines are for successive pulses. Intensity of the pulse is represented by the darkness of the character plotted. (b) Similar diagram for PSR 0809 + 74 showing the regular drift of subpulses from the trailing to the leading edge of the pulse.

sequently, over short intervals alternate pulses are strong and weak, respectively. These variations are not clearly associated with any drifting-subpulse behavior.

6.3.3 Nulls

Amplitude fluctuations of a different type have been observed in several pulsars. These variations, known as nulls (Backer, 1970), are characterized by a sudden decrease of the pulse energy to essentially zero for a few pulse periods, followed by an equally sudden increase to normal intensity. In some cases the occurrence of nulls may be related to a form of drifting subpulse behavior, but in general there is no clear correlation.

6.3.4 Long-Term Changes

Pulsars also show long-term intensity variations. A plot of the daily or weekly average intensity shows that over a period of several months the mean pulse energy can vary by an order of magnitude or more (Cole *et al.*, 1970). Long-term trends are frequently evident, with the pulse intensity steadily increasing or decreasing over a period of weeks. In some pulsars long-term quasi-periodic variations in intensity seem to occur, but more usually there is no obvious pattern.

6.4 Scintillation

6.4.1 Origin of Scintillation

All of the amplitude fluctuations discussed in the previous section are thought to be intrinsic to the pulsar. Observed pulse energies are, however, frequently affected by interstellar scintillation. Radiation from the pulsar is scattered by irregularities in the interstellar medium. Therefore at any given time an observer sees radiation from many different paths. These different signals interfere at the telescope, producing strong pulses at some times and weak pulses at others. The phase of the interfering signals varies continually because of relative motion of the pulsar and observer with respect to the scattering interstellar clouds.

6.4.2 Characteristic Fluctuation Times and Bandwidths

Scintillation introduces both a time and a frequency structure into the pulsar signals. This is illustrated in Figure 6.6, which shows a series of 40-second averages of the spectrum of radiation from PSR 0329+54. From this diagram it can be seen that if one observes this pulsar with a narrow bandwidth and integrates the pulse intensity over a period of about one minute, this averaged intensity will rise and fall with a characteristic time scale of about 20 minutes. These variations contrast with the pulse-to-pulse intensity variations which are frequently correlated over hundreds of megahertz. The characteristic decorrelation time (τ) and the width of the frequency bands ($\Delta\nu$) are dependent on the observing frequency and the dispersion measure of the pulsar.

A model consisting of a thin scattering region situated between the pulsar and the observer predicts that τ and $\Delta\nu$ should be proportional to the observing frequency and fourth power of the observing frequency, respectively (Salpeter, 1969). The effect of dispersion measure on these parameters must be treated using an extended screen model which shows that bandwidths should be inversely proportional to the square of the pulsar dispersion. Rickett (1970) and others have shown that these relations adequately describe observations of these parameters in most cases, although the scatter is frequently large.

6.4.3 Pulse Smearing

Scintillation has another important effect on the observed properties of pulsars. This is a smearing of the pulse shape which results from the variable delay of signals propagating along different paths to the observer. As the amount of smearing, *i.e.*, the characteristic delay time, $\Delta t \sim (\Delta\nu)^{-1}$, it is most pronounced in higher-dispersion pulsars at low frequencies. The principal consequence of this effect is that the trailing edge of the pulse is smeared out to an approximately exponential decay. In Figure 6.7 pulse shapes at several frequencies are given for PSR 0833−45 showing this effect. For short-period pulsars at a very low frequency, Δt can be of the same order as the period. In these circumstances the pulse shapes are drastically altered and the energy of the pulsed component of the radiation is reduced, resulting in a sharp low-frequency cut-off in the pulsar spectrum. This is clearly seen for the Crab pulsar, PSR 0531+21, in Figure 6.3.

The decorrelation time, τ, is inversely proportional to the relative velocity of the line of sight to the pulsar and the interstellar medium. Observed decorrelation times are consistent with relative velocities of more than 100 km sec^{-1} for several pulsars, showing that pulsars may be formed with a high velocity, *e.g.*, as runaway stars from binary systems.

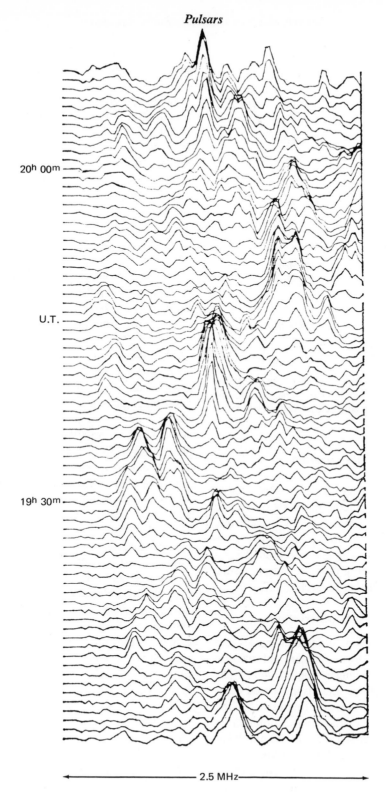

Figure 6.6 Plot of successive 40-sec averages of the spectrum of PSR 0329+54 showing the narrow-band scintillation structure. (Rickett, 1970. *Monthly Notices Roy. Astron. Soc.* 150:67.)

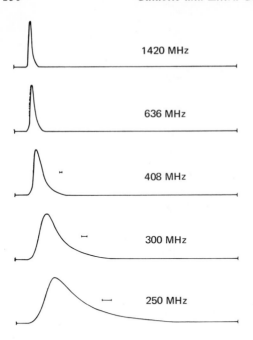

Figure 6.7 Profiles for PSR 0833−45 at several frequencies showing the smearing due to scintillation at low frequencies. (Ables *et al.*, 1970. *Astrophys. Lett.* 6:147.)

6.5 Polarization

Pulsar radiation is frequently highly polarized and, in general, there is a systematic variation of the polarization parameters across the pulse. Observations of these effects have been important in contributing to our understanding of pulsar mechanisms, as is discussed further in Section 6.7.

6.5.1 Individual Pulse Observations

In most pulsars individual pulses are highly polarized, with linear polarization generally being stronger than circular. The position angle of the linearly polarized component normally varies continuously through the pulse, and the circular polarization is often observed to change sense about the middle of the pulse (Clark and Smith, 1969). These polarization parameters usually vary greatly in succeeding pulses so that, in general, the integrated pulse is not as highly polarized as the individual pulses.

In most pulsars this variation appears to be random. However, for the pulsars in which regularly drifting subpulses are observed, the polarization of successive pulses is related to the drifting behavior. A plot showing the variation of polarization parameters for PSR 0809+74 is given in Figure 6.8. It can be seen that the observed position angle is strongly dependent on the subpulse phase and is consistent from one drifting band to the next.

6.5.2 Integrated Polarization

Observations show that in many pulsars the integrated pulse profile remains highly linearly polarized. Circular polarization is observed in some pulsars, generally near the center of the pulse, but it is usually rather weak. Polarization parameters are given for two pulsars in Figure 6.9, showing the strong linear and weak circular polarization. The position angle in almost every case varies in a smooth and continuous way through the pulse. The approximately linear variation shown for PSR 1133+16 in Figure 6.9 is observed in many pulsars (Manchester, 1971). The slope of the position angle variation varies in different pulsars from almost constant through the pulse to a rather large value. In PSR 1508+55 the position angle changes through the pulse by over 180° at an almost constant rate. In PSR 0329+54 the position angle varies continuously through the pulse by almost 360°. The form of position angle variation shown in Figure 6.9 for PSR 2045−16, with rapid variation near the center of the pulse and slow variation near the edges, is seen in several pulsars. The optical pulse from the Crab Nebula pulsar is also polarized in this way (Kristian *et al.*, 1970).

Observations at several frequencies have shown that the variation of position angle through the pulse is the same at all frequencies (Radhakrishnan and Cooke, 1969; Manchester, 1971). This shows that the variations observed are not the result of differing Faraday rotation across the pulse.

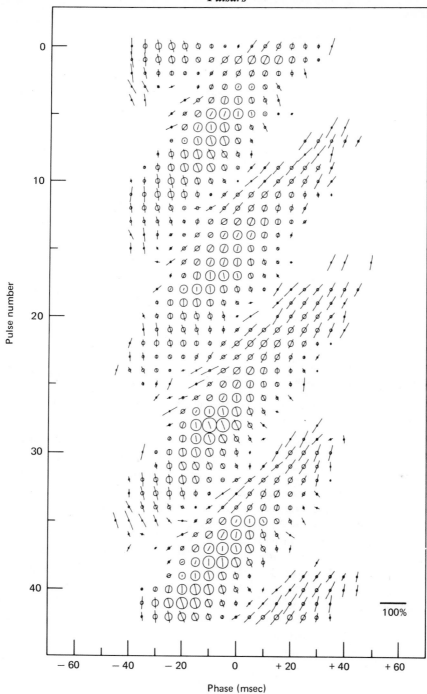

Figure 6.8 A phase-time diagram for PSR 0809+74 showing the polarization of individual sub-pulses. At each point, intensity is represented by the circle diameter, percentage polarization by the length of the line, and position angle by the orientation of the line. (Taylor *et al.*, 1971. *Astrophys. Lett.* 9: 205.)

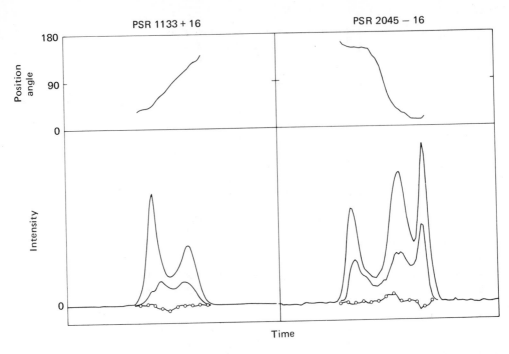

Figure 6.9 Integrated profiles and polarization parameters for two pulsars PSR 1133 + 16 and PSR 2045 − 16. In the lower part of the figure the upper line gives the pulse profile, the second line is the linearly polarized component, and the line with small circles is the circularly polarized component.

6.6 Correlation of Pulsar Properties

6.6.1 Pulsar Categories

The properties described in the previous sections can be used to divide pulsars into different categories. Relationships between the various observed properties provide a basis on which to build pulsar models. Taylor and Huguenin (1971) have divided pulsars into three categories on the basis of their integrated pulse profile and the presence or absence of drifting subpulses. Pulsars having a simple profile are called Type S; those with a more complex profile, having two or more components of comparable intensity, are Type C; and those possessing drifting subpulses are Type D.

Many of the properties of pulsars are related to these classifications. In Figure 6.10 the percentage linear polarization of the integrated profile is plotted against pulsar period for 23 pulsars, with different symbols for each pulsar class (Huguenin, Manchester,

and Taylor, 1971). This figure clearly shows that all Type S pulsars have periods of less than one second, whereas Type C pulsars have longer periods. Also most Type C pulsars are quite highly polarized, whereas Type D pulsars, without exception, have weakly polarized integrated profiles.

The quantity $P(dP/dt)$, where dP/dt is the period derivative, is an important parameter of pulsars. In the next section its relation to the energy loss mechanism which causes the observed secular increase in period will be given. As is also discussed there, the period divided by the derivative, $P/(dP/dt)$, is a characteristic age for the pulsar. In Figure 6.11(a), $P(dP/dt)$ is plotted against pulsar period P, and in Figure 6.11(b) $P(dP/dt)$ is plotted against the characteristic age $P/(dP/dt)$, with different symbols for the three pulsar classes in each case. These diagrams show that Type C pulsars tend to have large $P(dP/dt)$ values, whereas for Type D pulsars this quantity is always small. For

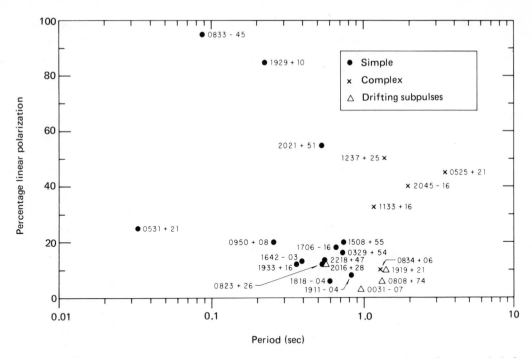

Figure 6.10 Mean percentage linear polarization plotted against period with different symbols for the three types of pulsars. (Huguenin *et al.*, 1971. *Astrophys. J.* 169: 97.)

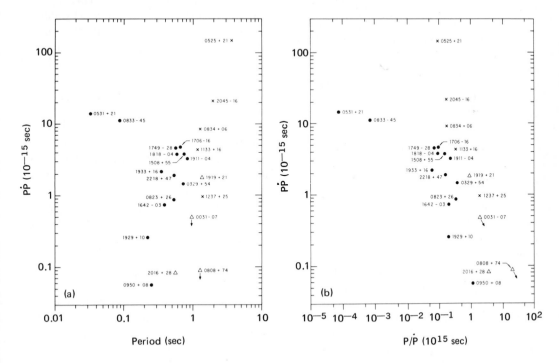

Figure 6.11 (a) The parameter $P(dP/dt)$ plotted against period P, and (b) $P(dP/dt)$ plotted against the characteristic age $P/(dP/dt)$, with different symbols for pulsar types. (Huguenin *et al.*, 1971. *Astrophys. J.* 169:97.)

Type D pulsars the quality of the drifting subpulses seems to be related to the $P(dP/dt)$ value, with the best examples of drifting subpulses mostly occurring in those pulsars with the smallest $P(dP/dt)$ values.

6.6.2 Period Dependence of Properties

Other properties of pulsars are apparently related to period. For example only the shortest-period pulsar, that in the Crab Nebula, has been observed at wavelengths outside the radio range. Only in this pulsar and that with the next shortest period, PSR $0833-45$ in Vela, have discontinuous period changes been observed. This may of course be partly a result of the longer baseline (in periods) available in which to observe these effects. Periodic fluctuations in pulse amplitude are not usually observed in pulsars with periods less than about 0.75 second. This is equivalent to saying that few Type S pulsars have periodic intensity fluctuations. There may be some correlation between distances of pulsars from the galactic plane and their period, with shorter-period pulsars nearer the plane. This could result, on the average, if pulsars are formed with a high velocity (as was suggested in Section 6.4). Longer-period pulsars are probably older and therefore have had more time to move away from the plane.

6.7 Pulsar Models

6.7.1 Rotating Neutron Stars

Following the discovery of pulsars, a large number of models involving pulsation, rotation, or orbiting of either neutron stars or white dwarfs were proposed. With accumulating observational data it soon became clear that only the rotating neutron star model (Gold, 1968) could account for the observed properties. The arguments which led to this conclusion were as follows. First, the great stability of the pulse repetition rate implies that pulsars have large inertia and that forces acting on the pulsar have very little effect on the period. Second, the wide range of periods

observed (100:1) is not consistent with any form of stellar pulsation, as it implies a 10^4:1 range in density, which is too large for one class of object. Third, the very short periods observed for the Crab (33 msec) and Vela (89 msec) pulsars are inconsistent with rotation of a white dwarf and also with orbiting models. Any star with density less than that characteristic of a neutron star could not resist the disruptive forces at these high rotational velocities. Periods in the range of tens of milliseconds are also too short for pulsation of white dwarfs, and the longer observed periods are too long for pulsation of neutron stars. Fourth, the observed steady increase in period is consistent with energy loss from a rotating system. The period of orbital models would steadily decrease with time owing to loss of energy through gravitational radiation. Fifth, the regular swing of position angle observed in integrated profiles can readily be explained in rotating star models, but is difficult to account for in pulsating or orbiting models. Finally, searches for optical identifications of nearby pulsars were not successful, ruling out white dwarf models, as these stars would have been easily visible at the expected distances. Also, it had been predicted that neutron stars would be formed as the collapsed remnant of a supernova explosion, and the two fastest, and presumably youngest, pulsars appear to be associated with supernova remnants. Three other pulsars, PSR $0611+22$, PSR $1154-62$ and PSR $1919+21$, may also be associated with supernova remnants. Thus we are left with rotating neutron stars as the only model (thus far) consistent with the observed properties of pulsars.

6.7.2 Magnetic Dipole Radiation

Neutron stars are expected to possess strong magnetic fields as a result of flux conservation during collapse of the star. The structure of this field is unknown, but it is generally assumed to possess a strong dipole component with a field strength at the stellar surface as large as 10^{12} gauss. The presence of this strong field on a rotating object has a number of interesting consequences.

In general there will be a component of the dipole moment perpendicular to the rotation axis. This model, of a rotating star with a magnetic axis at some angle to the rotation axis, is often called the oblique-rotator model for pulsars. As the star rotates it will radiate electromagnetic waves at the rotation frequency. This energy loss may be responsible for the secular slowdown of pulsars (Pacini, 1968). The rate of energy loss by this mechanism is given by

$$\frac{dE}{dt} = -\frac{2m^2\Omega^4}{3c^3} = I\Omega\frac{d\Omega}{dt} \qquad (6.3)$$

where m is the perpendicular component of the magnetic dipole moment, Ω is the rotation frequency, and I is the moment of inertia of the neutron star. As the period $P = 2\pi/\Omega$, Equation (6.3) shows that

$$P\frac{dP}{dt} = \frac{8\pi^2 m^2}{3Ic^3} \qquad (6.4)$$

That is, the product $P(dP/dt)$ is proportional to the square of the dipole moment and inversely proportional to the moment of inertia of the neutron star.

A characteristic age, τ, can also be defined:

$$\tau = P/\frac{dP}{dt} \qquad (6.5)$$

If the star was rotating rapidly when it was formed and magnetic dipole radiation is the dominant energy loss, then the time since formation is $\tau/2$ (Ostriker and Gunn, 1969a).

These results, in conjunction with the correlations discussed in Section 6.6, show that Type D pulsars probably have weak effective dipole moments, whereas strong magnetic fields are indicated for Type C pulsars. The data presented in Figure 6.11(b) show that pulsars with larger characteristic ages have smaller indicated dipole moments. Ostriker and Gunn (1969b) suggested that this indicates a decay of magnetic field with time in a given pulsar. However, recent calculations of neutron star conductivity indicate that,

except for low-mass stars, decay of the magnetic field would be negligible during the lifetime of a pulsar. It therefore seems more probable that different pulsars are formed with different field strengths, which then remain essentially constant during the pulsar lifetime.

Figure 6.11(a) then suggests the following evolutionary sequence for pulsars. Young pulsars are formed with a relatively short period and fall in the Type S category of pulsars. Then those with large magnetic moments will slow down relatively rapidly to become Type C pulsars when their period reaches about one second. Pulsars with small magnetic moments are also Type S when their period is short, but become Type D as their period increases. Of course, Type D pulsars may not have true ages as large as their characteristic age $P/(dP/dt)$. They may be formed with a relatively long period and may already possess the drifting subpulses, in which case their true age would be much less than their characteristic age.

6.7.3 Radio and Optical Emission Mechanisms

Another important consequence of the strong rotating magnetic field is that an enormous electric field is set up near the neutron star surface as a result of the dynamo effect (Goldreich and Julian, 1969). This strong electric field accelerates to very high energies charged particles which stream from the surface of the star along the field lines. Corotation of the plasma and field is possible inside a cylinder of radius $r = c/\Omega$ and axis the same as the rotation axis—the light cylinder. Continuous particle emission is possible only along open field lines, which penetrate the light cylinder.

A schematic diagram of this pulsar model—the oblique-rotator model—is given in Figure 6.12. Pulses are observed when an emission beam, rotating with the star, sweeps past the observer. A number of different schemes have been proposed to account for the formation of the beamed radiation. These

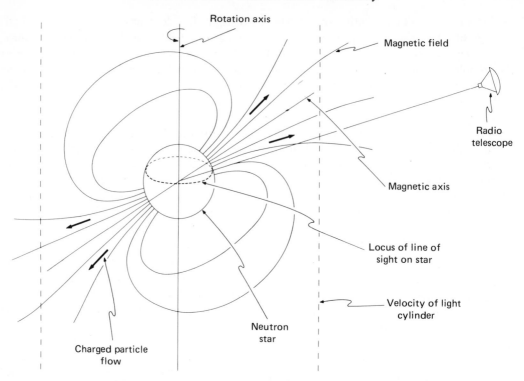

Figure 6.12 Schematic diagram of the oblique-rotator model for pulsars. Charged particles are thought to stream out from the magnetic poles along those field lines which pass through the light cylinder.

schemes divide into two classes: (1) those with the emission originating close to the light cylinder and (2) those with the emission originating close to the surface of the neutron star. Observations of pulse structure having very short time scales (Section 6.3) imply that the emission originates in a relatively small region. For example, a structure of 10 μsec width indicates that the extent of the emission region is less than 3 km. Observed pulse energies therefore require very high brightness temperatures ($\sim 10^{30}$°K) at the source, showing that the radio emission mechanism must be coherent.

In the first class of models the beaming results from the relativistic tangential velocity of corotating plasma near the light cylinder. Gold (1968) proposed that charged particle bunches near the light cylinder emit synchrotron radiation in the tangential direction. Smith (1970) has suggested a variation of this model in which particles radiate cyclotron radiation in the rotating coordinate frame. This radiation is then beamed and Doppler-shifted by the relativistic tangential velocity. Models of this type can account for the short duty cycle of individual pulses, but at present give no adequate explanation for the observed confinement of radiation to a small portion of the period or the stability of the mean pulse profile.

In the second class of models radiation is normally thought to arise in the vicinity of a magnetic pole. This is suggested by the observed variation of position angle through the integrated profile (Radhakrishnan *et al.*, 1969). Charged particles streaming along the open field lines near a pole will radiate in the direction of their velocity. As for light-cylinder models, coherent emission from charge bunches or sheets is required. Alternatively the coherent emission may arise from maser action. Pulsed optical radiation (as observed in the Crab pulsar) is attributed to normal synchrotron radiation from the same particles and need not be coherent. As the radiation is emitted only from the polar region, the observed narrow integrated pulse shapes can be explained. Interpulses may be accounted for by the beam from the opposite pole of an approximately dipolar field intercepting the observer. Normally this will happen only when the magnetic and rotation axes are nearly perpendicular.

The forms of position angle variation shown in Figure 6.9 can be accounted for if the radiation is polarized at some fixed angle to the projection of a vector which rotates with the star. In the magnetic pole model this vector is the magnetic axis. When the line of sight passes close to the magnetic axis the position angle varies rapidly; when it is further away the variation is less rapid. Calculations of the coherence of radiation from charge sheets flowing from the polar region indicate that the intensity will peak not at the pole but in a direction a few degrees from the magnetic axis (Tademaru, 1971). This effect results in double-peaked pulses, which are commonly observed.

No adequate explanation of the drifting-subpulse phenomenon exists. Intensity modulation by plasma waves in the pulsar magnetosphere and emission from the peaks of waves in the surface layers of the neutron star have been suggested. A difficulty with all models is that, to an outside observer, the disturbance must propagate around the star with an angular velocity slightly larger than the stellar rotational velocity to account for the regular drift of subpulses from the trailing edge to the leading edge of the profile. Also, as the drifting pattern typically persists for 10 or more pulse periods, the pulsar must possess some form of "memory" which lasts at least this long. The time for relativistic particles to travel from the star to the light cylinder is of the order of only one period.

6.7.4 Period Discontinuities

Discontinuous changes in the pulsar period may result from small changes in the effective moment of inertia of the neutron star–magnetosphere system. Several possible mechanisms have been discussed. In one, the change results from a cracking and readjustment of the neutron star surface—a starquake. This would normally result in a decrease in the moment of inertia and hence a decrease in the period. An alternative explanation is that plasma builds up in the region of closed field lines in the pulsar magnetosphere and then at some point when the stresses become too great it breaks out from the corotating region. This would also cause a sudden decrease in the moment of inertia and a corresponding decrease in period. Another possible mechanism involves non-uniform transfer of angular momentum from a rapidly spinning neutron star core to the crust of the star. If, as has been suggested, the interior of a neutron star is superfluid, then it would be rotating faster than the crust, as the external torques responsible for the secular slowdown act on the crust rather than the interior.

Clearly, considerable progress has been made in the understanding of pulsars since their discovery. However there are still many unsolved problems and the subject is far from closed. A more extensive review of the observational and theoretical material now available, including numerous references, has recently been published by ter Haar (1972), and a list of pulsar parameters (as known at the end of 1971) is given by Manchester and Taylor (1972).

References

Ables, J. G., M. M. Komesaroff, and P. A. Hamilton. 1970. *Astrophys. Lett.* 6:147.
Backer, D. C. 1970. *Nature* 228:42.
Clark, R. R., and F. G. Smith. 1969. *Nature* 221:724.
Cocke, W. J., M. J. Disney, and D. J. Taylor. 1969. *Nature* 221:525.
Cole, T. W., H. K. Hesse, and C. G. Page. 1970. *Nature* 225:713.
Gold, T. 1968. *Nature* 218:731.
Goldreich, P., and W. H. Julian. 1969. *Astrophys. J.* 157:869.

Hankins, T. H. 1971. *Astrophys. J.* 169:487.

Hewish, A., S. J. Bell, J. D. Pilkington, P. F. Scott, and R. A. Collins. 1968. *Nature* 217: 709.

Huguenin, G. R., R. N. Manchester, and J. H. Taylor. 1971. *Astrophys. J.* 169:97.

Kristian, J., N. Visvanathan, J. A. Westphal, and G. H. Snellen. 1970. *Astrophys. J.* 162:475.

Manchester, R. N. 1971. *Astrophys. J. Suppl.* 23: 283.

———, and J. H. Taylor. 1972. *Astrophys. Lett.* 10:67.

Ostriker, J. P. and J. E. Gunn. 1969a. *Astrophys. J.* 157:1395.

———, 1969b. *Nature* 223:813.

Pacini, F. 1968. *Nature* 219:145.

Radhakrishnan, V., and D. J. Cooke. 1969. *Astrophys. Lett.* 3:225.

———, D. J. Cooke, M. M. Komesaroff, and D. Morris. 1969. *Nature* 221:443.

Rickett, B. J. 1970. *Monthly Notices Roy. Astron. Soc.* 150:67.

Salpeter, E. E. 1969. *Nature* 221:31.

Smith, F. G. 1970. *Monthly Notices Roy. Astron. Soc.* 149:1.

Tademaru, E. 1971. *Astrophys. Space Sci.* 12:193.

Taylor, J. H., and G. R. Huguenin. 1971. *Astrophys. J.* 167:273.

———, G. R. Huguenin, R. M. Hirsch, and R. N. Manchester. 1971. *Astrophys. Lett.* 9:205.

ter Haar, D. 1972. *Phys. Rep.* 3C:57.

CHAPTER 7

RADIO STARS

Robert M. Hjellming

7.1 The Early Years

The detailed study of radio emission from stars is just beginning as this is being written in 1972. For this reason the reader should view this chapter as a description of the pre-history of the subject. Accordingly, the approach will be historical and will deal primarily with the observations.

7.1.1 The Search for the Sun

Soon after Heinrich Hertz, in 1888, detected the electromagnetic wave predicted by Maxwell's electromagnetic theory, Sir Oliver Lodge made the first attempt to detect radio emission from a star—the Sun. This attempt was made in his laboratory in Liverpool, England, using a crude detector, without even an antenna wire; however, only the radio interference from the industrial city of Liverpool was detected. Soon after that a graduate student at the University of Paris, Charles Nordmann, came to some conclusions that were far ahead of his time and attempted an experiment that might have worked if he had been patient. Nordmann realized that (1) an antenna was necessary to provide collecting area, even if it was only a long wire; (2) a quiet site away from radio noise was essential; (3) the radio emission from the Sun might vary with the solar cycle; and (4) he could expect to get radio emission, which today we would call nonthermal emission, from solar storms related to sunspots. Nordmann carried his apparatus to the top of a high glacier in France, tried to

detect the Sun with a 175-m antenna wire on a single day (September 19, 1901), obtained negative results, and did not try the experiment again. This is surprising, because he knew it was a time of minimum solar activity and he expected time variation. It has been speculated that with more patience his detection apparatus was capable of detecting the stronger bursts of solar radio emission. It was not until the 1940's that solar radio emission was finally detected, almost 10 years after Jansky and Reber had already detected (whether they knew it or not) radio emission from the galactic center, a supernova remnant, a radio galaxy, and ionized hydrogen complexes.

One of the major causes of the delay was the effect of a theory. In 1902 Planck announced his revolutionary radiation theory. Unfortunately, people became so convinced that the Sun must radiate like a blackbody at radio frequencies that they *knew* radio emission would not be detectable with available equipment. It is amusing to speculate whether detection of the nonthermal radio bursts from the Sun would have been used to argue that Planck's theory was wrong. In any case, after decades during which radio operators listened unknowingly to radio static from the Sun, Hey, Southworth, and Reber all independently detected the radio Sun.

7.1.2 Characteristics of the Radio Sun

Since the properties of the radio Sun are not covered elsewhere in this book, let us briefly mention some of them. The variety of

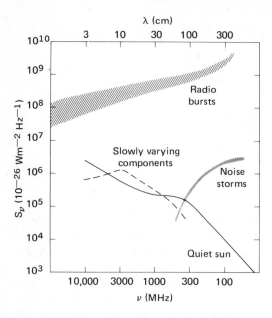

Figure 7.1 Schematic representation of radio flux density as a function of frequency for four types of radio emission from the Sun. The slowly varying component corresponds to that for sunspot maximum. (Adapted from Kundu, 1965. Copyright 1971 by the American Association for the Advancement of Science. R. M. Hjellming and C. M. Wade, 1971. *Science* 173:1087.)

solar radio events is so great that doubtless many phenomena found in other radio stars will be found to be more extensive versions of solar events.

For our brief discussion we shall divide solar radio events into five groups: (1) long-wavelength (meter, decimeter, and decameter) bursts; (2) centimeter-wave bursts; (3) noise storms; (4) the slowly varying component; and (5) the quiet Sun radio emission. The flux densities typical of these groups are schematically shown in Figure 7.1.

The strongest and most complex solar radio events are the bursts occurring at meter and longer wavelengths. They are generally associated with solar flares and attain brightness temperatures of up to $10^{12}°$K, with typically nonthermal spectra. The bursts are short-lived events with lifetimes typically of minutes, although some last up to a few hours. Most begin with the ejection of plasma from

active regions that are usually associated with sunspots. These radio bursts are not well understood, but complex phenomena involving synchrotron emission, gyromagnetic emission, Cerenkov plasma waves, shock waves, and plasma instabilities are frequently considered to be involved.

The bursts at centimeter wavelengths are somewhat better understood. They appear with a rapid rise in intensity followed by a slower decline. The radiation is frequently partially circularly polarized and appears as a smooth continuum. Three types of centimeter bursts are: (1) the *impulsive bursts*, in which a rapid rise in times of the order of a minute is followed by a decline lasting a few minutes; (2) the *post-burst*, lasting up to a few tens of minutes, in which there is a slow decay of intensity to a pre-burst level; and (3) bursts showing a *gradual rise and fall* of intensity, lasting up to a few tens of minutes. The centimeter bursts are empirically well correlated with X-ray flare events.

Noise storms frequently contribute the major part of the solar radio emission at 1- to 10-m wavelengths. They consist of a slowly varying broad-band continuum enhancement lasting from a few hours to a few days. Near the maximum of the 11-year sunspot cycle noise storms are in progress about 10% of the time. The radiation originates between 0.2 and 1 solar radius above the photosphere, and it is generally beamed almost radially outward from the Sun. The cause of noise storms is poorly understood, but they could be produced by magnetically trapped particles previously injected from a flare. Such particles can leak back to the site of the flare, exciting inward-moving plasma waves which produce radio waves that are subsequently reflected outward.

The so-called *slowly varying component* is probably thermal in origin and comes from bright regions with high-density coronal condensations, at temperatures of about $2 \times 10^{6}°$K, existing over sunspots and plage regions. Radio emission from a particular plage can persist for weeks. The amount of radiation from plages is well correlated with the overall level of solar activity. There is a possibility that gyromagnetic radiation produces the peak in the slowly varying component at about 10 cm (cf. Figure 7.1.).

Finally, there is the radio emission from the quiet Sun. This is a relatively stable radiation level that is always present. At a particular frequency, ν, the radiation is produced by thermal bremsstrahlung in or near the atmospheric layers where $\nu \approx \nu_p$, where ν_p is the plasma frequency given by (Kundu, 1965)

$$\nu_p = 0.009 N_e^{1/2} \text{ MHz} \qquad (7.1)$$

and N_e is the electron concentration in cm^{-3}. At meter wavelengths the radiation is emitted from the hot tenuous corona. At centimeter wavelengths it comes mainly from the much denser and cooler chromosphere. At short millimeter wavelengths it comes from the still denser and cooler photosphere.

As seen in the above discussion of the quiet Sun radio emission, there is some simple physics concerning the propagation of radio waves through a stellar atmosphere that should be understood. The index of refraction for radio waves of frequency ν in a completely ionized plasma with a plasma frequency ν_p is given by

$$n = \left[1 - \left(\frac{\nu_p}{\nu} \right)^2 \right]^{1/2} \qquad (7.2)$$

At any level in a stellar atmosphere where $\nu_p > \nu$, n will be imaginary and the radio emission is suppressed. Since radio waves propagate freely only when $\nu > \nu_p$, radiation of frequency ν can come from any height in the stellar atmosphere above the level where $\nu \approx \nu_p$. We can consider a stellar atmosphere to be opaque at the density level at which $\nu \approx \nu_p$. Therefore, since N_e decreases monotonically with increasing distance from the surface, observations at different frequencies tend to "see" different heights in the atmosphere. The higher the frequency, the deeper and more dense the layer at which the observed radiation can be produced. In the case of an optically thick, thermally emitting stellar atmosphere it is exactly the level at

which $\nu \approx \nu_p$ that is observed at frequency ν; and the observed brightness temperature will be the electron temperature at that level.

Because of the variety of radio phenomena associated with the Sun, eventually the explanations of the much more extensive radio emission from other stars will probably be in terms of processes known on the Sun. The reader can pursue this subject by referring to Kundu (1965).

7.1.3 When Is a Star a Radio Star?

Some very simple considerations will tell us the gross properties of any star likely to be observed as a radio source. We will consider any radio star to be represented by an average brightness temperature, T_B, and an average solid angle for the emitting region, Ω_s; then the flux density at frequency ν is given by

$$S_\nu = \frac{2k\nu^2}{c^2} T_B \Omega_s \qquad (7.3)$$

where k is the Boltzmann constant and c is the speed of light. It is useful to define an equivalent disk diameter, θ, by $\Omega_s = \pi \theta^2 / 4$; we can then write Equation (7.3) in the following useful form:

$$S_\nu = \frac{T_B \theta^2}{1970 \lambda^2} \qquad (7.4)$$

where S_ν is in flux units (1 f.u. = 10^{-26} Wm^{-2}Hz^{-1}), T_B is in °K, θ is in arc seconds, and λ is in cm.

If $S_{\min}$ is the lowest flux density that can be measured with a particular radio-telescope, this instrument can detect a particular radio star only if

$$T_B \theta^2 \geq 1970 \lambda^2 S_{\min} \qquad (7.5)$$

The parameters of a radio star that govern its detectability are the brightness temperature and the angular diameter. Of course, this is true of any radio source, but Equation (7.5) is very important in the radio star business because, of the two parameters, angular diameters affect detectability the most, and almost by definition, θ is very small for a star. With a minimum detectable flux density of

0.005 f.u., which is typical for eight hours of observing with the NRAO three-element tracking interferometer, $T_B\theta^2 \geq 135$ is needed to detect a radio star at 3.7 cm and $T_B\theta^2 \geq 1210$ is needed to detect a radio star at 11.1 cm. The normal stars with the largest known angular diameters, the red supergiants, subtend only 0.05 arc second; hence it would be necessary for $T_B \geq 54,000°K$ at 3.7 cm and $T_B \geq 484,000°K$ at 11.1 cm. Typical electron temperatures, which we denote by T_e, for photospheres or "normal" ionized plasmas are $10^{4}°K$, but brightness temperatures of 10^5 to $10^{6}°K$ can occur in stellar chromospheres and coronas. We can therefore turn the arithmetic around to say that a $10^{4}°K$ optically thick plasma $(T_B = T_e)$ requires $\theta \geq 0.12$ arc second at 3.7 cm and $\theta \geq 0.35$ arc second at 11.1 cm for detection with the NRAO interferometer. For optically thick coronas with temperatures of $10^{6}°K$, $\theta \geq 0.012$ arc second and $\theta \geq 0.035$ arc second are needed for detection at 3.7 and 11.1 cm, respectively.

One result of this simple arithmetic is the obvious conclusion that if any radio stars are detectable, particularly at high-flux density levels, physical processes involving unusual amounts of energy and unusually high surface brightnesses will be encountered. To see this in another way let us consider the maximum distance at which a star like the Sun would be observable at, say, 11 cm. If 0.01 f.u. is taken as a typical minimum detectable flux density for instruments in 1972, the quiet Sun could be seen out to 0.04 parsec; the slowly varying component could just be seen at 0.06 parsec; and very strong radio bursts could be detected out to 1.4 parsecs. Since the nearest star, Proxima Centauri, is 1.3 parsecs away, any stars beyond Proxima Centauri would have to involve much greater radio luminosity than the Sun at its strongest to be detectable. This is the prime reason why any observable radio stars must involve phenomena not encountered before in stellar astronomy. We should therefore not be surprised if the radio stars produce some surprises.

7.1.4 Red Dwarf Flare Stars

After the many false alarms of the late 1940's and 1950's, when almost every newly detected radio source* was called a "radio star," whether by way of interpretation or simple disregard for the English language, the 1960's produced the first reports of radio emission from true stars. Red dwarf stars of certain types were known to increase their optical brightness radically for brief periods of time, typically a few hours. These so-called flare stars exhibit phenomena that suggest analogs of solar flares. Since solar flares are accompanied by intense nonthermal radio emission, a pioneering effort by Lovell (1969) and other British and Australian radio astronomers was made to detect these objects as radio stars. During several thousand hours of observing time, these investigators found several cases of simultaneous radio and optical flares for a few stars, notably UV Ceti, YZ Canis Minoris, and V 371 Orionis. One of the best examples is the flare of YZ Canis Minoris, which is shown in Figure 7.2, in which the flux densities measured by Lovell at 240 and 408 MHz are plotted, together with simultaneous measurements at visual and UV wavelengths by Kunkel (1969) and Andrews (1969).

The interpretation of the radio emission from red dwarf flare stars is still quite uncertain. Most of the effort has been simply to point out possible analogies to solar events, but there is the difficulty that brightness temperatures are up to three orders of magnitude stronger than known on the Sun. This is an example of the point discussed earlier: Detection of almost any radio star will indicate the presence either of new phenomena or of known processes in unusually energetic forms.

7.1.5 The Twilight Years

From the initial detection of red dwarf flare stars until 1970 the radio astronomy of stars was in a sort of "twilight" phase. Very

---

* As late as 1963 it was still believed by some authors that quasars were stars.

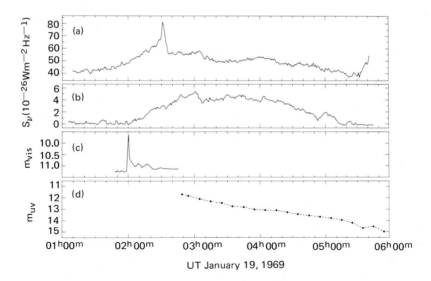

Figure 7.2 Radio flux densities at (a) 240 MHz and (b) 408 MHz as measured by Lovell (1969) as functions of universal time. Curves (c) and (d) show, respectively, the visual and UV magnitudes of this star. (Copyright 1971 by the American Association for the Advancement of Science. R. M. Hjellming and C. M. Wade, 1971. *Science* 173:1087.)

few radio astronomers were interested in observing the red dwarf flare stars because of the enormous investment in observing time on the biggest available instruments needed to observe a relatively few events.

In addition there were a number of reports of detections which, both then and now, are difficult to assess fairly. As pointed out earlier, the angular diameter of a star is so important in the detectability of an object that early efforts to observe other radio stars concentrated on the nearby red supergiants. The first report of possible detection of radio emission from a red supergiant was made by Kellermann and Pauliny-Toth (1966), who found an apparent signal of 0.11 ± 0.03 f.u. while observing α Orionis (Betelguese) at a wavelength of 1.95 cm on February 21, 1966. They found no signal, however, on the next 11 nights. Seaquist (1967) reported the possible detection of 2.85 cm radiation from α Orionis and another red supergiant, π Aurigae, at the 0.023 ± 0.006 and 0.031 ± 0.009 f.u. levels, respectively. The problem with evaluating these reports is simple, but insoluble: As with most scientific experiments, results which are not reproducible are assumed (sometimes by the experimenters and nearly always by everybody else) to be the result of problems with the equipment. Further, at weak levels the problem of confusion with background sources can arise, particularly with single antennas that have beam sizes of some minutes of arc. Both of these effects produce serious skepticism when reported detections are near the noise or confusion levels—even when such skepticism is unwarranted.

An example of unwarranted skepticism about results obtained during the twilight years is the early history of the radio emission from the peculiar blue star identified with the X-ray source Sco X-1. In 1968 Andrew and Purton (1968) found weak radio emission from the vicinity of Sco X-1 at the level of 0.021 ± 0.007 f.u. at 4.8 cm. Soon after that Ables (1969) reported time variations about a mean level of 0.022 ± 0.002 f.u. at 6 cm. It is fair to say that most radio astronomers did not believe the evidence for variability, and many felt the detection of a source was doubtful because of confusion effects. We now know that in some sense, nearly everybody was correct. As we shall discuss in connection with the radio counterparts of X-ray sources, the single dish measurements were seeing three radio sources, only one of which coincides with the X-ray star, and it is very variable. Because of this, the early work on the radio counterpart of Sco X-1 can be properly described as part of the twilight years.

7.2 The Dawn of Stellar Radio Astronomy

7.2.1 Radio Stars and Radio Interferometers

There are three main reasons why one can say the systematic study of radio stars dawned in 1970. These are the three reasons why radio stars are observed best with a phase-stable tracking interferometer, and observations of radio stars with such interferometers began in 1970.

The first and most significant reason for the importance of radio interferometers is the capability they provide to determine very accurate positions of radio sources. Anyone who can place a radio source on the position of a known star to within errors of the order of an arc second can claim to have clearly detected a radio star. The second reason is the high sensitivity attainable with a tracking interferometer. This is true because all of the information gained about a particular field during, say, 10 hours of tracking can be combined together to reduce the signal-to-noise ratio. This cannot be done reliably with continuum measurements by a single antenna, even if it has comparable noise figure, bandwidth, and collecting area; there is a limit to the amount of integration time a single antenna can usefully put on a single source. This limit does not exist for an interferometer where the useful sensitivity really does improve as the square root of the amount of time spent observing a particular field. The third advantage a multi-element interferometer has for the radio star business is now obvious, but only after the fact: *The proof that a time variable or transient event is real is most certain if independent pairs in an interferometer are detecting the same time-*

variable event. Since it turns out that all radio stars are variable, with typical time scales of hours, it is essential that one be able to prove the reality of unique events. In older radio astronomy texts one frequently sees versions of the statement, "thou shalt not observe a variable source with an interferometer." Just the opposite is true, for reasons that will be clear from the following example. Consider the problem of a tracking interferometer measuring the visibility function $V(t; \alpha, \delta)$, where t is the sidereal time, α is the right ascension, and δ is the declination of the center of the field of view being observed by the interferometer. If the normalized beam pattern of the (identical) telescopes is $f(x,y)$ and the brightness distribution is $B(x,y)$, with x and y being position displacements from the point (α, δ), then the visibility function is given by the Fourier transform.

$$V(t;\alpha,\delta) = \int\limits_{-\infty}^{\infty} f(x,y)\, B(x,y)\, e^{-i2\pi(ux+vy)}\, dxdy$$

$$(7.6)$$

where u and v are the usual projections of the interferometer baseline on the x,y plane (plane of the sky). Because of the form of Equation (7.6) a single variable point source with flux density $S(t)$ at a position (x_s,y_s) has a contribution to the visibility function, which is given by

$$f(x_s,y_s)\, S(t)\, e^{-i2\pi(ux_s+vy_s)}$$

Hence by observations of exactly the same field on two different sidereal days one can subtract the visibility function for the two days to obtain

$$f(x_s,y_s)\, e^{-i2\pi(ux_s+vy_s)}[S(t_1) - S(t_2)]$$
$$= V(t_1,\alpha,\delta) - V(t_2,\alpha,\delta) \quad (7.7)$$

With t_1 and t_2 differing by only an integral number of sidereal days, one obviously can solve for $S(t_1) - S(t_2)$, obtaining a perfect relative measurement of the time variation of the point source. In the case where the source is not present on one day, a measurement of the actual flux variation on a day when it is present is obtained. A deeper examination of the problem of variable sources viewed by a tracking interferometer shows that in principle, with only mathematical difficulties, relative variations of any number of variable sources in the field can be obtained. Fortunately, with present levels of observation, one can nearly always assume a single variable source. Finally, we note that there is no confusion problem due to steady background sources, because their effects will be subtracted out with proper differencing of the visibility functions. Apart from the effects of noise, the prime limitation of this technique is the stability obtainable for both amplitude and phase calibration of the interferometer in question.

The systematic study of radio stars began in 1970 because two interferometers were first used for this purpose at that time and in the following years: the NRAO interferometer in Green Bank, West Virginia and the Westerbork interferometer in Holland. Because of greater flexibility in observing and in evaluating and processing data, most of the initial successes were with the NRAO interferometer. However both instruments were capable of attaining detection limits of only a few milliflux units and both were capable of doing comparable work. The capability of the NRAO interferometer to simultaneously measure a variable source at 3.7 and 11.1 cm made it a valuable, but crude, radio spectrometer; this is particularly important since the radio spectrum for a variable source can be obtained only by simultaneous measurements. The initial Westerbork interferometer operated only at 21 cm so that it could detect radio stars and establish variability at low flux levels, but it could not determine radio spectra.

Within their limitations, single antenna observations of radio stars continued to play a role, but the early years were dominated by the interferometers, which will continue to dominate the subject in the future.

7.2.2 Radio Novae

There is one obvious exception to the general rule that radio stars will have angular diameters considerably less than an arc second. This will occur when the expanding atmosphere of an exploding star, a nova or supernova, is observed at the right time under the right circumstances. In June, 1970, Hjellming and Wade detected radio emission from Nova Delphini 1967 and Nova Serpentis 1970 with the NRAO interferometer. The variations of 2695- and 8085-MHz flux densities for these novae between June, 1970, and October, 1972 (Hjellming *et al.*, 1974), are plotted in Figure 7.3. The data are consistent with the evolution of an expanding thermal source, as would be expected for an ionized nova shell. The thermal nature of the radio emission is seen from the radio spectra shown in Figure 7.4 for measurements taken during 1970 (Wade and Hjellming, 1971). Figure 7.4 also establishes that the nova shells are complex structures, since the theoretical slopes (dashed curves) for an optically thick source are never attained at the lower frequencies; this effect is well known for complex HII regions to be the result of different emission measures for different parts of the source.

A first-order attempt to interpret the nova data can be made by solving empirically for the parameters of an assumed homogeneous source from data at two frequencies.

The flux density for such a radio source is given by

$$S_\nu = \frac{2k\nu^2}{c^2} T_e \Omega_s$$
$$\times [1 - \exp(-0.0823\, E T_e^{-1.35}\, \nu_{\text{GHz}}^{-2.1})]$$

$$(7.8)$$

where Ω_s is the solid angle subtended by the source, T_e is the electron temperature, and E is the emission measure in units of pc cm^{-6}. Simultaneous measurements at two frequencies allow the determination of $T_e \Omega_s$ and $E T_e^{-1.35}$ empirically whenever the source is slightly optically thick. Since optical data for the nova shells indicate that T_e is of the order of $10^4\,$°K, and for theoretical reasons is not likely to change too much as long as the source remains ionized, the radio data

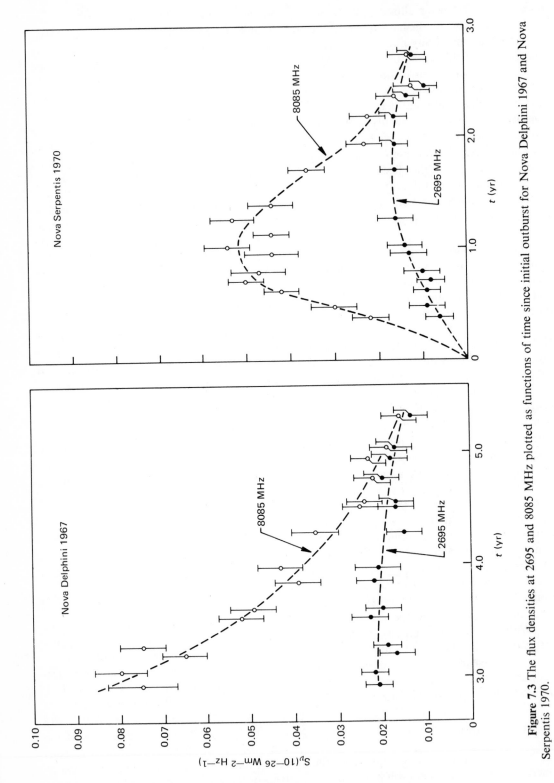

Figure 7.3 The flux densities at 2695 and 8085 MHz plotted as functions of time since initial outburst for Nova Delphini 1967 and Nova Serpentis 1970.

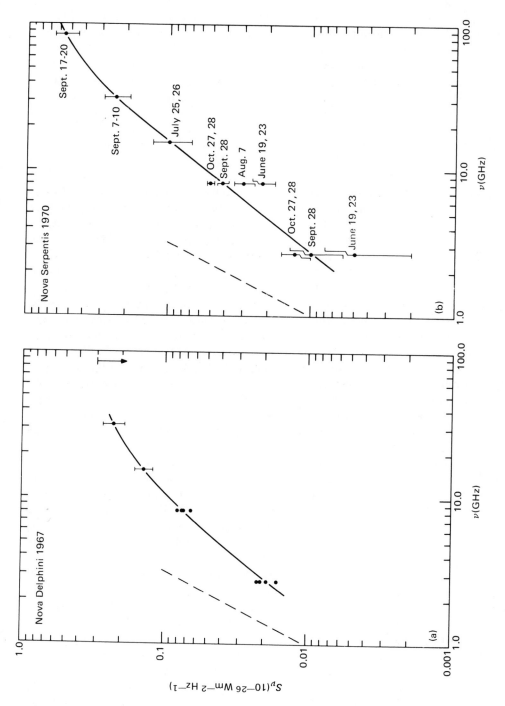

Figure 7.4 Radio spectra of Nova Delphini 1967 and Nova Serpentis 1970 as measured from June to October, 1970. (C. M. Wade and R. M. Hjellming, 1971. *Astrophys. J.* 163:L65.)

essentially determine the evolution of Ω_s and E as functions of time. Hjellming, Herrero, and Wade (1974) discuss the results of such an analysis in detail. A nova shell expanding with uniform velocities, however unsymmetrically, in all directions would show $\Omega_s \propto t^2$, where t is the time since initial outburst. However, the data indicate $\Omega_s \propto t^{0.94}$ for Nova Delphini 1967 and $\Omega_s \propto t^{0.74}$ for Nova Serpentis 1970. This indicates that both nova shells have undergone drastic deceleration. The apparent complexity in the variation of Ω_s and E provides severe constraints for any models of the expansion of these nova shells.

The observations of Nova Delphini 1967 and Nova Serpentis 1970 show that important new data about nova shells are obtained from radio observations for a few years after outburst (if the nova is close enough). Extensive observations of old novae indicate that their emission measures became so low that even though the angular diameters may exceed an arc second, they are undetectable with sensitivity limits at the 0.005 f.u. level. New bright novae, however, can now be observed in detail to determine the evolution of the shells from radio data; hence radio astronomy will play an important role in future studies. In particular, future novae detected with interferometers capable of resolutions of the order of 0.1 arc second will be able to easily resolve the nova shells at an early stage in their evolution. For all such cases the determination of θ at that time will allow determination of distance, one of the most important parameters for any object.

7.2.3 Radio Binary Stars

As of 1972, the most interesting radio stars appear to be associated with binary systems. Their numbers are, however, so few that the statistical significance of this result is unclear. In any case, let us now describe the early work on the radio stars α Scorpii B (Antares B), β Persei (Algol), and β Lyrae.

The first attempts by Wade and Hjellming (1971) to observe red supergiants included observations of α Scorpii; and on March 25, June 4, and November 12, 1970, a radio source at the 0.005 ± 0.001 f.u. level at 11.1 cm was detected near the red supergiant. Because the radio source was so weak and it was detectable only at 11.1 cm, the position uncertainty was about 3 seconds of arc. Even though the position uncertainty included both the position of α Scorpii A, the red supergiant, and α Scorpii B, a faint B 3V companion 3″.2 away, it was concluded that one of the long-sought radio supergiants had been detected. This conclusion was found to be premature when, on June 1, 1971, Hjellming and Wade (1971a) discovered a radio flare at 3.7 cm which allowed clear detection of a 0.011 ± 0.002 f.u. source with a position uncertainty of only 1 second of arc. The radio source clearly belonged to the B 3V companion star, as can be seen from Figure 7.5.

The prime importance of the α Scorpii B source was the tentative line of logic it introduced. When asking the simple question of what else is unusual about α Scorpii B, the answer is two things: Unusual emission lines in a nebulosity surrounding the B star, and the fact that it is near one of the red supergiants known to spew a considerable amount of matter into space. The two clues of mass motions and emission lines immediately suggested as likely candidates for radio stars the close binaries in which emission lines are clear indicators of mass motions in the system. The next star observed after this, on October 23, 1971, was β Persei (Algol), and it was clearly (Wade and Hjellming, 1972) a strong radio star. Several days later, on November 4, 1971, β Lyrae was also detected. During extensive observations between October 23 and November 10, 1971, it was shown that β Persei was clearly a variable radio source ranging from 0.019 ± 0.003 f.u. to 0.006 ± 0.005 f.u. at 3.7 cm, and 0.014 ± 0.003 f.u. to less than 0.004 f.u. at 11.1 cm, with spectral indices that were always thermal. At that time β Lyrae was not clearly variable, being at the 0.015 f.u. level at 3.7 cm to within the allowable errors, and was undetectable at 11.1 cm, again indicating a thermal spectral index.

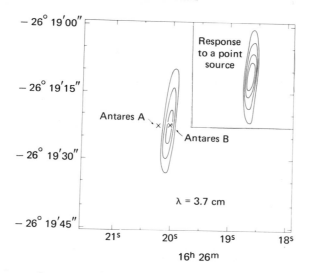

Figure 7.5 Radio map of Antares B (α Scorpii B) made during the 3.7 cm flare on June 1, 1971. [R. M. Hjellming and C. M. Wade, 1971. *Astrophys. J.* (*Lett.*) 163:L65.]

Extensive observations by Hjellming, Wade, and Webster (1972a) with the NRAO interferometer during January and February of 1972 established three results: (1) The β Lyrae radio source was weaker than previously detected, falling below a detection limit of 0.006 f.u. at 3.7 cm; (2) a search for radio emission in other Algol- and β Lyrae–type binary systems, many of which were comparable in distance and other properties to the prototype systems, produced only negative results; and (3) β Persei was found to flare up to levels a factor of 20 greater than the maximum observed in 1971. Figure 7.6 shows the variation of radio flux densities of β Persei at 8085 and 2695 MHz during January and February of 1972. Also shown in Figure 7.6 is the associated optical eclipse cycle; a comparison reveals no sign of correlation of radio and optical variability.

Observations of β Persei during April to July, 1972 (Hjellming *et al.*, 1972b), established three further unusual characteristics of the radio source: (1) sudden changes of radio flux by as much as a factor of two in less than 10 minutes; (2) a brief period between at least June 21 and July 5, during which the source behaved as a low-level optically thin thermal emitter, was followed by a resumption of extensive flaring; (3) existence of abrupt quenching of a nonthermal component which went from 0.25 f.u. to less than 0.01 f.u. in less than 45 min.

Before discussing further details of the β Persei radio source, it should be noted that unusual optical events were reported by Bolton (1972) at the time (February, 1972) that the radio flaring was becoming more frequent. Bolton noted the presence of both an absorption and an emission component in the K-line of CaII whenever the radio source was slightly optically thick between 3.7 and 11.1 cm; further the strength of the feature varied with time.

The unusual changes in behavior in β Persei, β Lyrae, and α Scorpii B, coupled with the negative results on other comparable binary systems, makes it fairly certain that a relatively rare state of stellar activity is involved.

Most of the time the β Persei radio source appears as a variable thermal source. Most of the radio flares are strikingly similar, particularly in their spectral index behavior. More detailed representation of the data for eight days is shown in Figure 7.7, in which seven of the stronger flares and an example of nonthermal behavior are shown. Examination

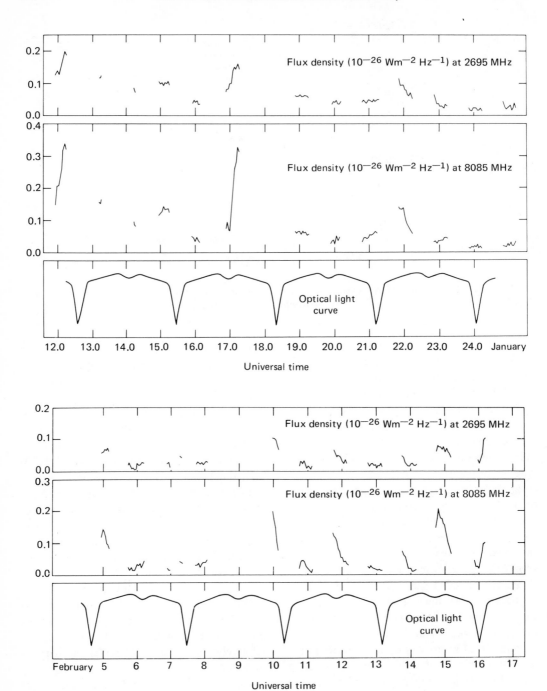

Figure 7.6 Radio flux densities of β Persei (Algol) as measured at 2695 and 8085 MHz during January and February, 1972, plotted as a function of universal time. Also plotted is a schematic representation of the optical light curve. (R. M. Hjellming *et al.*, 1972. *Nature Phys. Sci.* 236:43.)

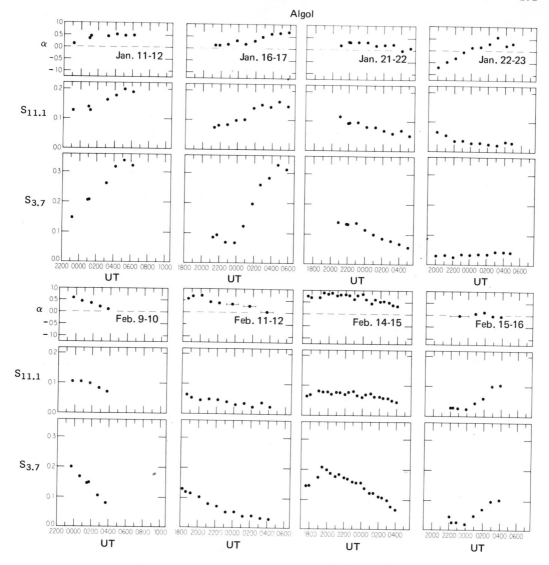

Figure 7.7 Flux densities for β Persei (Algol) at 2695 and 8085 MHz, and the associated empirical spectral index, plotted as a function of universal time for eight days during 1972.

of the data reveals striking consistencies: (1) Both before and after flaring events the spectral indices are essentially zero, as expected for optically thin thermal sources; (2) during the rise of a flare the spectral index changes steadily, as if the thermal source were becoming slightly optically thick; and (3) during the decay of a flare the spectral index slowly returns to zero again. The seven purely thermal-like events given in Figure 7.7 show this clearly. Somewhat longer time scale

variations are frequently seen in which the spectral index remains zero throughout.

Because the β Persei radio source has always appeared as a point source at 3.7 cm for the Green Bank interferometer, an upper limit of 0.''5 can be placed on the angular diameter. This means that a lower limit of $5 \times 10^{4}\,^{\circ}$K can be placed on the brightness temperature during the peak flaring in January, 1972. Furthermore, if one takes the dominately thermal appearance of the radio spectrum as proof that it is a thermal radio source, the electron temperature was greater than $2.5 \times 10^{5\circ}$K when flaring (Hjellming, 1972).

If one considers that the β Persei radio source is clearly located within the gravitational potential well of the binary system, the considerations discussed earlier for radio observations of the solar atmosphere are applicable to this case. If this is true, the β Persei radio source, when flaring, is dominated by emission from the level where $N_e \sim 10^{10}$ cm^{-3}, $T_e \sim 10^{8\circ}$K, $\theta \sim 0\overset{''}{.}01$, and the diameter of the source is 0.3 A.U., since β Persei has a distance of 25 pc. Under these brief conditions β Persei should be a transient X-ray source (Hjellming, 1972).

It is most interesting that the diameters of the stars in the β Persei system are about 0.028 A.U. and the separation of the close pair is 0.2 A.U.; therefore the size scale for the variable thermal source is that of either of the Roche lobes of the binary. Support for the argument that this is the typical size scale of the radio source is obtained from the unusual discontinuity in radio emission observed at both 3.7 and 11.1 cm on April 27, 1972 (Hjellming *et al.*, 1972b). In the space of 10 minutes, while a calibrations source was being observed, the radio flux at 3.7 cm went from 0.14 to 0.22 f.u. Clear variation on time scales as short as or shorter than a few minutes occurred at that time. Since an incoherent radio source with a size of the order of 0.3 A.U. could vary on time scales as short as 2.4 minutes, the occurrence of such a short time scale change is a confirmation of the estimate of 0.3 A.U. for the size scale of the variable source. It is further noteworthy that temperatures as high as 10^7 to $10^{8\circ}$K would be needed to be consistent with the extremely weak optical emission lines present in β Persei.

The above-mentioned values for the temperature, density, and size scale of the flaring radio source can be used to estimate a mass loss rate which is of the order of 10^{20} g/sec. With a duration of about a year, this would mean $\sim 10^{27.5}$ g mass loss. Since the mass in stars is a few solar masses, this is a few parts in 10^6 of the mass of the system. Changes in mass of this order should produce a comparable change in period. Indeed, Algol is known to undergo period discontinuities of the order of a few parts in 10^6 roughly once every 25 to 35 years (Herczeg, 1968). Since the last such change was in 1942, the occurrence of another in 1972 could be expected. It is therefore very tempting to associate major changes in stellar evolution, occurring roughly every 30 years, with the basic changes needed to produce such drastic radio flaring. One would think of the period 1972 as an uprising in stellar "activity." Occurrences during roughly one year out of 30 would explain the statistics whereby only three out of the approximately 60 binary systems searched during 1972 show radio emission.

As this is being written, it is not firmly established whether most of the flaring radio emission of α Scorpii B, β Persei, and β Lyrae should be interpreted as thermal, although this is clearly the simplest and most obvious conclusion. The situation is complicated by the fact that both α Scorpii B and β Persei seem to have nonthermal components at times. A proof of the thermal nature of the major flares would be detection of X-ray emissions. This tantalizing possibility would mean that stellar radio astronomy is inextricably linked with X-ray astronomy.

7.2.4 Radio Counterparts of X-ray Stars

The early radio observations by Andrew and Purton (1968) and Ables (1969) established that there was radio emission in the vicinity of the X-ray star Sco X-1 and suggested it might be variable. The unusual nature of this radio source was first revealed from observations with the NRAO interferometer by Hjellming and Wade (1971b). The Sco X-1 radio source was found to be a triple radio source, with all three sources placed on a line with position angle 29°. Further, the central component was found to be a nonthermal variable with a range between 0.26 f.u. and less than 0.004 f.u. at 11.1 cm.

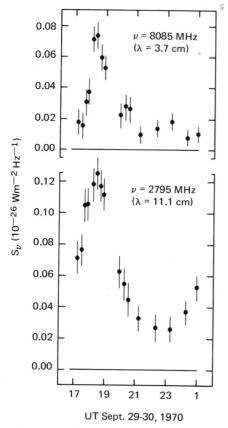

Figure 7.8 Radio flaring in Sco X-1 at 2695 and 8085 MHz seen on September 29–30, 1970. (Copyright 1971 by the American Association for the Advancement of Science. R. M. Hjellming and C. M. Wade, 1971. *Science* 173:1087.

The variable component coincided exactly with the peculiar blue star identified with the X-ray source (Sandage *et al.*, 1966), and the companion sources were both about 1.2 away. Figure 7.8 shows the spectacular radio flare of September 29–30, 1970. The radio emission from the central component of Sco X-1 always appears nonthermal in spectral index. Figure 7.9 shows maps of the triple radio source made with the NRAO interferometer (Wade and Hjellming, 1971).

The strong radio variability of the central component of Sco X-1 suggested that radio counterparts of other X-ray sources could be found by searching the X-ray positions, which always have large uncertainties of minutes of arc or more, for unusual sources

variable with a time scale of hours. The first successes resulted in the detection of radio counterparts to GX 17+2 and Cyg X-1 by Hjellming and Wade (1971d) with the NRAO interferometer. The Cyg X-1 source was independently detected by Braes and Miley (1971b) with the Westerbork array. The GX 17+2 source was a case where the radio emission rose above the detection limit to the level of 0.022 f.u. at 11.1 cm during only one day. Its radio properties are nonthermal and similar to what would be expected of Sco X-1 if it were about a factor of three further away. Even though it was observable for only one day, an accurate radio position was obtained. As pointed out by Tarenghi and Reina (1972), this position is coincident with that of a very faint, and unusually red, star.

The Cyg X-1 radio source is unusual in many ways. It was initially identified with the X-ray source, because it jumped from below a detection limit of 0.005 f.u. to a level of roughly 0.015 f.u. The accurate position obtained for the radio source resulted in identification of the X-ray and radio source with the spectroscopic binary, HDE 226868, which contains a normal ninth magnitude B0Ib supergiant and an invisible companion which is at least as massive as the supergiant. Because of the apparent violation of the mass luminosity law it may turn out that the invisible companion in the binary is a massive black hole. The Cyg X-1 radio source is also the only one, as of 1972, to show correlation between X-ray and radio variations. The radio source appeared sometime between March 22 and 31, 1972, remaining relatively steady thereafter, and it was during this period that the X-ray source (Tananbaum *et al.*, 1972) began a transition from one mean level of X-ray emission to another mean level.

The fourth detected radio counterpart of an X-ray source was GX 9+1 found by Zaumen *et al.* (1972) with the NRAO interferometer. The fifth was Cyg X-3, first detected by Braes and Miley (1972) with the Westerbork array. This radio star has the distinction of having been briefly one of the strongest compact radio sources in the sky.

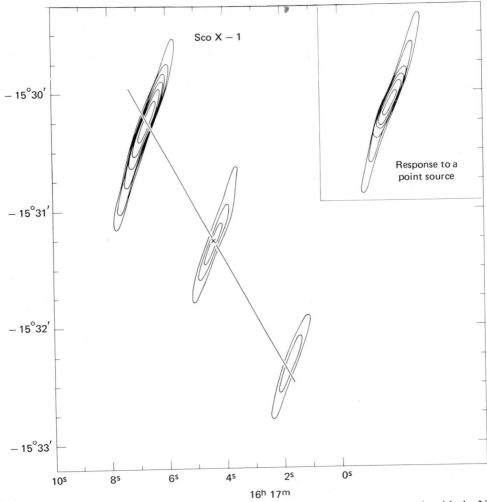

Figure 7.9 A synthesis map of the Sco X-1 triple radio source at 2695 MHz made with the NRAO interferometer. The cross marks the position of the peculiar blue star identified with the X-ray source, and the line indicates a position angle of 29°.

7.2.5 The Radio Counterpart of Cyg X-3

After the initial detection of the radio counterpart of Cyg X-3 by Braes and Miley (1972), further observations by Hjellming *et al.* (1972b) established that it was the strongest of the known radio counterparts to X-ray sources, reaching levels as high as 0.5 f.u. during flares. During most of the observations between May and August, 1972, Cyg X-3 appeared much more similar to β Persei than to any of the other radio counterparts of X-ray sources. In July, 1972, it was found (Hjellming and Balick, 1972a) that

Cyg X-3 underwent intense flaring at milli-meter wavelengths, with each event typically lasting an hour and attaining levels of about 1 f.u. at 31.5 GHz. On August 30–31, 1972, Cyg X-3 was at the lowest flux levels at which it had ever been observed, 0.01 f.u. at both 2695 and 8085 MHz. On September 2 a planned program of coordinated observations of β Persei with the NRAO interferometer, the Algonquin Park 150-foot antenna, and a number of optical observers was forestalled by the serendipitous circumstance whereby the 150-foot observers (Gregory *et al.*, 1972a) looked at Cyg X-3 because β Persei had not

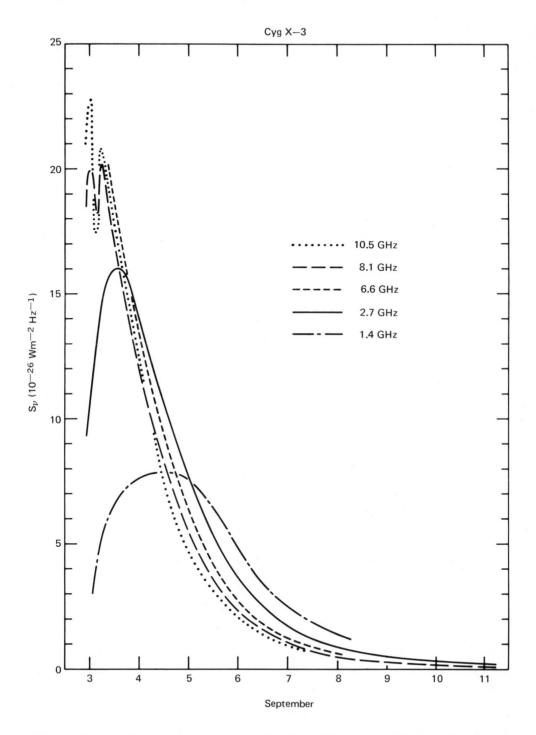

Figure 7.10 Curves drawn through the data at 10.5, 8.1, 6.6, 2.7, and 1.4 GHz for the first spectacular radio event detected in Cyg X-3 during early September, 1972. (R. M. Hjellming, 1971. *Astrophys. J.* 170:527.)

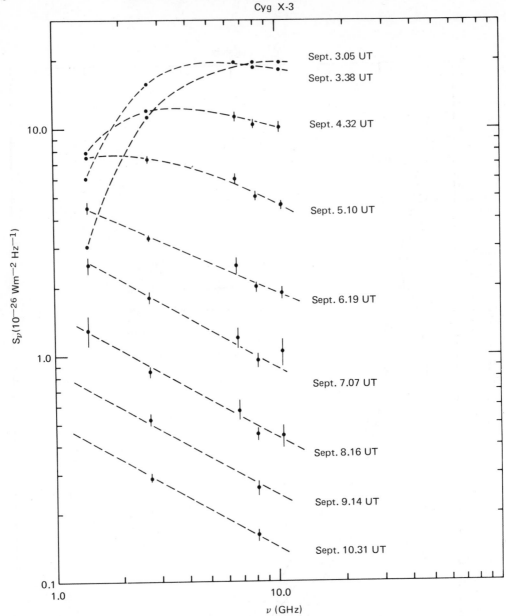

Cyg X-3

Figure 7.11 The evolving radio spectrum of Cyg X-3 between September 3 and 10, 1972.

yet risen, only to find it at the level of 22 f.u. at 10.6 GHz. Within moments the NRAO interferometer was also observing Cyg X-3 (Hjellming and Balick, 1972a), finding it to be 20 f.u. at 8085 MHz and 10 f.u. at 2695 MHz. Other observers were notified of this unique event by telephone, and these contacted others, until in a day or so most working radio telescopes in the world were observing Cyg X-3. The data taken from September 2–11, 1972, by Gregory *et al.* (1972a) at 10,630 MHz, Hjellming and Balick (1972a) at 8085 and 2695 MHz, and Shaffer *et al.* (1972) at 1420 MHz are schematically indicated in Figure 7.10. The most revealing aspect of this event is the evolution of the radio spectrum, as shown in Figure 7.11. We see in Figure 7.11 the decay of a flat spectrum at high frequencies which is initially self-absorbed at the lowest frequencies. As time increases the source expands, becomes less self-absorbed at the lower frequencies, and the radiating particles lose energy by a combination of adiabatic losses, synchrotron losses, and inverse Compton losses. One or both of the latter loss mechanisms cause the higher frequency spectrum to steepen. As time passes the various spectral regions attain a stable spectral index of -0.55. Furthermore, after the appropriate portion has reached the stable spectral index, the decay of the source is that of a power low in time which can be well approximated by

$$S_\nu = 4200 \, (t - t_o)^{-4.1} \, \nu^{-0.55} \, \text{f.u.} \quad (7.9)$$

where t is measured in days, ν in GHz, and t_o is taken to correspond to September 1.3 UT (Gregory *et al.*, 1972b). The result given by Equation (7.9) is just as expected for an expanding cloud of relativistic electrons losing energy by adiabatic losses.

The fluctuations observed at 10,630 and 8085 MHz on September 2–3 (see Figure 7.10) are real, and represent complications in the relativistic particle injection and/or evolution. In addition, the early stages of the event are not as expected from simple theory (Gregory *et al.*, 1972b).

After the decay of the event shown in Figures 7.10 and 7.11, Cyg X-3 underwent a complicated series of events beginning September 18, 1972, during which the source flared above the 10 f.u. level three times in the space of two weeks (Hjellming and Balick, 1972b), with very complicated behavior.

Given the impetus of the unusual radio outbursts it has been found that an extremely reddened star, visible only in the near infrared (Becklin *et al.*, 1972) is coincident with the source. From the observed properties of this star in the infrared it has been inferred that it could be similar to the blue supergiant in the binary system identified with Cyg X-1. The X-ray observers noted no unusual enhancement of X-ray emission during the radio events, but examination of old data revealed that the X-ray emission of Cyg X-3 is modulated with a 4.8-hour period (Parsignault *et al.*, 1972). The low-energy X-ray cut-off of Cyg X-3 has been used to infer a distance of 3 to 4 kpc (Giacconi, 1970); however, 21 cm absorption against the variable Cyg X-3 radio source (Lauque *et al.*, 1972) indicates a kinematic distance of 8 to 11 kpc.

As this is written, little is understood about the nature of the Cyg X-3 radio outbursts, except that they will provide a great deal of information about the production and re-injection of synchrotron radiation events. Sometime between August 31 and September 2, 1972, the object producing the X-ray source known as Cyg X-3 clearly initiated a series of events in which relativistic electrons and magnetic field were ejected at velocities of the order of a few tenths the speed of light, attaining an angular size of the order of 0.01 arc second and a physical size of the order of a light day a few days later. The events beginning September 18 were presumably of similar nature, but with a large number of ejection events to produce a very complicated evolution for the observed source.

7.3 The Future of Stellar Radio Astronomy

Even though the subject of stellar radio astronomy is just beginning as this chapter is being written, there are a number of things that indicate it will develop into a major area of astronomical research. First of all, there is an apparent correlation of the radio emission phenomena with binary stars that have been studied optically for decades. This cross-referencing between a venerable area of stellar astronomy and radio astronomy bodes well for future development of both fields. Secondly, the seeming relationship between radio stars and compact X-ray sources is of great interest. Finally, the simple fact that for a brief period of time a radio star was one of the strongest compact radio sources in the sky implies a great deal. It is possible that at any particular time such objects could dominate the radio sky; and it is possible that their numbers are large, since their chance of accidental detection is small. In addition, the clear implication of long time scale cycles of activity is suggested by the behavior of all the radio stars, but particularly Cyg X-3 and β Persei. Phenomena on this sort of time scale related to stellar evolution provide a major new dimension by which one can obtain data related to stellar evolution. Most astronomers have always been primarily interested in the study of stars. It is therefore interesting that in coming years the radio astronomers may have something new to contribute to this area.

References

Ables, J. G. 1969. *Astrophys. J.* 155:L27.

Andrew, B. H., and C. R. Purton. 1968. *Nature* 218:855.

Andrews, A. D. 1969. *I.A.U. Inform. Bull. Variable Stars, No. 325.*

Becklin, E. E., J. Kristian, G. Neugebauer, and C. G. Wynn-Williams. 1972. *Nature Phys. Sci.* 239:130.

Bolton, C. T. 1972. *I.A.U. Circ. No. 2388.* Feb. 25, 1972.

Braes, L. L. E., and G. K. Miley. 1971a. *Astron. Astrophys.* 14:160.

———, and G. K. Miley. 1971b. *Nature* 232:246.

———, and G. K. Miley. 1972. *Nature* 237:506.

Giacconi, R. 1970. *I.A.U. Symp. No. 37*, p. 107.

Gregory, P. C., P. P. Kronberg, E. R. Seaquist, V. A. Hughes, A. Woodsworth, M. R. Viner, and D. Retallack. 1972a. *Nature* 239:440.

———, P. P. Kronberg, E. R. Seaquist, V. A. Hughes, A. Woodsworth, M. R. Viner, D. Retallack. 1972b. *Nature Phys. Sci.* 239:113.

Herczeg, T. 1968. *Non-Periodic Phenomena in Variable Stars.* Dordrecht: Reidel.

Hjellming, R. M. 1972. *Nature Phys. Sci.* 238:52.

———, and B. Balick. 1972a. *Nature* 239:443.

———, and B. Balick. 1972b. *Nature Phys. Sci.* 239:135.

———, M. Hermann, and E. Webster. 1972. *Nature* 237:507.

———, V. Herrero, and C. M. Wade. 1974. (to be submitted to *Astrophys. J.*)

———, and C. M. Wade. 1970. *Astrophys. J.* 162:L1.

———, and C. M. Wade. 1971a. *Astrophys. J.* 168:L115.

———, and C. M. Wade. 1971b. *Astrophys. J.* 164:L1.

———, and C. M. Wade. 1971c. *Astrophys. J.* 170:523.

———, and C. M. Wade. 1971d. *Astrophys. J.* 168:L21.

———, C. M. Wade, and E. Webster. 1972a. *Nature Phys. Sci.* 236:43.

———, E. Webster, and B. Balick. 1972b. *Astrophys. J.* 178:L139.

Kellermann, K. I., and I. I. K. Pauliny-Toth. 1966. *Astrophys. J.* 145:953.

Kundu, M. R. 1965. *Solar Radio Astronomy.* New York: Interscience.

Kunkel, W. E. 1969. *Nature* 222:1129.

Lanque, R., J. Lequeux, and Nguyen-Quang Rieu. 1972. *Nature Phys. Sci.* 239:119.

Lovell, B. 1969. *Nature* 222:1126.

Parsignault, D. R., H. Gursky, E. M. Kellog, T. Matilsky, S. Murray, E. Schrier, H. Tananbaum, and R. Giacconi. 1972. *Nature Phys. Sci.* 239:123.

Sandage, A., P. Osmer, R. Giacconi, P. Gorenstein, H. Gursky, J. Waters, H. Bradt, P. Garmire. B. Sreekantan, M. Oda, K. Osawa, and J. Lugaku. 1966. *Astrophys. J.* 146:316.

Seaquist, E. R. 1967. *Astrophys. J.* 148:L23.

Shaffer, D. B., G. A. Shields, and B. Schupler. 1972. *Nature Phys. Sci.* 239:131.

Tananbaum, H., H. Gursky, E. Kellog, R. Giacconi, and C. Jones. 1972. *Astrophys. J.* 177:L5.

Tarenghi, M., and C. Reina. 1972. *Nature Phys. Sci.* 240:53.

Wade, C. M., and R. M. Hjellming. 1971. *Astrophys. J.* 163:L65.

———, and R. M. Hjellming. 1972. *Nature* 235:270.

Zaumen, W., G. T. Murthy, S. Rappaport, R. M. Hjellming, and C. M. Wade. 1972. *Nature* 235:378.

CHAPTER 8

THE GALACTIC MAGNETIC FIELD

Gerrit L. Verschuur

8.1 Introduction

Radio astronomical observations of such widely differing types of objects as extra-galactic nonthermal radio emitters, both quasars and galaxies, galactic supernova remnants, pulsars, and neutral hydrogen clouds, as well as observations of the non-thermal galactic background radiation, give information about the magnetic fields in interstellar space. This information is com-plemented by data on the optical polarization of starlight. Since rigorous models of the galactic magnetic field are difficult to formu-late, our discussion is necessarily qualitative.

All the various types of observations pertaining to the galactic magnetic field involve the existence of polarization of electromagnetic radiation. The continuum radiation from the so-called galactic "back-ground," presumably synchrotron emission, is intrinsically linearly polarized. The polar-ized signal propagating through the inter-stellar medium, which contains both thermal electrons and magnetic fields, can be de-composed into two oppositely circularly polarized components having different phase velocities. On emergence from the medium the waves can be recombined, but with relative phases which may be different from those which they had on entering the medium. The effect of this is a rotation of the plane of polarization, the amount of rotation varying as the square of the wavelength, and, as will be shown below, it is possible to estimate the magnetic field strength and the density of

thermal electrons in the line of sight by measuring this rotation. Furthermore, by extrapolating measurements at several wave-lengths to zero wavelength we can derive the orientation of the magnetic field component transverse to the line of sight in the region in which the radiation originates. This rotation of the plane of polarization, known as Faraday rotation, also affects the polarized emission from pulsars and extra-galactic radio sources. The amount of Faraday rotation is given in terms of a parameter, known as the rotation measure, which is one of the most important tools for studying the galactic magnetic field.

A different approach to studying the galac-tic magnetic field is by measuring the circular polarization of the 21-cm spectral line from interstellar neutral hydrogen. A spectral line in the presence of a magnetic field splits into several components (in frequency), each of which shows a specific polarization. For example, this so-called Zeeman splitting of the 21-cm line produces oppositely circularly polarized components, when viewed along the magnetic field lines, whose frequency difference is proportional to the field strength.

8.2 Observing the Galactic Magnetic Field

8.2.1 Specification of Polarization

In order to specify polarization of a wave completely we must quantify four param-eters. The first is the total intensity of the received signal; the second, the degree of polarization of this signal; the third, the degree of ellipticity (for an elliptically polarized wave); and the last is the plane of the major axis of the polarization ellipse measured from north through east. These four quantities are not directly measurable since an antenna can measure only the strength of the received signal and the antenna is either a linearly polarized dipole or a circularly polarized feed of some sort. However, the directly measurable quantities can be simply related to those needed to specify the received wave completely by using the Stokes parameters.

Equations for the Stokes parameters (I, Q, U, V) are derived in many sources (e.g., Chandrasekhar, 1960) and we shall quote only the results here. They are first the total intensity:

$$I = I_x + I_y = \langle E_x^2 \rangle + \langle E_y^2 \rangle = I_{RH} + L_{LH}$$
(8.1)

Here E_x and E_y are orthogonal components of the transverse electric field whose phase difference is δ, and I_x and I_y are the corresponding intensities. If two oppositely circularly polarized antennas are used, then the intensities in those two antennas, I_{LH} and I_{RH}, are used. The other parameters are

$$Q = \langle E_x^2 \rangle - \langle E_y^2 \rangle = I_x - I_y$$
$$= I_e \cos 2\beta \cos 2\chi = I_p \cos 2\chi \quad (8.2)$$

$$U = \langle 2E_xE_y \rangle \cos \delta = I_s - I_t$$
$$= I_e \cos 2\beta \sin 2\chi = I_p \sin 2\chi$$
$$= Q \tan 2\chi \quad (8.3)$$

I_s and I_t are the intensities measured at 45° to the original set of axes, and

$$V = \langle 2E_xE_y \rangle \sin \delta = I_{RH} - I_{LH} = I_e \sin 2\beta$$
(8.4)

I_e is the intensity of the polarized part of the wave, and I_p the linearly polarized component. Also $I = I_e + I_u$ for a partially polarized wave, where I_u is the unpolarized component. From these relationships we obtain the required parameters, the degree of polarization

$$p = \frac{(Q^2 + U^2 + V^2)^{1/2}}{I}$$
(8.5)

the axial ratio, r, from $\beta = \tan^{-1} r$ and $\sin 2\beta = (V/I_e)$, and finally the plane of polarization, χ, from $\tan 2\chi = (U/Q)$. The two parameters U and Q specify the linear polarization; in most galactic and extra-galactic work V is assumed zero. Recent observations show this assumption to be reasonable. Note that a wave whose electric vector rotates clockwise when viewed along the direction of propagation is defined as right-hand polarized in radio astronomy. This is opposite to the optical convention.

It is possible to measure linear polarization by making observations at two orientations of a pair of orthogonal antennas separated by 45° or at three orientations of a linearly polarized feed (0°, 45°, 90°). Alternatively, a measurement can be made with one set of orthogonal antennas and then a phase delay placed in one of them. Cohen (1958) has discussed how the product E_xE_y is proportional to the cross-correlation coefficients of the signals in the oppositely polarized antennas, so that one can obtain the desired polarization parameters. This is also discussed in Chapter 10.

8.2.2 The Faraday Effect

A beam of electromagnetic radiation incident on a magneto-ionic medium, *i.e.*, a medium consisting of electrons and positive ions under the influence of a magnetic field, propagates as two waves, known as the characteristic waves or characteristic modes of propagation, each with its own phase velocity and its own elliptical polarization. Owing to the difference in phase velocities of the two rays, called the ordinary and extraordinary rays, the phases alter relative to one another as they travel through the medium. When the two modes are recombined after leaving the medium, the resultant differs from the original incident ray. Provided the applied magnetic field is not transverse to the direction of propagation, the orientation of the polarization ellipse of the resultant wave rotates in the medium, an effect named after its discoverer, Michael Faraday. Polarized radio waves traveling through the interstellar medium will undergo Faraday rotation which can be measured.

The plane of polarization of a linearly polarized radio wave is rotated by an amount Ψ (radians) as it passes through a magneto-ionic medium given by

$$\Psi = 8.1 \times 10^5 \, \lambda^2 \int N_e \, B_{\parallel} \, dl \qquad (8.6)$$

where N_e is the electron density in cm^{-3}, l is the path length in parsecs, λ is the wavelength in meters, and $B_{\parallel}$ is the longitudinal component of the magnetic field in the medium in Gauss. A quantity called the rotation measure, $R(\lambda)$, is useful in describing the amount of rotation that has occurred in a given wavelength interval, and is given by

$$R(\lambda) = \frac{\partial \Psi}{\partial (\lambda^2)} = 8.1 \times 10^5 \int N_e B_{\parallel} dl \text{ rad m}^{-2}$$
$$(8.7)$$

The sign of this measure indicates the direction of the magnetic field. A positive value indicates a field toward the observer. Thus measurements of the orientation of the plane of polarization at many wavelengths suffice to determine the rotation measure. Furthermore, extrapolation to zero wavelength gives the intrinsic polarization angle, since there the rotation is zero.

Now consider a polarized beam of radiation rotated in a medium by an amount Ψ when observed at a wavelength λ. If the spectrum of the received signal is uniform across the bandwidth of the receiver, then the plane of polarization also varies across this band. The variation in orientation across the band is sometimes called the dispersion, $d\Psi$, and, from the fact that $\Psi \alpha$ constant $\times \lambda^2$ $\equiv$ (constant/ν^2) = (c/ν^2), where ν is the frequency, the dispersion across the band is given by

$$\frac{d\Psi}{d\nu} = -\frac{2c}{\nu^3} = -\frac{2\Psi}{\nu} \qquad (8.8)$$

If

$$\frac{\Delta \nu}{\nu} = \frac{\nu_1 - \nu_2}{\nu_0} \ll 1$$

then

$$\frac{d\Psi}{\Psi} = -2\frac{d\nu}{\nu} \equiv -2\frac{\Delta \nu}{\nu}$$

where $\Delta \nu$ is the bandwidth.

Therefore, observations of linearly polarized radio emission have to be made with a receiver with sufficiently small bandwidth, $\Delta \nu$, so that the dispersion in the band is kept to a small fraction of the rotation itself at a particular frequency. In solar radio burst observations, for which high Faraday rotations are observed, this is a very small bandwidth of the order of kHz, but for observations of galactic and extra-galactic radio sources bandwidths of the order of MHz are allowed. For pulsar observations bandwidths as small as 10 kHz are often required. If bandwidths are not sufficiently small, the measurements will be subject to depolarization by the differential Faraday effect across the band.

Several other phenomena occur which produce a depolarization of the radio signals being observed. Depolarization resulting from the finite beamwidth of the radio-telescope is one of these. If several regions of emission with different rotation measures (either in the source or in the intervening medium) simultaneously exist within the beam, the net polarization measured will be less than the straightforward sum of the component parts, and the rotation measure observed will be related to those in the beam in a complex way. Usually in such a case the plane of polarization observed is not a linear function of λ^2. The reader is referred to Burn (1966) for a discussion of this point.

Depolarization also results when the intrinsic polarization vectors vary across the source within the beam or when the components within a source have different spectral indices so that at one wavelength the polarization of one component might dominate and at another wavelength a differently polarized component might dominate. Again a λ^2 dependence of the polarization angle will be absent.

8.2.3 Background Polarization and Depolarization

Observations of the polarization of the continuum radio emission from the galaxy give information about the magnetic fields in those regions of space in which the radiation is emitted as well as about the intervening medium. The term "background polarization" is in fact a misnomer because it is the polarization of the foreground emitting material which we observe, the radiation from beyond several hundred parsecs being depolarized in its path through the medium. This can be readily shown as follows. Typically a rotation measure of 10 rad m^{-2} is found for continuum emission from the galaxy. At a wavelength of 1 m, the total rotation is 10 rad, and to avoid depolarization the differential rotation within the beam has to be less than 1 rad. For a typical interstellar magnetic field of 3 μ Gauss and electron density of 0.01 cm^{-3} we find that a path length of 40 pc gives a rotation measure of 1 rad m^{-2}. Structures within the beam with this scale size will each produce differential Faraday rotation of the order of a radian. It is therefore very likely that emission from beyond a few hundred parsecs will reach the observer unpolarized.

Polarization of the continuum emission from the galaxy was first clearly demonstrated by Westerhout *et al.* (1962) and followed up extensively by other workers in Holland who made observations at several different wavelengths. These indicated that Faraday rotation and depolarization of the radiation were occurring and that both were varying from point to point in the galaxy. These data, and those from subsequent surveys, provide information concerning the local magnetic field within a few hundred parsecs of the Sun and are discussed below.

8.2.4 The Effect of the Ionosphere

The ionosphere is an ionized medium containing a magnetic field and will therefore induce Faraday rotation on a radio wave passing through it. Typical rotations range from only a few degrees to 10°, at a wavelength of 21 cm, to about 60° at the longer wavelengths typically used in making background polarization observations (75 cm). Baker and Smith (1971), for example, found 30° of rotation at night. Since the total electron content of the ionosphere decreases at night, observations have typically been made then, particularly in the several hours before sunrise, when the short-term variations of the ionospheric electron content are smallest.

To correct for the ionospheric Faraday rotation one can either observe well-understood calibration sources at regular times throughout the day, or use ionospheric sounder data or satellite sounder data. Alternatively, the rotation of one plane of polarization of a signal from a geo-stationary satellite can be measured. A model for the ionosphere is then used to obtain the expected rotations for other azimuths and elevations.

8.2.5 Residual Polarizations

All polarimeters in radio astronomy are subject to spurious or residual polarization effects of some sort. Generally, in polarization observations, signals from two different antennas are combined in some way, and usually one is subtracted from the other. Therefore, unless one has identical polar diagrams for the two antennas, the difference is nonzero even if the source being observed is unpolarized. These differences can give an apparent polarization because the Stokes parameters Q, U, and V, instead of being zero, will have some finite value. Such spurious polarizations are sometimes called residual polarization, and great care has to be taken in correcting for these effects. This can be done, provided they are systematic in nature and repeatable.

There is a type of residual polarization which results from differences in the emission in the sidelobes of the two antennas; these sidelobes differ because of small differences

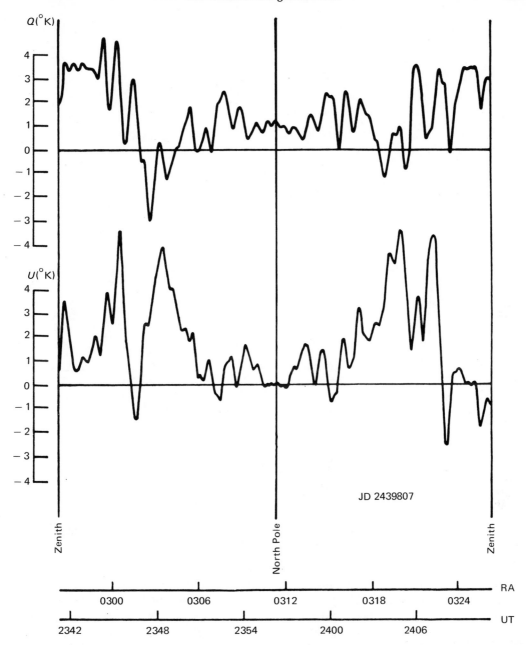

Figure 8.1 Observations of the Stokes parameters Q and U, in units of polarized brightness temperatures, from the paper by Baker and Smith (1971). The telescope was scanned from the zenith to the North Celestial Pole and back again while the data were taken.

in the actual construction or matching of the antennas, or because of irregularities in the parabolic reflector being illuminated. Since the continuum emission from the galaxy is confined primarily to the galactic plane and since this radiation can be substantially polarized, one finds that residual polarizations arising from differences of the level of emission in the distant sidelobes of a polarimeter can be serious. A variation of this affects observations made when the sidelobes of the antenna system are partly illuminated

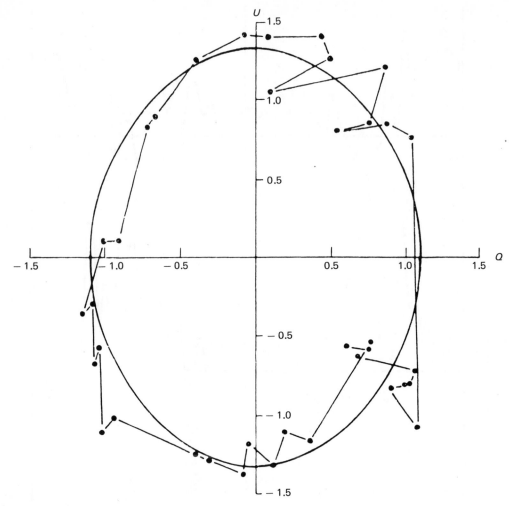

Figure 8.2 The residual polarization ellipse at the NCP in the Q–U plane from the data of Baker and Smith (1971), the units again being polarized brightness temperature (see text).

by the ground. Such ground effects obviously become more serious at lower elevations.

Several observing methods have been used to overcome these sidelobe problems. They consist basically of observing so-called standard regions. Unless the standard region or source is unpolarized or has a known polarization, however, the data are naturally modulated by the polarization vector of the source as well as by the vector representing the residual polarization.

We will illustrate the way background polarization observations are made and corrected for residual effects by taking the

recent observation of Baker and Smith (1971) as an example. In Figure 8.1 we show one of their scans from the zenith to the North Celestial Pole (NCP). The two Stokes parameters Q and U are shown. In order to find the residual polarization as well as the ionospheric correction they used their observations of the NCP as follows. Because of the rotation of the sky about the pole, the apparent position angle of the polarized emission from the NCP rotates during the day. Superimposed on this rotation is the varying Faraday rotation in the ionosphere. In Figure 8.2 we show the values for Q and U

at the NCP found by Baker and Smith (1971), plotted on rectangular coordinates. They note that the ellipticity of the data is a measure of the ratio of the sensitivities between the two recording channels and the center gives the zero levels for the channels. These data were combined with a knowledge of the ionospheric electron content and the sidereal time of the observations to obtain the separate contributions to the polarization ellipse at the NCP from ionospheric rotation and from the intrinsic polarization of the NCP. Thus the amount of ionospheric rotation can, in principle, be used to correct the observations made at nearly the same time of day. The reader should refer to the paper by Baker and Smith (1971) for further details. The polarization of the NCP at a 73-cm wavelength is found to be 1.2°K at a position angle of 170 ± 5°.

Another basic cause of residual polarization is the existence of cross-polarization lobes within the main lobe of the primary polar diagram. This is a serious problem in single dish observations at a primary focus. A plane wave striking a parabolic reflector excites currents in its surface which follow the curvature of the dish. Thus, a plane wave which is 100% polarized in the y direction will establish a current which has a small x component. Along the axis of the main beam the circular symmetry of the parabolic surface ensures cancellation of the unwanted component. Slightly off axis, however, a dipole placed at the focus and oriented parallel to the x-axis will detect a signal. The resultant sidelobe patterns are shown in Figure 8.3. These two patterns are for the case of two orthogonal dipoles measuring U and Q and are taken from the observations of Spoelstra (1972). Polarization measurements of point radio sources centered in the beam will be unaffected, but obviously measurements of extended sources will suffer seriously if the source structure varies across the cross-polarization lobes. Apparent polarizations will be produced which are difficult to correct. Obviously small differences in gain between the two antennas magnify the

effect. Not only does this difference in gain produce a residual polarization at the center of the main beam, but the cross-polarization sidelobes are no longer symmetrical.

An effective way of eliminating these sidelobes is to use a secondary reflector whose curvature cancels that of the primary reflector. Thus a Cassegrain antenna is highly desirable for making polarization measurements.

8.2.6 Rotation Measure Data

In 1964 Mayer *et al.* (1964) first measured the linear polarization of an extra-galactic radio source, Cygnus A. Since then the polarization of hundreds of extra-galactic sources and many galactic sources, including pulsars, have been measured. These observations obviously give information about the magnetic field structures and strengths in the sources, but the polarized signal itself serves as a useful probe of the interstellar medium. Faraday rotation and depolarization due to the interstellar medium are observed in the radiation from extra-galactic sources. The details of exactly what these probes are telling us about the interstellar medium is still the subject of controversy. The main problem is sorting out how much rotation is occurring respectively in the source, intergalactic space and interstellar space. In general, however, the rotation measures of large numbers of extra-galactic sources can be used to derive a general pattern for the magnetic field in the solar neighborhood because the sign of the rotation measure indicates the direction of the field.

The linearly polarized radiation from the pulsars is also rotated in the interstellar medium. The rotation measure is proportional to the integral $\int N_e B_\parallel dl$, which is the mean magnetic field along the line of sight weighted by the electron density. Another measurable quantity for pulsars is the dispersion measure, proportional to the integral of the electron density along the line of sight, $\int N_e \, dl$. Hence the ratio of the rotation measure and dispersion measure yields the

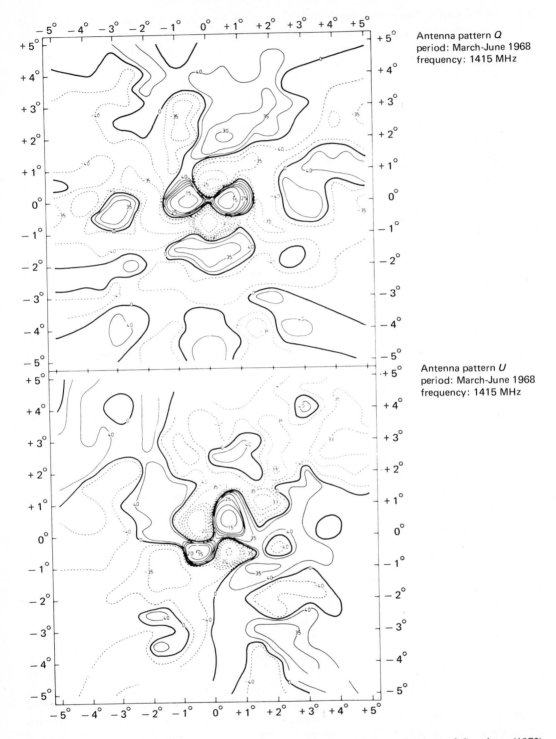

Figure 8.3 Antenna pattern for the Stokes parameters Q and U from the data of Spoelstra (1972). The scale is in dB below the main beam for unpolarized radiation. Dotted lines indicate negative values and full lines positive values for the Stokes parameters. The cross-polarization side-lobes are clearly illustrated.

weighted mean longitudinal magnetic field strength,

$$\langle B_{\parallel} \rangle = \text{constant} \times \frac{\int N_e B_{\parallel} dl}{\int N_e dl} \quad (8.9)$$

The plane of polarization of the pulses can be measured in several different ways. If a fixed linearly polarized feed is used, then the rotation of the plane of polarization causes a variation in the observed pulse amplitude with frequency. Provided the response of the antenna and receiver as a function of frequency is known, the magnitude of the rotation measure can be calculated from the spacing between maxima in the spectrum. The sign of the rotation measure cannot be obtained by this method because not enough information is available, if only one feed is used, to determine the Stokes parameters fully.

In order to obtain both the magnitude and the sign of the rotation measure one needs to observe the pulsar at a number of discrete frequencies using an orthogonal antenna system that does allow the Stokes parameters to be determined. Smith (1968) used this method, as did Manchester (1972). A number of pulses are integrated and observations are made at several frequencies, spaced sufficiently that the rotation of the plane of polarization significantly exceeds the noise. Manchester (1972) determined the plane of polarization on a point-by-point basis across a set of integrated profiles obtained at different frequencies and aligned in time. Corrections for ionospheric rotation at the observing frequency were derived from satellite observations of the ionosphere made at the Illinois Ionospheric Radio Laboratory. As is customary, a simple model had to be assumed in order to calculate the rotations expected at Green Bank, West Virginia. The resultant ionospheric rotation measures were found to be between 0.1 and 0.6 rad m^{-2}. In order to determine the rotation measures unambiguously, observations both at closely spaced and at widely separated frequencies are required. Manchester used up to seven different frequencies in the 250- to 500-MHz range.

8.2.7 The Zeeman Effect

After Bolton and Wild (1957) first suggested that the Zeeman effect should be observable in the 21-cm spectrum of neutral hydrogen, a great deal of effort was spent in trying to find it. If a cloud of neutral hydrogen radiating in its ground state is permeated by a magnetic field, the emission line should be split in frequency into two oppositely circularly polarized components. The amount of splitting depends on the strength of the line-of-sight component of the magnetic field and should be easily measurable if the interstellar field were 10^{-5} Gauss or greater, as was expected in 1960. Since the splitting is only 2.8 Hz per μGauss, it is not measurable directly in lines whose widths are usually ~20 kHz. However, a differential technique can be used in which the difference between the left- and right-hand circular components is measured as a function of frequency across the emission or absorption line. This difference profile will have the shape of a differentiated Gaussian profile in the case of a single spectral line, and therefore the greatest signals are expected for the most intense and the narrowest 21-cm lines. These are usually absorption lines where antenna temperatures of hundreds of degrees have been found, and it is only in such cases that the Zeeman effect has been detected. The observations give information only about the fields in the HI clouds themselves, and extrapolating to the remainder of the interstellar medium is an uncertain task.

The attempts to measure a Zeeman effect require long integrations while the telescope is tracking the source. The signals in two oppositely circularly polarized feeds have to be subtracted from one another, and since each of these is usually large (>100°K for absorption sources studied) and the difference profile (the Zeeman pattern) is very small, one has to be very careful in balancing the two feeds and being sure that their gains are perfectly stable within the switching cycle. Also, it appears that small-scale structures exist in the hydrogen-line emission from the galaxy and therefore the HI emission in the various cross-polarization lobes can be very different for the two feeds. Subtraction, to obtain the Stokes parameter, V, therefore also reveals the difference between the emission profiles in the two sets of oppositely polarized cross-polarization lobes. Residual signals due to this effect were found on all measurements made at $b \leq 1°$ by Verschuur (1970). Rotation of the feed system about the axis of the telescope meant that the cross-polarization pattern was rotated, and then the residual effects change, depending on the structure in the HI emission, whereas a true Zeeman effect would remain as a constant signal. The detection of the Zeeman effect in the direction of Orion A was certain only after many observations were made at different feed orientations on and around the source.

8.3 Discussion of the Observations

8.3.1 Background Polarization Surveys and Their Interpretation

The discovery that the galactic background was polarized, and its subsequent mapping at many wavelengths, has led to various interpretations of the observed distribution of the polarized emission in terms of the magnetic field structure in the local spiral arm. An example of such a map is given in Figure 8.4. The region of high polarization around $l = 140°$, $b = +10°$ has been of particular interest, and we will tentatively suggest a reason for its existence at the longer wavelengths. We will first outline the general properties that might be expected of the galactic polarized emission and then examine whether the observed features correspond to the predictions.

It is well known that the synchrotron process acts to produce polarized emission from the Milky Way which is mainly concentrated to the plane. If a uniform magnetic field pervaded the spiral arms, the degree of polarization observed would follow the intensity of the continuum emission. Since at the same time the plane of the galaxy also contains the greatest concentration of neutral and ionized matter, the Faraday rotation suffered by the polarized emission will be greatest there. It also follows that if the interstellar medium is not perfectly homogeneous, the differential Faraday effect over quite small regions of the sky will be greatest in the plane. Therefore, although the intrinsic polarization is largest in the plane, the depolarization is also greatest. A study of the way the degree of polarized emission varies with latitude at short wavelengths (10 to 20 cm) should therefore show high polarization near the plane, with the intensity decreasing with increasing latitude. At longer wavelengths (75 cm) the depolarization effects in the plane would lead us to expect relatively low polarizations there.

The observations (Berkhuijsen *et al.*, 1964; Mathewson *et al.*, 1966; Bingham and Shakeshaft, 1967) show that there are exceptions to this general picture. This is illustrated in Figure 8.4, where, besides the high polarization seen around $l = 140°$, one also notices patches of higher polarization well away from the plane. The high polarizations seen at high latitudes are associated with the galactic spurs (Mathewson and Milne, 1965; and Bingham, 1967), so we need to invoke a model that first accounts for the enhanced emission from the spurs and then for the high polarization. Mathewson and Milne (1965) suggest that the regions of high polarization at 75 cm are concentrated in a band about 60° wide centered on a great circle cutting the plane at $l = 160°$ and 340°. However, closer examination of this suggestion shows that the spurs themselves lie mainly in this band 60° wide. This band therefore may not, by itself, reveal the configuration of the local spiral arm field unless we wish to explain the spurs in terms of distortions of this field.

The only other notable exception to the general outline suggested above is the presence of an apparently atypical highly polarized region, around $l = 140°$, $b = +10°$ (see Figure 8.5). One of the lesser spurs

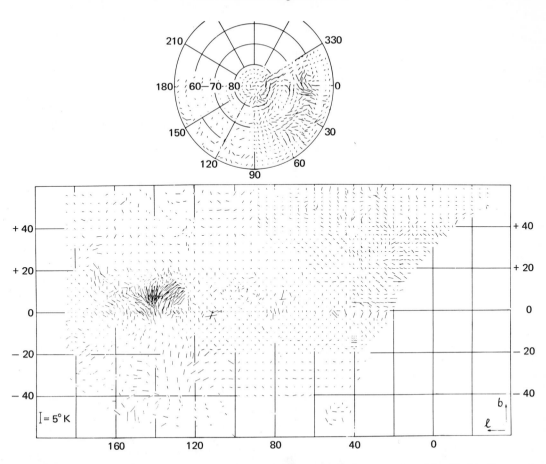

Figure 8.4 The 408-MHz polarization map of the northern sky as reported by Berkhuijsen and Brouw (1964). The magnitude and direction of the vectors in the diagram represent the magnitude and direction of the polarization vectors observed. Notice the region of high polarization around $l = 140°$. [Berkhuijsen and Brouw. *Bull. Astron. Inst. Neth.* (1964) 17:185.]

(Loop III) appears to start in this region and also shows high polarizations. It has been suggested (Berkhuijsen *et al.*, 1964; Bingham and Shakeshaft, 1967; Seaquist, 1967) that we are viewing the local spiral arm field normally at this point, and hence the intrinsic polarization appears high and the rotation measures are low. Such an interpretation does not account for the asymmetry of this region with respect to the plane nor the lack of a counterpart at $l = 320°$. It appears that we are here seeing through a local hole in the depolarizing medium in addition to viewing the local field normally in this direction.

The boundaries of this region of high polarization, when viewed at 75-cm wave-

lengths, occur roughly at $l = 125$ and $160°$, $b = 0$ and $+20°$ (Berkhuijsen *et al.*, 1964). Whereas the decrease in polarization around $l = 160°$ is rather gradual with longitude, the decrease around $l = 120°$ is relatively sharp. Examination of the 408-MHz map of Berkhuijsen and Brouw (1964) (Figure 8.4) shows this rather clearly. It could be accounted for by a boundary occurring in a region that is depolarizing the radiation which originates at longitudes $< 120°$ since there is a suggestion in the Cambridge (Bingham, 1967) and Dutch (Brouw, 1967) polarization maps at 1400 Mc/sec that the region of high intrinsic polarization extends to at least $l = 105°$.

We now note a remarkable similarity

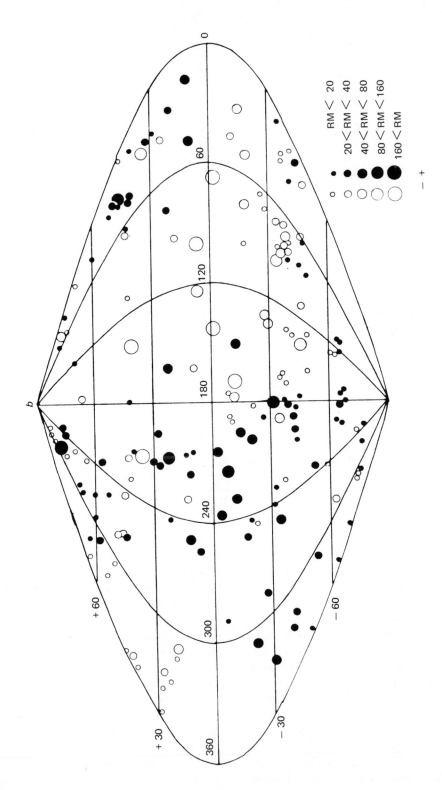

Figure 8.5 The rotation measures of 192 radio sources plotted in galactic coordinates. The filled circles indicate a field toward the observer, while the open circles indicate a field away from the observer, the size of the circle being representative of the magnitude of the rotation measure. From the data of Mitton (1972), *Monthly Notices Roy. Astron. Soc.* 155:373.

between this radio polarization cut-off at $l = 120°$ and a sudden change in optical polarization near this longitude. The results of Hiltner (1956), showing the polarization vectors of starlight in this region, reveal a definite discontinuity in the length and direction of the vectors at $l = 122°$. Possibly here we are viewing a boundary to a region that is capable of depolarizing both the radio background and the starlight at longitudes $< 122°$. We expect the former to occur in the presence of thermal electrons and line-of-sight magnetic field components so as to generate differential Faraday rotation in the medium. For the latter we might expect that a disorganized field structure within the polarizing clouds in the line of sight would help to keep the polarization low. Alternatively less polarizing material (*i.e.*, dust) might exist where $l < 120°$. There is, however, no evidence for this, and we can more easily explain the effect in terms of the lesser degree of field alignment inferred from the increased scatter in position angles.

Seaquist (1967) finds a correlation between starlight polarization and radio polarization in this region of space for stars in Hall's (1958) and Hiltner's (1956) catalogs; he concludes that since all the stars in the data are > 700 pc distant, the polarization must originate at distances > 700 pc in order to produce the correlation. However, the sample of stars he used is such that 90% are > 1 kpc and 75% > 1.6 kpc distant, so that unless re-examination of the data proves otherwise, the polarization could originate more than 1.6 kpc away.

8.3.2 The Spurs

At high latitudes we expect the magnitude of the polarization vectors to be lower, and if there are regions of very much larger polarization we should seek their origin. Those areas above the plane showing the highest polarization coincide with the so-called spurs of enhanced continuum emission. The nature of the spurs is not fully understood yet, but the polarization measurements are obviously important for any model that might be derived. Berkhuijsen (1971) has estimated that in regions of the North Polar Spur the percentage polarization reaches a value $> 70\%$, but this conclusion is reached after performing the difficult task of subtracting out a general background level of emission. Such a degree of polarization is as high as can be expected on the basis of synchrotron theory. Spoelstra (1971) also finds a very high percentage polarization in the North Polar Spur and a very close dependence of the degree of polarization on the structure in the Spur.

Originally Bingham (1967) presented evidence for magnetic field lines running along the ridge of the North Pole Spur (NPS) based on the analysis of 21-cm polarization measurements. The optical data also suggest a field generally parallel to the NPS. Mathewson and Milne (1965) reported a similar effect in the Cetus arc. However, Spoelstra (1971) has studied the NPS in greater detail, and using rotation measure data (typically 15 to 25 rad m^{-2}), finds that the intrinsic magnetic field in the NPS lies perpendicular to the ridges observed in the continuum radiation. In one model the NPS is part of a supernova shell and a circumferential magnetic field is expected. However, Spoelstra's conclusions make that somewhat doubtful.

8.3.3 Rotation Measure Data

Polarization observations of the synchrotron emission from large numbers of radio sources have been made at many wavelengths. These data give information about the source magnetic fields, and when rotation measures are obtained, about the magnetic fields and electron densities between the source and the observer. The interpretation of rotation measure data of discrete sources in order to learn something of the galactic magnetic field has been and still is controversial. It has been customary to examine the dependence of the sign of the rotation measures on galactic longitude and latitude in order to derive a model for the galactic magnetic field in the local spiral arm. In such models the lowest latitude sources have been left out because there are contributions to their rotation measures from distant spiral arms.

In Figure 8.5 we show the distribution on the sky from all the sources with known rotation measure (from Mitton, 1972). Below $b = -10°$ we get the impression that there is a certain degree of order in the way the signs of the rotation measures are distributed, but the pattern is not so obvious in the northern hemisphere at $b > 10°$. Sofue *et al.* (1969) have suggested that rotation measure signs are correlated with the red shift of the sources, implying perhaps that the rotation measures have an intergalactic contribution. Reinhardt (1972) has recently examined all the data in more detail, and although he cannot rule out the presence of an inter-galactic effect, he derives a direction and strength of the uniform component of the galactic magnetic field. A regression analysis for quaser data gives a mean field running toward $l = 110°$, $b = 10$ to $20°$. Radio galaxies did not fit this picture too well, and Reinhardt suggests that they might have larger intrinsic rotation measures which blur out much of the galactic effect. His similar analysis of the pulsar data gives the direction of the field to be toward $l = 120°$ and $b = -30°$.

It is more generally accepted that rotation measures of extra-galactic sources contain a large component introduced by the galaxy, because a fairly good dependence of rotation measure on latitude is found with some notable exceptions. The interpretation of the rotation measure data is therefore quite complicated. It is obvious from Figure 8.5 that small-scale variations exist. For example, rotation measure differences of 100 rad m^{-2} are seen to occur over 3° in the sky and some high-latitude sources have very large rotation measures (~ 100 rad m^{-2}). If these are galactic in origin, and path lengths there are supposed to be only ~ 100 pc, it seems obvious that at low latitudes very high rotation measures (> 1000 rad m^{-2}) might be common. Since these are not found, the existence of high-latitude, high-rotation-measure sources suggests that large intrinsic rotation measures exist in some extra-galactic sources.

Whether or not the polarization of extra-galactic sources shows a depolarization effect which is clearly dependent on galactic latitude is also controversial. No obvious dependence of the degree of polarization of sources as a function of latitude has been reported.

Pulsar rotation measures appear to be more likely to give reliable information concerning the local galactic magnetic field. Advantages of these data are an apparently insignificant contribution to the observed rotation measure from the pulsar itself (Manchester, 1972) and the knowledge that the pulsars are galactic and fairly close to the Sun. All the rotation is therefore occurring in the interstellar medium or in the ionosphere, and since the latter is well determined, the pulsar data apparently give the best estimates of the mean interstellar field vector. This contrasts with the data for extra-galactic radio sources, where we can never be certain about the amount of rotation occurring in the source as opposed to the galaxy. In Table 8.1 we show Manchester's (1972) results for the rotation measures and dispersion measures, as well as the value for the weighted mean field in the direction of 19 pulsars. In addition to Manchester's data, two other pulsars included in Table 8.1 are PSR $0628-28$ (Schwarz and Morris, 1971) and $0833-45$ (Komesaroff *et al.*, 1971).

8.3.4 Zeeman Effect Data

Only several discrete interstellar clouds have been found to contain magnetic fields strong enough to be detected in the 21-cm hydrogen-line wavelength. A complete description of the observational aspects of this experiment is given in Verschuur (1969a). In Table 8.2 we summarize the most significant Zeeman effect data that exist to date, taken from Verschuur (1969a, 1970, 1971). All the positive detections have been confirmed elsewhere, the most recent being the result for

Table 8.1 Rotation and Dispersion Measures for Pulsars

PSR	l (degrees)	b (degrees)	Dispersion measure (cm^{-3} pc)	Rotation measure (rad m^{-2})	Mean line-of-sight field* (μGauss)
0329+54	145	−1	26.7	−63.7	−2.93
0525+21	184	−7	50.8	−39.6	−0.96
0531+21	185	−6	56.8	−42.3	−0.92
0628−28	237	−17	34.4	+47.0	+1.6
0809+74	140	+32	5.8	−11.7	−2.5
0818−13	236	+13	40.9	−2.8	−0.08
0833−45	264	−3	69.2	+33.6	+0.59
0834+06	220	+26	12.9	+24.5	+2.3
0950+08	229	+44	2.9	+1.8	+0.7
1133+16	242	+69	4.8	+ 3.9	+0.99
1237+25	252	+87	9.2	−0.6	−0.07
1508+55	91	+52	19.6	+0.8	+0.05
1642−03	14	+26	35.7	+16.5	+0.58
1818−04	26	+5	84.4	+70.5	+1.0
1929+10	47	−4	3.1	−8.6	−3.3
1933+16	52	−2	158.5	−1.9	−0.01
2016+28	68	−4	14.1	−34.6	−3.0
2021+51	88	+8	22.5	−6.5	−0.36
2045−16	31	−33	11.5	−10.8	−1.15
2111+46	89	−1	141.4	−223.7	−1.95
2217+47	98	−8	43.5	−35.3	−1.00

* A positive field component is directed toward the observer.

Orion by Brooks *et al.* (1971) and for M17 by Murray (private communication).

8.4 The Structure of the Magnetic Field

8.4.1 General Considerations

In this and the subsequent sections we shall discuss some of the models that have been proposed for the magnetic field structure in our galaxy, and by the very nature of the available data, this means a discussion primarily of the models for the magnetic field in the solar neighborhood. A basic consideration is that the various phenomena discussed above probably occur in different volumes of interstellar space. For example, the Zeeman data apparently refer only to the fields in dense neutral hydrogen clouds, optical polarization data (not discussed in any detail here) are indicative of conditions in interstellar dust clouds, and rotation measure data refer to the ionized regions of space.

8.4.2 The Longitudinal Field

The theoretical expectation has always been that the field is aligned along the local spiral arm. In distant arms it was likewise expected to be aligned parallel to the arms and hence should appear to be aligned parallel to the plane. The optical polarization data, notably those of Hiltner (1956) and Hall (1958) for distant stars in the Perseus region (l = 40 to 140°), indeed indicate a high degree of alignment of the polarization vectors parallel to the plane. The background polarization data of Berkhuijsen and Brouw (1964) and Westerhout *et al.* (1962) supported this conclusion, because in the direction of l = 140° the radio polarization vectors were normal to the plane, irrespective of wavelength. This indicated zero rotation measure, which would result if the line of sight were normal to the field. The magnetic field was therefore thought to be longitudinal and directed toward $l \approx 50°$.

Table 8.2 Magnetic Fields Detected by Means of the Zeeman Effect at 21 cm

Direction	l (degrees)	b (degrees)	Velocity of cloud, km sec^{-1} (l.s.r.)	Field* (μGauss)
Taurus A	185	−6	+10	−3.5 ± 0.7
			+4	−1.5 ± 0.9
Cassiopeia A	112	−2	−38	+18.0 ± 1.9
			−48	+10.8 ± 1.7
Orion A	209	−19	+7	−50 ± 15
			+2	−70 ± 20(?)
M17	15	−1	+14	+25 ± 10
Cygnus A	76	+6	+3	+3.0 ± 2.2 ⎫ Doubtful
			−84	+4.0 ± 2.2 ⎭ detections

* A negative sign indicates a field toward the observer.

The early maps of the distribution of the rotation measures of extragalactic sources (Morris and Berge, 1964; Gardner and Davies, 1966; and Berge and Seielstad, 1969) also suggested that the magnetic field was longitudinal, but perhaps toward a slightly different longitude and having opposite directions above and below the plane. Gardner and Davies (1966) suggested that it was directed toward $l = 95°$ at low latitudes ($|b| < 20°$), toward $l = 90°$ for $−20 > b > −60$, and toward $l = 270°$ for $+20 < b < +60°$. Gardner *et al.* (1969) using more recent data suggested that the distribution of rotation measures is consistent with a uniform magnetic field directed toward $l = 80°$, with a localized distortion around $l = 20°$ in the shape of a loop of magnetic lines arching out from the otherwise uniform configuration. This would then give rise to the sign reversal in that part of the sky. This loop might be associated with the North Polar Spur in some way. The reader might draw his own conclusions from the more complete data collected by Mitton (1972), shown in Figure 8.5.

Manchester (1972) interprets the distribution of pulsar rotation measures on the sky in terms of a simple longitudinal field. His data are shown in Figure 8.6. He states that a field directed toward $l = 90°$ is clearly consistent with these observations. However, it appears that those pulsars indicating a field away from the observer lie between $l = 30$ and $190°$, and those of opposite sense lie between 210 and 30°. This, taken without reference to any other data, would indicate a field directed toward $l = 120°$, although the two positive rotation measures near $l = 30°$ might be associated with the Spur (Manchester, private communication).

It appears that the background radio polarization data and optical data, both indicating a field toward $l = 50°$, are at variance with the suggestion by Gardner *et al.* (1969) of a field toward $l = 90°$ and the data of Manchester (1972), which this author takes to indicate a field toward $l = 120°$.

Whatever the direction of a possible longitudinal field, along what is it aligned? There is little evidence of a local spiral arm directed toward $l = 90°$. The distribution of young stars serving as spiral arm tracers indicates that the local arm (or spur) is directed toward $l = 50°$ (see the map of Becker and Fenkart, 1970). The 21-cm-line galactic structure data are difficult to interpret near the Sun. A tangential point exists in the direction of $l = 80°$ in the continuum emission from the galaxy, but that is not necessarily the local arm. Simonson (1971), summarizing recent discussions on this problem, noted that many galactic structure workers were calling the local region a spur to a spiral arm. Perhaps the optical and radio continuum data indicate the field

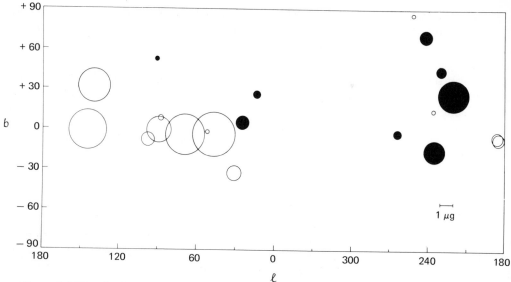

Figure 8.6 The distribution of pulsar rotation measures in galactic coordinates from Manchester (1972). Again filled circles indicate a field toward the observer and open circles a field away from the observer. From Manchester (1972). *Astrophys. J.* 172:43.

parallel to this spur. But then which field is being revealed by rotation measure data?

8.4.3 The Helical Field

Various theoreticians and observers have suggested that a helical magnetic field might be the expected configuration in the solar neighborhood. From the starlight polarization data it appears that there is a scale length of between 100 and 200 pc in the interstellar field (Jokipii and Lerche, 1969). Consequently the degree of polarization of starlight increases fairly rapidly for stars up to distances of this order, but increases only slowly beyond that. This is supported by the conclusions of Hall (1958) and Hiltner (1956), who found no distance dependence of the polarization of O or B in the Perseus spiral arm. The optical polarization data might therefore be the best indicator of the local field configuration.

Only Mathewson (1968) has so far tried to collect *all* the available optical polarization data and explain the direction of the observed vectors in terms of a single comprehensive model. He invoked a local (< 500 pc distance) helical field configuration. This helix was oriented with its axis toward $l = 90$ and $270°$, with a small pitch angle (7°), and sheared through 40° in an anti-clockwise sense viewed from the North Pole by differential rotation, so that in the direction of longitude 140° there is a high degree of alignment of the vectors. This had previously been interpreted as support for a longitudinal magnetic field directed normal to this direction, *i.e.*, to $l = 50$ or $230°$, which was roughly the direction of the so-called local spiral (spur) arm.

Mathewson's model explained the polarization vectors of all the stars observed to date irrespective of their distance. Although he did not state it explicitly, this implies that all the polarization must be produced in this local helical magnetic field region. Mathewson and Nicholls (1968) later added a possible longitudinal component to the model to account for the radio source rotation measure data.

Manchester (1972) has pointed out that several of the pulsar rotation measures contradict the helical field model as outlined by Mathewson. The main observational evidence for the helical field comes from the optical polarization data, but these interpretations may not be unique. However, a detailed discussion of these is beyond the scope of a book on radio astronomy. The radio data, on the other hand, are also difficult to interpret, and until rotation measures, and hence intrinsic polarization vectors, are available for the continuum emission over the whole sky, we are unable to state with certainty what the structure of the local magnetic field is.

8.5 The Magnetic Field Strength

8.5.1 General Remarks

As previously mentioned, different types of measurements of the interstellar magnetic field refer to different regions of space. Several assumptions must be made in order either to derive the field strength from a given set of data or to localize the region in space under consideration. Zeeman effect data give directly the line-of-sight field strengths only in HI clouds, although this author feels that a fairly useful estimate of the field in the more general interstellar medium can be derived from these data. The pulsar rotation measure data give the weighted mean field in the clouds of thermal electrons between the pulsar and the observer. Optical polarization data appear to be less helpful in providing estimates of the field strength, as these estimates have ranged from 3 to 30 μgauss, depending on the model assumed for the dust grains and their alignment.

8.5.2 The Pulsar Rotation Measure Data

Manchester (1972) has given the values for the weighted mean field in the direction of 21 pulsars, which are shown in Table 8.1. His conclusion is that a relatively uniform field of 3.5 μGauss is implied by these results. However, the three pulsars showing mean fields of 2.9, 3.0, and 3.3 μGauss lie in directions $l = 145°$, 68°, and 47° respectively, which, because of the 100° range of longitude they cover, argues against a very uniform field direction in the galaxy.

That the pulsar PSR 0329+54 at $l = 145°$ has such a large mean field value is surprising, since this direction is supposedly

normal to the local field derived from the radio background polarization and optical polarization data. If much of the rotation in this pulsar is occurring in the Perseus arm, we have to conclude that the field there is certainly not longitudinal either, since the optical arm there is also aligned normal to $l = 140°$.

On the simplest assumption, Manchester's model of a longitudinal field directed toward $l = 90°$ predicts some systematic variation of mean field strength as a function of longitude. However, a plot of the field strength data in Table 8.1 versus longitude shows no systematic pattern at all. One might therefore conclude that the magneto-ionic medium is not very uniform.

8.5.3 The Zeeman Effect Data

An unambiguous positive detection of the Zeeman effect was first made in 1968. Fields in the range 3 to 50 μGauss, listed in Table 8.2, have since been detected (Verschuur, 1969a, 1971); thus a simple answer to the strength of the interstellar field is not obvious. Verschuur (1969b) suggested that the field in a given cloud appeared correlated with the HI density in the cloud.

For a uniformly contracting cloud with a frozen-in field one expects the field strength, B, to vary as the radius squared, whereas the density varies as the radius cubed. Thus the field will vary as the density, n_H, to the two-thirds power. One can perhaps derive the value of the interstellar field before contraction of the medium into the clouds by selecting the value of the field at the density corresponding to the mean interstellar HI density (~ 0.7 cm^{-3}). If so, one attains a field of about 1 μGauss, bearing in mind that the line of sight is, on average, at about 45° to the field direction and taking the upper of the two lines as an upper envelope. However, of the five clouds in which fields have been measured, two lie in front of, and probably near, HII regions, and two lie in front of Cassiopeia A. It is the author's opinion that the clouds in front of Cas A are also associated with HII regions in the Perseus arm. Since there are a very large number of HII regions, clusters, and associations in that arm, it is likely that all the HI clouds there are being compressed by expanding ionization fronts. Therefore the clouds in question need not be at the state of gravitational collapse. They might have a larger pressure than the normal "intercloud" medium quoted by Field, Goldsmith, and Habing (1969). Even in this case the field strength might be expected to increase as $(n_H)^{2/3}$.

The remaining positive detection was in the direction of Taurus A, where a field of 3.5 μGauss is seen in a cloud that does not appear to be particularly dense. This might reflect the "normal" interstellar field more closely.

8.5.4 A Theoretical Approach

Parker (1970) has discussed the equilibrium of the galactic disk. He considers the strength of the field required to contain thermal gas in the galaxy. Treating the cosmic rays as a gas on a large scale, he shows that the thermal gas is effectively coupled to the field and can retain a field in the disk which is not too strong. A knowledge of the mean density of material leads to an upper limit of the field which can be trapped. The value so derived for the root-mean-square field is $\leq 5\,\mu$Gauss. However, various arguments suggest that the cosmic rays themselves spend only about 10^6 years in the disk of the galaxy. Parker estimates the minimum field required to contain them for this time period is 3 μgauss. Thus his theoretical arguments predict a field between 3 and 5 μGauss.

8.6 Summary

The data suggest that the mean field strength in the interstellar medium might be around 1 to 3 μGauss, but the field configuration is not yet clearly established. The field appears to be stronger in localized regions of space, such as dense HI clouds, and there is considerable small-scale structure in the field direction if one compares the Zeeman data with pulsar rotation measure data in the same directions.

The study of the distribution of radio source rotation measures on the sky shows that it is possible to derive very general impressions about the way the magnetic field lines are directed, but the picture is not yet so well understood that a clear model based on such data alone can be formed. Further data might help clarify this situation, although they are also likely to show that the field structure is much more complex than suspected.

Background polarization surveys now being carried out with high-resolution instruments should produce useful additional data concerning the local magnetic field, in particular if maps of rotation measures over the whole sky could ever be made. Lastly, it does not at present seem that the measurements of such a classical phenomenon as the Zeeman effect at 21-cm will add much to our knowledge of magnetic fields in interstellar space in the near future, unless very-low-noise switched receivers become available, which would allow the detection of fields of the order of 3 μGauss in emission features.

Recent review articles relative to this chapter have been prepared by Verschuur (1967, 1973a), van de Hulst (1967), and Davis and Berge (1968). The apparent confusion seen in the distribution of rotation measures on the sky has been discussed by Verschuur (1973b).

References

Baker, J. R., and F. G. Smith. 1971. *Monthly Notices Roy. Astron. Soc.* 152:361.

Becker, W., and R. Fenkart. 1970. *I.A.U. Symp. No. 38.* "Spiral Structure of Our Galaxy." W. Becker and G. Contopoulos, eds. New York: Springer-Verlag, p. 205.

Berge, G. L., and G. A. Seielstad. 1969. *Astrophys. J.* 157:35.

Berkhuijsen, E. M. 1971. *Astron. Astrophys.* 14: 359.

———, and W. Brouw. 1964. *Bull. Astron. Inst. Neth.* 17:185.

————, W. Brouw, C. A. Muller, and J. Tinbergen. 1964. *Bull. Astron. Inst. Neth.* 17: 465.

Bingham, R. G. 1967. *Monthly Notices Roy. Astron. Soc.* 137:157.

————, and J. R. Shakeshaft. 1967. *Monthly Notices Roy. Astron. Soc.* 136:347.

Bolton, J. G., and J. R. Wild. 1957. *Astrophys. J.* 125:296.

Brooks, J. W., J. D. Murray, and V. Radhakrishnan. 1971. *Astrophys. Lett.* 8:121.

Brouw, E. 1967. *I.A.U. Symp. No. 31, Noordwijk.* Private communication.

Burn, B. J. 1966. *Monthly Notices Roy. Astron. Soc.* 133:67.

Chandrasekhar, S. 1960. *Radiative Transfer.* New York: Dover, p. 24.

Cohen, M. H. 1958. *Proc. Institute of Radio Engineers,* 46:172.

Davis, L., and G. L. Berge. 1968. *Stars and Stellar Systems.* Vol. 7. Chicago: University of Chicago Press, p. 755.

Field, G. B., D. W. Goldsmith, and H. J. Habing. 1969. *Astrophys. J.* 155:L149.

Gardner, F. F., and R. D. Davies. 1966. *Australian J. Phys.* 19:129.

————, D. Morris, and J. B. Whiteoak. 1969. *Australian J. Phys.* 22:813.

Hall, J. S. 1958. *Publ. U.S. Naval Obs.*, 2, No. 17.

Hiltner, W. A. 1956. *Astrophys. J. Suppl.* 2:389.

Hulst, van de, H. C. 1967. *Ann. Rev. Astron. Astrophys.* 5:167.

Jokipii, J. R., and I. Lerche. 1969. *Astrophys. J.* 157:1137.

Komesaroff, M. M., J. G. Ables, and P. A. Hamilton. 1971. *Astrophys. Lett.* 9:101.

Manchester, R. M. 1972. *Astrophys. J.* 172:43.

Mathewson, D. S. 1968. *Astrophys. J.* 153:L47.

————, and D. K. Milne. 1965. *Australian J. Phys.* 18:635.

————, N. W. Broten, and D. J. Cole. 1966. *Australian J. Phys.* 19:93.

————, and D. C. Nicholls. 1968. *Astrophys. J.* 154:L11.

Mayer, C. H., J. P. McCullough, and R. M. Sloanaker. 1964. *Astrophys. J.* 139:248.

Mitton, S. 1972. *Monthly Notices Roy. Astron. Soc.* 155:373.

Morris, D., and G. L. Berge. 1964. *Astrophys. J.* 139:1388.

Parker, E. 1970. *I.A.U. Symp. No. 39.* "Cosmical Gas Dynamics." H. J. Habing, ed., p. 168.

Reinhardt, M. 1972. *Astron. Astrophys.* 19:104.

Schwarz, U. J., and D. Morris. 1971. *Astrophys. Lett.* 7:185.

Seaquist, F. R. 1967. *Astron. J.* 72:1359.

Simonson, S. C. 1971. *Astron. Astrophys.* 9:163.

Smith, F. G. 1968. *Nature* 220:891.

Sofue, Y., M. Fujimoto, and K. Kawabata. 1969. *Publ. Astron. Soc. Japan* 20:388.

Spoelstra, T. A. Th. 1971. *Astron. Astrophys.* 13:237.

————. 1972. *Astron. Astrophys.* 5:205.

Verschuur, G. L. 1967. *I.A.U. Symp. No. 31.* "Radio Astronomy and the Galactic System." H. van Woerden, ed. New York: Academic Press, p. 385.

————. 1969a. *Astrophys. J.* 156:861.

————. 1969b. *Astrophys. J.* 155:L155.

————. 1970. *Astrophys. J.* 161:867.

————. 1971. *Astrophys. J.* 165:651.

————. 1973a. *Fund. Cosmic Phys.* Journal in preparation.

————. 1973b. *Proc. I.A.U. Colloquium 23.* "Planets, Stars, and Nebulae Studied with Photopolarimetry." T. Gehrels, ed.

Westerhout, G., C. L. Seeger, W. N. Brown, and J. Tinbergen. 1962. *Bull. Astron. Inst. Neth.* 16:187.

CHAPTER 9

INTERSTELLAR MOLECULES

Barry E. Turner

9.1 Introduction

At the time of writing 24 interstellar molecular species have been identified and published, and all but three of them have been found at radio wavelengths. In addition, two unidentified lines have been found in the λ3-mm region, which are widely believed to be molecular in origin. Table 9.1 lists in chronological order of detection the presently known interstellar molecules at radio wavelengths. In addition, CH, CH$^+$, and CN have been detected at optical wavelengths, and H$_2$ and CO have been detected at UV wavelengths.

As Table 9.1 clearly shows, a revolution has taken place over the past three years in our understanding of the chemistry and physics of the clouds of gas and dust that fill a sizeable fraction of the interstellar medium. Some of these clouds, which were originally thought to have densities of order 10 cm^{-3}, temperatures of $\sim100°$K, and to contain mainly only hydrogen with small amounts of a few diatomic molecular species, are now known to contain a large variety of complex molecules. Total particle densities of at least 10^7 cm^{-3} are required to explain the observations of some of these molecules, while derived cloud temperatures range from close to 3° to $\sim200°$K. The densities of many of these molecular species are high enough to defy explanation on the basis of older theories of molecule formation such as gas phase radiative association or even early ideas of formation on grain surfaces. Although more efficient formation processes are required, the problem of explaining very large concentrations of complex molecules is at least partly resolved by the recent realization that the photo-destruction rates of molecules inside dense clouds may be much slower than previously believed. Molecules appear to exist largely in clouds dense enough to provide very efficient shielding against the interstellar UV radiation field.

The picture which has emerged over the past two years, then, is that molecules occur largely in dense cold clouds, many being identifiable in dark dust clouds. All molecules are probably associated with dust. All molecules are thought to occur in roughly the same regions and to have roughly the same velocity. The distribution is very non-uniform on a broad scale or even throughout individual clouds.

With the exception of only OH and H$_2$CO, the molecules in Table 9.1 have typically been studied in only about a dozen regions (in the directions of the brighter HII regions), and several are known only in the direction of the source Sgr B2 near the galactic center. In these very few sources the velocities of most molecules appear to be the same, within uncertainties of typically a few tenths of a km/sec. OH and H$_2$CO are sometimes exceptions to this rule, differing by up to 10 km/sec from the majority of molecules in some cases. In addition OH velocities appear generally to differ somewhat from those of H$_2$CO; other differences (see Section 9.2.2) suggest that OH and H$_2$CO

Table 9.1 Interstellar Molecular Abundances

Molecule	Formula	Date	Wavelength	Observed abundances (cm^{-2})			Abundances relative to CO			"Expected" abundance
				Sgr B2	W 51	Ori A	Sgr B2	W 51	Ori A	
Hydroxyl	OH	1963	18 cm	1×10^{18}	8×10^{15}	3×10^{14}	2×10^{-2}	1×10^{-4}	1×10^{-4}	3×10^{3}
Ammonia	NH$_3$	1968	1.3 cm	$>2 \times 10^{17}$	2×10^{15}	2×10^{15}	4×10^{-3}	2.5×10^{-5}	7×10^{-4}	4×10^{2}
Water	H$_2$O	1968	1.4 cm	?	?	?	?	?	?	3×10^{3}
Formaldehyde	H$_2$CO	1969	6.2 cm	2×10^{15}	2×10^{14}	3×10^{14}	4×10^{-5}	2.5×10^{-6}	1×10^{-4}	1
Carbon monoxide	CO	1970	2.6 mm	5×10^{19}	8×10^{19}	3×10^{18}	—	1	1	1
Cyanogen radical	CN	1970	2.7 mm	—	1×10^{15}	1×10^{15}	—	1.5×10^{-5}	3×10^{-4}	1.3×10^{-1}
Hydrogen cyanide	HCN	1970	3.4 mm	3×10^{14}	8×10^{14}	2×10^{15}	6×10^{-6}	1.0×10^{-5}	7×10^{-4}	1.3×10^{-1}
X-ogen	?	1970	3.4 mm	?	?	?	?	?	?	?
Cyanoacetylene	HC$_3$N	1970	Several	1×10^{16}	—	3×10^{14}	2×10^{-4}	—	1×10^{-4}	1.1×10^{-8}
Methyl alcohol	CH$_3$OH	1970	Several	3×10^{15}	—	$\leq 2 \times 10^{15}$	6×10^{-5}	—	$\leq 7 \times 10^{-4}$	1
Formic acid	HCOOH	1970	18 cm	10^{15} to 10^{16}	—	—	2×10^{-5} to 2×10^{-4}	—	—	7×10^{-4}
Carbon monosulfide	CS	1971	2.0 mm	$\sim 10^{14}$	$\leq 2 \times 10^{15}$	$\leq 5 \times 10^{14}$	2×10^{-6}	$\leq 2.5 \times 10^{-5}$	$\leq 1.7 \times 10^{-4}$	2.5×10^{-2}
Formamide	NH$_2$CHO	1971	6.5 cm	$\leq 2 \times 10^{16}$	—	—	$\leq 4 \times 10^{-4}$	—	—	9×10^{-5}
Silicon monoxide	SiO	1971	2.3 mm	4×10^{13}	—	—	8×10^{-7}	—	—	6×10^{-2}
Carbonyl sulfide	OCS	1971	2.5 mm	$>8 \times 10^{15}$	—	—	$>1.6 \times 10^{-4}$	—	—	8×10^{-6}
Methyl cyanide	CH$_3$CN	1971	2.7 mm	$>2 \times 10^{14}$	—	—	$>4 \times 10^{-6}$	—	—	4×10^{-5}
Isocyanic acid	HNCO	1971	3.4 mm	5×10^{14}	—	—	10^{-5}	—	—	9×10^{-5}
Hydrogen isocyanide?	HNC?	1971	3.3 mm	?	?	?	?	?	?	?
Methyl acetylene	CH$_3$C$_2$H	1971	3.5 mm	10^{15}	—	—	2×10^{-5}	—	—	1.3×10^{-4}
Acetaldehyde	CH$_3$CHO	1971	28 cm	5×10^{14}	—	—	1×10^{-5}	—	—	3×10^{-4}
Thioformaldehyde	H$_2$CS	1971	11 cm	Undetermined	—	—	Undetermined	—	—	2.5×10^{-2}
Hydrogen sulfide	H$_2$S	1972	1.8 mm	2×10^{14}	2×10^{14}	5×10^{14}	4×10^{-6}	2.5×10^{-6}	1.7×10^{-4}	7.5×10^{1}
Methyleneimine	CH$_2$NH	1972	5.7 cm	6×10^{14}	—	—	1.2×10^{-5}	—	—	1.3×10^{-1}

may not be in close physical association. The question whether molecules are physically associated with continuum sources, particularly HII regions, is harder to answer, since for all molecules other than OH and H_2CO the observations have been restricted almost entirely to directions of HII regions. OH and H_2CO are very likely not closely associated with HII regions; instead they appear to exist in clouds that occupy the same general regions of spiral arms as HII regions, namely, those where the density is enhanced. Most of the other molecules do appear to be associated with continuum sources in the sense that they reside in dense neutral clouds that surround or lie close to these sources; their velocities are generally quite close to the recombination-line velocities. The question how closely molecules are correlated with dust is harder to answer. Regions where most molecules are seen are completely opaque to visible light. OH and H_2CO are, however, seen in less dense regions in a few cases where extinction can be measured. Recent surveys in these regions indicate that OH and H_2CO are correlated with extinction, though not necessarily with total gas density. In other regions, however, OH does not appear to be correlated with extinction.

Present observations shed rather little information on the chemistry of interstellar molecules. Certainly most of the molecules are composed of the four most abundant atoms, hydrogen, nitrogen, carbon, and oxygen, while sulfur and silicon are next in order of abundance. Without further information, one might conclude from this that the chemistry of interstellar molecules is random —that the relative abundances of molecules are related to the relative abundances of their constituent atoms. When the actual abundances of the different molecules are considered, however, this conclusion is found to be untenable. Table 9.1 lists the molecules presently detected at radio wavelengths. Estimated column densities are given for the three sources Sgr B2, W 51, and Ori A for which molecules have been most extensively studied. Column densities normalized to the density of CO are also given and can be compared with the "expected" density, defined as the molecular abundance which would apply solely on the basis of the relative abundances of the constituent atoms. A comparison of the "expected" and observed ratios shows clearly that there are marked deviations from the random-chemistry model. Nearly all of the listed molecules are underabundant with respect to CO, but a few species (HC_3N, NH_2CHO, OCS) are overabundant. Organic molecules are highly favored over inorganic: All of the carbon-containing species are overabundant relative to OH, NH_3, and SiO (note that the ratios of the inorganic species $[NH_3]/[OH]$ and $[SiO]/[OH]$ are roughly as "expected"). There is also evidence that oxygen-containing molecules are favored over their thio- (sulfur-containing) analogues: $[CS]/[CO] \simeq 10^{-5}$, while $[S]/[O] \approx 1/40$, $[SH]/[OH]$ is less than $[S]/[O]$ by at least a factor of 4, and $[H_2CS]/[H_2CO]$ is apparently less than $[S]/[O]$. CO is, of course, vastly overabundant relative to all other interstellar molecules; not only is $[SiO]/[CO] \approx 10^{-6}$ while $[Si]/[C] \approx 1/17$, but also upper limits on the abundances of ClO, SO, and FeO indicate that they are much less abundant with respect to CO than the cosmic abundances of the atoms Cl, S, and Fe would suggest. There is also some suggestion from Table 9.1 that among the organic molecules the nitrogen-containing species may be slightly overabundant compared to those that do not contain nitrogen; confirmation of this possibility must however await more reliable abundance determinations. Table 9.1 also reveals a large variation in relative molecular abundances (up to a factor of 100 or more) from source to source. In Sgr B2, for example, $[HC_3N]/[HCN] \approx 10$, in itself a surprising result, while in Ori A the ratio is ~ 0.05.

Another interesting aspect of interstellar chemistry is the underabundance of free radicals as compared with stable molecules. At radio wavelengths, OH and CN are the only presently observed free radicals. Although OH is widespread throughout the galaxy, it appears to be less abundant than H_2O in the very dense sources where they are found together. CN is generally less abundant than HCN and HC_3N in sources where they are commonly observed. Many free radicals that might be expected on the basis of the known molecules in Table 9.1 are as yet undetected at radio wavelengths. The most conspicuous of these are HCO ([HCO]/[H_2CO] < 10^{-1}), NCO ([NCO]/[HNCO] < 1), and CH ([CH]/[OH] < 10^{-1}). The simplest explanation for their absence is that the densities of the clouds are high enough that these radicals react (with H or H_2) to form the stable molecules in a short time. Such reactions would most likely occur on grain surfaces, the collision rate of molecules and grains being very rapid in the densest clouds. In such clouds, even gaseous free radical reactions at a high rate cannot be ruled out, owing to their negligible activation energy.

It should be pointed out that several difficulties exist in determining abundances from spectral-line observations (see Section 9.4) which may typically produce uncertainties of an order of magnitude. Such uncertainties are not enough to violate the conclusion that interstellar chemistry is nonrandom. The variations in abundance from cloud to cloud, although large, also do not alter the conclusion that organic molecules are overabundant relative to inorganic ones. These cloud-to-cloud variations are particularly noticeable in the cases of OH and H_2CO, which have been studied in a large number of sources, but also apply to NH_3, CO, CN, and HCN, which have been observed in roughly a dozen sources. Such variations have several possible explanations. First, although atomic abundances are probably reliable to within a factor of 2 or 3 on the average, they may well vary substantially from region to region. For example, the ratio [CO]/[H] in W 51 appears to exceed (by a factor of $\sim$3) the value which would apply if all C were in the form of CO, assuming that the average cosmic ratio of [C]/[H] $\simeq 3 \times 10^{-4}$ applies in this region. And as indicated by recent studies of NH_3, HC_3N, HNCO, CH_3OH, and SiO, the

Sgr B2 source consists of two distinct clouds, in one of which the nitrogen-containing molecules appear to be relatively more abundant. Regions of high obscuration may well have different atomic abundances than regions of low obscuration. Second, different molecular formation processes may occur in different regions, *e.g.*, owing to differing dust-to-gas ratios or differing grain compositions. Destruction rates will also vary from region to region, depending on the strength of the UV radiation field and the wavelength dependence of the absorption by dust that presumably must shield the molecules against rapid photo-destruction.

It would appear, then, that interstellar chemistry is very specific and does not follow the patterns of conventional laboratory reactions, which are typically conducted under LTE conditions. Laboratory chemistry may be characterized, for example, by temperatures of $\sim$300°K and densities of $\sim 10^{19}$ cm^{-3}. Interstellar chemistry operates under temperatures of $\sim$3 to 200°K and densities of $\sim$10 to 10^8 cm^{-3} corresponding to mean free paths of $\sim 10^8$ to 10^{16} cm. Over most of this density range, collision rates ($\sim 10^{-9}$ to $\sim 10^{-2}$) are much slower than electronic decay rates (10^3 to 10^9 sec^{-1}) or vibrational decay rates ($\sim 10^{-1}$ to 10^2 sec^{-1}) of typical molecules. Thus all interstellar molecules must undergo chemical reactions while in ground electronic, vibrational, and (in the case of certain diatomic molecules) even rotational states. By contrast, laboratory reactions occur with molecules populated according to a Boltzmann distribution of rotational and vibrational states. Considerable evidence, of both a theoretical and an experimental nature, indicates that collision cross-sections and reaction rates depend very strongly upon the specific mode of excitation of the reacting species. Unfortunately, the known thermodynamic parameters (*e.g.*, activation energy, cross-sections) that are used to predict reaction rates apply only to the LTE case in which all relevant excitation modes of a molecule are populated by a Boltzmann distribution. The corresponding parameters of the interstellar case are not known. These difficulties apply to gaseous reactions. In the event that surface reactions dominate interstellar chemistry, even more formidable difficulties hamper interpretation. These stem from the fact that virtually nothing is known of the detailed processes that occur in surface reactions, not only under interstellar conditions (where the nature of the surface is unknown) but under laboratory conditions with controlled surfaces as well.

Although little of a definite nature can as yet be inferred from interstellar chemistry, the observations so far have provided clues to physical conditions and processes—some rather unexpected in nature. A few examples, discussed more fully later, will illustrate.

The observation of several different lines from a given molecule is useful not only in deriving excitation temperatures (equivalent to kinetic temperatures in certain, specifiable, cases), but also in indicating density distributions. H_2CO lines at 6-, 2-, and 1-cm wavelengths indicate, for example, that the distribution of molecules may be very clumped in several regions, much like a polka-dot pattern with perhaps quite small densities in the regions between the clumps. The strong maser-type emission from OH and H_2O similarly has been shown to arise in very small regions, only a few A.U. in size.

Some molecules, such as CO, CN, HCN, and OCS, appear to be quite normally excited (that is, to be nearly in thermal equilibrium with their surroundings) and are also seen in emission. The collision rates of these molecules, which provide their excitation, must therefore exceed the molecular interaction rate with the 3°K cosmic background radiation. This allows a lower limit to be placed on the total particle densities in the molecular clouds, which is typically 10^3 cm^{-3}, although it is much higher in some cases.

Recently, certain short-lived states of NH_3 have been observed in the source Sgr B2. These states must be collisionally excited at rates comparable to their rapid rate of decay to the ground state, $\sim 10^{-2}$ sec^{-1}. The total particle densities required for this are found to exceed 10^8 cm^{-3}.

In at least some molecular clouds, differences in the excitation temperatures of molecules such as CO, CS, and HCN can be shown to imply that the excitation must be due to collisions with neutral particles rather than with electrons. In turn it is found that the electron-to-neutral number density ratio is far smaller than is usually taken to apply to HI regions. Thus carbon atoms in these dense clouds must be largely neutral, indicating that insignificant amounts of UV radiation, soft X-rays, and low-energy cosmic rays penetrate these clouds.

The anomalous excitation of interstellar molecules is not discussed in detail in this review. The best-known cases, reviewed elsewhere (Turner, 1970a, 1970b) are those of OH and H_2O, both of which in some sources display powerful maser emission corresponding to brightness temperatures as high as 10^{13}°K and originating in regions no larger than a few A.U. in size. Densities as high as 10^{12} to 10^{13} cm^{-3} and kinetic temperatures of at least several hundred Kelvins are inferred in these regions, which must be proto-stellar in nature. In other types of sources, which display a different kind of OH maser action, the existence of powerful infrared radiation fields is indicated; these may be produced by large-scale shock fronts in the clouds, or by nearby "IR nebulae" which are observed in some cases. Yet another type of maser emission from OH and H_2O arises in the atmospheres of some cool "IR" stars of variable nature; the physical and chemical conditions in the atmospheres are found to vary in interesting ways with light-phase.

Molecules observed at optical wavelengths (CH$^+$, CH, CN) reveal quite different physical conditions from those observed at radio wavelengths. The "optical" molecules occur in relatively rarefied clouds of densities no larger than 10 to 10^2 cm^{-3} and apparently devoid of the more complex "radio" molecules listed in Table 9.1. Radiative recombination and gaseous exchange reactions appear to be the dominant formation mechanisms for these species, rather than surface reactions on grains. The interaction of these molecules with the 3°K cosmic background results in the excitation of certain transitions and not others; from the observations of these the intensity of the 3°K background can be derived at wavelengths at which the Earth's atmosphere is opaque and for which the accuracy is much greater than can be obtained by rocket observations. The results obtained in this way are so far consistent with a blackbody interpretation of the cosmic background to a wavelength of 0.55 mm.

9.2 Observations of Interstellar Molecules at Radio Wavelengths

9.2.1 Types of Objects Where Molecules Are Found

Of the 23 radio molecules listed in Table 9.1, 12 have been seen in two or less sources (seven in only one source—Sgr B2), and another eight in fewer than 10 sources. Only OH, H_2CO, and CO (and, in far fewer sources, H_2O) have been detected in sufficiently many regions to allow some idea of the overall galactic distribution of molecules to emerge.

Table 9.2 summarizes the types of objects in whose directions molecules are found. The entries refer to the percentage of the total number of each type of object searched that has yielded detectable molecular abundances. Entries such as "1 of 4" indicate cases where the number of sources searched is so small, or the sensitivity of the search is so poor, that meaningful percentages cannot as yet be given. A dash (—) means that no adequate search has yet been made.

The ordering of the objects in Table 9.2 is not accidental. It is thought that the most highly evolved objects are to the right in the table and that this also corresponds to increasing total particle densities.

The "extended obscured regions" refer to large ($\sim 3° \times 4°$) areas in the Taurus and Serpentis-Scorpius sections of the sky, which are characterized by small, somewhat non-uniform extinction at visual wavelengths (~ 1 to 5 magnitudes). The extinction has been measured accurately by means of star counts (Bok, 1956); we shall refer to these regions as Bok regions. Both OH and H_2CO have been detected and mapped throughout these regions (Turner and Heiles, 1973; Kutner, 1973), and their strengths are found to be linearly correlated with the localized extinction. They do not correlate, however, with the density of atomic hydrogen, as deduced from 21-cm line observations (Knapp, 1972). The important question is whether molecules like OH and H_2CO are fundamentally correlated with the presence of grains (*i.e.*, extinction) or with total gas density. This question is at present unanswered, because it is not known whether gaseous hydrogen is in the form of H or H_2 in these regions. OH and CO appear to be thermally excited in these regions, thus in principle allowing a reliable estimate of their abundances to be made. Preliminary results for OH indicate column

Table 9.2 Galactic Objects Where Interstellar Molecules Are Found

Molecule	Extended obscured regions	Dust clouds	Globules	Continuum sources	"Optical" emission nebulae	IR stars
OH	2 of 2	100%	2 of 5 ?	80%	5%	~20%
H_2CO	2 of 2	100%	5 of 7	50%	—	0
H_2O	0	0	0 of 3	10%	—	~20%
NH_3	—	1 of 4	—	10 of 14	—	0 of 3
CO	1 of 1	8 of 11	—	9 of 9	—	1 of 1

densities only slightly smaller than those found in dust clouds, and comparable excitation temperatures. H_2CO is observed in absorption in these regions, and since there is no measurable excess background continuum, the absorption of the 6-cm transition must occur against the 3°K cosmic background, corresponding to an excitation temperature for the 6-cm transition of $< 3°K$. Because the kinetic temperature in these regions exceeds 3°K, as deduced from the OH data, the 6-cm transition of H_2CO is evidently being anomalously excited (the energy level populations are "anti-inverted"). The evidence favours collisional, rather than radiative pumping for H_2CO.

a) *Dark Dust Clouds* (Heiles Clouds)

These clouds are visually seen as regions of very heavy obscuration, with sizes ranging from a few arc minutes to several degrees. They are distinguished from the Bok regions both by smaller sizes and by much higher extinctions, which at visual wavelengths range from lower limits of 8 magnitudes to ~100 magnitudes. The physical properties of these clouds have been reviewed by Heiles (1971). Only OH, H_2CO, NH_3, and CO have been detected in these clouds. Based primarily on the OH and CO emission, kinetic temperatures from 5 to 30°K are found in these clouds; from less direct arguments total gas densities from 10^2 to 10^4 cm^{-3} are inferred from the same data. These inferences are possible because the CO and OH molecules appear to be thermally excited in these clouds (the small deviations from LTE that are observed in some OH lines do not affect these conclusions). At these temperatures and densities it is expected that nearly all hydrogen is in the form of H_2, and this conclusion is borne out by 21-cm observations which typically show a depletion of atomic hydrogen in these clouds. As in the Bok regions, H_2CO is seen only in absorption via the 6-cm transition in these clouds, and again because the only measurable continuum radiation in these directions is the 3°K cosmic background, the 6-cm transition must be anoma-

lously cooled. The velocities of OH, CO, and H_2CO are usually identical within measurement uncertainties in these clouds, suggesting that they coexist and are reasonably homogeneously mixed in these clouds. However, in a few cases OH and H_2CO are each seen in two different velocity features in the same cloud, and in these cases the ratio of abundances appears to differ from one feature to the other.

An interesting comparison of the molecular properties in Bok regions and Heiles clouds is furnished by Heiles cloud 2, which is situated in the middle of one of the Bok regions. Although the boundary of cloud 2 is sharply defined in terms of the visual extinctions, the boundary at best may be characterized as "fuzzy" in terms of the molecular line radiation. The abundance of the OH (and probably also of the H_2CO) is a factor of only ~2 less in the Bok regions than in the Heiles cloud, while the extinction is at least 10 times less. Thus the linear relation between OH (and H_2CO) abundance and extinction, which applies for extinctions between 1 and 5 magnitudes, apparently ceases to apply for extinctions greater than a value somewhere between 5 and 50 magnitudes. This suggests that these molecules are not destroyed primarily by UV radiation at these higher extinctions.

b) *Globules*

Most of the Heiles clouds have been inferred from the molecular data to be gravitationally unstable and in a state of collapse. It seems logical to suppose that the next stage of evolution is the globules. These objects are even smaller (~1 to 10 arc minutes in size) and more opaque than the Heiles clouds. Since molecules act as efficient coolants in interstellar clouds, by means of their line radiation, one might expect globules to be cooler as well as more dense than the Heiles clouds. (Although globules are thought to evolve into proto-stars, the collapse would not have proceeded as far as to allow gravitational heating to compete with molecular cooling.) Partly because of their small size and also partly because of their low excitation conditions, it has proved difficult to detect molecules in these objects. Only H_2CO has been definitely detected in globules, by means of the 6-cm line (Palmer *et al.*, 1973), and as in Bok regions and Heiles clouds, it is seen only in anomalous absorption. Because radiation at visual (and shorter) wavelengths cannot penetrate globules, and because no IR sources seem to be associated with globules, the H_2CO pumping mechanism would appear either to be collisional or to involve microwave radiation, possibly as a result of non-blackbody characteristics of the $3°K$ background. The tentative detection of OH in globules is consistent with a thermal excitation at a very low temperature.

For these globules of average projected size ~ 10 arc minutes it appears that the OH abundance is less than that found in the dark dust clouds, despite the fact that the particle densities in the globules are likely much higher. On the other hand the H_2CO abundance in globules seems comparable to or greater than that in dust clouds, if we assume that the anomalous pumping of the 6-cm H_2CO transition is of comparable efficiency in globules and in dust clouds. It is possible that free radicals such as OH are converted to stable molecules at the higher globule densities; H_2O would be a likely endpoint, and searches have been made for it in globules without success. However, the presence of H_2O in globules cannot be ruled out on this basis, because the only accessible transition is the high-energy (~ 450 cm^{-1}) 6_{16} to 5_{23} one, which is almost certainly incapable of being excited in such low-energy environments as globules.

c) Continuum Sources

Every one of the 23 radio molecules so far discovered has been observed initially in the direction of continuum sources (HII regions and supernova remnants). In the directions of at least a few of these sources (Sgr B2, Sgr A, W 51, DR 21OH, W 3OH, Ori A) molecules are found with larger column densities than in any other type of object.

In the case of the two most widely observed molecules, OH is usually—and H_2CO is always—seen in absorption; all other molecules are always observed in emission. At wavelengths of order 3 cm and shorter, emission is expected because the continuum brightness temperature is small compared with typical molecular excitation temperatures. This situation is seen to apply to most of the molecules listed in Table 9.1. In the case of CH_3OH, $HCOOH$, NH_2CHO, and CH_3CHO, the observation of emission lines is consistent with probable anomalous excitation which produces population inversions and at least some degree of maser amplification in the observed transitions.

Are molecules physically associated with the continuum sources in whose directions they are seen? The answer to this question depends on the species of molecule. Many of the molecules in Table 9.1 are observed in only Sgr B2, W 51, and Ori A, with seven of them appearing only in Sgr B2. In these sources, all these molecules have the same velocity within the uncertainties, and the velocity tends to be very similar to, but not identical with, the recombination-line velocities that characterize the ionized region of the continuum source. It is generally assumed that the molecules reside in the neutral cloud that surrounds the HII region and out of which the HII region and its ionizing star originally formed. This picture appears to be directly confirmed in the case of Ori A, for which several of the molecules (CO, CS, HCN, X-ogen) are observed in extended clouds which center on the HII region. It is in this sense that the continuum source category in Table 9.1 is considered to be a later evolutionary stage than the dust clouds and globules. This picture is supported further by the fact that the molecular densities (and inferred total gas densities) are larger in these sources than in dust clouds.

While this picture seems to apply to most of the molecules in Table 9.1, it is too simple to describe the two most widely observed species, OH and H_2CO. OH has now been observed in emission, absorption, or both in some 400 directions in the galaxy, nearly all of them either corresponding to continuum sources or lying within $\pm 2°$ of the galactic plane, where there is significant background continuum radiation (Goss, 1968; Turner, 1970c, 1972; Robinson *et al.*, 1970; Goss *et al.*, 1970; Manchester *et al.*, 1970; Robinson *et al.*, 1971). H_2CO has been detected in some 150 directions, virtually all of which include the OH positions (Zuckerman *et al.*, 1970; Turner and Heiles, 1974). In most directions both OH and H_2CO reveal line features at several velocities. Often one of these velocities is close to the continuum source velocity, and it may also agree with the velocity observed for the other molecules in the few sources where they are seen. However, it is equally common for OH and H_2CO velocities to differ noticeably from the velocities of the other molecules, except for the case of H_2O, whose velocity is always the same as that of the anomalous OH emission with which it apparently coexists. There is no clear tendency for anomalous OH emission (observed in $\sim 25\%$ of the OH sources) to lie closer to the HII region in velocity than does the OH absorption when seen in the same direction. Those OH and H_2CO features which are unrelated to the continuum source in velocity presumably arise from dust clouds or perhaps even Bok-type regions which happen to lie in front of the source. In the case of H_2CO the unrelated features could also lie *behind* the continuum sources if the 6-cm transition is everywhere anomalously cooled; this could explain some cases in which H_2CO is observed where OH is not, despite the fact that OH appears more widespread throughout the galaxy than H_2CO. There is growing evidence that OH, and possibly H_2CO as well, may exist rather uniformly throughout the entire regions of spiral arms and possibly in the space between; that the OH clouds are in large part observed only against continuum sources is a selection effect, imposed both by the design of the surveys undertaken so far and by the sensitivity of present equipment. These two limitations are also responsible for the small sample of directions in which the other presumably widespread molecules such as CO and NH_3 have so far been searched; the sample includes only the sources exhibiting the largest abundances of OH and H_2CO. In the next section we discuss the large-scale distribution of OH and H_2CO, as derived from observations against regions of galactic background continuum.

d) *"Optical" Emission Nebulae*

These refer mostly to the objects given in Sharpless' Catalogue (1959), and similar ones in the Index catalogue (see also Terzian, 1970). These emission nebulae are from 1 minute to a few degrees in size and evidently are close to the Sun, judging from the lack of obscuration. Most of these objects, particularly those of large angular size, are characterized by low surface brightness at optical wavelengths and by little or no radio continuum. Although many have furrows or knots of dust associated with them, the total amount of dust and obscuration appears to be much less than is typical of most HII regions. These properties suggest that the optical emission nebulae are highly evolved HII regions. Most of the 13 cases of detected OH in these objects are of the anomalous emission variety known as Type I, which is also found primarily in the directions of younger HII regions. Although the OH is probably associated with the objects (on the basis of velocities and spatial distribution), there seems to be no correlation between the presence of OH and the presence of dust or other distinctive optical characteristics in these objects.

e) *IR Stars*

Anomalous emission from OH and H_2O has been found to be associated with specific types of IR stars, principally late M-type stars, the majority of which are Mira variables (W. J. Wilson *et al.*, 1970; Schwartz and Barrett, 1970). These stars exhibit a very large 5- to 10-μ flux, possibly due to circumstellar shells. The OH emission is of the type that is generally explained by IR radiative pumping (Litvak, 1969). Systematic time variations in the maser emission are commonly found for H_2O and are usually in phase with the variations in the total light curve of the star; OH time variations are often seen as well, and these tend to be in antiphase with the variations in the total light. The OH emission is thought to arise in circumstellar shells which derive their structure via mass loss from the star. The H_2O emission probably arises within the stellar atmosphere itself. VLBI observations of a few of these OH sources establish that the OH maser action indeed arises in regions immediately around or in the atmospheres of these stars (Hardebeck and Wilson, 1971).

One star, IRC 10216, is found to be a source of several other molecules: CO, CN, CS, and HCN (R. W. Wilson *et al.*, 1971; Morris *et al.*, 1971). This object is a very late-type carbon star that seems to possess an expanding or a turbulent envelope, with a high brightness temperature at $5\,\mu$ wavelengths, and a large angular size (0.7″ at $5\,\mu$). Both CO and HCN indicate a small value for the $^{12}C/^{13}C$ isotope ratio, consistent with a carbon-star interpretation. All observed molecules appear to be thermally excited in this object. Molecules such as NH_3 and HC_3N have not been detected and must have much smaller abundances than HCN in IRC 10216. This object is one of very few that contain several molecules of which OH is not one.

H_2CO has never been detected in any stellar objects; this applies to the 1_{01} to 0_{00} transition at 72 GHz, as well as to the 6-cm transition, so that the absence cannot be attributed to peculiarities in the excitation of the 6-cm line. The stellar objects that lack H_2CO include many cool IR stars (Mira variables, M supergiants, and a few carbon stars such as IRC 10216), whose temperatures are $\lesssim 1000°K$. Thus the absence of H_2CO cannot be explained by thermal dissociation $[D_o(H_2CO) \approx 3.5 \text{ eV}]$ and probably not by photo-dissociation. It is interesting that CO is present in IRC 10216, while H_2CO is not.

f) Galactic Objects Not Containing Molecules

Objects searched without success for OH include planetary nebulae, Wolf-Rayet stars (with and without associated nebulosity), clusters and associations of young stars with very high reddening (high dust content), reflection nebulae, and many types of stellar objects, including T-Tauri stars, p-Cygni stars, and emission B stars, some of which appear to have circumstellar shells of dust and gas (Turner, 1969; Grasdalen *et al.*, 1973). Absence of OH in planetary nebulae and Wolf-Rayet stars is probably due to high excitation in the atmospheres, which destroys the molecules. The difference between those stellar objects that show OH and those that do not is not understood. Possibly stellar objects not containing OH are carbon-rich stars with a low O/C ratio; in such objects all of the O may be in the form of CO. Observations have not been made in these stars for CO, CN, HCN, and CS, which should be expected if there is an overabundance of carbon.

9.2.2 The Galactic Distribution of OH and H_2CO

Extensive observations (Turner, 1972; Zuckerman *et al.*, 1970; Turner and Heiles, 1974) of OH and H_2CO have provided a picture of the overall distribution of molecules in the galactic plane and have allowed a statistical approach to the questions of how well OH and H_2CO are correlated with continuum sources and with each other. A statistical approach is useful because the spatial resolution with which OH and H_2CO have been observed is inadequate to determine the extent of these correlations on the basis of individual sources, which are usually smaller in angular size than the observing beam.

The large-scale distributions of OH and H_2CO can be characterized by apparent opacities, $\langle \tau \rangle$ (defined as antenna temperature in the line divided by antenna temperature of the background continuum), which on the average decrease monotonically with increasing distance from the galactic plane. The scale height for OH is about one-fifth that of atomic hydrogen, and for H_2CO it is still less.

As a function of galactic longitude l, $\langle \tau \rangle$ for OH has a well-known peak within $\pm 5°$ of the galactic center, assumes intermediate values in the regions $337° \leq l \leq 355°$ and $5° \leq l \leq 35°$, and shows a sharp decrease, by a factor of nearly 3, at $l \approx 35°$ (in the direction of the tangent to the Sagittarius spiral arm); for all $l \gtrsim 35°$, $\langle \tau \rangle$ for OH assumes the rather low values ($\lesssim 0.02$ typically) that seem to characterize the entire anti-center region of the galaxy. Most of the OH seen in the anti-center appears, from its velocities, to reside in the local spiral arm. Evidence of the local OH gas is also seen in the form of weak emission in a few high-latitude regions; it appears to be thermally excited.

H_2CO distribution is not so completely studied as that of OH, but appears to be as widespread throughout the galactic plane. $\langle \tau \rangle$ for H_2CO also shows a peak in the galactic center region and a general decrease with increase in l. The failure to detect H_2CO absorption from galactic gas at high latitudes indicates not only a high degree of concentration toward the galactic plane, but also that possibly there is no significant concentration of H_2CO close to the Sun. Alternatively, if the H_2CO/OH abundance ratio close to the Sun is as high as elsewhere in the galaxy, the local H_2CO may be unobservable because it is in equilibrium with the 3°K cosmic background.

a) Relation of OH and H_2CO with Continuum Sources

Maps of OH and H_2CO in the vicinity of a few continuum sources of large angular size typically show that the opacity does not correlate well with the distribution of continuum brightness, T_c. By itself, this proves little, because of the limited sample of such maps and because of probable variations in local physical conditions in the vicinity of the source. However, it is also found that the opacity does not correlate with T_c in a statistical sense, when all of the known OH and H_2CO data are considered. This conclusion remains valid, even when one includes only those molecular clouds whose velocity is within 5 km/sec of the HII region velocity.

Another approach is to look for a correlation of OH and H_2CO opacities with the difference in velocity, $|v(\text{molecule}) - v(109\alpha)|$, because $v(109\alpha)$ is directly associated with the continuum source. Again there is no correlation. An actual physical association might not be revealed by this method if, for example, the molecules were formed in shock fronts which were moving rapidly with respect to the nebula but at different relative velocities from nebula to nebula. However, because such shocks would generally move outward from the nebulae, a bias toward negative values of $v(\text{molecule}) - v(109\alpha)$ would be expected in the case of OH absorption, because the OH always lies in front of the nebula. No such bias is observed.

A third approach is to examine the number of OH and H_2CO clouds as a function of $|v(\text{molecule} - v(109\alpha)|$. Here there is a definite correlation: More molecular clouds have velocities close to HII region velocities than do not. For either molecule this correlation does not necessarily indicate a physical association with the continuum sources, because a significant fraction of all observations are made in the region $0° \leq l \leq 25°$. In this interval the velocities expected for either molecules or HII regions on the basis of galactic rotation vary slowly with distance from the Sun over most of the regions contained in the line of sight. Molecular clouds and HII regions lying in this vicinity would be expected to have small differences in velocity, even though they might well not be physically associated at all. It is interesting that the velocity correlation is tighter for OH than it is for H_2CO. Since the OH and H_2CO surveys cover very similar regions, the somewhat closer correlation of OH and HII region velocities may well indicate that proportionately more OH sources are indeed physically associated with continuum sources than is the case for H_2CO. It is equally possible, however, that H_2CO clouds may be seen behind as well as in front of continuum sources if the excitation temperature of the 6-cm transition is less than 3°K in most clouds. In any case, any part of the apparent correlation not explained by galactic rotation probably indicates only that OH, H_2CO, and HII regions exist in the same general regions of spiral arms where the gas and dust densities are highest. Therefore the velocities should be similar but not identical. It should be pointed out that no greater fraction of the molecule clouds have velocities similar to HII region velocities when the sample of sources is restricted to the strongest HII regions. Thus conditions for forming OH and H_2CO appear no more favorable near the strong HII regions than near weak ones. This conclusion is consistent with the idea that there is no close physical association of these molecular clouds and HII regions beyond the fact that both tend to occur in the regions of largest density in spiral arms.

Molecules other than OH, H_2CO, and possibly CO do appear, at present, to occur primarily in the neutral clouds immediately surrounding HII regions. This trend can, of course, largely be explained in terms of observational selection—many of these species have been sought only in the directions of the 10 or fewer brightest HII regions. In addition, many of the molecules require higher excitations than OH or H_2CO to be observable. One reason is that they are not capable of the anomalous pumping such as occurs in H_2CO, rendering it observable where it otherwise would not be. A second reason is that these molecules are largely seen at short wavelengths where, because of negligible background continuum radiation, they must be observed in emission; in this case, at least the first excited rotational state must be populated. The necessary excitation may be available for many of these species only in the vicinity of HII regions. Until more extensive observations are made with much improved sensitivity, one cannot rule out the possibility that all of the presently known molecules are as widespread as OH and H_2CO. However, it is interesting that the velocities of OH and H_2CO appear often to differ significantly from the velocities of other molecules, in the vicinity of those HII regions where all molecules are seen. This constitutes the only

evidence so far that OH and H_2CO may have a fundamentally different distribution, and hence chemistry, than some of the other species. If this is so, then the next obvious question is how well OH and H_2CO are themselves related.

b) *Relation of OH and* H_2CO

Several approaches indicate that OH and H_2CO themselves are not closely associated over much of the galaxy. First, $\langle \tau \rangle$ for OH does not appear to be correlated with $\langle \tau \rangle$ for H_2CO, even when one includes only those OH and H_2CO clouds whose velocities differ by less than 2 km/sec. Second, one can attempt to derive the thermal and turbulent contributions to the overall line widths by combining data respectively for OH and H, H_2CO and H, and OH and H_2CO. This procedure generally leads to inconsistent and unreasonable results, indicating that the basic assumption required in this method—that the H, OH, and H_2CO coexist and are homogeneously mixed in the cloud—is apparently invalid. Third, of a total of over 100 clouds observed for both OH and H_2CO, about 30% show OH but no H_2CO, and about 10% show H_2CO but no OH. And in another 6% of the cases the velocity differences are so large ($\gtrsim 30$ km/sec) that the OH and H_2CO clouds cannot be physically related.

Certainly some of these discrepancies can be explained if the 6-cm H_2CO transition is "refrigerated" throughout the galaxy. Then H_2CO will be observed in some regions where OH cannot be seen. Also, the excitation temperature of the 6-cm H_2CO line, and hence $\langle \tau \rangle$, may depend critically on local conditions such as radiation fields. This would tend to randomize any actual correlation between the apparent opacities of OH and H_2CO.

Some statistical evidence for a "universal" refrigeration of the 6-cm H_2CO line indeed exists. If the number of clouds, N, observed in both OH and H_2CO is plotted as a function of the difference in velocity $\Delta v \equiv v(H_2CO) - v(OH)$, the distribution in N is found to be strongly skewed toward positive values of Δv. This behavior applies equally well to those clouds seen in the directions of the strongest continuum sources as to the entire data sample. The most probable explanation for this asymmetry in the velocity distribution is that more H_2CO is seen at large distances than OH, as would tend to be the case if the H_2CO excitation temperature were less than the 3°K background while that of OH were not. In the galactic northern hemisphere this corresponds to observing more positive velocities in H_2CO than in OH. Based as it is on a statistical argument, this is the first indication that the anomalous "refrigeration" of the 6-cm H_2CO line (observed directly in the dark dust clouds) may be a universal phenomenon.

9.2.3 Distribution of Other Molecules

Other molecules have not been observed in as many directions as OH and H_2CO, so that by comparison their large-scale distribution is poorly known. However, because they are observed at shorter wavelengths, where observing beams are small, they have provided unique information about the small-scale distribution of molecules and their interrelationships, on the size scale of the individual sources. In this regard we discuss the distributions of NH_3, CO, H_2CO (λ2-mm lines), H_2O, CS, and HCN.

Observations of the galactic center region in H, OH, and H_2CO (6-cm line) had earlier shown that the distribution of different gases was very chaotic and complicated; for example at $l = 0°$, $b = 0°$, strong OH absorption is seen in the velocity range -140 to -200 km/sec, where no H or H_2CO is observed. With the fairly high spatial resolution ($\sim$6 minutes) available for H_2CO, the galactic center region is found to consist of 15 fairly discrete clouds within the region $359°5 \leq l \leq 1°7$, $-0°2 \leq b \leq 0°2$ (Scoville *et al.*, 1972); some but not all of these clouds appear to have OH counterparts. Observations of the (1,1) transition of NH_3 have been made in these clouds also with 6-minute resolution (Morris *et al.*, 1971); neither the intensities nor the velocities of the NH_3 appear to be related

to those of the H_2CO. Observations of the $1 \to 0$ line of CS in a few of these clouds with 2-minute resolution indicate a correspondence with NH_3 velocity features in some cases but with H_2CO features in other cases. CO-line radiation from the entire galactic center region is highly saturated, owing to the very large abundance; therefore no information is available on the detailed distribution of CO.

In contrast with the center of the galaxy, molecular sources observed elsewhere are more discrete spatially and have many fewer clouds in the line of sight. Nevertheless, if the three HII regions so far studied in detail (Sgr B2, W 51, Ori A) are typical, these sources are also characterized by complicated molecular distributions. Figure 9.1 shows spectra of NH_3 and CS in Sgr B2. The (2,1) line of NH_3 requires much higher densities for excitation than does the (1,1) line, because of its rapid decay to lower levels. The smaller range of velocity covered by the (2,1) line is indicative, therefore, of a higher density core in the Sgr B2 cloud from which the (2,1) radiation arises. Similarly, the $2 \to 1$ line of CS requires higher densities for excitation than the $1 \to 0$ line, yet in this case the $2 \to 1$ CS radiation seems to come from two discrete clouds (velocity ranges), whereas the $1 \to 0$ CS radiation does not. Further evidence for two discrete clouds within the Sgr B2 source comes from recent high-resolution ($\sim$1 arc minute) observations of CH_3OH, HC_3N, and HNCO (Turner *et al.*, 1972; Snyder and Buhl, 1972). These clouds are separated by only $\sim$2 arc minutes and appear to have different chemical compositions; the northern cloud (at 70 km/sec) is strong in NH_3, HNCO, and HC_3N, while the southern cloud (at 54 km/sec) appears stronger in CS, SiO, and CH_3OH, and seems to lack nitrogen-containing molecules. Different transitions of CH_3OH are seen in the two clouds, indicating that there are differences in excitation conditions as well. Finally, interferometric studies with a resolution of 20″ (see Fomalont and Weliachew, 1973) show that the H_2CO distribution bears no resemblance to that of the other molecules. In Table 9.1 no account

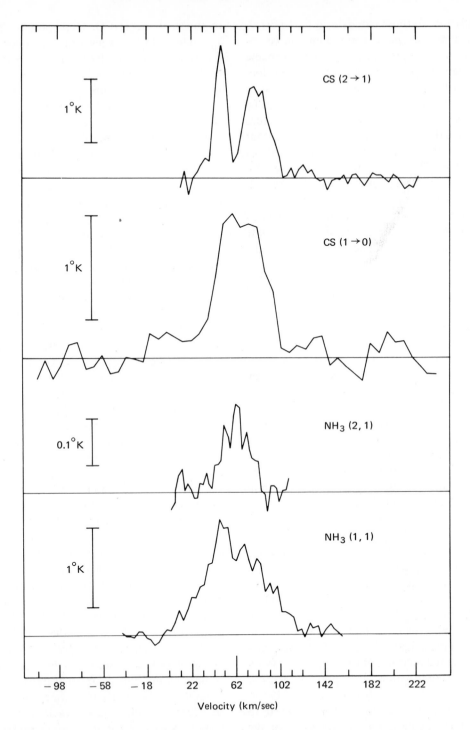

Figure 9.1 Spectra of NH_3 and CS in the Sgr B2 molecular source. The observation of the (2,1) line of NH_3 indicates that densities as high as 10^8 cm^{-3} may exist in the core of the source. The striking difference in the $2 \rightarrow 1$ and $1 \rightarrow 0$ lines of CS indicates that the Sgr B2 source may consist of two separate clouds or density condensations.

has been taken of these complexities. The molecular abundances given there are derived from earlier lower-resolution observations, which represent an average of the two clouds. The example of Sgr B2 makes clear the problems to be anticipated in attempting models of interstellar chemistry based upon current data. Both chemical composition and excitation apparently can be very different between clouds that typically may lie close enough together that they are not usually resolved with currently used telescopes.

In Ori A, all the molecular clouds observed in CO, HCN, X-ogen, CS, and H_2CO ($\lambda 2$-mm lines) are centered on the position of an infrared point source (Becklin and Neugebauer, 1967) but they show very different angular sizes. The CO cloud is about 1 degree in size, those of HCN and X-ogen are ~ 6 minutes in size, and those of CS and H_2CO (which appear to have identical distributions) are ~ 3 minutes in size. These differences are probably indicative of the different densities required to produce the favorable balance between formation and destruction rates which is needed to yield observable abundances.

The association of molecules with infrared sources has recently become more widely recognized. Such an association has never been clear for OH or for H_2CO (with the exception of anomalous OH emission); however, molecules such as CS and HCN, which are observed at mm wavelengths only under conditions of appreciable excitation, are nearly always observed in the directions of strong infrared sources, particularly those whose flux increases strongly toward larger wavelengths. In most, but not all, cases these objects are sources of anomalous OH emission, and may well be proto-stellar in nature.

The complicated distributions of molecules shown by Sgr B2 and Ori A are undoubtedly typical of other molecular clouds associated with continuum and IR sources, although few others have as yet been studied in much detail. NH_3 has been observed in several continuum sources outside of the galactic center, all of which are strong OH

emission sources, and in a dust cloud (Morris *et al.*, 1973). The spatial distribution of NH_3 within these sources is unknown. However, in some sources the NH_3 velocity agrees with that of OH and H_2CO but not with the HII region velocity, while in other sources the NH_3 velocity is that of the HII region and differs from the other molecular velocities. Even in those cases where there is agreement of NH_3 and OH absorption velocities, there appears to be no correlation in NH_3 and OH abundances. There is perfect agreement of all molecular velocities in cloud 4.

CO has been observed in emission at the 2.6-mm wavelength from a fairly wide variety of objects including HII regions, an IR star, dark dust clouds, and throughout the galactic center region. Where it is observed, it is the most abundant interstellar molecule so far detected (Table 9.1), and it is the only molecular constituent of the interstellar medium (except, by inference, H_2 in dense cold clouds) which appears to contain a large fraction of the available atoms. CO has been studied extensively in a few continuum sources with high spatial resolution (~ 1 minute) (Penzias *et al.*, 1971a). The CO-line radiation from these molecular clouds is always heavily saturated, owing to the very large abundance. Thus, even in the lowest-density regions, spatial variations on a small scale within these clouds tend to be washed out observationally by the presence of CO, in amounts sufficient to produce significant brightness in the CO-line radiation. Observations of the ^{13}CO isotope can allow optical depth effects to be accounted for (by the assumption of a fixed intrinsic abundance ratio $[^{13}CO]/[^{12}CO]$; although this method has been applied at single positions in several sources, the small-scale distribution of the ^{13}CO isotope has not yet been determined within these molecular regions. As further evidence that CO is much more abundant than other molecules, it is typically observed in clouds which, although centered at the same positions as the other molecular clouds, are much more extended in angular size. In addition to the case of Ori A, mentioned above, the CO cloud in both W 51 and Sgr B2 is also several times larger than most other molecular clouds, which appear to be centered on the HII regions. Only OH has comparable extent in these regions. The widespread and highly abundant nature of CO is observed in dark dust clouds as well. Despite much lower total densities, the CO lines are highly saturated in these clouds, as they are in the continuum source clouds. In each of the three dark clouds in which CO has been mapped over an extended region, it is found (Penzias *et al.*, 1972) that the CO extends beyond the optical cloud boundaries; in one case (cloud 2) the CO extends outside the cloud into an area of high star density, where the visual extinction is only ~ 2 magnitudes. The generally widespread distribution of CO is probably at least partly explained by the fact that it is more stable against photo-destruction by UV radiation than are the other molecules. Both the large extent and the high abundance of CO may also be due to efficient formation mechanisms which include a number of processes that destroy other molecules leaving CO as a residue, as well as formation by radiative association, a process that is possible for only a very few of the observed interstellar molecules.

Interstellar H_2O molecules are observed only via very powerful maser emission from the 6_{16} to 5_{23} transition at 1.35 cm; this emission is associated with only two types of galactic objects: (1) with Type I anomalous OH emission sources, which usually are observed in the directions of HII regions (Knowles *et al.*, 1969; Meeks *et al.*, 1969; Turner *et al.*, 1970; Sullivan, 1971; Turner and Rubin, 1971); and (2) with late-type IR stars, specifically M supergiants and Mira variables, which usually, but not always, also exhibit Type II(b) anomalous OH emission (Schwartz and Barrett, 1970). Nineteen H_2O sources are known in the first category and about a dozen in the second category. Positions have been determined by VLBI techniques to an accuracy of $\sim \pm 2''$ for several of the OH/H_2O emission sources in HII regions; although the OH and H_2O regions coincide, there appears to be little or no rela-

tion with the small knots of bright continuum emission which have recently been found to exist in nearly all HII regions (Webster and Altenhoff, 1970a, 1970b). Because considerable energy, by interstellar standards, is needed to excite the 6_{16} to 5_{23} transition of H_2O (~ 450 cm^{-1} above ground), it is not surprising that H_2O emission always coincides with anomalous OH emission. It is probable that the physical association of H_2O and OH emission is due to more than the presence of high energy. For at least the H_2O sources associated with Type I OH emission, it appears that the OH is formed by the collisional dissociation of H_2O into specially excited states which lead to the anomalous OH emission (Gwinn *et al.*, 1973). For the H_2O/IR star sources the excitation of both H_2O and OH emission is more likely caused by infrared pumping, and it is not clear whether the two molecules have any chemical relationship.

It will be difficult to observe H_2O in absorption in the interstellar medium, because it has no lower-energy transitions at wavelengths where there is any appreciable background continuum radiation from radio sources. At present H_2O can be observed only in very special high-energy regions. However, H_2O may well be at least as abundant as OH in most interstellar regions, because at temperatures below 2000°K the process $H_2 + OH \rightleftarrows H_2O + H$ is in equilibrium at large $[H_2O]/[OH]$ ratios.

9.3 Observations of Interstellar Molecules at Optical and Ultraviolet Wavelengths

9.3.1 The Observations

Only three interstellar molecules have been identified in the visual part of the spectrum, namely, CH^+, CH, and CN. In the far- or rocket-UV, CO and H_2 have recently been found. We discuss these two groups of molecules in turn.

In the more than 30 years since their discovery, CH^+, CH, and CN have been detected in the spectra of 62 stars out of approximately 300 stars which exhibit the H and K lines of Ca^+ as well as the sodium D lines in some cases (cf. McNally, 1962). These stars, and hence the molecules, appear to be widely distributed. The molecular velocities always agree with the Ca^+ velocities, but the molecular lines do not show the multiplicity that appears in the Ca^+ lines when observed with high-velocity resolution. Thus the molecules appear to be more localized than the Ca^+ ions.

The existing information on these "optical" molecules is very limited. At present a total of 16 lines is attributed to CH^+, CH, and CN but all except four of the lines have been observed in fewer than six stars. CN has been detected in a total of 14 stars; and CH^+ and CH in a total of 88 and 60 stars, respectively. A strong selection effect is introduced by the fact that most optical molecules are observed in the spectra of hot stars. The immediate question is whether the molecular lines are formed in interstellar space remote from the star, or in the neighborhood of the star as a direct result of the effects of the stellar radiation or of enhanced local densities. The latter possibility is suggested by analogy with comets which, when they arrive within ~ 3 A.U. of the Sun, produce line spectra of CH^+, CH, and CN (as well as other molecular lines). Hence it is possible to speculate that interstellar grains volatilize and produce the observed molecular lines only when they approach hot stars or other strong sources of heating. However, against this hypothesis is the observations of molecular lines in the spectra of a few later-type stars. The radio observations of CN (as well as of other molecules) indicate that molecular lines can be formed remote from any star, but radio observations cannot refute the possibility that the optical molecular lines may also be formed near stars. This is because radio and optical observations of molecules never refer to the same regions; wherever the extinction is low enough to permit optical observations, the apparent abundance of molecules is too low to be detected at radio wavelengths, since the radio transition probabilities are much smaller than those at optical wavelengths.

In the relatively few stars where CH and CN are seen together, their abundances are comparable despite the fact that H is cosmically about 10^4 times more abundant than N. The column densities are typically 10^{12} cm^{-2} for both CH and CN. This represents an intermediate density in the hierarchy of interstellar clouds examined for CH and CN. On the low-density end are most molecular line stars that show CH and CH^+ but no CN. Typical column densities for CH in these sources are $< 10^{12}$ cm^{-2}. On the high-density end are the interstellar clouds that reveal CN at radio wavelengths with typical column densities of 10^{15} cm^{-2}; recent searches for CH at the radio wavelength of 8.9 cm have placed upper limits on its column density of $\sim 10^{14}$ cm^{-2}, in these denser clouds. Apparently the relative chemistry of CH and CN is a strong function of density in interstellar clouds.

Further evidence of a very specific interstellar chemistry for molecules observable at optical as well as radio wavelengths comes from the failure to detect several expected molecules at optical wavelengths. Among these are OH, OH^+, NH, NH^+, CO^+, CN^+, NH_2, C_2, and C_3 (Herbig, 1960, 1968). The absence of NH in appreciable quantity is particularly surprising, since it has a large oscillator strength and a transition that lies at longer wavelengths than some detected transitions of CH. One explanation may lie in the fact that NH cannot form by radiative association (see Section 9.6.2), while CH and CN can form in this way. Both NH and NH_2 are photo-dissociation products of NH_3, with NH_2 dominating at wavelengths larger than 2000 Å (Stief *et al.*, 1972). The absence of these molecules in the ζ Oph cloud therefore implies a lack of NH_3 there also, although this has not been verified by a search at the 1.3-cm wavelength. Little or no NH_3 would be expected in these regions, however, because there is little protection from dissociating UV radiation, and only a low density of grains, on whose surfaces molecules such as NH_3 may possibly form (see Section 9.6.2). It is noteworthy that both NH and NH_2 are seen in comets, where they are both produced (by evaporation of NH_3 ice on solid comet material) and excited by the solar radiation field. If, as seems likely, any interstellar NH and NH_2 originates in the dissociation of NH_3, these molecules should be observed preferentially near hot stars. That they are not so observed might indicate that interstellar molecules are generally not found in the vicinity of hot stars. This conclusion receives slight support from the absence of CN^+; although not observed in comets, CN^+ might be expected to be produced from CN in the vicinity of the hottest stars.

The absence of C_2 and C_3 bears on the question whether graphite exists in the interstellar medium or in stars. If, as some theories suggest, graphite molecules are formed in superheated carbon vapor in the stellar atmospheres of cool carbon stars and then ejected into space, some C_2 and C_3 molecules should be expected to go with them. Their absence cannot be taken as strong evidence against this picture, however, because such molecules may be altered or destroyed by interstellar radiation fields after ejection (see Section 9.6.1).

Unfortunately, data on the two molecules H_2 and CO seen at rocket-UV wavelengths are presently limited. Theoretical studies of photo-dissociation processes and several rocket observations have shown that H_2 is not present in appreciable amounts in the general interstellar medium. However, H_2 appears to be the major gaseous constituent of the dense interstellar clouds that contain most of the molecules. Theoretical work indicates that $[H_2]/[H]$ increases with increasing extinction and has a large value for extinctions exceeding ~ 1.5 magnitudes in the visual. Carruthers (1970) recently detected H_2 for the first time, by means of its Lyman resonance bands, in the direction of the star ξ Per. Here $[H_2]/[H] \approx 0.3$ and the visual extinction is ~ 1 magnitude. In the nearby star ε Per, $\sim 4°$ away, $[H_2]/[H] \approx 0.2$ while the extinction is ~ 0.3 magnitudes. These results

are consistent with theories of catalyzed recombination of H atoms on grains to form H_2 (Hollenbach *et al.*, 1971). Another interesting result is that the ratio of total hydrogen to dust is significantly higher in the ξ Per cloud than in the lower-extinction ε Per cloud. Many more observations of H_2 are needed to determine the significance of this result.

CO has been detected by UV absorption in the direction of the star ζ Oph (Smith and Stecher, 1971), where CH^+, CH, and CN have also been detected. The estimated density of CO is $\sim 7 \times 10^{15}$ cm^{-2}, much larger than the density of CH and CN ($\sim 2 \times 10^{13}$ cm^{-2}) but about 4 orders of magnitude less than the density of CO observed in other sources at radio frequencies. If the abundance ratio [CH]/[CO] were similar in ζ Oph and in the much denser sources in which CO is observed at radio frequencies, CH would also have been detected in the searches made for it in the "radio" sources. Hence both [CH]/[CN] and [CH]/[CO] appear to decrease as the density of sources increases. Thus, as is found in the laboratory, CH appears to be a much more reactive radical than either CN or OH in the interstellar medium.

9.3.2 Physical Conditions Derived from Optical and Ultraviolet Molecules

With one exception, all of the absorptions from molecules in this category are due to electronic transitions from the ground state to various excited electronic and vibrational states. This indicates the general low level of excitation in these relatively rarefied interstellar clouds. The particular types of physical parameters we can derive from these molecules do not require a study of the equation of transfer and so will be discussed here. The physical parameters derivable from the radio observations are described in the next section, along with the necessary aspects of radiative transfer theory.

The most thoroughly studied molecular cloud at optical wavelengths is the one in front of the star ζ Oph (Herbig, 1968). A number of arguments, based on the Na lines seen in the same direction, establish the lines as arising in a thin dense HI region rather than in the HII region immediately surrounding ζ Oph. The radiation field is relatively well studied in this region, and there are 21-cm observations of H, which yield total densities of 50 to 1000 cm^{-3}, depending on the thickness of the region and assuming that $[H_2]/[H]$ is small. The molecular abundances are found to be 4×10^{13} cm^{-2} for CH, 3×10^{13} for CH^+, 8×10^{12} for CN, and 5×10^{13} for Na. Upper limits on abundances of molecules sought but not detected are $<7 \times 10^{13}$ for OH, $<8 \times 10^{12}$ for NH, and $<3 \times 10^{13}$ for CO^+. NH^+ and OH^+ are also not observed. CO was detected in the UV with an abundance of $\sim 7 \times 10^{15}$ cm^{-2}. These abundances are typical of those found in the spectra of other stars. In general, therefore, the molecular abundances are 2 to 3 orders of magnitude smaller in regions where molecules are seen optically than in those regions where they are observed at radio wavelengths. CO appears to be much the most abundant diatomic species in both regions (by a factor of 10^3 to 10^4). The small abundances of NH and OH relative to the other diatomics is especially significant. It is in direct contradiction to the predictions made by the Stecher and Williams (1966) theory for the formation of diatomic molecules on grain surfaces by means of chemical exchange reactions. On the other hand, just such a suppression of NH and OH with respect to CH and CN is to be expected if diatomic molecules are predominately formed by two-body radiative recombination in interstellar space. In Section 9.6.2 we discuss how this process yields the observed quantities of CH and CH^+ and how, by subsequent gaseous exchange reactions, the observed amounts of CN and CO are also produced. The physical conditions this theory implies in order to fit the observations are characterized by total densities of $10 \lesssim n_H \lesssim 100$ cm^{-3}, temperatures between 15 and $50°K$, electron densities of $\sim 6 \times 10^{-2}$ cm^{-3}, and opacities of ~ 2 at wavelengths of 1000 Å. The temperature is quite similar to that found

in the radio-molecule clouds, but the optical cloud densities and extinction are much lower, while the n_e/n_H ratio is significantly higher.

Although both rocket observations and theoretical studies indicate little H_2 in the general interstellar medium, much evidence has accumulated that H_2 is the major gaseous constituent in the dark dust clouds and denser regions, where the catalyzed recombination of H atoms on grains predominates over photodissociation of the H_2. Theoretical work (Solomon and Wickramasinghe, 1969; Hollenbach *et al.*, 1971) indicates that the ratio $[H_2]/[H]$ should be large whenever the total visual extinction in dust clouds exceeds ~ 1.5 magnitudes. The recent detection of H_2 in the direction of 15 stars, with extinction varying from 0.2 to 1.5 magnitudes (Spitzer, *et al.*, 1973) tends to confirm these predictions at least qualitatively. $[H_2]/[H]$ varies from 0.08 to 0.67 and is correlated with the extinction.

The UV spectrum of CO in ζ Oph produces a column density of 7×10^{15} cm^{-2}, which is more than 100 times the amount of CH found by Herbig (1968) in the same star. This is probably explained by the large (11.1 eV) binding energy of CO, which makes chemical destruction unlikely and reduces the photo-destruction rate under a typical radiation field to $\sim 10^{-3}$ that of CH. Measures of the atomic hydrogen abundance in front of ζ Oph (by means of the Ly α lines) show that $[CO]/[H] \simeq 2 \times 10^{-5}$, so that $\sim 7\%$ of all carbon is in the form of CO if usual cosmic abundance ratios ($C/H \simeq 3 \times 10^{-4}$) apply. This is in contrast with the much denser clouds observed at radio wavelengths, in which nearly all of the carbon appears to be tied up in molecules.

It should be pointed out that the abundance of CO cannot be determined independent of some assumptions about its excitation. The abundances mentioned above were obtained by assuming that the CO is in equilibrium with the 3°K cosmic background radiation, and it should be noted that a satisfactory fit to the curve of growth for the CO line could not be obtained under any other reasonable assumption. While this does not prove the existence of the 3°K background, it corroborates the firmer evidence which other molecules are able to provide.

9.3.3 Optical Molecules and the 3°K Cosmic Radiation

An interesting feature of most of the light diatomic molecules observed in the interstellar medium at optical wavelengths is that their ground-state rotational energy level schemes are characterized by $E/k \sim \hbar^2/2Ik \sim 3°K$ (I is the moment of inertia of the molecule). Hence these molecules are

extremely sensitive indicators of the density of the 3°K field. Figure 9.2 shows observed spectra and relevant energy levels of CH, CH^+, and CN used for sampling the 3°K field. The relative intensities of the lines shown depend on the temperature of the radiation field, assuming that the rotational level populations are in equilibrium with this field. Those lines which are not observed [R(2) in CN, R(1) in CH^+, and $R_1(1)$ in CH] indicate that only the lowest levels are populated under interstellar conditions. The energy levels show that CN samples the radiation field at 2.64 and 1.32 mm by means of the intensity ratios R(0)/R(1) and R(2)/R(1); similarly CH^+ samples the field at 0.359 mm

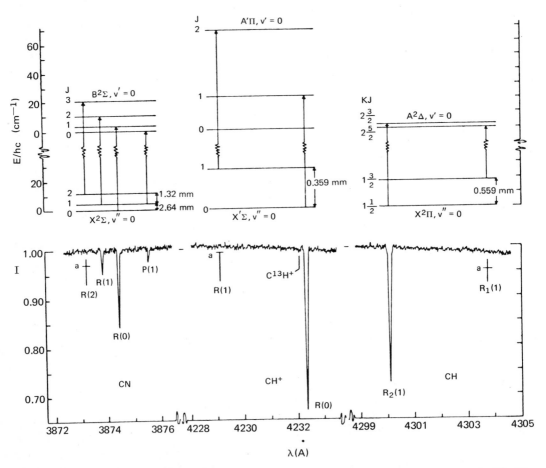

Figure 9.2 Observed spectra in the UV of the interstellar molecules CN, CH^+, and CH. The observed transitions are indicated on the energy level diagrams, as are the mm-wavelength transitions whose excitation temperatures are inferred to be that of the cosmic background radiation. (Figure reproduced by kind permission of P. Thaddeus.)

and CH at 0.559 mm. Because the CH^+ [R(1)] and CH [R_1(1)] lines are not detected, only upper limits to the equivalent cosmic radiation temperature are established: $\leq 5°K$ at 0.559 mm and $\leq 7°K$ at 0.359 mm. The CN results give T_{cosmic} (1.32 mm) $\leq$ 3.96°K and T_{cosmic} (2.64 mm) = 2.80°K (Thaddeus, unpublished), in excellent agreement with the average value 2.7°K measured at several longer wavelengths. Although these values apply to the star ζ Oph, agreement within 0.1°K has been obtained for at least two other stars, and although only one direct confirmation has been provided by these molecules, the several upper limits are also valuable, since they are much more stringent than those derived so far in the same spectral region by rocket and balloon observations. It should be noted that the spectral resolution provided by the molecular results is also very much higher than is possible by the other methods; it is typically $\Delta\nu/\nu \simeq 10^{-5}$ and is limited by the Doppler broadening of the absorption lines.

We noted above that the CO spectrum in ζ Oph was consistent with excitation by the 3°K background also. Since the dipole moment of CO is some 14 times smaller than that of CN, its excitation is some 200 times less strongly coupled to the 3°K radiation field than is CN. Collisions with CO must therefore be very infrequent in the ζ Oph cloud, and in fact imply that the particle densities are less than 100 cm^{-3}. This is consistent with the densities required to explain the observed abundance ratios of CH, CH^+, CN, and CO according to theories of formation by radiative recombination (see Section 9.6.2).

9.4 Microwave Molecular Observations As Probes of Interstellar Clouds

Studies of interstellar molecules at radio wavelengths can in principle reveal a wider range of physical parameters in interstellar clouds than is possible at optical wavelengths. This is because a relatively large number of energy levels, connected by radio wavelength transitions, have sufficiently small energies to be appreciably excited under interstellar conditions, either by collisions or by radiation. At radio wavelengths molecules may therefore be observed either in absorption or in emission. Except in HII regions, where interstellar molecules have very short lifetimes and do not appear to exist in significant quantities, optical transitions cannot be excited by collisions and can be observed only by absorption of optical photons. Under such conditions there is usually no derivable information on the physical conditions in the molecular environment, other than the abundance of the molecule itself. The only exceptions were discussed above, in connection with the 3°K background, and even this information could not have been derived had there been appreciable excitation of the mm-wavelength transitions of CN, CH^+, and CH by collisions.

9.4.1 Radiative Transfer Theory

A study of the equation of transfer is required to relate the observed quantities to the physical parameters of the source producing the spectral radiation. We shall outline only those aspects essential to deriving the parameters of interest.

The one-dimensional, time-independent equation of transfer relates the volume absorption coefficient κ_ν, volume emission coefficient ε_ν, and resultant specific radiation intensity $I_\nu d\nu$ in frequency range $d\nu$ by

$$dI_\nu/ds = -\kappa_\nu I_\nu + \varepsilon_\nu \qquad (9.1)$$

If there are population inversions (resulting in amplification by maser action), the sign in front of $\kappa_\nu I_\nu$ is positive; additional modifications must be made to this term if the maser amplification is saturated (cf. Turner, 1970a, 1970b). Equation (9.1) may be integrated to give

$$I_\nu(s_0) = I_\nu(0)\, e^{-\tau_\nu(s_0)} + e^{-\tau_\nu(s_0)} \int_0^{s_0} \varepsilon_\nu\, e^{\tau_\nu(s)}\, ds$$

$$(9.2)$$

where $\tau_\nu(s) = \int_0^s \kappa_\nu \, ds$ is the optical depth. The integration path s is taken along the line of sight from an initial point $s = 0$ to the position of the observer at s_0.

Although Equation (9.2) may be evaluated for a large number of specific cloud configurations (*e.g.*, see Goss, 1968), it is usual, and consistent with the presently limited observational information, to evaluate Equation (9.2) for the case of a uniform homogeneous cloud having an absorption coefficient $\kappa_0(s)$ corresponding to an atomic or molecular transition of frequency ν_0. In this case, Equation (9.2) becomes

$$I_0(s_0) = I_0(0) \, e^{-\tau_0(s_0)} + \left(\frac{\varepsilon_0}{\kappa_0}\right)\left(1 - e^{-\tau_0(s_0)}\right)$$
$$(9.3)$$

when evaluated at the transition frequency ν_0. The first term represents the attenuation of the incident radiation as it passes through the cloud, and the second term represents the emission generated by the molecules in the cloud and corrected for self-absorption. The quantity of interest in determining molecular processes is the excess intensity caused by the molecular transitions at ν_0; this is given by subtracting from $I_0(s_0)$ the intensity $I_\nu(s_0)$ at a frequency ν that lies close to the transition frequency ν_0. From Equation (9.3), $I_\nu(s_0) \approx I_0(0)$. The excess intensity due to the spectral line is then

$$\Delta I_0 \equiv I_0(s_0) - I_0(0)$$

$$= \left[\frac{\varepsilon_0}{\kappa_0} - I_0(0)\right]\left[1 - e^{-\tau_0(s_0)}\right] \quad (9.4)$$

As is well known, the quantities ε_0, κ_0, and τ_0 may be written in terms of the parameters that describe the molecular properties and excitations as follows:

$$\kappa_\nu = \frac{\lambda^2 A f(\nu)}{8\pi}\left(\frac{g_u}{g_l}n_l - n_u\right) \quad (9.5)$$

$$\varepsilon_\nu = \frac{h\nu A f(\nu) \, n_u}{4\pi} \quad (9.6)$$

where n_u and n_l are the molecular densities in the upper and lower levels of the transition of interest, g_u and g_l are statistical weights of these levels, and $f(\nu)$ is the line shape function for which $\int f(\nu)d\nu \equiv 1$; if the line shape is gaussian, $f(\nu) = 2\sqrt{ln2} \,/ \sqrt{\pi} \, \delta\nu$ is used, where $\delta\nu$ is the full width of the line at half-maximum intensity.

It is customary to define two temperatures in terms of which the radiation and molecular parameters are expressed. The brightness temperature, T_B, is an equivalent blackbody temperature used to describe the specific intensity by the Rayleigh-Jeans relation

$$I = \frac{2kT_B}{\lambda^2} \quad (9.7)$$

where k is Boltzmann's constant. The excitation temperature, T_s, of the molecular transition of interest is defined by the Boltzmann relation

$$\frac{n_u}{n_l} = \frac{g_u}{g_l}e^{-h\nu_0/kT_s} \quad (9.8)$$

If the molecules are in local thermodynamic equilibrium (LTE) with their surroundings, then the entire ensemble of energy levels is described by the same T_s, and $T_s = T_k$, where T_k is the kinetic temperature of the environment. In most astrophysical situations the molecular energy levels are not populated according to a Boltzmann distribution law. In most cases deviations from LTE will occur because of the low densities, resulting in collision rates which are too small relative to spontaneous decay rates to provide sufficient population in the higher-energy levels. More spectacularly, molecular interactions with non-blackbody radiation fields or with a variety of collision processes which are not in LTE often cause marked deviations of even the lowest energy-level populations from those described by LTE. In these cases, population inversions resulting in maser action may occur (*e.g.*, OH, H_2O, CH_3OH, NH_2CHO, CH_3CHO) or population anti-inversions, resulting in anomalous absorption (*e.g.*, H_2CO).

Combining Equations (9.4) through (9.8) with the assumption $h\nu \ll kT_s$ gives the solution of the radiative transfer equation, which is usually used as a starting point for the analysis of spectral line data:

$$\Delta T_B = (T_s - T_c)(1 - e^{-\tau_0(s_0)})$$

$$(9.9)$$

where $T_c = \lambda^2 I_0(0)/2k$ is the brightness temperature of the continuum radiation which lies behind the molecular cloud; $T_c \gtrsim 3°K$ owing to the cosmic background. ΔT_B is positive for an emission line ($T_s > T_c$) and negative for an absorption line ($T_s < T_c$).

In Equation (9.9) the determined quantities are ΔT_B and T_c (often with limited accuracy), and the physical information is contained in T_s and τ_0; τ_0 is related to the column density by

$$\tau_0 = \frac{2\sqrt{ln2}\,\lambda^2 A}{8\pi^{3/2}\,\delta\nu}\left(\frac{g_u}{g_l}N_l - N_u\right)$$

$$= \frac{2\sqrt{ln2}\,\lambda^2 A}{8\pi^{3/2}\,\delta\nu}\frac{g_u}{g_l}N_l(1 - e^{-h\nu/kT_s}) \quad (9.10)$$

where N_l and N_u refer to column densities. The total column density of molecules (summed over all states) can be found only by assuming a distribution for the energy level populations; if a Boltzmann distribution characterized by a temperature T_k is adopted, then N_l is related to the total population column density N by

$$\frac{N_l}{N} = \frac{g_l e^{-E_l/kT_k}}{Q}$$

$$(9.11)$$

where Q, the (rotational) partition function, is given by

$$Q = \sum_{J=0}^{\infty}\sum_{K=0}^{J}(2J+1)e^{-E_{K,J}/kT_k}$$

for symmetric or asymmetric top molecules (J, K are the usual quantum numbers specifying the energy levels). For linear molecules the sum over K is omitted, and with $E_J = J(J+1)hB$ and the sum replaced by an integral (valid for $hB \ll kT_k$), one finds that $Q = kT_k/hB$. B is the rotation constant of the linear molecule. For asymmetric tops Q is given by the expansion (Gordy *et al.*, 1953)

$$Q = \sqrt{\frac{\pi}{ABC}\left(\frac{kT_k}{h}\right)^3}\,e^{h(B+C)/8kT_k}$$

$$\times\left[1 + \frac{1}{2}\left(1 - \frac{B+C}{2A}\right)\frac{h(B+C)}{2kT_R} + \cdots\right]$$

where A, B, C are the rotational constants; for most cases the expansion may be truncated after the first term.

From the foregoing, one can see that there are several strong limitations on the reliability with which abundances can be determined.

First, if the molecular line source is optically thick, $\tau_0 \gg 1$, then Equation (9.9) shows that no information is obtained about τ_0, and hence about the abundances from observations of a single line; however, $T_s - T_c$, and hence T_s, is measured directly. In practice, in order to make progress the source is always *assumed* to be optically thin (unless there is information to the contrary), in which case expanding the exponentials in Equations (9.9) and (9.10) and combining these equations, one finds

$$\Delta T_B = \frac{2\sqrt{ln2}\,\lambda^2 A}{8\pi^{3/2}\,\delta\nu}\frac{h\nu}{k}\frac{g_u}{g_l}N_l\frac{T_s - T_c}{T_s} \quad (9.12)$$

For cases where $T_s \gg T_c$ there is no dependence on (or information about) T_s, but the abundance N_l is directly related to ΔT_B. This indicates why abundances are more reliably determined if the molecule is seen in emission than in absorption. At longer wavelengths where $T_s \ll T_c$ typically, Equation (9.12) indicates that only the ratio N_l/T_s may be measured.

Second, if the molecule is anomalously excited, not only is the relation between N_l and N unknown, but N_l itself will not be generally obtainable. In the case of an optically thin cloud, N_l can in fact be determined about as well as in the case of normal excitation. For example, in the presence of maser amplification under the unsaturated conditions that would correspond to optical thinness, the factor $e^{\tau_0} - 1$ replaces the factor $1 - e^{-\tau_0}$ in Equation (9.9), so that in the case $\tau_0 \ll 1$ we have $\Delta T_B \approx \tau_0(T_s - T_c)$ for both cases. However, when τ_0 is not small, it is clear that, in the example of unsaturated amplification, the value of τ_0 deduced from Equation (9.9) will be very different from that deduced with the factor $e^{\tau_0} - 1$. Use of Equation (9.9) for a source that is actually masering has the effect of greatly overestimating the abundance and/or excitation temperature.

Third, the accurate measurement of ΔT_B itself contains several problems. If the actual distribution of molecular clouds is a polka-dot pattern within a telescope beam whose size is much larger than the dots, then ΔT_B for the dots is underestimated, and the corresponding abundances likewise. Often the observation of two or more transitions of a given molecule can suggest such effects, for if the optical depth in the small clouds is large, then much the same apparent optical depths could occur for both strong and weak resonant transitions. The term "clumping" is usually used to describe this class of problems. Quite equivalent in its effect to the clumping problem is the common occurrence in which the molecular cloud, even if uniform in distribution, is badly underresolved by the telescope beam. This occurs nearly always at longer wavelengths. The usual assumption that the line source fills the antenna beam then leads to an underestimation of ΔT_B, and hence of N and/or T_s. For emission line sources the factor of underestimation is the ratio of antenna beam size to cloud size, Ω_B/Ω_s. For absorption line sources the factor is $\Omega_B/R\Omega_c$, where Ω_c is the continuum source size and R is the so-called "filling-factor"— the fraction of the continuum source covered by the molecular cloud.

From the foregoing it can be surmised that there are more unknowns than available data. For instance it can never be determined from the observation of a single line whether the cloud is optically thick or not. In principle, the observation of two lines, say two different hyperfine components, can resolve this problem: If the two lines have a ratio of strengths characteristic of LTE, then the cloud is

probably thin, whereas if the ratio approaches unity, the optical depth is probably large. However, it is well known that IR trapping effects can also alter the ratio of two hyperfine lines in either direction, as it does in certain types of anomalous OH emission (Litvak, 1969). Different isotopes of a given molecule can be used in the same way: If, for example, the $^{12}C/^{13}C$ ratio in molecules were known to have the terrestrial value, then the observed relative strengths of ^{13}CO and ^{12}CO could be used to deduce the optical depth. However, the interstellar isotope abundance ratio is not known *a priori*, and furthermore it is necessary to assume that the two isotopic molecular species have the same excitation temperatures in the observed transitions.

In practice one uses observations of many different transitions of many different molecules in many different sources (where possible) to build a "consistent picture" for the values of isotopic ratios, typical kinetic and excitation temperatures, total particle densities, etc. For example, in several sources it is found that $T_B(H_2^{12}CO)/T_B(H_2^{13}CO)$ has the terrestrial value of $^{12}C/^{13}C$ (≈ 89)*. It is then assumed that these sources are optically thin in both species and have the same excitations for both, so that the abundance ratio $[H_2^{12}CO]/[H_2^{13}CO] = 89$. Next, $[^{12}C]/[^{13}C] = 89$ might be assumed to apply for all molecules throughout the interstellar medium, the ratio then being used to deduce optical depths in those clouds which are not thin. Another example applies to molecules such as OH and H_2CO, which are most often observed in absorption against continuum sources. In these cases only the quantity N/T_s can be determined; in order to derive N, a value for T_s is sometimes actually assumed, which may represent a "typical" value for OH as determined in other galactic regions from studies of normal emission, or it may be the value of T_s found to apply to other molecules in the same source. In fact, the value for N_{OH}

---

* Whenever atoms in molecular formulae are written without superscripts, they refer to the normal isotopes ^{12}C, ^{16}O, ^{14}N, and ^{32}S.

in Sgr B2 used in Table 9.1 is derived assuming $T_s = 100°K$.

9.4.2 Excitation of Interstellar Molecules

As has been indicated, the excitation of interstellar molecules often does not correspond to thermal equilibrium. Instead, energy level populations are determined by several competing processes which include collisions with electrons or neutral particles and interaction with radiation fields which seldom have blackbody characteristics. Often, from other information it is possible to deduce the nature of the particles and possibly also of the radiation field in specific types of regions; however, the analysis of molecular data yields independent and useful information on these aspects.

It is generally impossible to relate in a simple way the excitation temperature T_s of a pair of energy levels with various temperatures which characterize the particles and radiation fields. Such relations often must be deduced by detailed numerical evaluation of the set of statistical equilibrium equations, which include a large number of the energy levels of the molecule. These calculations form the basis of so-called pumping theories which have been applied to the obvious cases of anomalous excitation of OH (Litvak, 1968, 1969, 1972; Turner, 1970a), H_2O (Hills, 1969), and H_2CO (Thaddeus, 1972; Townes and Cheung, 1969; Solomon and Thaddeus, 1970). In certain cases and with certain approximations, T_s can be related quite simply to the temperatures of the various exciting agencies, and useful insight is gained. Let transitions between the upper (u) and lower (l) levels be produced by collisions with particles at temperature T_K, by the cosmic background radiation field at blackbody temperature T_R, and by some other mechanism (*e.g.*, other radiation fields), which produces $u \to l$ transitions at rate $r_{ul}{}^L$ and $l \to u$ transitions at rate $r_{lu}{}^L$; this mechanism may, for example, operate via several other energy levels as intermediate steps. Then from the

statistical equilibrium equation

$$n_u(A_{ul} + B_{ul} I_v + n\sigma v + r_{ul}{}^L$$

$$= n_l\left(\frac{g_u}{g_l} B_{ul} I_v + \frac{g_u}{g_l} n\sigma v e^{-h\nu/kT_K} + r_{lu}{}^L\right)$$

$$(9.13)$$

one finds that

$$T_s = \frac{(\tau_K \xi + \tau_R \xi + \tau_R \tau_K \xi r_{ul}) T_R T_K}{\tau_R \tau_K T_R T_K + \tau_R \tau_R \xi + \tau_K T_K}$$

where $\tau_K = (n\sigma v)^{-1}$, $\tau_R = (1 - e^{-h\nu/kT_R})/A_{ul}$, and $\xi = (h\nu/k)(1/\tau_{ul}{}^L - 1/\tau_{lu}{}^L)^{-1}$. The various τ's are the lifetimes (reciprocal rates) for the respective processes. T_s is thus a rather complicated weighted mean of the temperatures T_R and T_K. In case the rates $r_{ul}{}^L$ and $r_{lu}{}^L$ can be related by a radiative temperature T_L in the form $(r_{lu}{}^L/r_{ul}{}^L) = (g_u/g_l) \exp(-h\nu/kT_L)$, then

$$T_s = \frac{(\tau_K \tau_L + \tau_R \tau_L + \tau_R \tau_K) T_R T_K T_L}{\tau_R \tau_K T_R T_K + \tau_R \tau_L T_R T_L + \tau_K \tau_L T_K T_L}$$

$$\xrightarrow{\tau_L \to \infty} \frac{(\tau_K + \tau_R) T_R T_K}{\tau_R T_R + \tau_K T_L} \quad (9.14)$$

In obtaining these results, it is assumed that $(h\nu/kT) \ll 1$, where T is any of the above temperatures. This assumption is quite poor when $T = T_R$ and $\nu \gtrsim 10^{10}$ Hz.

9.4.3 Cloud Densities and Temperatures Deduced from Molecular Observations

Whenever molecules are observed in emission, well removed from any apparent discrete sources of radiation, it can be surmised that collisions elevate T_s well above the cosmic background value $T_R \simeq 2.7°K$. Then, according to Equation (9.14), $\tau_R \gtrsim \tau_K$ is required. The density of particles that makes τ_K just equal to τ_R is given by (cf. Rank *et al.*, 1971)

$$n = \frac{A_{ul}}{\langle \sigma v \rangle (1 - e^{-h\nu/kT_R})} \quad (9.15)$$

where $A_{ul} = 64\pi^4 \nu^3 |\mu_{ul}|^2/3hc^3$ and where $\langle \sigma v \rangle$ is an average over the velocity distribution of the particles. It is assumed that hydro-

Table 9.3 Lower Limits for H_2 Densities in Molecular Clouds

Molecule	ν (GHz)	A_{ul} (sec^{-1})	n_{H_2} (cm^{-3})
OH	1.667	7.7×10^{-11}	2.6×10^{1}
NH$_3$ (1,1)	23.694	3.6×10^{-8}	1.0×10^{3}
H$_2$CO (2_{11}–1_{10})	140.839	5.3×10^{-5}	5.8×10^{5}
CO	115.271	7.5×10^{-8}	8.6×10^{2}
CN	113.492	1.3×10^{-5}	1.5×10^{5}
HCN	88.632	2.4×10^{-5}	3.0×10^{5}
HC$_3$N (1 → 0)	9.098	4.0×10^{-8}	2.7×10^{3}
CS	146.969	6.1×10^{-5}	6.6×10^{5}
SiO (3 → 2)	130.268	1.1×10^{-4}	1.2×10^{6}
HNCO (4_{04}–3_{03})	87.925	9.0×10^{-6}	1.1×10^{5}
CH$_3$C$_2$H (5_0–4_0)	85.457	1.9×10^{-6}	2.4×10^{4}
OCS	109.462	3.7×10^{-6}	4.3×10^{4}

gen makes the only significant contribution to the collision rate (either as H or as H_2), although in some regions electrons may also contribute significantly. For lack of information, it is usual to use $\langle \sigma v \rangle = \sigma_{geom} \bar{v}$, where σ_{geom} is the geometric cross-section ($\approx 10^{-15}$ cm^2) and $\bar{v}$ is the mean relative speed of the two colliding particles, given by $\bar{v} = (8kT_K/\pi m)^{1/2} \approx 10^5$ cm/sec for $T_K = 100°$K and for $m \equiv m_{H_2}$.

The value of n given by Equation (9.15) is a strong lower limit, since it would tend to produce a value of T_s roughly midway between T_K and T_R; in many regions there is evidence that $T_s \approx T_K$. Table 9.3 gives some of these lower limits for n, for a few specific molecules that are observed in emission and which do not give evidence of anomalous excitation.

These lower limits for n_{H_2} assume that virtually all colliding particles are H_2 molecules. Whether electron or neutral hydrogen collisions play a significant role in exciting many of the observed molecular transitions is not entirely clear. Certainly in most of the dark dust clouds, observations of the 21-cm line of atomic hydrogen indicate considerably less hydrogen inside than outside the clouds, and in many cases no hydrogen is detected at all. At the low temperatures ($<30°$K) and high densities ($\gtrsim 10^3$ cm^{-3}) found in these clouds, both considerations of thermodynamic equilibrium and also theories

of the recombination of hydrogen atoms on grains indicate that all H should have been converted to H_2. Arguments suggesting very low densities of electrons ($n_e/n_{H_2} \ll 10^{-4}$) have been summarized by Rank *et al.* (1971). They rest on the fact that at the densities characteristic of molecular regions, not only UV radiation but also soft X-rays and low-energy cosmic rays would fail to penetrate into the clouds in sufficient amounts even to ionize carbon. In addition, electron densities would be very low because of the fast recombination rate of molecules, in the absence of appreciable ionizing rates in the molecular clouds.

Further evidence of much higher densities n_{H_2} and of insignificant electron densities comes from the study of individual molecules or of comparisons of small groups of molecules in given sources. In discussing these implications it will be clear that they rest on consistent, but not necessarily unique, pictures. For instance, one assumption used throughout is that the cosmic isotopic abundance ratios for the elements carbon, nitrogen, and oxygen are the same as the terrestrial values (in Section 9.4.4 we discuss how these are derived).

To illustrate how the various problems in deducing densities, temperatures, and molecular abundances are interrelated, we recall that the two steps in calculating the total abundance N of a molecule are (1) to determine N_l, the density in the lower level of the observed transition, by estimating the opacity τ in the observed line, and (2) to estimate the partition function Q. Although N_l can be derived independently of T_s if $\tau \ll 1$, one generally cannot establish from a single-line observation whether $\tau \ll 1$ and T_s is high or whether $\tau \gg 1$ and T_s is low. And to estimate Q, not only T_s but also the total particle density must be known with reasonable accuracy in order to estimate up to which energy level the populations are thermalized. In addition to these problems, it must be established that there is no anomalous excitation, so that N_l or N can be effectively estimated at all.

Most of these problems have been overcome in the case of the CO observations in both dust clouds and in molecule sources associated with continuum sources. In both types of sources, but especially in dust clouds, the CO is spatially extended so that ΔT_B (see Equation 9.9) can be reliably determined free of beam dilution effects. $\Delta T_B(^{13}\mathrm{CO})$ is typically found to be only three or four times smaller than $\Delta T_B(\mathrm{CO})$, which indicates that the opacity in the CO line is very large. If it is assumed that

$$T_s(\mathrm{CO}) = T_s(^{13}\mathrm{CO}) = T_s(\mathrm{C}^{18}\mathrm{O})$$

then by Equation (9.9) the ratios $\Delta T_B(\mathrm{CO})/\Delta T_B(^{13}\mathrm{CO})$ and $\Delta T_B(\mathrm{CO})/\Delta T_B(\mathrm{C}^{18}\mathrm{O})$ give two independent values for $\tau(\mathrm{CO})$, assuming terrestrial isotopic ratios. As derived in this manner, values for $\tau(\mathrm{CO})$ are always very large—between 50 and 95 for most continuum sources and 15 to 20 in dust clouds. These large opacities are indicated also by the observation of larger line widths for CO than for $^{13}\mathrm{CO}$ and evidence of saturation over much of the central part of the CO lines. To determine the total CO abundance, $N(\mathrm{CO})$, from the opacities, the excitation must be considered. Because the opacities are large, $T_s = \Delta T_B$ for the observed transition. For many molecules including CO, one needs no further details of the excitation to calculate N fairly accurately; for instance, if the observed value of T_s is assumed to apply to all

pairs of energy levels of CO, the derived value of $N(\mathrm{CO})$ is only ~ 2 times larger than if it is assumed that all levels are unpopulated except the $J = 1 \to 0$ levels in the observed transition. In this way typical values of $N(\mathrm{CO}) \simeq 10^{19}$ cm^{-2} in continuum sources, and $\sim 10^{18}$ cm^{-2} in dust clouds, are derived.

Because of its small dipole moment $(0.1\,\mu)$, CO is also useful in establishing stringent upper limits on electron densities. Collision cross-sections for electrons $(\sigma_e \approx \mu e/\hbar\nu)$ are generally much higher than for neutrals $(\sim 2 \times 10^{-15}$ cm$^2)$, but the abundance ratio, e/H_2, is very small $(< 10^{-4})$. However, if the electron kinetic temperature were considerably higher than the neutral particle temperature, relatively few electron collisions could produce a sizeable effect on the CO excitation. The best argument against significant electron excitation, at least in the dark dust clouds, comes from the failure to detect the $J = 1 \to 0$ transition of HCN in these objects. If electron collisions were significant, HCN and CO should have roughly the same excitation temperature, despite the much larger dipole moment of HCN, because for electron excitation the ratio of excitation rate to radiative decay rate is independent of the dipole moment. Then HCN should be observable, assuming $[\mathrm{HCN}]/[\mathrm{CO}] \simeq 10^{-4}$, as is found in other sources. Taking the dominant excitation as due to collisions with molecular hydrogen, the hydrogen density necessary to thermalize CO is $n_{\mathrm{H}_2} \gg A_{10}/\sigma v \simeq 10^3$ cm^{-3}.

Such densities are typically also implied by the derived values of $N(\mathrm{CO})$, assuming that $[\mathrm{CO}]/[\mathrm{H}] \sim [\mathrm{C}]/[\mathrm{H}]$, and so the deduction that $T_s \approx T_K$ is not inconsistent. It should be noted that the usual criterion for thermalization of the molecules, namely, $n_{\mathrm{H}_2}\sigma v > A_{ul}$, applies only for an opacity $\tau \ll 1$ in the molecular line. When $\tau \gg 1$, the appropriate criterion for thermalization is $\tau \gg \eta^{1/2}$, where $\eta = (A_{ul} + C_{ul})/C_{ul}$ is the mean number of scatterings undergone by a resonant photon generated in the cloud before it is destroyed by a downward collision (C_{ul} is the collision rate). For $n_{\mathrm{H}_2} = 10^3$ cm^{-3} we have $C_{10} \approx A_{10}$ for the CO $1 \to 0$ line, so that $\eta \approx 1$ while $\tau(CO) > 15$ in all cases. Thus CO is thermalized in all dust clouds and continuum sources. Hence $T_K = T_s$, which may be derived directly from the observations by

$$\Delta T_B = \frac{h\nu}{k} [(e^{h\nu/kT_s} - 1)^{-1} - (e^{h\nu/kT_R} - 1)^{-1}]$$

$$(9.16)$$

where $T_R \approx 2.7°K$ is the cosmic background. In this way values of T_K between 4 and 18°K are derived for dark dust clouds.

The values of T_K in dust clouds as derived from CO observations are strikingly confirmed by OH measurements in the same clouds. The two principal ground state lines of OH, at 1667 and 1665 MHz, have relative strengths of 9:5 under LTE conditions and optical thinness. In dust clouds OH is observed weakly in emission with a ratio of principal line brightnesses that is typically slightly less than 9:5. This slight reduction is assumed to result from some saturation of the 1667-MHz line as a result of optical thickness, and is used to derive the opacity of the 1667-MHz line and hence the OH abundance. From Equation (9.9), with $T_c = 0$ for dust clouds, T_s may then be derived directly from the observed ΔT_B. The fact that these values of T_s agree quite well with the values of T_K established from CO observations indicates that $T_s(OH)$ may be identified with T_K also. In turn one could then derive values of n_{H_2} necessary to thermalize OH, but these values are much lower than are required for CO, owing to the much smaller value of A_{ul} for the 18-cm lines of OH.

Densities and temperatures derived from molecules other than CO and OH, and in sources other than dust clouds, are somewhat less reliable. Often even by comparing the observations of several molecules, it proves difficult to establish exact conditions. As an illustration, consider CO, CS, and HCN in Ori A. The angular extent of the CO source is $\sim 1°$, while those of HCN and CS are a few arc minutes. By use of isotopic ratios it is found that $\tau(CS) < 6$ (^{13}CS is not detected), while both CO and HCN are optically thick,

as established from observations of their ^{13}C isotopes. Then the observed fact that ΔT_B(HCN) $\ll \Delta T_B$(CO) indicates that T_s(HCN) $\ll T_s$(CO). In addition, the fact that ΔT_B(CS) $\ll \Delta T_B$(CO) suggests that T_s(CS) $< T_s$(CO). Now Table 9.3 shows that CS and HCN have large and comparable A_{ul} values, while A_{ul} for CO is much smaller. A lower value of T_s for the short-lifetime species indicates that the excitation rate is too slow to thermalize the observed levels. It also indicates that the excitation is taking place via collisions with neutral particles rather than with electrons. This is because the electron collision rate is proportional to A_{ul} (this is not so for neutral particles) and hence electrons should produce roughly equal excitations for CO, CS, and HCN.

When neither T_s nor τ can be determined, as in the case of CS in Ori A, a lower limit for the column density N_l in the lower level of the observed transitions can be established by using Equation (9.12) with the assumption that $\tau \ll 1$ ($T_s \gg T_c$ for mm-wavelength transitions such as CS). A *lower limit* for the total column density can then be obtained by maximizing the fraction of molecules in the N_l level, assuming that levels below l are populated according to their statistical weights, that the u level is populated at the minimum value of T_s consistent with the observed ΔT_B, and that all higher levels are unpopulated. An upper limit to N_l can be calculated by assuming the largest value of τ consistent with observations of other isotopes. Equation (9.10) now applies. In this case use of Equation (9.16) provides a *minimum* value for T_s. If this value of T_s is applied to all energy levels, we obtain an upper limit for N/N_l. This is because for collisional excitation at limited densities, and $\tau \gg 1$, the excitation temperature for each pair of levels decreases with increasing energy of the level. Therefore if the minimum T_s is assigned to all levels, the excitation of levels above u is overestimated and the excitation of levels below l is underestimated. Both of these result in overestimating the population of the other levels relative to the l level. For sources where these techniques have been applied, the upper and lower limits for N differ by factors between 10 and 100 (Penzias *et al.*, 1971b). Only a minimum value of T_s is obtained. A lower limit for the total density n_{H_2} is also obtained by use of Equation (9.13) when the terms in I_v and r^L are omitted. If we use values of T_K as deduced from CO data, and also adopt a minimum value for n_u/n_l consistent with the observed ΔT_B, then we obtain $n_{H_2} \geq 4 \times 10^5$ cm^{-3} for Ori A and W 51. These values are much higher than those deduced from observations of CO (because of its small A_{ul} value) or those presented in Table 9.3.

As one proceeds to molecules whose observed transitions lie at increasingly high energies, the demands on the total densities become correspondingly greater. At present, the properties of NH$_3$ and H$_2$O emission imply the most spectacular limits. Even though H$_2$O is observed only under the condition of maser action, it is possible to deduce limits on temperatures and densities in these regions. The more intense H$_2$O sources, if they are radiating more or less isotropically, as appears likely, must be emitting total powers as large as 10^{30} to 10^{33} ergs/sec in the microwave line. Adopting values for n_{H_2O}, as estimated by Sullivan (1971), and source sizes of ~ 1 A.U. leads to probable lower limits on n_{H_2} of $\sim 10^9$ cm^{-3}. Upper limits for total densities of both OH and H$_2$O maser sources are $\sim 10^{13}$ cm^{-3}, imposed by the lack of observed pressure broadened line shapes.

The NH$_3$ spectrum, now observed in a total of seven different lines, probably provides the most accurate indication of temperatures and excitation conditions in interstellar clouds. An analysis of five of these lines, all of which arise in metastable states, is well summarized by Rank *et al.* (1971). These lines have been used to deduce "current" kinetic temperatures that range from 20 to 80°K in various directions toward the Sgr B2 cloud. They may also be used to deduce the equilibrium between ortho- and para-ammonia; transitions between these two species require as long as 10^6 years, whether they occur

under collisions with H$_2$ or H in gaseous phase, or by interaction with grains. The "temperature" derived from this equilibrium is generally several times higher than the "current" temperature and represents an average temperature over the past 10^6 years.

Two recently detected NH$_3$ lines, the (2,1) and (3,2) inversion doublet lines, arise from nonmetastable states (Zuckerman *et al.*, 1971b) (see Figure 9.1). These connect directly to the (1,1) and (2,2) levels via far-IR transitions with lifetimes of only 78 and 29 sec, respectively. Nevertheless, analysis of these two lines along with the five lines arising from the metastable doublets shows that collisions rather than large IR radiation fields populate these levels. In the absence of any IR trapping, $n_{H_2} \gtrsim 10^9$ cm^{-3} is required to produce the observed excitation of the nonmetastable doublet radiation. Some IR trapping may occur, but the maximum effect has been estimated from the fact that $\tau(1,1) \leq 8$, inferred from upper limits for the [^{15}NH$_3$]/[NH$_3$] ratio. A conservative lower limit in the presence of IR trapping is found to be $n_{H_2} \geq 10^7$ cm^{-3}. These densities occur only in the case of the Sgr B2 cloud. A cloud mass $\geq 10^6$ M$\odot$ is inferred, in which a fraction $\leq 10^{-5}$ of the N atoms appear to be contained in the NH$_3$ molecules.

In summary, derived temperatures in the molecular clouds range from as low as $\sim 4°K$ in some dust clouds to as high as 150°K in the Sgr B2 source (deduced from CH$_3$CN excitation; Solomon *et al.*, 1971). There is no correlation of temperature with density within a given class of source. The dust clouds appear to have lower temperatures than the molecule clouds lying near continuum sources, despite the higher densities in the latter type. Presumably the HII regions heat the surrounding neutral clouds, an effect which doubtless promotes certain chemical processes that do not occur in the lower-excitation dust clouds. Whether this effect, or the higher densities, is responsible for the wider variety of molecules found in the continuum source type of molecular cloud is not at present known.

Compared with preconceived notions,

the densities as revealed by molecular data constitute a much greater surprise than do the temperatures. Very large masses and gravitational instability are implied by the large densities. A cloud collapses under free fall in a time $t \simeq (\rho G)^{-1/2}$, where ρ is the density and G the gravitational constant. For a density of $n_{H_2} = 10^8$ cm^{-3}, as apparently applies at least to the Sgr B2 cloud, the collapse time is only 3000 years. In this particular cloud there is as yet no evidence of adequate turbulence or rotational motion to prevent such rapid collapse; clearly this will be an important feature to look for when the spatial resolution of telescopes is improved. It is likely, however, that many local instabilities occur within such clouds. Star formation may already be active within these clouds. As is discussed in Section 9.6.2, so-called pre-solar nebulae have been considered for some time to be prime sites for the formation of interstellar molecules. Generated within such nebulae, of characteristic density $\sim 10^{15}$ cm^{-3}, molecules would be dispersed throughout the cloud by IR radiation pressure from the proto-star, never being exposed to the destructive influence of UV radiation. Such a picture appears more likely than, say, an increased molecular formation rate on grains in such dense clouds, at least in the case of OH and H_2CO. This follows from the observation that OH and H_2CO abundances do not differ significantly in the Bok region and in cloud 2, which is immersed in this region, despite a difference of at least a factor of 10 in the extinction. However, the merits of the pre-solar nebula hypothesis in molecular clouds such as Sgr B2 will be demonstrated only by high-resolution IR observations, which should indicate the presence of such nebulae by their expected high IR radiation. Alternatively, high-resolution studies of the molecular lines themselves may eventually reveal whether they arise in several very small regions of much larger density than is indicated by present measurements.

A proto-stellar origin for the maser emission of OH and H_2O is already strongly indicated. Brightness temperatures as high as 10^{13}°K characterize the emission of both molecules, which are found by long-baseline interferometry to emit from regions no larger than a few A.U. in size. Furthermore, several compelling arguments (cf. Turner, 1970b; Rank *et al.*, 1971) indicate that these maser sources emit more or less uniformly into a solid angle of 4π ster. As many as 6×10^{48} photons/sec are then generated in the 1.3-cm H_2O line in the strongest sources; this is equivalent to more than a solar luminosity concentrated into a single microwave spectral line. Such strong maser emission has proved incapable, on energetic grounds, of being explained in terms of radiative excitation. Collisional pumping is necessary. Especially in the case of the H_2O emission, which undergoes rapid time variations in intensity (on a time scale of a few days), the total particle densities are inferred to be very high, possibly as high as 10^{12} to 10^{13} cm^{-3}. Such regions must be proto-stellar in nature. Most of the strongest OH masers (of the Type I variety that radiate principally in the 1665- and 1667-MHz lines) appear to arise in the same sources as the H_2O masers, although they may possibly lie in the outer regions of the proto-stellar objects, since the densities in the OH-emitting regions do not, on theoretical grounds, appear to exceed $\sim 10^8$ cm^{-3}. Ambient temperatures must be moderately high to pump the H_2O masers (at least a few hundred degrees Kelvin) because the states involved in these masers lie ~ 450 cm^{-1} above the ground rotational state. The high densities and temperatures, along with the observed sizes, all point to proto-stellar regions of a few solar masses as the sites of the more intense OH/H_2O maser sources.

Other types of OH maser emission arise in quite different types of regions. The so-called Type II(a) variety, which emits in only the 1720-MHz line and is not associated with H_2O emission, appears to be excited by strong infrared radiation fields at $\sim 100\ \mu$, and seems to reside in clouds that are probably quite cold and have densities intermediate between those of the dark dust clouds and those of the strong OH/H_2O

maser sources. In some cases the sources of IR radiation are actually observed (as "IR" nebulae), while in other cases the presence of large-scale shock fronts in interstellar clouds is indicated. A third type of OH maser emission, Type II(b), which occurs mainly in the 1612-MHz line, seems always to arise in the atmospheres of cool IR stars, often of the Mira-variable type, and is often accompanied by H_2O maser emission. A strong near-IR radiation field at $2\,\mu$ and presence of temperatures in excess of $\sim2000°K$ appear to be the necessary and sufficient requirements for exciting the OH in this way (Litvak, 1969).

Other interstellar molecules which exhibit population inversions in their observed transitions include CH_3OH, NH_2CHO, HCOOH, and CH_3CHO (Zuckerman *et al.*, 1972; Turner *et al.*, 1972; Rubin *et al.*, 1971; Zuckerman *et al.*, 1971a; Gottlieb *et al.*, 1971). The emission, and probably the population inversions, are very weak for these molecules ($\lesssim10\%$). The inversions in these cases appear to arise simply from the effects of hard collisions, spontaneous decay, and energy level configurations that result in slightly higher net decay rates into the upper level of the observed transition than into the lower level. The collision rates must not be high enough to thermalize the levels, which implies total densities $\lesssim10^7$ cm^{-3} in typical cases. The population inversions appear to be somewhat larger for CH_3OH than for the other species in this category.

9.4.4 Isotopic Abundance Ratios

As discussed above, the role of isotopic abundance ratios has been central in the deduction of physical conditions in interstellar molecular clouds. In this section we describe how these ratios are obtained and summarize the current results (see Zuckerman, 1973).

Two practical problems arise in deriving isotopic abundance ratios from ratios of antenna temperatures. They are (1) large optical depth and (2) different excitation temperatures for the two species. The problem of large optical depth is probably the more severe; the optical depth can sometimes be deduced by combining data from several different molecules, but the only reliable way to avoid the problem is to deal with molecules whose radiation is known to be optically thin. The problem of different excitation temperatures may arise either because of nonequilibrium pumping, which affects the two isotopic species of a molecule differently, or because of different degrees of trapping of microwave or IR resonant radiation in the cloud, owing to different optical depths of the two isotopic species.

Table 9.4 summarizes the results of isotopic abundance ratio determinations for several different molecules and sources. The $^{12}C/^{13}C$ ratio has been most reliably obtained from optical CH$^+$ lines in the direction of ζ Oph (Bortolot and Thaddeus, 1969) and also from the UV spectrum of CO in the same direction (Smith and Stecher, 1971). Such determinations are free of the above problems, but have been possible in only this one direction. Both measurements give values close to the terrestrial one. Among radio molecules, CO and H_2CO have been mostly used for the $^{12}C/^{13}C$ ratio. Although it may suffer from differential pumping effects, H_2CO is observed to be optically thin in many sources, as judged from the apparent LTE ratio of the hyperfine components in the 6-cm lines. In these sources the $^{12}C/^{13}C$ ratio is found to scatter about the terrestrial ratio. In optically thick regions, such as in the galactic center region, additional information is needed.

This information is in the form of simultaneous observations of $H_2{}^{12}C^{18}O$ and $H_2{}^{13}C^{16}O$ in Sgr B2 (Gardner *et al.*, 1971). These species are apparently optically thin, from which it may be determined that $(^{13}C/^{12}C)/(^{18}O/^{16}O)$ is about 1.8 times the terrestrial value of 5.5. If $^{16}O/^{18}O$ has the terrestrial value, then $^{12}C/^{13}C \approx 50$ in Sgr B2 —about half the terrestrial value. This choice is considered more likely than the alternative one—that $^{12}C/^{13}C$ is terrestrial and $^{16}O/^{18}O$ is 1.8 times the terrestrial value—for several

Table 9.4 Interstellar Isotope Ratios

Atom	Terrestrial ratio	Interstellar ratio	Interstellar molecule and region
$^{12}C/^{13}C$	89	$\sim$50 if $^{16}O/^{18}O = 488$	H_2CO, galactic center
		$\sim$89 if $^{16}O/^{18}O = 870$	H_2CO, galactic center
		$\sim$89 $\pm$ 15	H_2CO, outside galactic center
		82 $\pm$ 15	CH^+, ζ Ophiuchus
$^{16}O/^{18}O$	488	488 if $^{12}C/^{13}C = 50$	H_2CO, galactic center
		488 if $^{12}C/^{13}C = 50$	CO, galactic center
		870 if $^{12}C/^{13}C = 89$	H_2CO, galactic center
		390 $\pm$ 100	OH, galactic center
$^{16}O/^{17}O$	2700	$\geq$ 2700	OH, galactic center
$^{14}N/^{15}N$	270	> 70	NH_3, galactic center
		230 $\pm$ 70	HCN, Orion
$^{32}S/^{34}S$	22.5	24 $\pm$ 5	CS, galactic center and elsewhere
$^{32}S/^{33}S$	125	> 100	CS, Orion

reasons. First, although determined from OH with quite large uncertainty, the ratio $^{16}O/^{18}O$ appears to be less than or equal to the terrestrial value, rather than greater (Gardner *et al.*, 1970; Wilson and Barrett, 1970b). Second, the CNO nuclear bicycle reaction in stars decreases the $^{12}C/^{13}C$ ratio (to a value as low as 4, under equilibrium), while no known nuclear process alters the $^{16}O/^{18}O$ ratio from the terrestrial value.

The use of CO and HCN to determine isotope ratios is somewhat less useful than H_2CO because of problems of optical thickness and poor sensitivity. The best determination from CO is for W 51 and gives $^{13}C/^{18}O$ about 1.5 times the terrestrial value, or a ratio $^{12}C/^{13}C = 60$ if the $^{16}O/^{18}O$ ratio is terrestrial. H_2CO data give the value 63 $\pm$ 20 for W 51. In other sources outside the galactic center the CO results corroborate the H_2CO data (although with inferior sensitivity) in suggesting terrestrial ratios.

Further information on the oxygen isotope ratios has been obtained from attempts to observe ^{17}OH. A preliminary upper limit for the $^{17}O/^{18}O$ ratio (Gottlieb *et al.*, 1971) appears to be consistent with the terrestrial ratio, indicating that the oxygen isotopes are unperturbed in general from their terrestrial values.

The nitrogen isotope ratio is as yet diffi-cult to obtain; the best existing value is $^{14}N/^{15}N \approx 230 \pm 70$, derived from HCN observations in Ori A (R. W. Wilson *et al.*, 1972). This is consistent with the terrestrial value of 277. $^{14}N/^{15}N \geq 70$ is found for the NH_3 (3,3) inversion transition in Sgr B2 (Zuckerman *et al.*, 1971b).

Values for the $^{32}S/^{34}S$ ratio have been obtained from the $J = 2 \rightarrow 1$ transition of CS in W 51, W 30H, and Ori A (Turner *et al.*, 1973). Assuming that the $^{12}C/^{13}C$ ratio is terrestrial in these sources, a value of 24 $\pm$ 5 is found for $^{32}S/^{34}S$. This is consistent with the terrestrial value of 22.5.

In summary, it appears that interstellar isotope ratios are the same as the terrestrial ratios, with the possible exception of a slight overabundance of ^{13}C in the galactic center region. Since any such enhancement of ^{13}C is likely a result of stellar nuclear processes followed by mass loss to the interstellar medium, it would not be surprising to find an overabundance of ^{13}C in the galactic center where, compared with other regions, there are proportionately more stars than interstellar material. That such enhancement of the $^{13}C/^{12}C$ ratio should have stellar origin is supported by the isotope ratios found in the carbon star IRC 10216 (Morris *et al.*, 1971). Here, observations of $H^{13}C^{14}N$ and $H^{12}C^{15}N$ establish that the ratio $[^{13}C][^{14}N]/[^{12}C][^{15}N]$

is larger than the terrestrial value. Since both ^{13}C and ^{14}N increase and ^{12}C and ^{15}N decrease in abundance in the CNO nuclear cycle, this is consistent with the idea that stars can alter the interstellar $^{13}C/^{12}C$ ratio.

9.5 "Missing" Molecules and Their Implications

9.5.1 Free Radicals

OH and CN are the only free radicals known at radio wavelengths in the interstellar medium. Radicals which have not been found at microwave frequencies despite sensitive searches are CH, SH, NO, NCO, HNO, and HCO. Yet HCO, when combined respectively with H, OH, NH_2, and CH_3, produces the known interstellar molecules H_2CO, HCOOH, NH_2CHO, and CH_3CHO. Similarly NCO is part of the interstellar molecule HNCO, SH is part of H_2S, and CH of course may be regarded as a basic building block of all organic molecules. The very low abundances of HCO, CH, SH, and NCO must be due either to low stability against photo-dissociation or to high reactivity, even at interstellar densities. It should be noted that the transitions searched in SH, HCO, and CH involve the ground states, so that excessive densities are not required to populate the observed levels. In NCO the relevant $\Pi_{3/2}$, $J = \frac{7}{2} \to \frac{5}{2}$ transition has a lower state some 2 cm^{-1} above ground. One can attempt to explain the apparent relative abundances of OH, CN, CH, SH, and HCO in terms of their photo-dissociation potentials and re-activities with H_2. HCO is most unstable to radiation, decomposing in the red; CH can dissociate at $\sim 3000 \text{ Å}$, while CN and OH require $\lambda \lesssim 1650 \text{ Å}$ and $\sim 940 \text{ Å}$, respectively. The reactivities of these radicals may be compared in terms of the reactions $H_2 + HCO \to H_2CO + H$, $H_2 + CH \to CH_2 + H$, $H_2 + OH \to H_2O + H$, and $H_2 + CN \to HCN + H$, which have heats of reaction of $+15$, $+3$, -15, and -19 kcal/mole, respectively (Herbst, 1972). Assuming that the acti-

vation energies are not decisively different for these processes (there are no existing data in some cases), it might appear that the relative UV stabilities are the controlling factors, and would largely explain the presence of OH and CN, but not of CH and HCO.

The case of SH in interstellar molecules is interesting. Although H_2S has been detected, it is almost certain that $[H_2S]/[H_2O] < [S]/[O]$. It has definitely been established that $[SH]/[OH] < [S]/[O]$, at least in dust clouds (Heiles and Turner, 1971).* Because the thermal dissociation energy, and probably the photo-dissociation energy, are not significantly less for SH than for OH, it would appear that at least for this case the formation mechanism is responsible for the under-abundance of SH.

The missing radicals NO, HNO are not themselves contained in any known (stable) interstellar molecules. In fact the absence of the NO bond has become a virtual "rule" in describing interstellar molecules; all attempts to find HCNO, H_2NOH, CH_3NO, and CH_2NOH have failed. The latter two molecules are isomers of NH_2CHO and are known to be less stable. Similarly, HCNO is an isomer of HNCO and is less stable under laboratory conditions. It is interesting, however, that NO has recently been shown to be more stable against interstellar UV radiation than several known interstellar molecules such as NH_3, CH_3C_2H, H_2O, H_2CO, and OCS (Stief, 1973).

Several free radicals have been sought but not found by optical means in the relatively low-density clouds in front of ζ Oph. Among these are NCN, CCN, CNC, NH_2, and NH (Herbig, 1960, 1968). Unless these polyatomic fragments resulted from the breakup of larger stable molecules, a low abundance would be expected, for they cannot form via gas-phase two-body reactions, as can diatomic radicals (see Section 9.6.2), and there would appear to be insufficient amounts of dust in the ζ Oph region to allow a very rapid rate of

---

* Preliminary unsuccessful searches for CH_3SH are not yet conclusive enough to establish a similar relationship with respect to CH_3OH.

formation on surfaces. Shielding from destructive UV radiation is also very limited.

9.5.2 Stable Molecules

Of the large number of stable molecules so far sought without success, we mention but a few—those which have the possibility of revealing some aspects of interstellar chemistry.

Relatively unstable isomers of CH_3CN and HNCO which are not found in space are CH_3NC and HCNO, respectively. In each of these unstable isomers, the nitrogen atom has an effective valence of 5, which means that a formal charge separation occurs in the molecules in the sense $CH_3^-NC^+$ and H^+CN^-O. The work involved in separating the charge is at the expense of weakening the bonding. Under inelastic collisions with grains, rearrangement reactions tend strongly to occur, producing the stable CH_3CN and HNCO molecules. Such reactions would be sufficiently exothermic to cause release of the reformed molecules from the grains. If surface reactions on grains are the principal formation mechanism for molecules in general, then it might also be expected that CH_3NC and HCNO would never be formed in appreciable quantity to begin with.

When two free radicals join to make a stable molecule, the strength of the joining bond is strong if one of the radicals is electronegative and the other electropositive, but is weak if they are both of the same sense. This principle is reflected in the relative abundances of interstellar molecules. HCO is strongly electropositive, and CH_3 less so, while $HC \equiv C$, NH_2, and OH are electronegative. Thus NH_2CHO and HCOOH are very stable, CH_3OH and CH_3NH_2 quite stable, and NH_2OH and $HC \equiv C - CHO$ relatively unstable. The first three molecules have been found in space, while the latter three have not been found, although all have been sought. This might suggest that collisions with grains are quite frequent in interstellar molecule sources, since such collisions would produce rearrangement reactions to the more

stable species on their surfaces. Collisions with gaseous H_2 or H could disrupt the molecules, but they could not reform via two-body recombination (see Section 9.6.2) and would require grain surfaces to form into stable molecules. Free radical recombination will always occur on grains, regardless of how low the temperature is, because such recombination has zero activation energy.

The pattern of molecules which are observed and which are not may also provide information on the temperature conditions which prevailed when these molecules were originally formed (see the Addendum at the end of the chapter). Under conditions of chemical and thermal equilibrium, unsaturated (multiple) carbon bonding dominates saturated (single) carbon bonding at high temperatures, and vice versa at low temperatures. Equilibrium occurs either under conditions of high gas density or under conditions of frequent collisions with grains whose surfaces are at the same temperature as their surroundings. Table 9.5 indicates eight sequences of molecules, all but one of which contains a known interstellar molecule.

The symbol $\langle\quad\rangle$ indicates that the molecule has been detected, while $\boxed{\quad}$ indicates that it has been sought but not found. The remaining species either have no known microwave spectra or have not yet been conclusively searched for. In sequences (2) through (4), only the unsaturated species have been found. In sequences (1), (7), and (8), only the saturated species have not been found. Sequences (5) and (7) show two different possible sequences based on the C—C—O frame. It should be pointed out that other factors, such as the bonding stability, may well be more decisive in determining the abundances than is the temperature; thus the unstable species $HC \equiv C - OH$ might be less abundant than $CH_3 - CH = O$, even at high temperatures. $CH_3 - OH$ (sequence 6) is the only known fully saturated organic molecule; however, it is less abundant than the "higher temperature" species $CH_2 = O$ and $C = O$.

The conclusion, then, is that interstellar

Table 9.5 Patterns of Observed Interstellar Molecules

Sequence		Increasing temperature———→	
(1)	$CH_3—NH_2$	$CH_2=NH$	$HC\equiv N$
(2)	$CH_3—CH_2—NH_2$	$CH_3—CH=NH$	$CH_3—C\equiv N$
(3)	$CH_3—CH_2—CN$	$CH_2=CHCN$	$HC\equiv C—C\equiv N$
(4)	$CH_3—CH_2—CH_3$	$CH_3—CH=CH_2$	$CH_3—C\equiv CH$
(5)	$CH_3—CH_2—OH$	$CH_2=CHOH$	$HC\equiv C—OH$
(6)	$CH_3—OH$	$CH_2=O$	$C=O$
(7)	$CH_3—CH_2—OH$	$CH_3—CH=O$	$CH_2=C=O$ (?)
(8)	$NH_2—CH_2—OH$	$NH_2—CH=O$	$HN=C=O$

molecules could have been formed under conditions of relatively high temperatures. Just what temperature range may be involved is quite uncertain, since it depends on the density, element abundances, accurately determined molecular abundances, and an assurance of strict equilibrium conditions during formation. Equilibrium calculations by Eck *et al.* (1966) at 1000°K and 10^{-6} atm (a pressure applicable to the pre-solar nebula) show that $[CH_3-NH_2]/[HC\equiv N] \approx 10^{-17}$ for $C:H:O = 4:94:2$. Such examples indicate that impossibly low limits may have to be achieved on the negative results if they are to be quantitatively useful in testing equilibrium theories.

The idea of interstellar molecules being formed at quite high temperatures is also suggested by the NH_3 observations in Sgr B2, which indicate that the average temperature over the past 10^6 years is considerably higher than the present temperature (see Rank *et al.*, 1971). Given the probable rate of collapse of a cloud as massive as Sgr B2, it is likely that the molecules must largely have been formed in a time no longer than 10^6 years.

The presently determined temperatures of molecular clouds are very much lower than those suggested above for the formation conditions. Why have not the relative abundances of the species adjusted to the lower temperature, by collisions with grains for instance? The answer is that a significant activation energy is probably required to initiate these surface reactions, even though the net reaction from high-temperature to low-temperature species is exothermic. At the present temperature of the grains this activation energy cannot be overcome; hence hydrogenation of the unsaturated bonds cannot be accomplished.

In Section 9.6.2 we discuss evidence that favors the formation of interstellar molecules in pre-solar-type nebulae, where densities attain $\sim 10^{15}$ cm^{-3} and temperatures reach ~ 1500°K. The evidence discussed here, based not only on detected, but on "missing" interstellar molecules, tends to support this idea. The picture is that in these nebulae the grains and high-density gas create equilibrium conditions at high temperatures and produce the unsaturated molecular species. These molecules may or may not be ejected into the interstellar medium from these nebulae, but their environment eventually cools down and "freezes in" the unsaturated species, which we presently observe at least in those types of molecular sources (those associated with continuum sources) where there is good reason to believe that star formation has taken place.

9.6 Formation and Destruction of Interstellar Molecules

The equilibrium abundance of a given molecule is of course determined by both its formation rate and its rate of destruction. If we are to clarify interstellar chemistry through an understanding of these processes, we must overcome the following problems. First, it must be established whether or not different processes are responsible for the observed molecular abundances in different regions. In Section 9.1 it was noted that molecular abundances appear to differ from source to source. This could be caused by different grain or gas compositions, or by the same compositions under different physical conditions. Thus one of the current goals of observational work is to correlate the observed abundances with the specific physical conditions found in different interstellar regions. Second, one must propose formation schemes which have adequate rates, even though physical conditions (for example, the nature of grain surfaces) are not always known. Laboratory experimentation is required to determine whether these uncertainties are decisive. Third, one must propose formation theories and environments which produce the patterns of observed and unobserved interstellar molecules, and their isotope ratios. This ties the molecular problem to a wider body of astronomical knowledge and allows us to virtually rule out some previously proposed sites of molecular formation.

9.6.1 Destruction of Interstellar Molecules

In exposed regions of interstellar space, unshielded by dust, the radiation field is dominated at short wavelengths by bright O and B stars; it peaks in intensity shortward of 2000 Å, in the photolytic UV, and at a typical photon energy of $\sim 10\,\text{eV}$ has a corresponding flux of $\sim 3 \times 10^8$ photons/cm²/ sec. Most molecules, both known and expected, dissociate very readily in such an environment. The questions we must answer are: (1) do the relative UV stabilities of known and expected molecules correlate with their observed abundances; (2) how efficient a protector is the dust that apparently is associated with interstellar molecules; (3) can the interstellar UV, appropriately attenuated and reddened by the dust, act as an energy source

for surface reactions on interstellar grains, thus aiding in the formation rather than the destruction of interstellar molecules.

The answer to question (1) is unknown for most of the molecules of interest, since no photochemical data exist for them. The best that can be done is to use the thermal dissociation energies; for a number of polyatomic species these tend to be linearly related to the photo-dissociation energies, although this tends to be less true for diatomic species (Herzberg, 1966). Figure 9.3 represents these thermal dissociation energies in a manner similar to that given by Rank *et al.* (1971), although we have included different species to account for the unobserved molecules. Within the rather severe limitations imposed by using thermal dissociation energies, there does appear to be a tendency for the observed molecules to have greater stability than the unobserved molecules.

For a few of these molecules the photo-dissociation cross-sections are known accurately as a function of wavelength, as are the quantum yields Φ_λ for the various paths of decomposition (Stief *et al.*, 1972). In conjunction with a knowledge of the spectral distribution, F_λ, and the transmissivity, T_λ, of interstellar dust, the lifetime τ of a molecule against photo-dissociation can be calculated from

$$1/\tau = \int_{912\text{Å}}^{\lambda_{\text{max}}} F_\lambda T_\lambda \sigma_\lambda \Phi_\lambda d\lambda$$

where the lower limit of integration assumes an HI environment. The results of such calculations, taken from Stief (1973), are shown in Figure 9.4. They show that the dust is remarkably efficient in screening the molecules from the most destructive UV. They also indicate that the ordering of relative stabilities of the molecules as presented in Figure 9.3 is qualitatively correct.

Several important points emerge from these calculations. First, the notable absence of NO and the presence of OCS and CH₃CHO clearly cannot be explained on the basis of destruction rates caused by UV or other known processes. The interstellar

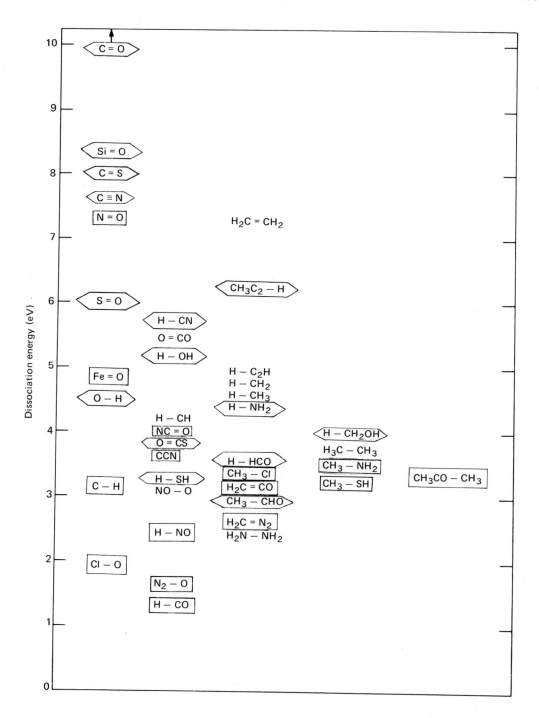

Figure 9.3 Thermal dissociation energies for molecules of astrophysical interest. Molecular complexity increases to the right. The symbol ⬡ indicates molecules that have been detected in the interstellar medium, while ▢ indicates molecules that have been sought but not found.

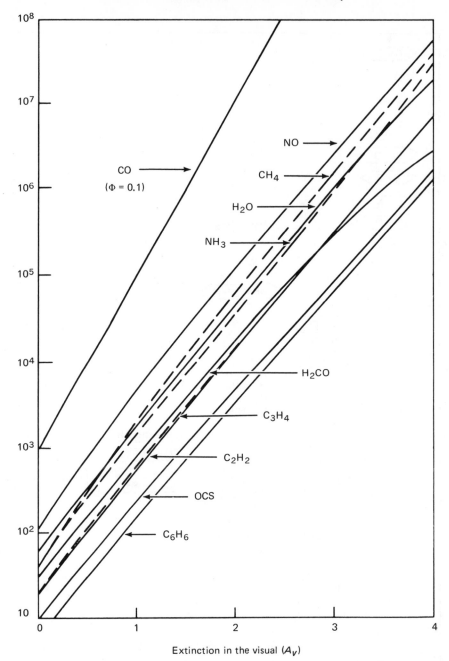

Extinction in the visual (A_v)

Figure 9.4 Lifetimes of molecules in interstellar clouds as a function of the visual extinction. The lifetimes are assumed to be determined only by photo-destruction by visual and UV photons.

chemistry of these species is evidently highly nonequilibrium and highly specialized. Second, all of these molecules dissociate in $\lesssim 100$ years in unprotected regions of space. Only CO has a lifetime of ~ 1000 years in exposed

HI regions. Unless implausibly large formation rates can be demonstrated, these results indicate that the observed molecules must either have been formed in the dense opaque clouds where they are now seen, or have been transported there in some protected way, as by being imbedded on grain surfaces.

Observations of the spatial distributions of these molecules tend to demonstrate the importance of the UV destruction mechanism in some cases, but to pose interesting questions in other regions. In Ori A, the molecular clouds of CO, CS, HCN, CN, and H_2CO are all centered on the IR nebula; however, the angular sizes of these clouds vary from $\sim 1°$ for CO, $\sim 7'$ for CS, $\sim 4'$ for HCN, to $< 1'$ for H_2CO. This is exactly as expected from the relative UV stabilities if the shielding effect decreases uniformly from the center to the edge of the large cloud. Similarly, despite the comparable cosmic abundances of N and O, observations indicate $[CN]/[CO] \sim 10^{-4}$ to 10^{-3}, which is consistent with the relative stabilities of CN and CO. On the other hand, CO, OH, and probably H_2CO are all found with nearly the same abundance in the Bok region surrounding cloud 2 as inside cloud 2, despite a decrease in extinction by a factor of at least 10 in passing outside cloud 2 (see Section 9.2.1). Perhaps the formation rate of OH and H_2CO increases outside cloud 2, possibly because of higher temperatures or the presence of UV radiation in the Bok region. The widespread distribution of H_2CO in the galaxy is itself somewhat puzzling (only OH is more widespread), since H_2CO has nontrivial absorption features to wavelengths as long as 3800 Å, where dust is an ineffective shield [this explains why $\tau(H_2CO)$ tails off toward larger values of A_v in Figure 9.4]. It would appear that H_2CO must have both a very general and a very rapid means of formation compared with most interstellar molecules.

Some specific properties of the photolysis of molecules are subject to observational test, in order to help decide whether—as appears very likely—UV dissociation is the principal destruction mechanism for interstellar molecules. For example, H_2CO decomposes into $H_2 + CO$ and $2H + CO$ very efficiently at wavelengths $\lesssim 2000$ Å, whereas at longer wavelengths it decomposes with considerably less efficiency into $H_2 + CO$ and $H + HCO$. It is quantitatively found

(Stief *et al.*, 1972) that HCO would be a minor product of the UV decomposition of H_2CO in the interstellar medium, although it could be formed by other means. This may partly explain why HCO is as yet undetected. NH_3 can decompose into $NH_2 + H$, $NH + H_2$, or $NH + 2H$, with relative importances that depend on the wavelength of light absorbed. In the interstellar field the net result is that NH_2, and to a lesser extent NH, may be present as a result of NH_3 photo-decomposition. Since NH cannot be formed by radiative association (see Section 9.6.2), it should be present only where it can be formed by surface reactions or where NH_3 is abundant; either case probably means regions of high extinction where optical detection would be difficult. Hence NH is as yet undetected. H_2O decomposes to either $H + OH$ or $H_2 + O$, with the former dominating at all wavelengths. This is probably not the principal source of interstellar OH, however. CH_4 is found to produce primarily CH_2 and CH_3 radicals upon decomposition, with very little CH. OCS decomposes much more readily to $CO + S$ than to $CS + O$; this is one of several examples that indicates the role of CO as a "residue" of many molecular processes and helps to explain its large interstellar abundance.

By comparison with photo-dissociation, photo-ionization can be ignored as an efficient destruction mechanism in space, because ionization energies lie several volts higher than do the dissociation energies. However, molecules can also be destroyed by interaction with high-energy radiation (X-rays and γ-rays) and energetic particles. These processes are much less effective than UV radiation in unshielded areas, but inside dense clouds where the UV is highly attenuated, the energetic radiation and particles may well still persist. Thus molecular lifetimes in clouds may depend upon these processes and be shorter than those shown in Figure 9.4. A quantitative treatment of these processes is not possible at present.

Finally, molecules may be lost from the interstellar gas by freezing out on grains. In this process, molecules strike and adhere to grains by van der Waal's or other forces, with a probability of about unity at the low temperatures that characterize most molecular clouds. As pointed out by Rank *et al.* (1971), if the usual assumptions are made that $n_{grains}/n_{hydrogen} \sim 10^{-12}$ and that the grains have radius 2×10^{-5} cm, then the molecular lifetime before striking a grain is

$$\tau_g \approx 10^{21}/vn \quad \text{sec}$$

where v is the molecular velocity and n the total particle density ($= n_{H_2}$). For a molecule of molecular weight 25, one finds

$$\tau_g \text{ (dust clouds)} \qquad \simeq 10^5 \text{ yr } (T = 10°\text{K}, \\ n = 10^4)$$

$$\tau_g \text{ (continuum sources)} \simeq 30 \text{ yr } (T = 100°\text{K}, \\ n = 10^7)$$

In dust clouds this rate is comparable to, or slightly less than, the UV dissociation rate. However, in the denser hotter sources such as Sgr B2, the freezing-out rate exceeds the UV rate even in free space. In either case the lifetime in gaseous form is much less than the lifetime of the cloud, so that molecules must be continuously formed at high rate in the gas phase, or regenerated or re-emitted from the dust grains many times during the lifetime of the molecular cloud. If chemical processes take place on the grains, particularly those involving free radicals which have no activation energy, then the molecular species observed in these clouds may depend strongly in the nature of the grain surfaces and bear little or no relation to those species that may originally have been formed by some other process (*e.g.*, in presolar nebulae; see Section 9.6.2). More likely, the reactions that take place on the grain surfaces are only those very few whose activation energies, A, can be overcome at temperatures $\lesssim 100°\text{K}$. This includes all free-radical reactions (and ion-molecule reactions), and may explain why free radicals are observed so rarely in interstellar space. Another interesting reaction with $A \simeq 120°\text{K}$ is $NO + N \rightarrow N_2 + O$, which may explain why NO is absent in

space. By comparison, most reactions between atoms and molecules or between molecules and grains have $A \sim 2000°\text{K}$ and may be ruled out in interstellar sources.

An interesting question is whether the very different grain interaction rates between dust cloud sources and continuum sources should produce differences in the observed chemical species, or whether any such differences might also arise from different original formation environments. A closer look at formation mechanisms for interstellar molecules is needed to answer this question.

9.6.2 Formation of Interstellar Molecules

There is still uncertainty concerning the method of production of interstellar molecules, though various mechanisms have been discussed. These are (1) radiative association (Solomon and Klemperer, 1972); (2) chemical reactions between atoms and molecules chemically adsorbed on grains (Stecher and Williams, 1966); (3) reactions on grain surfaces between weakly adsorbed atoms (Brecher and Arrhenius, 1971; Williams, 1968; Hollenbach and Salpeter, 1971); (4) breakdown of larger molecules evaporated from grains (cf. McNally, 1968); (5) formation under thermodynamic equilibrium in cool stellar atmospheres or pre-solar nebulae (Herbig, 1970; Shimizu, 1973). (See also the Addendum to this chapter.) We shall describe only those features of these processes that are essential in choosing among them, based on the observational information now available. Details can be found in the references. All of these processes may be applicable under different conditions in the interstellar medium.

a) Radiative Association

Two gas-phase atoms can combine to form a molecule during a binary collision, although with only very small probability ($\sim 10^{-6}$) per collision. Two conditions are necessary: (1) The created molecule must be formed in an excited state by two atoms which, under interstellar conditions, will be in their ground states; (2) the excited state must be connected to the ground molecular state by a permitted transition, to allow the excess collisional energy to be radiated away. The first requirement is violated by OH, the second by H_2. Also, CO, NH, and NO fail one or the other criterion. Among the simple diatomics of astrophysical interest, only CH^+, CH, and CN meet these requirements. However, if sufficient quantities of these can be formed, other diatomics may also be formed by highly exothermal gaseous exchange reactions with these species. This is the basis of the Solomon and Klemperer (1972) theory; in a complete study of all exchange reactions involving H, N, C, and O that are possible at low temperatures and densities, they find that, starting with CH^+ and CH, the diatomics CN, CO, H_2, and C_2^+ may be formed at adequate rates via collisions of CH^+ with N, O, H, and C^+, respectively. By radiative association and gaseous exchange reactions alone, it is clear that only diatomics can be efficiently formed. Nonexchange collisions between gaseous diatomics or between diatomics and atoms would not in general lead to triatomics or higher, but rather would dissociate the existing diatomic molecules, in a time equal to that of the dissociation vibration period, $\sim 10^{-13}$ sec; or $\sim 10^7$ times shorter than the excess energy would be radiated in a typical optical transition. Thus the Solomon-Klemperer theory is intended to apply only to the relatively rarefied interstellar clouds of low extinction in which diatomic species alone are seen by absorption of starlight at optical wavelengths. In the single case of ζ Oph, for which comprehensive observations of CH^+, CH, CN, and CO exist (Herbig, 1968), the theory gives adequate quantitative agreement. OH, NO, and NH are not predicted to form at all, and CO^+ in only minute quantity; all these species are undetected by Herbig to low limits in the direction of ζ Oph.

Two additional cases of radiative association not studied by Solomon and Klemperer appear to be possible in interstellar space. OH, CO, CN, NO, and C_2 may be formed by binary collisions and an inverse pre-dissociation process (Julienne *et al.*, 1971; Julienne and Krauss, 1973). The predicted efficiency is quite low, and possibly may not explain the observed OH abundances, even in clouds where OH is seen in absorption. The theory is not intended to apply to the much-higher-density maser sources. It may possibly explain the observations of OH in some regions where there is no visible extinction (Turner, unpublished data). However, in most of the highly obscured regions where fairly normal OH is observed, a formation mechanism involving ion-molecule reactions or grain surfaces probably dominates, because most of the hydrogen is in the form of H_2 rather than the atomic form necessary for the radiative association process. Radiative association may also be possible between two large free radicals to form a large stable molecule (Williams, 1971). Large radicals possess many internal degrees of freedom among which the excess energy of collision can be dispersed via vibrational de-excitation. However, calculation shows that such dispersal will occur with adequate efficiency only if each radical contains roughly five or more atoms; molecules with 10 or more atoms are as yet unknown in space, but should they be found, free-radical association might well explain their formation better than surface reactions, since the latter entail the problem of escape from the grain after formation. However, molecules with from 3 to about 10 atoms cannot be formed efficiently by binary collisions in the gas phase, and require one of the following processes.*

b) *Surface Reactions Between Fully Chemically Adsorbed Species*

In all surface reactions the grain acts as the third body to absorb the excess energy of collision of the incident species. Additionally, surfaces may weaken bonds of molecules that

---

* Herbst (1972) has shown that polyatomic molecules may also be formed rapidly by ion-molecule reactions; such processes probably dominate when the temperature is very low (<20 to $50°K$). See Addendum to this chapter.

chemically adsorb on them, so that they are easily dissociated into highly reactive fragments (atoms or radicals) which react readily with other incident particles to form new molecules. This is the basis of catalysis. Chemically adsorbed species, as distinct from physically (weakly) adsorbed species, form actual bonds with the surface molecules, and thus any surface reaction which forms a new molecule must be sufficiently exothermic to overcome this bond. This strongly limits under low-temperature conditions the variety of product molecules which can be formed and escape from the surface. On the other hand, molecules or atoms adsorbed only by weak physical forces (such as van der Waal's forces) tend to evaporate very quickly, before incident particles can impinge on the grain and interact with them. In the early studies of this type (Gould and Salpeter, 1963; McNally, 1962) it was found that, for ideal surfaces, the grain temperature had to be $\lesssim 7°$K to allow formation of H_2 and other diatomics; this temperature was considered to be excessively low, and led to the temporary abandonment of physical adsorption mechanisms in favor of chemi-adsorption mechanisms.

The only quantitative study of surface reactions involving fully chemi-adsorbed species is that of Stecher and Williams (1966). (See Addendum to the chapter.) Unfortunately, it is restricted to diatomic species. However, because chemical bonding forces dominate all others and are well known for several different types of surfaces, the predictions of such a theory (which of course depend strongly on the surface) are quite definite and may be compared with observations. Thus, either the process itself or at least the presence of certain types of grain surfaces may be tested on the basis of present observations. For this reason we summarize the results of the theory of Stecher and Williams.

For graphite grains the molecules are created by the chemical exchange reaction $\langle\rangle \cdot CX + Y \rightarrow \langle\rangle \cdot C + XY$, where $\langle\rangle \cdot C$ indicates that the carbon atom is a member of the last hexagon ring of the grain. The allowed reactions are those which are sufficiently exothermic to overcome the chemical bonding to the grain and allow the newly formed diatomic molecule to escape. The following reactions are possible. If $X \equiv H$ and $Y = H$, C^+, O, N, respectively, then H_2, CH, OH, and NH may be formed, respectively. If $Y = H$, only H_2 can form. If $Y = C^+$, CO and CN may be formed. If $X = C$, again CO and CN can form. Finally, if $X = N$, only N_2 may be formed (NO does not form).

If the grain surfaces consist of dirty ice, only H_2 and CO are capable of formation.

Restricted as it is to diatomic species, the Stecher and Williams theory was intended to describe the optically observed molecules in regions of low density. In this regard it fails on several aspects. First, for graphite surfaces, and in regions where CH occurs, it predicts the formation of both OH and NH, but not of CO. Yet in ζ Oph, CO, CN, and CH are observed, while OH and NH are not. The predicted abundance of OH is considerably higher than that of CH. For grains of dirty ice, CH and CN are not predicted—again, contrary to observations in ζ Oph. Second, the efficiency of the Stecher and Williams process is inadequate to produce the observed abundances of any of the diatomic molecules under ordinary HI conditions. The problem is that the activation energy of these reactions is typically $\gtrsim 0.2$ eV ($T \geq 2000°$K), while the available kinetic energy of the incident particles in HI regions is only $\sim 10^{-3}$ eV. The necessary energy can be produced in principle by passage close to hot stars or by cloud-cloud collisions. ζ Oph may possibly have been an adequate source of energy for the diatomic molecules seen in that direction. However, molecules in general are not observed preferentially near hot stars, and the presently observed temperatures in molecular clouds are not consistent with the occurrence of cloud-cloud collisions within the time that the molecules would freeze out on grains. In the highly dense clouds that harbor the polyatomic molecules seen at radio frequencies, the incident hydrogen particles would be in the form of H_2 rather than H; then the predicted pattern of diatomic molecular species, which fails to agree with the optical observations, may not apply. Furthermore, there are no predictions for polyatomic species. However, the problem of efficiency must continue to be a severe difficulty. It appears likely, therefore, that surface reactions involving fully chemically adsorbed molecules do not produce gaseous interstellar molecules in general, and certainly do not produce the observed diatomic species if the grain surfaces are either graphite or dirty ice.

c) *Surface Reactions Involving Weakly Chemisorbed Species*

Whereas fully chemically adsorbed surface reactions involve adsorption energies equivalent to $T_{ad} \gtrsim 2000°K$, weak chemical adsorption can occur on some kinds of surfaces, particularly those of transition metals (Fe, Ni, etc.) and their oxides, with $100°K \leq T_{ad} \leq 500°K$ (Brecher and Arrhenius, 1971). This is high enough to prevent excessively rapid evaporation of the surface atom or molecule before it can react with an incident particle, but not so high as to require excessive energy to desorb the newly formed molecule. Iron (Fe) is considered an important and abundant early refractory condensate in circumstellar dust clouds (Larimer, 1967; Hoyle and Wickramasinghe, 1968) and a likely component of interstellar dust (Wickramasinghe and Nandy, 1970). The importance of transition metals and their oxides lies both in the suitable range of T_{ad} values and in the fact that no activation energy is required for chemi-adsorption on their surfaces. This is because these species possess incomplete atomic d shells and hence have free valence electrons available for bonding. By contrast, this "semi-chemi-adsorption" is not possible on graphite, which requires a large activation energy for bonding (~ 30 kcal/mole, or $T_{act} \approx 15,000°K$) and has a large adsorption energy ($T_{ad} \simeq 2200°K$). It has been shown that H_2, N_2, CO, O_2, etc., will all chemi-adsorb rapidly on Fe and Ni at temperatures as low as $4°K$ (for H_2 and CO). Chemi-adsorption is accompanied by dissociation of the incident molecules into extremely reactive atomic species, which recombine on the surface in a variety of ways to form new molecules.

(i) *Nonactivated Adsorption*

This term refers to those cases in which no additional sources of energy are present to drive the catalytic surface reaction except the ambient kinetic energy, which for transition metals must be ~ 100 to $500°K$ to produce efficient rates. Such a range of temperatures is consistent with those found for the molecular sources associated with continuum sources. In addition, the surfaces can be expected to act as efficient catalysts indefinitely, rather than becoming contaminated with many surface layers of nondesorbed molecules. This is because the heat of formation of H_2 alone (~ 100 kcal/mole) will maintain the grains at elevated temperatures and ensure that the molecules evaporate quickly after association.

Under nonactivated conditions, H, H_2, N_2, O_2, and CO are known to adsorb rapidly. CO is chemisorbed without dissociation, but the other species dissociate with adsorption, leaving highly reactive atoms which recombine to form various molecules. The production rate of H_2 can be shown (Brecher and Arrhenius, 1971) to be faster than that considered for other types of surface reactions, even when the latter are accompanied by high-energy input (Stecher and Williams, 1966; Hollenbach and Salpeter, 1971; Hollenbach et al., 1971). Similarly, as shown by industrial catalytic processes on Fe, NH_3 would be rapidly produced by $3H_2 + N_2 \xrightarrow{Fe} 2NH_3$ and CH_3OH by $2H_2 + CO \xrightarrow{Fe} CH_3OH$. In general, the Fischer-Tropsch catalysts (Fe, Ni, Co) are used in industry for the synthesis of hydrocarbons, aldehydes, alcohols, and acids from CO and H_2. Finally, it has been convincingly argued (Anders, 1973; Studier et al., 1968) that at $T < 500°K$ Ni or Fe grains catalyze, by

Fischer-Tropsch reactions, the synthesis from H_2, CO, and NH_3 of various organic compounds present in carbonaceous meteorites. These are generally the same group of molecules found in the interstellar medium (Table 9.1).

(ii) Activated Adsorption

The foregoing discussion indicates that surface reactions on transition metals and their oxides may produce the observed molecules at suitable rates under the physical conditions found in the continuum source molecular clouds. However, the much lower temperatures (and densities) in the dark dust clouds suggest the need for an additional energy source in those clouds for these surface reactions. Because of the much lower (although still large) extinction of dust clouds compared with continuum sources, UV radiation might provide an energy source to overcome the low temperatures.

A series of laboratory experiments using transition metals (Ni) and relatively long-wavelength UV (2537 Å) has been performed in the presence of various gases (H_2, N_2, CO) by Breuer and his associates (Breuer *et al.*, 1970a, 1970b; Breuer, 1973). All of the adsorbed molecules in these experiments (H_2, N_2, CH_4, CO) dissociate upon adsorbing; in the case of CO the dissociation is precipitated by the UV, which is energetic enough at 2537 Å, as a result of the weakening of the CO bonds in the adsorbed state. This same UV is, however, of insufficient energy to dissociate most of the astrophysically interesting molecules in the gas state. If CO and CH_4 are adsorbed alone, a variety of aldehydes including H_2CO and CH_3CHO are formed. If CO, H_2, and N_2 are coadsorbed, all of the nitrogen-bearing molecules observed in space are formed. When compared with observed relative abundances in Sgr B2, the laboratory results seem to predict too much H_2CO and not enough HC_3N, although relative abundances of the other molecules (HCN, HNCO, CH_3CN, NH_2CHO) seem consistent. However, these experiments are appropriate mainly to dust cloud molecular sources,

where adequate amounts of long-wavelength UV might penetrate, or arise from stellar sources embedded in the clouds. In dust clouds H_2CO is observed in far greater abundance relative to the nitrogenous molecules than it is in continuum sources, in agreement with the UV-induced laboratory work.

Another UV-activated experiment of possible astrophysical relevence has been performed by Khare and Sagan (1971), who irradiate under high vacuum a mixture of H_2O, NH_3, $(H_2CO)_n$, and C_2H_6 that has been frozen out on silicate glass at 77°K. The UV is monochromatic, at 2537 Å. The only gases that are found to remain uncondensed under these conditions are CO, CH_4, and traces of H_2. Most of the hydrogen has reacted to form a variety of other molecules which remain condensed at 77°K on this type of surface. Upon heating, the condensate evaporates and is found to contain CH_3OH, HCN, CH_3CN, CH_3CHO, and HCOOH, all known interstellar molecules, and C_2H_5OH (ethanol), $(CH_3)_2CO$ (acetone), and CH_3OOCH (methyl formate), which have been sought in vain in the interstellar medium. The vapor pressures of both groups are very similar, while the UV stability is lowest for acetone and the aldehydes. Neither of these considerations, therefore, seems to explain why some of the products are found in space and why some are not. In addition, the formation of these molecules appears to produce too little energy to facilitate their efficient release to the gas phase. These considerations may serve to indicate the very specialized physical conditions required in order to produce the observed molecular species.

d) Dust Clouds versus Continuum Source Molecular Clouds: A Different Chemistry?

We have seen (Table 9.5) that in continuum source clouds "high temperature" molecules rather than "low temperature" species seem to be formed; also there is a rapid interaction with grains ($\tau_g \sim 30$ yr) and probably no appreciable UV radiation field. In dust clouds, present temperatures and densities are lower, $\tau_g \sim 10^5$ yr, and some UV may be present. Do these different conditions imply a different chemistry in the two types of source?

Four possibilities may explain the observation of "high temperature" molecules in the continuum source clouds: (1) Molecules are formed in LTE at high temperature and are not strongly modified by continual interaction with grains thereafter, even though the grains and gas cool considerably with time. (2) The observed unsaturated molecules are not necessarily formed at high temperatures, but are favored in subsequent surface reactions with the grains because the activation energies for their formation happen to be low. (3) The observed unsaturated molecules happen to be those which are most volatile (have the largest vapor pressures) in contact with the interstellar grains. (4) Molecules are formed primarily by ion-molecule reactions (see the Addendum to the chapter). Among these choices, it appears possible to tentatively* rule out only (3). An examination of known vapor pressures for organic molecules shows no correlation with the observed relative abundances in space. For example, the "low temperature" unobserved molecules propane and propylene have higher vapor pressures than CH_3C_2H (see sequence 4, Table 9.5). Similarly HCN and H_2CO have lower vapor pressures than CH_3NH_2 and CH_3OH, respectively (sequences 1 and 6, Table 9.5). There is no definite information relating to possibility (2), although it may be mentioned that an activation energy of ~ 0.5 eV is required to hydrogenate some simple double-bonded hydrocarbons such as ethylene. Also, as described above, surface reactions involving CO, H_2, N_2, CH_4, and UV activation do not by experiment seem to form unsaturated species preferentially (HCN and HC_3N are formed with low efficiency). However, the action of the UV may be de-

cisive in selecting the formation species, and may not represent the interstellar conditions. The available evidence therefore points toward possibility (1) or (4) for the continuum source molecular clouds, but with considerable uncertainty.

In the dark dust clouds, only OH, NH_3, H_2CO, and CO are so far observed. OH and NH_3 are fully saturated species, while both CO and H_2CO tend to form CH_3OH under collisions with H_2 (on surfaces) at low temperatures. CH_3OH is unobserved in dust clouds but is observed in continuum source clouds. This is consistent with the higher grain-interaction rates in the latter sources but should not be taken to indicate that grain interactions or polyatomic-ion buildup are negligible in dust clouds. This is because of the highly anomalous excitation displayed by CH_3OH in the continuum source clouds, an excitation that probably is lacking in the lower-temperature dust clouds. In a similar vein, the lack of detected HCN and HC_3N in dust clouds does not necessarily indicate that the formation chemistry was at low temperature or does not involve ions; the reason is that, because of the relatively low densities in these clouds, the excitation of the transitions searched in these molecules may be very low, rather than the abundance being necessarily low. Further observational work, in more transitions and at greater sensitivity, is needed to decide whether the relative abundances of interstellar molecules differ in dark dust clouds and in continuum source clouds. If such a difference is found, it will be an important clue to the formation mechanisms in both types of sources.

e) Formation of Grains and Molecules in Stars

Although interstellar molecules may be produced on grains surfaces, the question of the origin of the grains themselves has never been answered. One of the oldest ideas is that they form in the atmospheres of cool late-type stars which have sufficiently high densities and low temperatures to allow carbon and other substances to condense; these stars are also observed to eject material into the

---

* One must be cautious in relating measured vapor pressures with the tendency to freeze out on interstellar grains. Vapor pressures refer to the gas phase of a substance in equilibrium with its own liquid or solid, whereas in the interstellar case the gas phase would be in interaction with some other (unspecified) surface.

interstellar medium. Molecules may of course form under these conditions as well, in many-body collisions.

Detailed theories for the condensation of carbon into graphite in these stellar atmospheres (Hoyle and Wickramasinghe, 1962; Fix, 1969; Donn *et al.*, 1968) require temperatures $\lesssim 2000°K$ and stars with a large [C]/[O] ratio; this restricts the possible classes of stars to spectral types C and N. These types generally have $^{12}C/^{13}C$ ratios which are substantially smaller than the terrestrial value. It can be estimated (Hoyle and Wickramasinghe, 1962) that all the N stars observed could produce enough graphite particles to account for interstellar extinction, assuming that most of the carbon in these stars forms graphite (as indicated by calculation) and that most of the graphite is ejected from the stars. Although the formation process for graphite appears to be efficient, there is still considerable doubt as to whether most of the graphite can escape from the star in solid form. The portion which escapes can carry adsorbed molecules with it for eventual release to the interstellar medium.

Solid material other than graphite will form in the atmospheres of cool stars which are not carbon-rich. For example, the refractory silicates Al_2SiO_5 and Mg_2SiO_2 will likely predominate in cool ($T \lesssim 2000°K$) oxygen-rich stars, although significant amounts of $CaMgSi_2O_6$ and Fe are also expected (Gilman, 1969). For transition-type stars with [C]/[O] ≈ 1, the most abundant condensate is SiC. Whether these grains would typically become coated with ices or other surface-adhering molecules during or after ejection from the star is not known. It is to be expected, however, that stars of varying composition will produce grains of varying composition also. This in turn might be expected to result in different molecule formation (if grain surfaces are involved) as well as variations in the extinction law throughout the interstellar medium. The wavelength-dependence of the extinction is, in fact, found to be strikingly similar in all directions, which might indicate that if grains originate in stars

at all, those stars must have a very narrow range of properties.

Although molecules can form efficiently in the cool stellar atmospheres, there is much more doubt about their ability to escape from the star intact than is the case for grains. It would seem very difficult to shield the molecules from photo-dissociation by the stellar radiation or from thermal dissociation in passing through the hot coronas of some of these stars. Nonetheless, it is worthwhile to summarize the predicted relative abundances of molecules as calculated under LTE conditions for several types of stellar atmospheres. Some interesting patterns emerge, which may be compared with observations.

Tsuji (1964) has calculated the equilibrium composition of 44 molecules composed of H, C, N, and O for various abundance ratios of these elements, temperatures, and pressures appropriate to stellar atmospheres. In general, hydrocarbons larger than C_4H_3 are unimportant. Among nonorganic molecules, only H_2O, NH_3, and NH_2 are stable enough to form in noticeable amounts, but they do so over a wide range of conditions or types of stars. However, species like OH, NO, NH, H_2O_2, N_2O, and NO_2 are always negligible. The relative element abundances are much more critical in determining organic molecule production. For small C/O ratios (F to M stars), the dominant species are H_2O, NH_3, HCN, CO_2, CH_3, and CH_4, although H_2, N_2, and CO are quite abundant. For carbon stars with C/O $\simeq 5$ and N/C $\simeq 10$, polyatomics such as HCN, C_2N, C_3N, C_3, and a variety of small hydrocarbons are principally formed along with CO, which contains most of the oxygen; if these species successfully escape from the star, the radicals C_2N and C_3N will be hydrogenated, producing CH_3CN and HC_3N. These polyatomic species are produced in more abundance than diatomics (C_2, CH, CN) especially for temperatures below 2800°K. For carbon stars with N/C $\simeq 0.2$, more C_2 and C_3 are produced than C_2N and C_3N. Molecules containing other elements, such as SiO, SO, and CS are also predicted by other calculations

(Shimizu, 1973; Vardya, 1966) in various types of stars.

Thus many of the observed interstellar molecules could in principle be formed in stellar atmospheres (H_2O, NH_3, CO, CN, HCN, CH_3CN, HC_3N, SiO, CS), but others are conspicuously absent (H_2CO, CH_3OH, HCOOH, HNCO, NH_2CHO, etc.). Direct searches have failed to detect H_2CO in a variety of cool stars (see Table 9.2), yet it is one of the most widespread interstellar molecules. Among the predicted molecules that have been directly observed in stellar atmospheres are HCN and H_2O (at both microwave and IR wavelengths) and C_3 (at optical wavelengths). HCN and H_2O are predicted as among the most abundant species in carbon-rich and oxygen-rich stars, respectively.

Because of their short lifetimes against the interstellar UV field, and the absence of some of the most abundantly observed species, polyatomic interstellar molecules likely do not originate in cool stars in significant amounts. It is more difficult to dismiss a stellar origin for diatomic molecules. Not only are they more stable against UV radiation, but also they are in theory produced in great abundance not only in cool carbon stars but in M stars. CO is predicted to be more abundant than CH, CN, or H_2, in agreement with the optically observed interstellar species. In addition, many M stars (although not all) have a roughly terrestrial $^{12}C/^{13}C$ ratio, as is observed in both CH and CO among the "optical" interstellar species. Although the predictions of radiative association (Solomon and Klemperer, 1972) are in agreement with all observed aspects of interstellar diatomic molecules, including CH^+, one cannot rule out at least a small contribution to these interstellar species from M stars, which are far more numerous than all other types of stars.

f) Formation of Grains and Molecules in Pre-solar Nebulae

As an interstellar cloud collapses under its own gravity at low temperature, it even-tually takes on the form of a dense disk-shaped cloud centered on a denser core, out of which the proto-star forms. As the proto-star approaches the top of the Hyashi track the particle densities in the surrounding disk reach 10^{14} to 10^{15} cm^{-3}, and the temperatures reach $\sim 2000°K$. Thereafter the temperatures decline. Under these cooling conditions the interstellar gas is depleted in those elements (Ca, Ti, Fe, Al) that are the first to condense from a cosmic gas. Under equilibrium conditions, Fe condenses as a metal rather than as a compound. Thus all of the conditions needed for production of molecules are present, namely, high densities and temperatures and the presence of catalysts (Herbig, 1970). Laboratory experiments (Studier *et al.*, 1968; Anders, 1973) indicate, for example, that a large number of complex hydrocarbons and nitrogenous molecules will form rapidly from H_2, CO, and NH_3 in the presence of Fe catalysts at temperatures $\lesssim 1300°K$. After the molecules form in the pre-solar nebula and the proto-star reaches the Hyashi track, the dust and nebula are dispersed, largely by IR radiation from the proto-star, but also because of the excess angular momentum of the system. The molecules then disperse into interstellar space, or concentrate in large HI clouds that characteristically surround the sites of star formation.

A considerable quantity of information is available on the chemistry of pre-solar nebulae, in connection with attempts to explain the composition of the carbonaceous chondrites. These carbon-rich meteorites, as well as the inner solar system generally, are strongly depleted in carbon relative to the cosmic abundance scale. This carbon, along with other volatiles (H, O, N, noble gases) apparently were lost to interstellar space when the solar nebula dissipated. The strongest likelihood is that this material was lost in the form of polyatomic organic molecules, dust grains, and uncondensed gas. Taking our own solar system as typical, for which $\sim 3 \times 10^{-3} M \odot$ of volatiles escaped, calculations show that the dispelled mass from all pre-solar nebulae evolved through the lifetime of the galaxy can easily account for the presently estimated mass of interstellar dust (Herbig, 1970).

Anders (1973) has reviewed the molecular chemistry which likely occurred as the pre-solar nebula cooled from $\sim 2000°K$. One prominent process is the Fischer-Tropsch type (FTT) reaction:

$$n\, CO + \frac{n+m}{2} H_2 \overset{Fe}{\Rightarrow} C_n H_m + n\, H_2O$$

which makes use of the most prevalent un-condensed gases, CO and H_2, under these conditions to produce a wide variety of hydro-carbons. A comparison of the most efficiently produced hydrocarbons by this mechanism with those found in meteoric compounds shows a remarkable similarity, both for ali-phatic and for aromatic forms. Another form of FTT reaction includes NH_3 as a reactant and a wide variety of nitrogenous organic molecules as products. It is difficult to com-pare predictions of this latter reaction with the nitrogen molecules found in meteorites, but there is a suspicion that insufficient NH_3 may have been present to produce the observed meteoric nitrogen molecules under the temperatures and densities necessary to fit the hydrocarbons. Additional NH_3 may have been synthesized by shock waves in the pre-solar nebulae, to make up any shortage.

Having established that FTT reactions were dominant in the pre-solar nebulae, one can check how well they account for the interstellar molecules identified so far. Of the 21 identified molecules in Table 9.1, all but seven (HC_3N, NH_2CHO, CS, CH_2NH, SiO, H_2CS, and H_2S) are known to be produced by FTT reactions, and for these seven there is no available laboratory information. Thus there is qualitative agreement. To decide whether there is quantitative agreement, with respect to both positive identifications and "missing" species, better abundance determi-nations are needed for the interstellar mole-cules, and some allowance will have to be made for photochemical reactions during dissipation of the pre-solar nebulae. These

would affect mainly gaseous molecules, less so those protected on grain surfaces. Another point in favor of an FTT origin for inter-stellar molecules is that the abundances of these molecules relative to CO in the inter-stellar medium are at least 2 orders of magni-tude lower than yields in FTT synthesis (Anders, 1973). Unless this low abundance is ascribed entirely to their short photolytic lifetimes, which seems unlikely in the dense continuum source molecular clouds, even a small degree of star formation and CO pro-cessing by FTT reactions would be enough to account for the interstellar molecules.

FTT reactions are generally nonequi-librium processes which are usually quenched before they have gone to completion. In the pre-solar nebulae, most of the FTT synthesis occurred under nonequilibrium conditions, while the nebulae were cooling (rapidly) from ~ 2000 to $\sim 700°K$; in this range certain pro-cesses, such as conversion of the hydrocarbons to alkanes, are inhibited. Thus the composi-tion of the molecules as they are ejected from the nebulae depends in principle upon the temperature and density of the nebulae at that point in time, as well as on the way in which these parameters varied with time during the molecular synthesis. Unless most solar nebu-lae evolve very similarly, or unless the FTT mechanism is quite insensitive to differences in physical conditions among different nebu-lae, any similarity in FTT molecular abund-ances and those observed in interstellar space might be fortuitous.

As an alternative possibility to FTT reactions, molecules may have formed under LTE conditions in pre-solar nebulae. Very little work has been done to test this possi-bility adequately. Existing calculations have either not made use of the cosmic elemental abundances or not considered ranges of temperature and pressure probably applicable to the pre-solar nebular environment. Table 9.6 gives a comparison of predicted and ob-served molecular abundances under three sets of LTE conditions. LTE(1) refers to the work of Shimizu (1973), who calculates the equi-librium concentrations of diatomic and tri-

Table 9.6 Observed versus Predicted Molecular Abundances*

Molecule	Observed abundances relative to CO			Predicted abundances relative to CO		
	Sgr B2	W 51	Ori A	LTE(1)	LTE(2)	LTE(3)
OH	2×10^{-2}	1×10^{-4}	1×10^{-4}	3.3×10^{-2}	—	—
NH_3	4×10^{-3}	2.5×10^{-5}	7×10^{-4}	8.2×10^{-5}	5×10^{-4}	2.3×10^{-9}
H_2O	?	?	?	6.2×10^{-1}	1	1.4×10^{-10}
H_2CO	4×10^{-5}	2.5×10^{-6}	1×10^{-4}	—	7×10^{-7}	1.2×10^{-12}
CO	1	1	1	1	1	1
CN	—	1.5×10^{-5}	3×10^{-4}	1.1×10^{-5}	—	—
HCN	6×10^{-6}	1×10^{-5}	7×10^{-4}	4×10^{-5}	3×10^{-6}	1.1×10^{-1}
X-ogen	?	?	?	?	?	?
HC_3N	2×10^{-4}	—	1×10^{-4}	Very small	—	—
CH_3OH	6×10^{-5}	—	$\leq 7 \times 10^{-4}$	—	1.7×10^{-8}	4.9×10^{-20}
HCOOH	2×10^{-5} to 2×10^{-4}	—	—	—	1×10^{-7}	3.8×10^{-24}
CS	2×10^{-6}	$\leq 3 \times 10^{-5}$	$\leq 2 \times 10^{-4}$	6.4×10^{-6}	—	—
NH_2CHO	$\leq 4 \times 10^{-4}$	—	—	—	—	—
SiO	8×10^{-7}	—	—	—	—	—
OCS	$>1 \times 10^{-4}$	—	—	2.9×10^{-7}	1.2×10^{-4}	5.9×10^{-7}
CH_3CN	$>4 \times 10^{-6}$	—	—	—	—	—
HNCO	10^{-5}	—	—	—	—	—
HNC?	?	?	?	?	?	?
CH_3C_2H	2×10^{-5}	—	—	—	—	—
CH_3CHO	1×10^{-5}	—	—	—	—	—
H_2CS	Undetermined	—	—	—	1.4×10^{-8}	9.7×10^{-12}
H_2S	4×10^{-6}	2.5×10^{-6}	1.7×10^{-4}	—	1.4×10^{-2}	4.4×10^{-4}
CH_2NH	1.2×10^{-5}	—	—	—	—	—

* The dash (—) in columns 2 to 4 indicates that the corresponding molecule has not yet been observed; the same symbol in columns 5 to 7 indicates that the molecule was not included in the calculations.

atomic molecules composed of H, N, C, O, and S, using cosmic abundances for these elements. In attempting to account for the ejection of molecules from many proto-stars of various masses and evolutionary stages, Shimizu adopts "typical" proto-star conditions of 1500 to 3000°K and 10^{-2} to 10 atm pressure. Pre-solar nebulae are probably better described by pressures of 10^{-6} to 10^{-2} atm. LTE(1) fits the observed OH, NH_3, CN, and HCN abundances quite well, and possibly the H_2O also, but appears to underestimate CS and OCS. It is interesting that a good fit is obtained for OCS when the temperature is taken as 1500°K. Although these comparisons seem encouraging, HCO and SO are overestimated in LTE(1) by a factor of at least 100 with respect to the observed upper limits. And there is extreme difficulty in forming the more complex molecules in

adequate amounts. HC_3N could form only in observed quantities, for example, if carbon were overabundant by a factor of at least 3 relative to the cosmic scale; the recent observations of HC_3N in appreciable quantities outside the galactic center region (Morris *et al.*, 1972b) make this possibility less likely.

LTE(2) and LTE(3) are calculations of Eck *et al.* (1966). In LTE(2) the temperature is 1000°K, pressure is 1 atm, and N:C:H:O = 1:1:7:2; that is, roughly cosmic except that hydrogen is underabundant by a factor of ~400. Nevertheless, where direct comparison is possible, the agreement with the observed abundances seems qualitatively about as good as for LTE(1). LTE(3) is for a temperature of 1000°K, pressure 10^{-6} atm, and N:C:H:O = 2:2:47:1. Here, O/C is four times less than cosmic, although H/C is now only 60 times less than cosmic. Clearly LTE(3) drastically underestimates the abundances of nearly all of the molecules except HCN. Additional calculations by Eck *et al.* (1966) show that this underestimation occurs for a wide range of element abundance ratios, as long as the pressure is low (10^{-6} atm). On the other hand, the predictions of LTE(1) and LTE(2), both at pressures of ~ 1 atm, are much more similar, despite large differences in both temperature and element abundance ratios. If formation of interstellar molecules occurs under conditions approaching LTE, a pressure of 10^{-2} to 1 atm is therefore indicated. This might shift the locale of molecule formation in pre-solar nebulae closer to the proto-stellar core itself rather than in the outer disk.

It is probably reasonable to conclude that studies of both FTT reactions and LTE processes in pre-solar nebulae environments are as yet too preliminary to provide an unambiguous test of this means of interstellar molecule formation. Quantitative laboratory studies of FTT reactions under assumed nebular conditions and directed specifically at the observed interstellar molecules are needed. Similarly, more complex LTE calculations are needed that include the more complex molecules.

g) *Possible Formation Schemes for Interstellar Molecules versus Observations*

We may summarize the various possible formation schemes for interstellar molecules as follows:

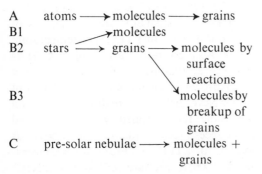

A   atoms ⟶ molecules ⟶ grains
B1
B2   stars ⟶ grains ⟶ molecules by surface reactions
B3   molecules by breakup of grains
C   pre-solar nebulae ⟶ molecules + grains

Most of the relevant observational evidence is useful in arguing against some of these schemes, rather than in favor of them.

All of the schemes fall into either of two categories: (1) "steady state," where the interstellar molecules were formed where they are now observed, and the abundances are determined by a balance between formation and destruction rates; (2) molecules formed in special regions where formation rates were temporarily very high, and were then dispersed throughout the interstellar medium. This avoids problems of formation but encounters problems of destruction rates during the dispersal phase.

Category B1 appears very unlikely. Since the ejected molecules are not sequestered on grains, they could not disperse far from the stars ($\lesssim 0.01$ pc at an ejection velocity of 10 km/sec) before being photo-dissociated. Yet molecules are not observed to occur preferentially near these cool stars, which characteristically do not possess surrounding nebulosity. In fact, searches for several molecules (other than OH and H_2O) in the directions of cool stars have been negative. The apparent terrestrial isotopic ratios observed in interstellar molecules also tend to rule out B1. All of the stellar types considered suitable for forming molecules in their atmospheres and ejecting them have definitely nonterrestrial $^{12}C/^{13}C$ ratios. The ratio for N stars ranges from 2 to 20, and for R stars it is ~ 5. Some less direct arguments also disfavor B1: Both calculations and observations suggest that the molecules C_2 and C_3 are formed in cool stellar atmospheres in greater abundance than other polyatomic molecules. Yet C_2 and C_3 have never been observed in the general interstellar medium, although their optical spectra are well known. There is also a difficulty in ejecting molecules intact, even from cool stars. Tsuji (1964) shows that mass loss from cool stars can produce the observed abundance of interstellar molecules (at optical wavelengths), assuming that all molecules escape these stars intact. However, outer convection zones of considerable extent are apparently required to produce the mass loss in a way similar to the solar wind; such convection zones probably also produce stellar coronae, the temperatures of which would rapidly destroy molecules passing through them. M giants and supergiants also lose mass but do not seem to have coronae; however, McNally (1968) has shown that all of the mass lost from all of these stars would be insufficient to yield the observed quantity of interstellar molecules, even if it were in the form of free molecules.

Several arguments may also be made against B3. The terrestrial value of $^{12}C/^{13}C$ for interstellar molecules rules out the cool carbon stars (types N and R) as sources for the grains, although their physical conditions are most conducive to grain formation. And grains from cool oxygen-rich stars, if such are produced, would almost certainly not contain carbon (Gilman, 1969) and therefore could not explain any of the organic molecules. There are other difficulties with the breakup of any such grains to form molecules. Breakup by hot stars is unlikely because molecules are not generally observed near hot stars. Furthermore, diatomic species would be far more abundant than polyatomics in the harsh UV field of any stars energetic enough to disrupt grains. The optically observed molecules CH, CH^+, and CN are also unlikely to be explained by B3 because the observed $^{12}C/^{13}C$ ratio appears to be terrestrial. Also, only a drastic (and unlikely) variation in grain composition from region to region could explain the large variations in [CN]/[CH] from region to region. Grains could be broken up by shocks or by cloud-cloud collisions; however, not only is there no observational evidence for such phenomena, but also the energy required for grain disruption would be very rapidly dissipated by the cooling effect of the molecules that are formed, leaving only a very short time for their formation. Finally, aside from the carbon-isotope problem, it would be impossible to explain the absence of such relatively stable molecules as NO, NH, etc., by breakup of grains.

Category B2 is not so easily refuted. The observed terrestrial ratio of $^{12}C/^{13}C$ for interstellar molecules is not a problem as long as the carbon atoms in the observed molecules come from the interstellar gas and not

from the grains. However, this appears to be unlikely if the grains are graphite. For example, in the Stecher-Williams (1966) theory the carbon in CH comes from the interstellar gas but the carbon in CO and CN comes from the grains. One might also expect to find some preference for molecules near some of the 100 or so N-type stars observed within 1 kpc of the Sun, but such is not the case. The complex organic molecules are instead concentrated in the neutral clouds surrounding the brightest HII regions, *i.e.*, are in the vicinity of the earliest and hottest stars—those least capable of producing grains of any sort. Finally, the observed (terrestrial) isotopic ratios appear quite uniform throughout the interstellar medium, implying that the interstellar chemistry has remained unchanged over the lifetime of the solar system ($\sim 5 \times 10^9$ yr). If either the gas or the grains which form molecules had been significantly processed by stars with highly variable and nonterrestrial isotopic ratios, one would expect these vagaries in the isotopic ratio to show up in the molecules themselves.

Category A refers to the formation of both molecules and grains by spontaneous accretion from the free atomic interstellar gas. The first step in this process, that of formation of diatomic molecules by radiative association, is efficient enough to produce the observed quantity of diatomic molecules [see Section 9.6.2(a)]. And *n*-atom free radicals can radiatively associate to form 2*n*-atom molecules with high association probability per collision if $n \gtrsim 5$ (Williams, 1971). If 10-atom particles exist, there should be no difficulty in forming by accretion larger solid particles, with *n* as large as 500 or more, as proposed by Platt (1956) to explain the interstellar extinction law. The difficulty lies in forming particles with $3 \lesssim n \lesssim 10$. Direct association cannot produce these particles in the observed abundances of the polyatomic molecules. However, it is not clear that over the lifetime of the galaxy the overall process cannot have produced a significant fraction of the interstellar grains. Sufficient grain formation by direct accretion has been found possible by ter Haar (1943), using effective sticking probabilities that, compared with those of Williams (1971), were overestimated by many orders of magnitude for $3 \lesssim n \lesssim 10$, but that were underestimated by large factors for $n \gtrsim 10$. Once n exceeds a value of roughly 10, the UV destruction rate of the particles is greatly reduced, explaining why after sufficient time there would be many more grains than small polyatomic molecules formed by direct accretion. The number of large particles tends to accumulate with time, while the number of small polyatomic molecules does not. Of course small polyatomics in turn can be produced at greatly enhanced rates on the surfaces of grains once these have accumulated in sufficient number.* As yet, the mechanism of direct accretion is not on firm theoretical footing, but it would appear to be a possible explanation for the presence of dust grains in dark dust clouds well removed from any evident sites of star formation.

Ion-molecule reactions (see Addendum to this chapter) offer a more likely explanation for formation of molecules with $2 \leq n \leq 10$ directly from atoms, provided that H_2 is already present (H_2 can form only on grains). Ion-molecule reaction rates exceed rates of other formation mechanisms and compete with loss rates of molecules by adsorption on grains. The gas is therefore cycled continually through the molecular phase and into the accretion of grains. The equilibrium abundance of molecules is predicted to be as high as that observed in general; moreover, the pattern of observed molecules appears to be produced by these reactions (Table 9.5).

Category C appears at present to be consistent with observations of interstellar molecules at microwave frequencies. Isotopic ratios will not have been perturbed by nuclear reactions in the central star, because the

---

* It is interesting that the space abundance of diatomics relative to hydrogen is $\sim 10^{-8}$ to 10^{-5}, and of grains is $\sim 10^{-12}$. The abundance of grains appears abnormally large, given the typical number of atoms that they contain. It should also be noted that in general the abundance of polyatomic molecules decreases as their complexity increases.

molecular formation occurs long before the stars become self-luminous. The main radiation field is IR, which aids the ejection of molecules and grains, rather than destructive UV. And all of the most complex observed molecules are found only in dense neutral clouds surrounding continuum sources, which are sites of recent star formation. Other indirect evidence favors process C for the continuum source molecular clouds. The observed molecules in comets (OH, OH^+, CO, CO^+, NH_2, NH, CN, CH, CH^+, C_2, and C_3) look very much like the dissociation fragments of the identified interstellar molecules. Comets themselves are apparently remnants of the pre-solar nebulae, bearing possible generic relations to some Apollo asteroids (Opik, 1964) and the Type I carbonaceous chondrites.

Although process C appears to explain molecules observed in clouds associated with continuum sources, the origin of molecules in dark dust clouds is still an interesting question. There appear to be three possibilities for the origin of molecules in dark dust clouds. (1) Both grains and molecules are ejected from pre-stellar nebulae, but only the grains can traverse large regions of interstellar space without disruption. It is possible that these grains come to rest preferentially in interstellar clouds that have already started to form, because the larger densities in these clouds tend to slow or halt the grains as they pass through. Diatomic molecules would presumably already be present in such clouds, and the arrival of grains would precipitate formation of more complex molecules by surface reactions, thus cooling the cloud and increasing its density further. (2) Alternatively, direct accretion, probably via ion-molecule reactions (process A) may account for most of the molecules and grains in dark dust clouds. (3) Finally, dark dust clouds may characteristically contain sites of star formation (as yet not revealed by observation) which would supply molecules and grains by process C.

Because the temperature of dark dust clouds is, and must always have been, much lower than the temperatures in pre-solar nebulae during the molecular formation stage, it would be expected that molecules formed under possibility (2) (*i.e.*, at low temperatures) would exhibit a chemistry different from that observed in the continuum source molecular clouds unless the latter are also produced by ion-molecule reactions. The chemistry would not be expected to differ under possibility (3). Possibility (1) is unclear on this point; assuming the grains that reach the dust clouds have the same composition as those in the continuum source clouds, the chemistry in the two types of sources should be similar if the pattern of molecules in the continuum source clouds is controlled, not by the primary process (formation at high temperatures), but by the rapid interaction with grains at the current lower temperatures. (This possibility assumes the presence of suitable UV-activation.) Whether or not grain chemistry has imposed its characteristics on the primary formation chemistry in the continuum source clouds probably depends on the nature of the grains. Some progress in understanding should be possible through laboratory experiment with different catalysts. Meanwhile one of the most important goals for the radio astronomer is to determine whether there are differences in relative molecular abundances in dark dust clouds and in continuum cloud sources. It is equally important to improve the reliability of abundance estimates, in order to determine what range of chemical properties may be attributed to a given class of molecular region, such as those associated with continuum sources.

9.7 Conclusions

In this review we have emphasized, on the observational side, the types of regions where interstellar molecules are found and the physical conditions that may be derived from those species that are apparently normally excited. On the interpretive side, possible chemical processes are considered that might be expected in the types of regions harboring these molecules and which might produce the particular patterns of molecules observed. In emphasizing these aspects, we have necessarily omitted some areas of current research on interstellar molecules. The most obvious of these is the anomalous excitation of interstellar molecules, as most clearly revealed by OH, H_2O, and H_2CO. This choice was made on the grounds that, while of intrinsic interest, the understanding of these processes probably reveals little of a general nature about processes in the interstellar medium (as distinct from those in proto-stars, for example). The anomalous excitation of OH and H_2O has been reviewed elsewhere (Turner, 1970a, 1970b). We have also omitted listing all of the observed transitions of the known interstellar molecules and the details of their characteristics in individual sources. The two recent reviews of interstellar molecules by Rank *et al.* (1971) and by Snyder (1972) cover these aspects.

Although they cover a wide range of physical conditions, all molecular regions associated with the interstellar medium may be characterized as cool and dense by usual interstellar standards. In these clouds most of the gas appears to be in the form of simple, relatively stable molecules, with atoms and free radicals much less abundant. Through a study of the temperatures, densities, isotope ratios, and associations with various types of galactic objects that characterize molecular sources, a preliminary pattern is starting to emerge. Molecules appear to be associated with prestellar objects, starting with the most tenuous of interstellar clouds and ending with the proto-stellar stages immediately before stars become self-luminous. (A few molecules are also found in the atmospheres of cool stars.) The unexpectedly high abundance and complexity of interstellar molecules in regions of low temperature has prompted a revision of classical ideas on formation and destruction of interstellar molecules. Largely through the recent observational clarification of isotope ratios and of the physical nature of regions with which these molecules are associated, some progress has been made in choosing among the possible formation processes—at least in terms of the fundamental processes involved. The details of these processes, however, are little understood and require clarification in the laboratory. In addition to this requirement is the observational need for increasingly detailed information on the spatial relationships of the various molecules with one another and with other interstellar phenomena, such as extinction, infrared sources, small continuum sources, etc. In particular, a better idea of the degree of uniformity of cosmic element abundances and of the molecular formation processes among different molecular clouds must be gained in order to interpret the observations in light of improved laboratory work. The stage now appears to have been reached in which progress is retarded more by lack of a detailed picture of the known interstellar molecular clouds than by a limitation in the number of interstellar molecules that are known.

Addendum

Since June 1972, when this article was completed, some new developments have occurred which are important enough to summarize here.

The first is the investigation of gas phase (positive) ion-molecule reactions as a formation mechanism for interstellar molecules (Herbst and Klemperer, 1973; Watson, 1973a,b). In these processes, the primary required ions arise from the ionization of H_2 by cosmic rays (H_2 is still regarded as forming solely on grain surfaces). Resultant ions H^+ and H_2^+ react rapidly with other neutrals to build up polyatomic ions which are eventually neutralized by recombination with electrons. Although the degree of ionization is low ($n_e/n_{H_2} \ll 10^{-4}$), these formation rates are shown to dominant those of other processes in dense cold clouds (in warmer clouds, surface reaction rates would be competitive) and to be comparable with rates of molecular destruction by adsorption onto cold grain surfaces. Quantitative predictions of relative molecular abundances by ion-molecule formation are at present highly uncertain, owing to lack of information on several reaction rates and to the possibility of omission of a few important reactions. However, qualitative predictions of the types of molecules preferentially formed are in close agreement with the observational situation described in Table 9.5 (*cf.* Turner, 1974). That is, molecules with unsaturated carbon bonding should be formed more readily than those with saturated bonds. This comes about because the final reaction producing each molecule is electron-recombination, a process that is much more likely to be dissociative for single (saturated) carbon bonds than for multiple (unsaturated) carbon bonds.

Another important development is the theoretical investigation of the types of surface reactions which may occur under conditions of very low temperature (Watson and Salpeter, 1972a,b). The sticking probability for heavy (C, N, O) atoms is found to be close to unity (but is uncertain for C^+ ions). Whether or not chemisorption or quantum-mechanical diffusion is possible, each heavy atom remains on a grain surface long enough for another particle to stick and react with it. Thus a particle that sticks will form some molecule with high efficiency, mainly OH, H_2O, CH to CH_4, and NH to NH_3. Some of the unreactive products (CH_4, NH_3, H_2O, H_2CO, etc.) remain on the surface and form more complex molecules (in a reaction with an excited radical or after activation by absorption of UV photons). Two important results of this study indicate that surface reactions are probably inadequate to form appreciable amounts of interstellar molecules at temperatures under ~ 20 to $50°K$. First, it is shown that thermal evaporation, as well as several non-thermal evaporation processes (involving IR radiation and cosmic rays) are insufficient to eject the molecules that remain on the surface after reactions. Only UV photons (1000–2000Å) are adequate, and only if the quantity $\xi = \frac{n}{100} \cdot e^{2.5 \, \tau_v}$ is $\lesssim 10^4$. Here n is the gas density and τ_v the visual opacity of the cloud as seen from the center. Thus, for $\xi > 10^4$ molecules will be depleted by adsorption onto grain surfaces, unless the formation by some other mechanism, e.g., ion-molecule reactions, whose rate is higher than that of surface reaction at low temperature. Observationally, ξ easily exceeds 10^4 in both dark dust clouds and "black" clouds which exist near HII regions unless these clouds have unobservable sources of UV imbedded in them. (However, in the "black" clouds the temperature might be high enough to allow appreciable thermal evaporation from grain surfaces.) The second important result, although somewhat less certain, is that surface catalysis at low temperatures probably forms molecules more efficiently with saturated than with unsaturated carbon bonding. This contradicts the observed pattern described in Table 9.5.

A third important development is the discovery of interstellar deuterium, in the form of DCN (Jefferts *et al.*, 1973) and HD (Spitzer *et al.*, 1973). The apparent abundance ratios are [DCN]/[HCN] $= 6 \times 10^{-3}$ (in Orion A), and [HD]/[H$_2$] $= 2.1 \times 10^{-3}$ to 2.1×10^{-2} for 9 stars observed in the rocket UV. Although these ratios are much larger than the terrestrial ratio [D]/[H] $= 1.48 \times 10^{-4}$, it has been shown that formation on grain surfaces (as probably applies to HD and H$_2$) would favor production of the deuterated species (Solomon and Woolf, 1973) and also that formation by ion-molecule reactions (DCN, HCN) would favor the deuterated species (Watson, 1973c). In some cases, such as NH$_3$ and H$_2$CO, deuteration would not be favored; this conclusion seems borne out by failures to detect NH$_2$D and HDCO despite sensitive searches.

References

Anders, E. 1973. *Molecules in the Galactic Environment.* M. A. Gordon and L. E. Snyder, eds. New York: Wiley Interscience.

Becklin, E. E., and G. Neugebauer. 1967. *Astrophys. J.* 147:799.

Bok, B. J. 1956. *Astron. J.* 61:309.

Bortolot, V. J., and P. Thaddeus. 1969. *Astrophys. J.* 155:L17.

Brecher, A., and G. Arrhenius. 1971. *Nature-Phys. Sci.* 230:107.

Breuer, H. D. 1973. *Molecules in the Galactic Environment.* M. A. Gordon and L. E. Snyder, eds. New York: John Wiley.

————, H. Moesta, and N. Trappen. 1970a. *Naturwissenshaften* 9:452.

————, H. Moesta, and N. Trappen. 1970b. *Naturwissenschaften* 9:453.

Carruthers, G. R. 1970. *Astrophys. J.* 161:L81.

Cheung, A. C., D. M. Rank, C. H. Townes, S. H. Knowles, and W. T. Sullivan. 1969. *Astrophys. J.* 157:L13.

Donn, B., N. C. Wickramasinghe, J. P. Hudson, and T. P. Stecher. 1968. *Astrophys. J.* 153:451.

Eck, R. V., E. R. Lippincott, M. O. Dayhoff, and Y. T. Pratt. 1966. *Science* 153:628.

Fix, J. D. 1969. *Monthly Notices Roy. Astron. Soc.* 146:37.

Fomalont, E. B., and L. Weliachew. 1973. *Astrophys. J.* 181:781.

Gardner, F. F., R. X. McGee, and M. W. Sinclair. 1970. *Astrophys. Lett.* 5:67.

————, J. C. Ribes, and B. F. C. Cooper. 1971. *Astrophys. Lett.* 9:181.

Gilman, R. C. 1969. *Astrophys. J.* 155:L185.

Gordy, W., W. V. Smith, and R. F. Trambarulo. 1953. *Microwave Spectroscopy.* New York: John Wiley.

Goss, W. M. 1968. *Astrophys. J. Suppl.* 15:131.

————, R. N. Manchester, and B. J. Robinson. 1970. *Australian J. Phys.* 23:559.

Gottlieb, C. A. 1973. *Molecules in the Galactic Environment.* M. A. Gordon and L. E. Snyder, eds. New York: John Wiley

————, H. E. Radford, J. A. Ball, and B. Zuckerman. 1971. Private communication.

Gould, R. J., and E. E. Salpeter. 1963. *Astrophys. J.* 138:393.

Grasdalen, G., C. E. Heiles, and B. E. Turner. 1973. Unpublished.

Gwinn, W. D., B. E. Turner, W. M. Goss, and G. Blackman. 1973. *Astrophys. J.* 179:789.

Hardebeck, E. G., and W. J. Wilson. 1971. *Astrophys. J.* 169:L123.

Heiles, C. E. 1971. *Ann. Rev. Astron. Astrophys.* 9:293.

————, and B. E. Turner. 1971. *Astrophys. Lett.* 8:89.

Herbig, G. H. 1960. *Astron. J.* 65:491.

———. 1968. *Z. Astrophys.* 68:243.

———. 1970. *Mem. Soc. Roy. Sci. Liegé* 19:13.

Herbst, E. 1972. Doctoral dissertation. Harvard University.

———, and W. Klemperer. 1973. *Astrophys. J.* 185:505.

Herzberg, G. 1966. *Electronic Spectra and Electronic Structure of Polyatomic Molecules.* Princeton, N.J.: Van Nostrand.

Hills, R. E. 1969. Unpublished manuscript.

Hollenbach, D. J., and E. E. Salpeter. 1971. *Astrophys. J.* 163:155.

———, M. W. Werner, and E. E. Salpeter. 1971. *Astrophys. J.* 163:165.

Hoyle, F., and N. C. Wickramasinghe. 1962. *Monthly Notices Roy. Astron. Soc.* 124:417.

———, and N. C. Wickramasinghe. 1968. *Nature* 218:1126.

Jefferts, K. B., A. A. Penzias, and R. W. Wilson. 1973. *Astrophys. J.* 179:L57.

Julienne, P. S., M. Krauss, and B. B. Donn. 1971. *Astrophys. J.* 170:65.

———, and M. Krauss. 1973. *Molecules in the Galactic Environment.* M. A. Gordon and L. E. Snyder. eds. New York: John Wiley.

Khare, B. N., and C. Sagan. 1971. Cornell University preprint, CRSR 454.

Knapp, G. 1972. Doctoral dissertation. University of Maryland.

Knowles, S. H., C. H. Mayer, A. C. Cheung, D. M. Rank, and C. H. Townes. 1969. *Science* 163:1055.

Kutner, M. 1973. *Molecules in the Galactic Environment.* M. A. Gordon and L. E. Snyder, eds. New York: John Wiley.

Larimer, J. W. 1967. *Geochim. Cosmochim. Acta* 31:1215.

Litvak, M. M. 1968. In *Interstellar Ionized Hydrogen.* Y. Terzian, ed. New York: W. A. Benjamin, p. 713.

———. 1969. *Astrophys. J.* 156:471.

———. 1972. Lecture notes. 1971. Scottish Universities' Summer School in Physics. T. R. Carson and M. J. Roberts, eds. New York: Academic Press.

Manchester, R. N., B. J. Robinson, and W. M. Goss. 1970. *Australian J. Phys.* 23:751.

McNally, D. 1962. *Monthly Notices Roy. Astron. Soc.* 124:155.

———. 1968. *Adv. Astron. Astrophys.* 6:173.

Meeks, M. L., J. C. Carter, A. H. Barrett, P. R.

Schwartz, J. W. Waters, and W. E. Brown. 1969. *Science* 165:180.

Morris, M., B. Zuckerman, P. Palmer, and B. E. Turner. 1971. *Astrophys. J.* 170:L109.

———, B. Zuckerman, P. Palmer, and B. E. Turner. 1973. *Astrophys. J.* 186:501.

———, P. Palmer, B. E. Turner, and B. Zuckerman. 1972. *Proc. I.A.U. Symp. 52.*

Opik, E. J. 1964. *Adv. Astron. Astrophys.* 2:219.

Palmer, P., B. Zuckerman, and D. Buhl. 1974. In preparation.

Penzias, A. A., K. B. Jefferts, and R. W. Wilson. 1971a. *Astrophys. J.* 165:229.

———, P. M. Solomon, R. W. Wilson, and K. B. Jefferts. 1971b. *Astrophys. J.* 168:L53.

———, P. M. Solomon, K. B. Jefferts, and R. W. Wilson. 1972. *Astrophys. J.* 174:L43.

Platt, J. R. 1956. *Astrophys. J.* 123:486.

Rank, D. M., C. H. Townes, and W. J. Welch. 1971. *Science* 174:1083.

Robinson, B. J., W. M. Goss, and R. N. Manchester. 1970. *Australian J. Phys.* 23:363.

———, J. L. Casswell, and W. M. Goss. 1971. *Astrophys. Lett.* 9:5.

Rubin, R. H., G. W. Swenson, R. C. Benson, H. L. Tigelaar, and W. H. Flygare. 1971. *Astrophys. J.* 169:L39.

Schwartz, P. R., and A. H. Barrett. 1970. *Proceedings of the Conference on Late-Type Stars, Kitt Peak National Observatory.* October 27–28.

Scoville, N. Z., P. M. Solomon, and P. Thaddeus. 1972. *Astrophys. J.* 172:335.

Sharpless, S. 1959. *Astrophys. J. Suppl.* 4:257.

Shimizu, M. 1973. *Prog. Theor. Phys.* 49:153.

Smith, A. M., and T. P. Stecher. 1971. *Astrophys. J.* 164:L43.

Snyder, L. E. 1972. International Series on Physical Chemistry, Vol. 3, Spectroscopy. London: Butterworth.

———, and D. Buhl. 1972. *Astrophys. J.* 177:619.

Solomon, P. M. 1973. *Molecules in the Galactic Environment.* M. A. Gordon and L. E. Snyder, eds. New York: John Wiley.

———, and N. C. Wickramasinghe. 1969. *Astrophys. J.* 158:449.

———, and P. Thaddeus. 1970. Unpublished manuscript.

———, K. B. Jefferts, A. A. Penzias, and R. W. Wilson. 1971. *Astrophys. J.* 168:L107.

———, and W. Klemperer. 1972. *Astrophys. J.* 178:389.

————, and N. J. Woolf. 1973. *Astrophys. J.* 180:L89.

Spitzer, L., J. F. Drake, E. B. Jenkins, D. C. Morton, J. B. Rogerson, and D. G. York. 1973. *Astrophys. J.* 181:L116.

————, J. F. Drake, E. B. Jenkins, D. C. Morton, J. B. Rogerson, and D. G. York. 1973. *Astrophys. J.* 181:L116.

Stecher, T. P., and D. A. Williams. 1966. *Astrophys. J.* 146:88.

Stief, L. J. 1973. *Molecules in the Galactic Environment.* M. A. Gordon and L. E. Snyder, eds. New York: John Wiley.

————, B. Donn, S. Glicker, E. P. Gentieu, and J. E. Mentall. 1972. *Astrophys. J.* 171:21.

Studier, M. H., R. Hayatsu, and E. Anders. 1968. *Geochim. Cosmochim. Acta* 32:151.

Sullivan, W. T., III. 1971. *Astrophys. J.* 166:321.

ter Haar, D. 1943. *Bull. Astron. Inst. Neth.* 10:1.

Terzian, Y. 1970. *Astron. J.* 75:1155.

Thaddeus, P. 1972. *Astrophys. J.* 173:317.

Townes, C. H., and A. C. Cheung. 1969. *Astrophys. J.* 157:L103.

Tsuji, T. 1964. *Ann. Tokyo Astron. Obs.* (2nd ser.) 9:1.

Turner, B. E. 1969. *Astron. J.* 74:985.

————. 1970a. *J. Roy. Astron. Soc. Can.* 64:221 and 64:282.

————. 1970b. *Publ. Astron. Soc. Pacific* 82:996.

————. 1970c. *Astrophys. Lett.* 6:99.

————. 1972. *Astrophys. J.* 171:503.

————. 1974. *J. Roy. Astron. Soc. Can.* In press.

————, and R. H. Rubin. 1971. *Astrophys. J.* 170:L113.

————, D. Buhl, E. C. Churchwell, P. G. Mezger, and L. E. Snyder. 1970. *Astron. Astrophys.* 4:165.

————, and C. E. Heiles. 1974. In preparation.

————, M. A. Gordon, and G. T. Wrixon. 1972. *Astrophys. J.* 177:609.

————, B. Zuckerman, P. Palmer, and M. Morris. 1973. *Astrophys. J.* 186:123.

Vardya, M. S. 1966. *Trieste Colloquium on Late-Type Stars.* M. Hack, ed. Dordrecht: Reidel. p. 242.

Watson, W. D. 1973a. *Astrophys. J.* 182:L73.

————. 1973b. *Astrophys. J.* 183:L17.

————. 1973c. *Astrophys. J.* 181:L129.

————, and E. E. Salpeter. 1972a. *Astrophys. J.* 174:321.

————, and E. E. Salpeter. 1972b. *Astrophys. J.* 175:659.

Webster, W. J., and W. J. Altenhoff. 1970a. *Astron. J.* 75:896.

————, and W. J. Altenhoff, 1970b. *Astrophys. Lett.* 5:233.

Wickramasinghe, N. C., and K. Nandy. 1970. *Nature* 227:51.

Williams, D. A. 1968. *Astrophys. J.* 151:935.

————. 1971. *Astrophys. Lett.* 10:17.

Wilson, R. W., P. M. Solomon, A. A. Penzias, and K. B. Jefferts. 1971. *Astrophys. J.* 169:L35.

————, A. A. Penzias, K. B. Jefferts, P. Thaddeus, and M. L. Kutner. 1972. *Astrophys. J.* 176:L77.

Wilson, W. J., A. H. Barrett, and J. M. Moran. 1970. *Astrophys. J.* 160:545.

————, and A. H. Barrett. 1970a. *Proceedings of the Conference on Late-Type Stars, Kitt Peak National Observatory,* October 27–28.

————, and A. H. Barrett. 1970b. *Astrophys. Lett.* 6:231.

Zuckerman, B. 1973. *Molecules in the Galactic Environment.* M. A. Gordon and L. E. Snyder, eds. New York: John Wiley.

————, D. Buhl, P. Palmer, and L. E. Snyder. 1970. *Astrophys. J.* 160:485.

————, J. A. Ball, and C. A. Gottlieb. 1971a. *Astrophys. J.* 163:L41.

————, M. Morris, B. E. Turner, and P. Palmer. 1971b. *Astrophys. J.* 169:L105.

————, B. E. Turner, D. R. Johnson, P. Palmer, and M. Morris. 1972. *Astrophys. J.* 177:601.

CHAPTER 10

INTERFEROMETRY AND APERTURE SYNTHESIS

Edward B. Fomalont and Melvyn C. H. Wright

10.1 Basic Concepts

10.1.1 The Need for High Resolution in a Telescope

The resolution of a telescope has a diffraction limit of $\sim \lambda/D$, where λ is the observing wavelength and D is the aperture diameter. The resolution of large optical telescopes is limited to about a half arc second by atmospheric fluctuations, although the diffraction limit is much smaller. However, for radio frequencies a diffraction limit of about 1 arc minute at the highest radio frequencies is currently reached using filled-aperture telescopes. Further significant increase of the resolution of filled-aperture telescopes is not likely.

Many sources of radio radiation are confined to small angular extents, and a resolution much higher is needed to help understand the physical processes in radio sources. The majority of extra-galactic radio sources are smaller than 1 arc minute and some radio components have not yet been resolved ($<3 \times 10^{-4}$ arc second) using intercontinental baselines. Radio emission in our galaxy, though resolved with filled-aperture telescopes, nevertheless shows fine-scale structure in the second of arc range. The study of spectral-line radiation associated with molecules in space also requires high-resolution studies. Some maser emission regions are extremely small, and some of the larger clouds of molecular emission show small spatial clumps in their associated emission and absorption profiles. The mapping of hydrogen using the 21-cm transition frequency is limited to the very nearest galaxies with a filled-aperture telescope.

The sensitivity for observations of continuum radiation is often limited, not by receiver noise, but by confusion. There is little usefulness in increasing the sensitivity beyond the point where more than one source is likely to be within the antenna beam. The density of sources is such that to reach the confusion limit of a large antenna with the best receivers requires a few minutes or less of integration time. Better sensitivity can be fully utilized only in connection with increased resolution.

The need for high resolution was recognized with the first radio interferometers used in the late 1940's (Ryle and Vonberg, 1946; McCready *et al.*, 1947). The work of Stanier (1950) showed that an interferometer could be used to measure the Fourier components of a brightness distribution, and developments by Ryle and co-workers in England and by Christiansen and Mills and co-workers in Australia (Ryle, 1952; Christiansen, 1953; Mills and Little, 1953) extended interferometry to multi-elements and movable elements. The detailed principle of aperture synthesis was formulated by Ryle and Hewish (1960).

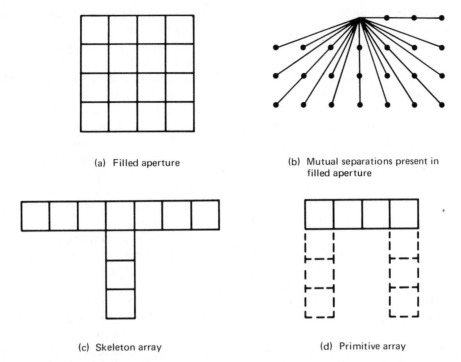

(a) Filled aperture

(b) Mutual separations present in filled aperture

(c) Skeleton array

(d) Primitive array

Figure 10.1 Principle of aperture synthesis. (a) Filled aperture consisting of 16 elemental areas. (b) Vectors indicating the 24 separations present in the filled aperture. (c) A "Tee" skeleton array consisting of 10 elemental areas and spanning the same mutual separations as the filled aperture. (d) A primitive array consisting of four stationary elements and six locations for one or more additional moveable elements.

10.1.2 Aperture Synthesis Technique

There are many methods of obtaining high resolution, and all involve the placing of receiving elements, suitably connected, over a large area. The speed at which a large aperture is synthesized depends upon the arrangement, the number, and the mobility of the elements. In the study of radio sources whose emission is not time-variable, it is not necessary that the whole of the telescope aperture be present at the same time. Following Ryle and Hewish (1960), the principle behind aperture synthesis may be understood by considering the operation of a conventional, filled-aperture telescope, which may be regarded as composed of N elemental areas as in Figure 10.1(a). For convenience a square aperture with $N = 16$ is taken. The signal in the nth area due to a source of emission is

$$I_n \cos (wt + \phi_n) \qquad (10.1)$$

where I_n is the amplitude of the signal and ϕ_n is the relative phase of the radiation. The relative phase over the aperture depends on the source direction. If the signals from the elements are added together vectorially (as, for example, by the feed at the focus of a parabolic reflector) and time-averaged, the power output is

$$P \propto \frac{1}{2} \sum_{j=1}^{N} \sum_{k=1}^{N} I_j I_k \cos(\phi_j - \phi_k)$$

$$= \frac{1}{2} \sum_{j=1}^{N} I_j^2 + \sum_{j=1}^{N-1} \sum_{k=j+1}^{N} I_j I_k \cos(\phi_j - \phi_k) \qquad (10.2)$$

The first term is proportional to the sum of the powers received by the elementary areas. The resolving power, which is related to how P changes with the direction of the source (and hence with ϕ_n), derives from the cross-product terms. Each individual term can equally well be measured with just two elementary areas in positions j and k. *All the terms can accordingly be measured sequentially with only two elementary areas which can be moved about on the ground. The summation of the cross products thus measured can be performed later, e.g., in a digital computer.* This is the principle of aperture synthesis: The two elementary areas can be used to synthesize the result of a measurement with the much larger area (the "synthesized aperture").

The term $\phi_j - \phi_k$ can be written as $(2\pi/\lambda)\,\mathbf{B}_{jk}\cdot\mathbf{s}$, where $\mathbf{B}_{jk}$ is the separation of the two elemental areas, $\mathbf{s}$ is a unit vector defining the source position, and λ is the wavelength of the radiation.

Although there are $N(N-1)/2$ cross-product terms, many occur with redundancy. The aperture in Figure 10.1(a) with 120 cross-product terms has only 24 independent mutual separations, with highest redundancy occurring in the close separations. These basic separations are shown in Figure 10.1(b).

An array which contains all of the relative positions of a filled aperture is called a skeleton array. The "Tee" array in Figure 10.1(c) is an example. Only 10 elemental areas are needed to span the filled rectangular aperture. More generally, if we decompose a square aperture into an $(n \times n)$ elemental area, the corresponding "Tee" skeleton array contains only $3n - 2$ elemental areas. There are many examples of skeleton arrays; *e.g.*, a ring-shaped configuration is a skeleton array for a filled circular aperture (Wild, 1967).

The redundancy of the mutual separations for a skeleton array is, however, different than that for an equivalent filled aperture. This leads to a different beam pattern and sidelobe level. Also, the discrete nature of the mutual separations in a skeleton array leads to grating sidelobes. These properties are discussed in more detail in section 10.3.

An aperture may be synthesized by physically moving elements on the ground to occupy in turn all relative positions that occur in the aperture. Thus even with a two-element interferometer, it is possible to span completely a large aperture. Such arrays are called primitive arrays. For example, as seen in

Figure 10.1(d), the array of four fixed elements and one moveable element is able to synthesize the desired aperture with six locations of the moveable element. Alternatively, with two moveable elements, only three configurations are necessary. The Cambridge 4C survey (Ryle, 1960) used a scheme where one arm of a "Tee" array was constructed and the other arm was obtained by moving a small element in a perpendicular direction. For these arrays the addition of the separate configurations are usually performed in a digital computer after the set of observation has been completed.

One method, developed at Cambridge (*e.g.*, Ryle, 1962), uses the rotation of the Earth to change the aspect of the array. This method is illustrated in Figure 10.2. As viewed from the radio source, the array (assumed to be a two-element interferometer oriented in the east-west direction) rotates through all possible orientations over a period of 12 hours. With Earth rotation an aperture can be synthesized using a relatively small number of elements aligned in one direction. A line aperture is obtained instantaneously with the array, and this line

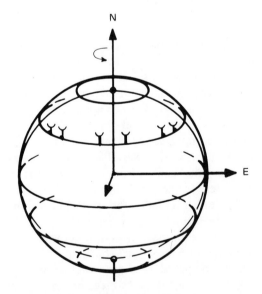

Figure 10.2 The rotation and foreshortening of a two-element east-west interferometer due to the rotation of the Earth as seen by an observer at the source.

aperture is rotated by the Earth's rotation.

The redundancy of the separations for a primitive array, as with a skeleton array, is different from that of the equivalently filled aperture. However, when performing the summation in a digital computer, an advantage is that the different separations may be combined with any desired weighting. One can in this way compensate for the differences in redundancy and so make the resultant weighting correspond more closely to that of a filled aperture (see Section 10.3).

For the most part we will concentrate on primitive arrays which utilize the Earth's rotation. Most of the newer or planned arrays are of this type. Often the arrays are built with elements aligned in one or in several directions, with moveable telescopes to fill in spacings along an array axis or to change the resolution of the array.

The building block of all arrays, the two-element interferometer, is discussed in the remainder of Section 10.1. The geometry needed for Earth rotation synthesis, the measurement of the visibility function, effects of bandwidth, polarimetry, and a tabulation of useful formulae in the Appendix are discussed. In Section 10.2 a working interferometer is briefly discussed. Variations of the usual techniques are also covered. Aperture synthesis techniques and problems are reviewed in Section 10.3. Sensitivity problems are also discussed. Finally in Section 10.4 the inversion processes for obtaining maps of the source brightness from interferometric data are outlined.

10.1.3 Two-Element Interferometer

a) Response to a Point Source

The two-element interferometer provides high resolution by correlating the signals of the two antennas. The correlation is normally achieved by the multiplication or addition of the signals, which produces a spatial modulation of the primary beam of the antennas with interference fringes. In this way fine structure is introduced into the primary beam to increase the resolution.

The response of the system to a point source of monochromatic radiation of frequency ω or wavelength λ is shown in Figure 10.3. A voltage E proportional to the electric field caused by the source is generated at the feed of each telescope at slightly different times. This time difference is called the geo-

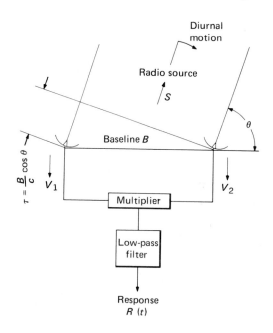

Figure 10.3 A simple two-element correlating interferometer.

metric delay and is denoted by τ. The voltages at the multiplier input are

$$V_1 \propto E \cos (\omega t)$$

$$V_2 \propto E \cos [\omega(t - \tau)]$$

$$= E \cos \left(\omega t - \frac{2\pi B}{\lambda} \cos \theta \right)$$

(10.3)

where B is the separation of the two antennas, λ is the wavelength of the radiation, and θ is the angle between the point source and the line joining the two antennas. The expression $(B/\lambda) \cos \theta$, the interference term, gives the phase path-length difference of the radiation travel along the two possible paths. The diurnal motion of the Earth causes θ to vary with time. The output, $R(t)$, of the multiplier, after a high-frequency term is rejected by a low pass filter, is

$$R(t) \propto S \cos\left(\frac{2\pi B}{\lambda} \cos\theta(t)\right) \quad (10.4)$$

This is the basic equation of interferometry. The flux density, or power S, of the source has replaced E^2. The fringe spacing is given by the angle which produces a change of one wavelength in the path-length difference.

The phase path-length difference (B/λ) $\cos\theta$ can be more generally written as $\mathbf{B}\cdot\mathbf{s}$, where $\mathbf{B}$, the physical spacing, is equal to the element separation in wavelengths and its direction is that of the line joining the elements.* The direction to the source is given by the unit vector $\mathbf{s}$. Equation (10.4) then becomes

$$R(t) \propto S \cos(2\pi \, \mathbf{B}\cdot\mathbf{s}(t)) \quad (10.5)$$

The interference response of a two-element interferometer can be pictured as a set of *fixed* quasi-sinusoidal fringes in the sky. The fringes are a function only of the interferometer separation. As the Earth rotates, a radio source travels through the fringe pattern (*i.e.*, θ is a function of time), producing a quasi-sinusoidal response with a period determined by how quickly the term $\mathbf{B}\cdot\mathbf{s}(t)$ varies with time. Usually the antennas track the radio source (*i.e.*, follow the radio source in its diurnal motion) so that maximum sensitivity is obtained from the desired radio source. A graph of fringe sizes and other parameters associated with an interferometer are given in Appendix I at the end of this chapter.

b) Fringe-Source Geometry

In order to describe aperture synthesis which makes use of the Earth's rotation, it is necessary to determine the fringe-source geometry. The forms of the necessary formulae are complicated and depend on the adopted coordinate system. The general relations will be derived here, and a list of the useful, working formulae are given in Appendix II.

* In order to avoid ambiguity, the baseline direction will be defined as a vector from telescope 2 toward telescope 1.

A useful quantity in interferometry is the projected spacing $\mathbf{b}$ of the physical baseline $\mathbf{B}$ as viewed from a radio source. The change of the projected spacing due to the rotation of the Earth is illustrated using Figure 10.2. The Earth is pictured as viewed by an observer at the radio source, and the projection of the baseline clearly changes as the Earth rotates. In this example of an interferometer oriented in the east-west direction the projected baseline appears to rotate 180° in a 12-hour period.

Mathematically, the projected spacing $\mathbf{b}$ is given by

$$\mathbf{b} = \mathbf{s}x(\mathbf{B}x\mathbf{s}) = \mathbf{B} - (\mathbf{s}\cdot\mathbf{B})\mathbf{s} \quad (10.6)$$

and is equal to the physical baseline $\mathbf{B}$ less the component of $\mathbf{B}$ in the direction of the source, *i.e.*, $\mathbf{b}$ is the projection of the physical baseline perpendicular to the source direction. Generally, the projected spacing is resolved into components along directions to the east and north, which are commonly denoted u and v, respectively. The resultant path described in the "$(u\text{-}v)$" plane produced as a source is tracked and is useful in understanding the methods of aperture synthesis. Equations and examples of $(u\text{-}v)$ paths are shown in Appendix II.

c) Response to an Extended Source

The response of a two-element interferometer to an extended source can be obtained by considering the source to be a collection of point images and summing their individual responses. Let $\mathbf{B}$ be the physical spacing of the interferometer and $\mathbf{s}$ a convenient coordinate near the source. This point is denoted as the phase center. Any other point can be denoted by $\mathbf{s} + \boldsymbol{\sigma}$. If $I(\boldsymbol{\sigma})$ describes the brightness distribution (angular distribution of power), then the response to the extended source is

$$R(t) = \int d\boldsymbol{\sigma} \, I(\boldsymbol{\sigma}) \cos[2\pi \, \mathbf{B}\cdot(\mathbf{s}(t) + \boldsymbol{\sigma})] \quad (10.7)$$

which is the response to a point source, Equation (10.5), integrated over the source. The flux density S of the source is equal to $\int d\boldsymbol{\sigma} \, I(\boldsymbol{\sigma})$. In most astronomical applications the radio source is followed in its diurnal motion by each antenna (tracked) so that $I(\sigma)$ is constant.

The term $I(\sigma)$ is the brightness distribution as modified by the primary response (or beam pattern) of the individual antennas. If $A(\sigma)$ is the primary response and $I'(\sigma)$ the real brightness distribution, then

$$I(\sigma) = I'(\sigma) \cdot A(\sigma) \qquad (10.8)$$

See Section 10.3.1.

Since the angular size of the region observed is limited by the extent of the antenna response (typically less than one degree), the phase term in Equation (10.7) can be expanded to first order for sufficient accuracy.

$$\mathbf{B} \cdot (\mathbf{s} + \sigma) \simeq \mathbf{B} \cdot \mathbf{s} + \mathbf{B} \cdot \sigma$$
$$= \mathbf{B} \cdot \mathbf{s} + \mathbf{b} \cdot \sigma \qquad (10.9)$$

Since σ is nearly perpendicular to $\mathbf{s}$, only the projected spacing $\mathbf{b}$ is used in the second term of the cosine. The response becomes

$$R(t) = \int_{-\infty}^{\infty} d\sigma\, I(\sigma) \cos\left[2\pi\mathbf{B} \cdot \mathbf{s} + 2\pi\, \mathbf{b} \cdot \sigma\right] \qquad (10.10)$$

This may be expanded into

$$R(t) = \cos(2\pi\mathbf{B} \cdot \mathbf{s}) \int_{-\infty}^{\infty} d\sigma\, I(\sigma) \cos(2\pi\mathbf{b} \cdot \sigma)$$

$$-\sin(2\pi\mathbf{B} \cdot \mathbf{s}) \int_{-\infty}^{\infty} d\sigma\, I(\sigma) \sin(2\pi\mathbf{b} \cdot \sigma)$$

However, it is much easier to work with the more compact complex form

$$R(t) = \exp\{i2\pi\mathbf{B} \cdot \mathbf{s}(t)\} \int_{-\infty}^{\infty} d\sigma\, I(\sigma) \exp\{i2\pi\mathbf{b} \cdot \sigma\}$$

$$= V \exp\{i2\pi\mathbf{B} \cdot \mathbf{s}(t)\} \qquad (10.11)$$

where

$$V = \int_{-\infty}^{\infty} d\sigma\, I(\sigma) \exp\{i2\pi\mathbf{b} \cdot \sigma\}$$

and where the real part of the right-hand side is implied. The exponential term outside of the integral is identical to Equation (10.5), the response of a point source located at the phase center. The integral, denoted as the visibility function V, is a complex number and gives the interference of the source. The amplitude of V is proportional to the amplitude of the fringe pattern and the argument of V equals the phase shift in the fringe pattern from that of the response to a point source at the phase center.

From the form of Equation (10.11) the visibility function is obviously the Fourier transform of the brightness distribution. For a simple extended source, such as a double, the visibility function has a simple dependence on the projected spacing, and the separation and relative strengths of the two components of the double may be found by fitting such a model to the observed visibility function. Visibility functions corresponding to simple models are given in Appendix III. For more complicated sources the emission may be recovered from the observed response by performing the inverse Fourier transform

$$I(\sigma) = \int_{-\infty}^{\infty} d\mathbf{b}\, V(\mathbf{b}) \exp\{-i2\pi\mathbf{b} \cdot \sigma\} \qquad (10.12)$$

Such a transform would be straightforward if the visibility function could be measured at all projected spacings. This is impossible, and the methods for extracting I from V for incomplete and discrete coverage are discussed in Section 10.4. Since the brightness distribution is a real function, $I^*(\sigma) = I(\sigma)$, where (*) denotes the complex conjugate. It is easily shown that $V(-\mathbf{b}) = V^*(\mathbf{b})$. That is, we need only measure the visibility function over half of the projected spacing plane.

It is common to use a Cartesian coordinate system, moving at the diurnal rate, in the neighborhood of the source. Let $\mathbf{s} = (\alpha, \delta)$ and $\sigma = (x, y)$, where x is an eastward displacement from α, and y is a northern displacement from δ. With such a coordinate system, and the convention that the phase increases for a source displacement toward the north and east, we get

$$V(u,v) = \int dx \int dy\, I(x,y) \exp\{+i2\pi(ux + vy)\} \qquad (10.13)$$

$$I(x,y) = \int du \int dv \; V(u,v) \exp\{-i2\pi(ux+vy)\}$$
$$(10.14)$$

The effect of the curvature of the sky plane is generally insignificant.

d) Effect of Bandwidth

For most astronomical applications wide-frequency bandwidths are desired to increase the signal-to-noise ratio. The response of a two-element interferometer with a frequency response given by $\alpha(\omega)$, the bandwidth function, can be obtained from Equation (10.4):

$$R(t) \propto S \int d\omega \; \alpha(\omega) \cos(\omega\tau) \quad (10.15)$$

where $\omega\tau = 2\pi \mathbf{B}\cdot\mathbf{s} = 2\pi(B/\lambda)\cos\theta$. If the bandwidth function extends over $\Delta\omega$ where $\Delta\omega > 1/\tau$, then the path-length change across the frequency band is sufficient to cause a loss of correlation.

Coherence across the frequency band is achieved by the insertion of time delay $\tau_D \approx (B/c)\cos\theta$ in one or both arms of the interferometer to approximately equalize the path-length over both paths to the multiplier. The amount of delay must be changed to compensate for the variation of the geometric delay as the source is tracked. Each change of delay produces a phase jump in the output. In the simple interferometer in Figure 10.3 a continuously varying delay which exactly compensates for the geometric delay would lead to a constant response. A more detailed description of delay tracking and its effects is given in Section 10.2.

e) Interferometric Polarimetry

In general, radiation is polarized and the measurement of the polarization parameters is important in the understanding of the emission mechanisms. A description of polarized radiation is given by Chandrasekhar (1950), Cohen (1958), and Kraus (1966), and the application to interferometry is elucidated by Morris, Radhakrishnan, and Seielstad (1964), Conway and Kronberg (1969), and Weiler (1973). The electric field associated with a beam of monochromatic radiation of angular frequency ω can be written as

$$\mathbf{E} = \hat{\mathbf{e}}_x \; Ex \cos(\omega t) + \hat{\mathbf{e}}_y \; Ey \cos(\omega t - \delta)$$
$$(10.16)$$

where Ex and Ey are the electric field amplitudes in the x and y directions ($\hat{\mathbf{e}}_x$, $\hat{\mathbf{e}}_y$) and δ is the phase difference between the two orthogonal modes. Any receptor of radiation (feed) is sensitive to only one of the two orthogonal modes of radiation [Equation (10.16) is only one of a possible set of these modes] and thus intercepts only that component of the radiation.

The radiation is coherently polarized if Ex, Ey, and δ remain constant over a long period of time. However, the radiation from most celestial objects is produced by a large number of independent radiators, and the resultant electric field from the ensemble varies randomly with time.

However, the fluctuations of Ex and Ey, although random, may be correlated. Such a beam of radiation is said to be polarized. For example, the synchrotron radiation emitted by a collection of electrons confined to planar orbits is partially polarized. The electric field for a noncoherent partially polarized beam of radiation written in vector complex form can be described by four parameters:

$$\mathbf{E} = \{\hat{\mathbf{e}}_x \; Eo \; (\cos\beta\cos\chi - i\sin\beta\sin\chi)$$
$$+ \hat{\mathbf{e}}_y \; Eo \; (\cos\beta\sin\chi + i\sin\beta\cos\chi)$$
$$+ Eu\} \; e^{i\omega t} \quad (10.17)$$

where Eo is the amplitude of the electric field of the polarized radiation, β is the ellipticity,* and χ is the position angle. The amplitude of the unpolarized part of the radiation is Eu; its phase and direction are random. The parameters Eo, β, and χ can be written in terms of Ex, Ey (here Ex and Ey refer only to the polarized part of the radiation), and δ in Equation (10.16) (see Chandrasekhar, 1950, p. 26).

The characteristic of partially polarized noncoherent radiation is most commonly

---

* If $\beta > 0$, the radiation is said to be left-hand elliptically polarized. If $\beta < 0$, it is right-hand elliptically polarized. Left-hand polarization means a clockwise rotation of the electric vector with the wave approaching.

described in terms of the four Stokes intensity parameters I, Q, U, and V. These are defined by the following relations:

$$I = \text{total intensity of radiation}$$
$$= \langle Eo^2 \rangle + \langle |Eu^2| \rangle$$
$$= \langle Ex^2 \rangle + \langle Ey^2 \rangle + \langle |Eu^2| \rangle$$

$$Q = \text{linearly polarized intensity}$$
$$= \langle Eo^2 \rangle \cos 2\beta \cos 2\chi$$
$$= \langle Ex^2 \rangle - \langle Ey^2 \rangle$$

$$U = \text{linearly polarized intensity}$$
$$= \langle Eo^2 \rangle \cos 2\beta \sin 2\chi$$
$$= \langle Ex\, Ey \rangle \cos \delta$$

$$V = \text{circularly polarized intensity}$$
$$= \langle Eo^2 \rangle \sin 2\beta$$
$$= \langle Ex\, Ey \rangle \sin \delta$$

(10.18)

The Stokes parameters are defined in terms of r.m.s. time averages of the electric field components.

The response to the radiation will depend on the orientation and ellipticity of the feed. A dipole parallel to the x-axis will respond only to the x-component of the radiation and measures $\frac{1}{2}(I + Q)$. Rotation of the dipole to the y-direction will change the response to the y-component of the radiation and measure $\frac{1}{2}(I - Q)$.

The response of an interferometer with arbitrary feed characteristics can also be expressed in terms of Stokes parameters, and specific examples are also given in Appendix IV. Since all the Stokes parameters are intensity parameters, the brightness distribution of $Q(\sigma)$, $U(\sigma)$, and $V(\sigma)$ can be found in the same way as the total intensity, $I(\sigma)$. In most cases we will discuss explicitly the interferometry of the total intensity, but the term $I(\sigma)$ can be replaced by a sum of the various Stokes parameters $F(I,Q,U,V;\sigma)$ where F depends on the feed configuration.

10.2 A Working Interferometer

10.2.1 A Complete System

A block diagram of a working interferometer and the response of the system are shown in Figure 10.4. The diagram is complex and includes the major components of a complete system. The response in mathematical terms to incoming radiation at various stages is shown. Only the phase behavior of the system is incorporated into the equations. The gain factors for the system are neglected. Error terms related to cable length changes, temperature effects, etc., have not been included.

The major difference from the simplified system in Figure 10.3 is the conversion of the observed radio frequency (RF) to an intermediate frequency (IF) using a standard heterodyne process with a coherent local oscillator (LO). With this conversion most of the path over which the signals are joined is at a moderately low frequency, where cable losses are smaller, path-length changes due to external variations are smaller, and electronic components are less expensive. In addition, a change of observing frequency requires only a change of components in front of the heterodyning stage and in the LO frequency.

A description of the technical aspects of an interferometer has been given by Read (1963) and Swenson (1969) and will not be repeated here. Some obvious requirements and the function of various components are given below:

(1) A low-noise RF amplifier. The sensitivity of most systems is limited by the noise generated in the RF amplifier. A large gain is also desirable to decrease the effect of noise generated by following stages.

(2) A stable heterodyne process. The RF signal is converted to an IF signal by mixing (multiplying) the RF signal with a strong LO signal. In order to preserve the RF signal characteristics in the IF signal, the LO signal must be monochromatic and the phase relationship of the LO signal at each mixer must be coherent. Coherence is difficult to obtain over a distance of more than ~ 10 km. A term $\phi(t)$ has been included in the LO signal sent to the mixer of antenna #2. Phase errors in the interferometer system and system modifications such as lobe rotation can be

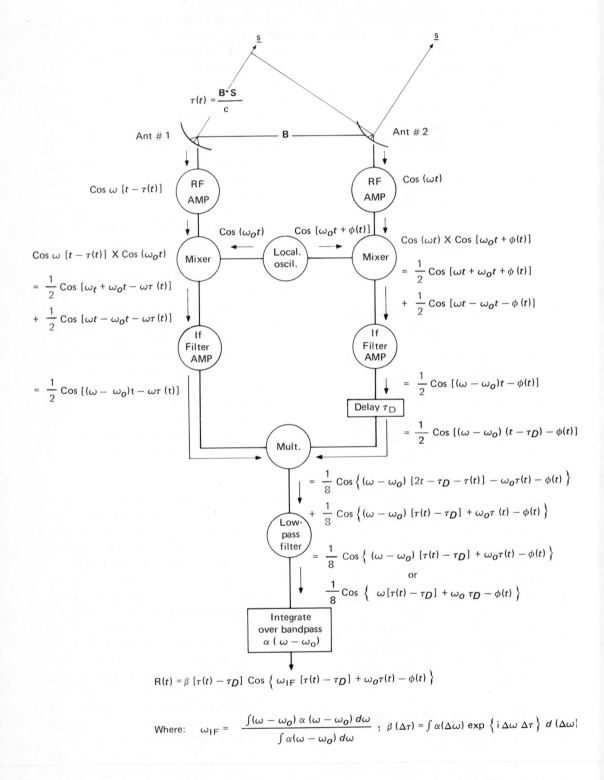

Figure 10.4 The schematic working diagram of the response of an interferometer.

understood in terms of the behavior of $\phi(t)$. With a heterodyne process two RF frequencies (sidebands) are superimposed when converted to an IF frequency ω_{IF}. The upper sideband has a frequency of $\omega_o + \omega_{IF}$, and a lower sideband has a frequency of $\omega_o - \omega_{IF}$. For many astronomical applications both sidebands are used; however, one sideband may be rejected by placing an appropriate filter in the RF line.

(3) IF system. Amplification is needed to increase the signals for subsequent processing. A filter is also needed to reject the high-frequency signal and to limit the IF frequency range within the specifications of the multiplier, delay lines, and IF amplifier bandpass. For spectral line work, narrow-band filters may also be present. A variable delay line, usually consisting of a binary series of lengths of cable, is placed in one or both of the IF lines to equalize the travel time of the radiation via the two possible paths. The insertion of delay τ_D is shown in the response of antenna #2.

(4) Multiplication and terminal processing. The signals from each antenna are multiplied and the high-frequency response $(\omega - \omega_o)2t$ is rejected by a low-pass filter. The subsequent monochromatic response is given in two forms in Figure 10.4.

10.2.2 Bandwidth Effects

a) Response

The response after multiplication and filtering is that for a monochromatic signal within the system bandwidth. The bandwidth is usually determined by IF filtering, but can be affected by a limited bandwidth in any part of the system. If the bandwidth function of the response is denoted by $\alpha(\omega - \omega_o)$* (a complex quantity), the resulting response is given at the bottom of Figure 10.4. The average IF frequency weighted by the bandwidth function is ω_{IF}. The term $\beta(\tau_D - \tau(t))$ is called the fringe washing function and is the complex Fourier transform of $\alpha(\omega - \omega_o)$; see Appendix V.

b) Delay Tracking

We have already alluded to the loss of coherence over a large bandwidth in Section 10.1. The difference in the time of travel from the source via the two possible paths $\Delta\tau \equiv \tau_D - \tau(t)$ must be made smaller than the reciprocal bandwidth; otherwise $\beta(\Delta\tau)$ will be significantly less than unity. At present there is no technically feasible method for obtaining a continuously variable delay line; a binary array, usually driven by a computer, sets $\tau_D \approx \tau(t)$ within the stepping interval of the delay. Unfortunately, in a single sideband system a discrete change of τ_D produces a phase jump equal to $\omega_{IF}\Delta\tau$ in the response, which disturbs the sinusoidal output. Three solutions are possible: (1) Change τ_D in steps of $2\pi/\omega_{IF}$ so that all of the phase jumps are 2π at the center of the IF band. However, significant losses occur unless the bandwidth is much less than ω_{IF}. (2) Change τ_D in very small steps so that each phase jump is small. (3) Compensate for the phase change due to the delay step in the IF line by a phase-adding device.

For observations of continuum radiation where there is no fundamental difference between the two sidebands, the use of a double sideband system avoids the problem of phase jumps caused by delay tracking. The phase jump for each sideband is in the opposite direction and hence they cancel each other for a double sideband system, i.e., $\omega_{IF} = 0$ for a balanced double sideband system.

10.2.3 Calibrations

The complete response $R(t)$ of a two-element interferometer to an extended source of radiation (see Equation 10.11) is

$$R(t) \sim Re\{\beta(\Delta\tau)\,V(\mathbf{b})\exp i\{2\pi\mathbf{B}\cdot\mathbf{s}(t)+\phi(t)\}\}$$

$$(10.19)$$

where β is the fringe washing function, $\Delta\tau$ is the time delay difference in the two paths

---

* The term $\alpha(\omega - \omega_o)$ is the bandwidth of the system in terms of the input RF frequencies. For a double sideband system $\alpha(\omega - \omega_o)$ extends above and below the LO frequency ω_o, although the sidebands are folded together in the IF conversion.

$[\tau_D - \tau(t)]$, V is the visibility function, **b** is the projected baseline, **B** is the physical baseline, **s**(t) is the unit vector to the source, and $\phi(t)$ is a phase term containing lobe rotation and other deliberate phase modifications as well as random phase errors. In this section we shall assume that the delay is accurately tracked and that β is unity.

The response and the precise time as well as other useful parameters are sampled by a computer, and the apparent visibility function is determined by fitting the response to the form $\exp i\{2\pi\mathbf{B}\cdot\mathbf{s}(t)\}$. The apparent visibility function must then be corrected for certain instrumental effects.

Nearly all corrections to the observed response are determined by the use of radio sources known as calibrators. Generally these are point sources (a source of emission with an angular size much less than a fringe separation) of known flux density, position, and polarization. The calibrators are observed periodically, and certain system parameters are adjusted so that the correct values of flux density, polarization, and zero phase are obtained for these sources. The same (or interpolated) adjustments are made to all observations, *i.e.*, the apparent visibility functions are corrected in the same way. There are four basic calibrations for continuum measurements to determine the instrumental parameters: gain, baseline, phase, and polarization.

The gain calibration is the determination of the ratio of the flux density of the calibrators to the correlated amplitude of the response. All of the electronic components shown in Figure 10.4 have some variation of gain with time, and the observation of a source of known flux density every few hours or days (depending on the overall gain stability of the system) is needed to follow the variations. Often the system is equipped with an automatic level control which holds the IF power constant at some late stage of the system. An increase in the system noise temperature then leads to a reduction in the amplitude of the signals, and observation of calibrators is still required to monitor ampli-

tude fluctuations. Other systematic gain changes must be applied before a satisfactory time dependence of gain can be obtained, such as the effects of the filter responses, atmospheric attenuation, antenna efficiency as a function of elevation, added system noise temperature due to a strong source or ground radiation pick-up, and loss of coherence with various delay elements. Determination of the gain to an accuracy of several percent is frequently achieved.

An *a priori* baseline **B** is used to fit the interferometer response to a computed fringe pattern. The baseline cannot be surveyed accurately enough [better than $(1/100)$ of a wavelength] and must be determined by the calibration process. Point sources of accurately known position are observed over a wide range of hour angle and declination, and the systematic variation of the visibility phase response with hour angle and declination is used to determine a more accurate baseline value. Phase drifts in the system response and errors in the assumed source positions are detrimental to a good baseline determination.

The baseline separation between two telescopes is equal to the vector joining their respective phase centers, which are usually near the focal points of the telescopes. Most telescopes are built with intersecting axes of motion so that the phase center remains at a fixed distance from the intersection regardless of the pointing direction. Thus, the baseline separation **B** is independent of the pointing position. Even if the antennas are moved slightly off position, the phase center defined by the source radiation does not move. The amplitude of the response to the source will decrease, but the phase of the sinusoidal response is unaltered. For nonidentical antennas with nonintersecting axes, the baseline separation is a function of source direction. The geometry has been worked out by Wade (1970) for a nonintersecting equatorial mount.

For telescopes with intersecting axes, the baseline error is due to the uncertainties in the mutual positions of the antennas. Since many arrays use moveable antennas, exact positions cannot be determined beforehand.

From Equation (10.19) the expected variation in phase over the sky is $\Delta\mathbf{B}\cdot\mathbf{s}$, where $\Delta\mathbf{B}$ is a constant. Observations of at least four sources, and preferably several tens of sources for adequate redundancy, are needed to determine the three coordinates of $\Delta\mathbf{B}$ and a phase offset.

Once an accurate baseline separation has been obtained, the phase behavior (after known phase modifications are considered) of the system $\phi(t)$ can be followed by observing calibrators as often as necessary.

Before the above phase calibrations can be made, systematic phase variations due to the delay in the atmosphere must first be considered. If the antennas are at different elevations, the two radiation paths go through different amounts of atmosphere, causing a phase change as a function of source position and ground weather conditions. The resultant phase correction, which depends on the refractive index, can be calculated. In the plane-parallel approximation the troposphere produces no additional phase change. Although the position of a source can be changed by many arc minutes by refraction, the phase path length between the two telescopes is unaffected. For observations at low elevation, the spherical nature of the troposphere must be considered, and a phase correction dependent on gross atmospheric properties is needed. This correction is significant for element separations greater than ~ 1 km (Hinder and Ryle, 1971), corresponding to a typical tropospheric scale size. Also, short-term phase fluctuations occur, mainly due to water vapor, which gives rise to path-length changes of a few millimeters. These short-term fluctuations are due to small-scale irregularities of size $\lesssim 1$ km and cannot be effectively calibrated.

The polarization response $F(I,Q,U,V;\boldsymbol{\sigma})$ of an interferometer is given in Appendix IV. The feed alignment and ellipticity cannot be accurately measured, so observations of calibrators with known polarization values are used. The details of the calibration process depend on the type of polarization desired and the feed configurations. The polarization

response also varies over the primary beam and can be calculated in some cases and may vary somewhat with the telescope pointing.

10.2.4 Modifications to the Basic Interferometer

There are many modifications to the basic interferometer system. Some systems use several intermediate frequencies, each produced by an independent local oscillator. This is often done for convenience (*e.g.*, using a 300-MHz IF system with delays designed for 10 MHz), but is also used to adjust the output in a prescribed manner. Many systems use complicated LO chains for greater frequency versatility and phase-compensating systems. Four modifications—lobe rotation, spectral-line interferometry, very-long-baseline (VLB) interferometry, and intensity interferometry—will be discussed in some detail.

a) Lobe Rotation

The frequency of the interferometer response (the fringe frequency) can be considered as the carrier frequency for the signal (the visibility function). The rate of change of the visibility function is proportional to the rate of change of the projected spacing, whereas the fringe rate is proportional to u, the east-west projected spacing. At times the fringe frequency is less than that associated with the visibility change and information can be lost. Also, the variation of the fringe frequency with baseline separation and source position is a nuisance in the data analysis; the frequency dependence of the system must be accurately calibrated and the data must be sampled at least two times a fringe period.

One method of arbitrarily changing the fringe frequency is the addition of a varying phase, usually controlled by a computer, to one side of the interferometer. The additional phase can be added in the RF, LO, or IF lines using a phase rotation capacitor device or by a deliberate offset of the local oscillator signal to one antenna (Read, 1963). The response then has the time dependence

$$\cos(2\pi \mathbf{B} \cdot \mathbf{s}(t) + \phi(t)) \qquad (10.20)$$

Often $\phi(t)$ is set equal to $-2\pi \mathbf{B} \cdot \mathbf{s}(t) + \Omega t$, where Ω is the constant fringe frequency, independent of baseline or source position. For multi-element arrays it is not possible to obtain the same fringe frequency for all correlated pairs, although the time-dependent term can be subtracted. The Westerbork synthesis telescope in the Netherlands uses a system in which the fringe rate is zero. Special switching techniques are then necessary to avoid a slow drifting of the response and to measure the real and imaginary parts of the visibility function (Casse and Muller, 1973).

b) *Spectral-Line Interferometry*

The discovery of many radio-frequency lines has made spectral-line interferometry necessary in order to achieve high angular resolution, especially for low-frequency lines. The frequency (velocity) dependence of many sources of line radiation is complicated, and complete observations require many channels of narrow bandwidths (1 kHz to 1 MHz) to adequately cover the relevant radiation.

Two methods are used to obtain high spectral resolution. First, two sets of matched filters are introduced, one into each IF line (see Figure 10.5a). Each filter produces a separate output, which is individually multiplied with the other matched filter (Rogstad *et al.*, 1967). The response of each filter pair is identical to that of a broad-band interferometer and is a direct measure of the visibility function averaged over the filter bandwidth.

Another method of obtaining high spectral resolution uses the multiplication of the broad-band signal with many large delay intervals (Baldwin *et al.*, 1971). The analogue procedure is shown in Figure 10.5(b). The output from each correlator is proportional to $\beta(k\Delta\tau)$, $k = -N \leq 0 \leq N$, where $\Delta\tau$ is the delay interval step and $N\Delta\tau$ is the total delay range in the correlation. Since the bandpass is the Fourier transform of the fringe washing function β, the bandpass with respect

to ω_{IF} may be recovered as

$$\alpha(l\Delta\omega) = \sum_{k=-N}^{N} \beta(k\Delta\tau)\exp\{i\pi lk/N\};$$

$$-N \leq l \leq N \quad (10.21)$$

The resolution is $\Delta\omega \approx (N\Delta\tau)^{-1}$ and the aliasing frequency is $\omega_A \approx \Delta\tau^{-1}$; so that the response is given by

$$\alpha'(\omega) = \sum_{n=-\infty}^{\infty} \alpha(\omega + n\omega_A)$$

where n is an integer. The aliasing can be avoided by placing filters in the IF lines to limit the IF bandwidth range to less than ω_A. The filter shape corresponding to the simple sum in Equation (10.21) is

$$\frac{\sin(N\Delta\tau\omega)}{(N\Delta\tau\omega)} \qquad (10.22)$$

This filter has the undesirable property of high sidelobes. Convolution with a Gaussian function [produced by multiplying $\beta(k\Delta\tau)$ by a Gaussian] or hanning will reduce the sidelobe level but increase the resolution size $\Delta\omega$.

The same procedure can be obtained from a digital cross-correlator. Several seconds of IF signals from each element are stored and subsequently multiplied with various time-delay offsets $\Delta\tau$ to obtain $\beta(k\Delta\tau)$.

Normal operation of the spectral-line interferometer is single sideband. The rejection of one sideband is usually obtained by a bandpass filter in the RF line of each antenna. The bandpass of $\alpha(\omega)$ is the product of the system bandpass and the radiation from the source. The system bandpass can be measured experimentally or determined from observations of a calibrator source with no line radiation in the frequency range.

An interesting technique for spectral-line observations is that of a double sideband system operating with a deliberate delay offset so that the two sidebands cancel (Radhakrishnan *et al.*, 1971). This occurs when $\beta(\tau_D - \tau(t)) = 0$. If the upper and lower sidebands are identical in frequency characteristics and the delay tracking is accurate, the

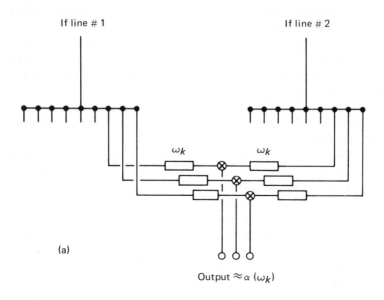

(a)

Output $\approx \alpha\ (\omega_k)$

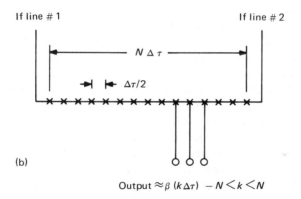

(b)

Output $\approx \beta\ (k\Delta\tau)\ \ -N < k < N$

Figure 10.5 Spectral-line receivers. (a) A conventional filter bank receiver. (b) Cross-correlation spectrometer with correlation at various delay intervals.

output response is proportional to the difference of radiation between the two sidebands. The result is a null experiment, unless there is line radiation in either sideband.

c) Very-Long-Baseline Interferometry

As baselines are increased, the major problem is the maintenance of LO stability between the elements. Over moderate distances of about 100 km, radio links have been used to transport the local oscillator signals and IF responses. However, unless sophisticated path-length compensation schemes are

used, the phase change $\phi(t)$ of the response can vary unpredictably (Elgaroy *et al.*, 1962; Basart *et al.*, 1970).

For very-long-baseline (VLB) interferometry independent local oscillators are used at each element. This technique is now possible, with the advent of very accurate frequency standards which use the discrete atomic energy transitions (cf. Cohen, 1969). In order to obtain a phase stable response, the error in the LO phase, $\phi(t)$, must remain constant within a radian. For observations at 10 cm, for example, stability over 15 minutes requires an oscillator stability of 1 part in 3 $\times$ 10^9 Hz $\times$ 10^3 sec $\approx$ 1 part in $10^{12.5}$. Such stability can be obtained using hydrogen masers as frequency standards.

The IF output of each element is recorded separately on a high-speed analogue or digital recording device. Systems are currently available in which 10^7 bits of information can be recorded each second, thereby permitting bandwidths of several megahertz to be properly sampled. The two IF responses are then multiplied in a digital computer or on a special analogue device. The IF responses must be lined up with the proper time delay before multiplication. This requires a timing accuracy of about 10^{-6} sec for a bandwidth of 1 MHz. Otherwise, the fringe washing function $\beta(\tau_D - \tau(t))$ will be much less than 1.

For the reduction of VLB data the observed response is fitted to the form of Equation (10.19) in the same way as a conventional interferometer. Because of baseline inaccuracies, source position errors, timing errors, and oscillator drift, the delay may be significantly in error, causing loss of correlation. Also, error in the calculated fringe rate may limit the integration time. The response is therefore fitted using a range of delay offsets and fringe frequencies around the nominal values. The maximum amplitude obtained is proportional to the visibility amplitude of the source. The visibility phase can be measured only when baseline errors and clock drifts are also accurately determined.

A systematic analysis of the fringe frequency and the time-delay measurements for observations of sources can be used to obtain a more accurate baseline separation and improved source positions. Such an analysis is described by Cohen and Shaffer (1971) using an equation similar to that given in the Appendix for the fringe frequency and time delay. Two modifications are needed. Firstly, since the time delay is many milliseconds between two distant telescopes, there is a substantial change of the baseline separation during this time. A "retarded" rather than an instantaneous baseline must be used. Secondly, a relativistic correction caused by the large mutual velocity of the telescopes is necessary. Source positions of 0.1 arc second accuracy are now being obtained by using the fringe frequency and time delay.

One method for improved accuracy of VLB interferometry is by the observation of several sources or frequencies simultaneously. The relative visibility functions among the sources or frequencies can be obtained even with inaccurate baselines and source positions, and sizeable oscillator drifts. Accurate maps of complicated regions of small-diameter OH and H_2O emission regions have been obtained in this way (cf. Moran *et al.*, 1968). For continuum observations four-antenna experiments using two distant observatories, each with two antennas, have been successful. Two different sources are observed with each VLB interferometer, but the same local oscillator is used at each location, and the two responses are recorded simultaneously on the same device. Most of the gross baseline errors, oscillator errors, and position errors are cancelled by the technique.

The ultimate measurement of positions in the order of milliseconds of arc will require a careful calibration, understanding, and measurement of many effects: continental drift, Earth tides, precision rotation of the Earth, atmospheric and ionospheric path-length changes, and relativistic bending of radio waves. All of these effects are significant at the millisecond of arc level.

d) Intensity Interferometry

Hanbury Brown and Twiss (cf. Hanbury Brown, 1968) recognized that it is not necessary to have a phase stable system for an interferometer. In an intensity interferometer the RF signals from each antenna are mixed with two incoherent LO signals, with the resulting lack of phase stability in the IF signals. The separate signals are each detected and then multiplied together. Although no phase relationship exists between the two outputs, some correlation is obtained because of common intensity fluctuations in each output.

An analysis of the response for an intensity interferometer has been given by Bracewell (1958) and MacPhie (1966). The response is proportional to the square of the visibility amplitude and all phase information is lost, although some phase behavior can be recovered with multi-element intensity interferometry.

The signal-to-noise ratio for an intensity interferometer is much lower than that for a correlating interferometer. The general formulae have been considered by Clark (1968). The loss of signal-to-noise is related to the bandwidth of the fringe power. For a correlating interferometer with perfect phase stability [*i.e.*, $\phi(t) = 0$], all of the correlated power falls within a narrow frequency range $(1/T)$, where T is the length of the observation. Only the noise falling within that passband need be associated with the measurement. For an intensity interferometer the fringe power is spread over the entire frequency range of several megahertz determined by the detector and the resulting noise is thus much larger. Before the use of hydrogen masers as oscillator standards, the sensitivity of VLB observations was also limited because of random fluctuations in the phase stability, which broadened the fringe frequency range considerably.

10.3 Aperture Synthesis

10.3.1 Filled and Unfilled Aperture Beams

In Section 10.1 we alluded to some of the different characteristics between the beam of a filled aperture and the beam of a synthesis instrument. Most of the differences are caused by incomplete and discrete coverage of the (u,v) plane with aperture synthesis instruments.

Most telescopes can be described in terms of an aperture plane on which currents are induced by the incoming radiation (Kraus, 1966, Chapter 6; Christiansen and Högbom, 1969, Chapter 3). The voltage pattern $\mathscr{V}(\boldsymbol{\sigma})$ of the aperture plane is given by

$$\mathscr{V}(\boldsymbol{\sigma}) = \int_{\text{aperture}} d\mathbf{x} \exp\{-i2\pi\mathbf{x}\cdot\boldsymbol{\sigma}\} \, g(\mathbf{x}) \quad (10.23)$$

where $\boldsymbol{\sigma}$ is the direction of the response with respect to the direction of the pointing axis. The term $g(\mathbf{x})$ is the aperture distribution (grading) and gives the relative response over the aperture. The aperture distribution depends on the characteristics of the antenna surface and the illumination by the feed.

The power pattern $A(\boldsymbol{\sigma}) \equiv \mathscr{V}|(\boldsymbol{\sigma})|^2$ is given by

$$A(\boldsymbol{\sigma}) = \int d\mathbf{x} \exp\{-i2\pi\mathbf{x}\cdot\boldsymbol{\sigma}\} \, h(\mathbf{x}) \quad (10.24)$$

with

$$h(\mathbf{x}) \equiv \int d\mathbf{x}' \, g(\mathbf{x}') \, g^*(\mathbf{x} - \mathbf{x}') \quad (10.25)$$

where $h(\mathbf{x})$ is called the transfer function. It is a measure of the density of mutual spacings in the aperture. An example of the above quantities for a filled-aperture radio-telescope is shown in Figure 10.6.

The above analysis is identical for an array; however, the aperture distribution is now produced by many spatially separated antennas, each associated with an element of the array. In the case of an array with a total set of mutual separations $\mathbf{b}_j$, $j = 1, \cdots, N$ and with identical antennas each of grading $g_A(\mathbf{x})$ and transfer function $h_A(\mathbf{x})$, the transfer function of the array is

$$h(\mathbf{x}) = \sum_{j=1}^{N} h_A(\mathbf{x} - \mathbf{b}_j) \quad (10.26)$$

and

$$P(\boldsymbol{\sigma}) = A(\boldsymbol{\sigma}) \sum_{j=1}^{N} \exp\{-i2\pi\mathbf{b}_j\cdot\boldsymbol{\sigma}\} \quad (10.27)$$

where $P(\boldsymbol{\sigma})$ is the power pattern for the array and $A(\boldsymbol{\sigma})$ is the power pattern for the elemental antenna.

The aperture functions for two linear arrays are also shown in Figure 10.6. The first array consists of seven identical equally spaced elements with an overall size equal to that of the filled aperture. The envelope of the shape of the transfer function is similar to that of the filled aperture and the central peaks of the power patterns are about the same width. The discrete spacings (stepping interval) of the array produce grating responses (images) that are attenuated by the

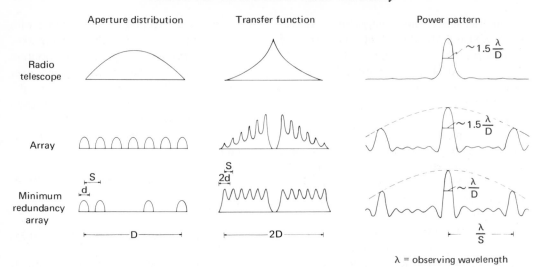

Aperture distribution　　　Transfer function　　　Power pattern

λ = observing wavelength

Figure 10.6 The aperture distribution, transfer function, and power pattern for: (a) a radio telescope, (b) an array with equally spaced elements, (c) a minimum redundancy array (placing four elements in such a way so as to obtain six nonredundant spacings).

elemental power pattern. For a multiplying interferometer the zero spacing (self-correlating pairs) is not obtained.

The second array contains only four elements but they are placed in such a way that the same six mutual separations of the previous array are measured. This type of an array is called a minimum redundancy array (Moffet, 1968). The envelope of this transfer function is more uniform and corresponds to a filled aperture strongly illuminated at the edge. The central beamwidth is narrower than the previous array, but the sidelobe level [wiggles in $P(\sigma)$] is larger because of the sharper cut-off in the transfer function. The grating response, a function of the stepping interval, is similar to the previous array.

Neglecting the power pattern of the individual elements, the synthesized beam of an array is

$$P(\sigma) = \sum_{j=1}^{N} \exp\{-i2\pi\mathbf{b}_j\cdot\sigma\}\, w_j \quad (10.28)$$

or

$$P(x,y) = \sum_{j=1}^{N} \exp\{-i2\pi(ux+vy)\}\, w_j \quad (10.29)$$

in rectangular coordinates. The weighting w_j of each data point has been incorporated into

the equations. If the data for each telescope pair are individually combined in a computer, the weighting may be chosen to modify the envelope of the transfer function. The optimum weighting depends on sidelobe level and signal-to-noise considerations.

The output of a synthesis array is a set of visibility functions $V(u_j,v_j)$, and the resultant brightness distribution (cf. Equation 10.14) is

$$I(x,y) = \sum_{j=1}^{N} V(u_j,v_j) \exp\{-i2\pi(u_jx+v_jy)\}\, w_j \quad (10.30)$$

The synthesized beam is thus the brightness distribution obtained by observing a point source. The observed brightness distribution is the convolution of the true brightness distribution with the synthesized beam.

A skeleton array contains all of the desired mutual separations at any instant so that it is not necessary to track a radio source in order to obtain a reasonable synthesized beam. The visibility functions are then transformed to obtain the brightness distribution. For a primitive array the necessary baselines are obtained by moving some of the elements and/or utilizing the rotation of the Earth. For both types of arrays the properties of the synthesized beams are discussed in the next section. Explicit illustrations of skeleton arrays and their responses are given in Christiansen and Högbom (1969).

The visibility functions are obtained by averaging the quasi-sinusoidal response of each inteferometer pair. The averaging duration should not exceed the time for which the visibility function significantly changes. This depends on the angular size of the radio emission. The finite angular size imposed by the primary response of the individual elements limits the averaging time to several minutes for a 1-km baseline and a 25-m element size. The averaging time for observations of intense small-diameter sources may be considerably increased. A general practice is to average the visibility function of all baselines at equal intervals. Each sampled point then has about the same signal-to-noise although the smaller spacings in the array have been sampled more often than necessary.

10.3.2 Properties of a Synthesized Beam

A more detailed discussion of the properties of a synthesized beam is given in conjunction with Figures 10.7 and 10.8. In Figure 10.7 the coverage in the $(u-v)$ plane is shown for a set of observations of CAS A. Each point represents an observation of a unit time interval—hence approximately unit weight. The interferometer baseline is not aligned east-west, causing a displacement of the $(u-v)$ ellipses from the origin.

a) Beam-Weighting and Its Effects

The natural weighting (effective aperture distribution) for the array is proportional to the density of sampled points weighted in accordance to the signal-to-noise of each point. However, for most synthesis instruments we are at liberty to weight each observation as desired, since the summation of Equations (10.29) and (10.30) is usually done in a digital computer.

The synthesized beam and brightness distribution for Cas A with natural weighting are shown in Figure 10.8(a). If the outer spacings are more heavily weighted, the cor-

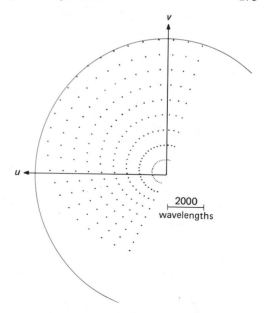

Figure 10.7 The coverage in the $(u-v)$ plane for observations of CAS A, using the NRAO interferometer. The source was tracked from hour angle -6^h to $+6^h$ at baselines 100 to 900 m in 100-m intervals. Each point represents a 30-minute average of the data. In practice the averaging interval used is $\frac{1}{2}$ minute.

responding beam and brightness distributions are shown in Figure 10.8(b). With this weighting the transfer function is $u(\mathbf{x}) \approx 1$, corresponding to a uniformly sampled aperture. The beamwidth is smaller but the sidelobes are higher. In Figure 10.8(c) the beam and brightness distribution with the inner spacings weighted more heavily is given. The synthesized beam width is increased but there is a decrease of the inner sidelobes.

b) Sidelobe Levels

At some angles signals from parts of the aperture add in phase to produce ripples in the power pattern. These ripples are similar to those of filled-aperture telescopes. Two major causes of sidelobes are (1) diffraction at the edge of the aperture and (2) gaps in the synthetic aperture coverage in the (u,v) plane. The inner sidelobes caused by the aperture edge can be decreased by lowering the weighting of the outer spacings so that the edge-like diffraction of the aperture extremities is

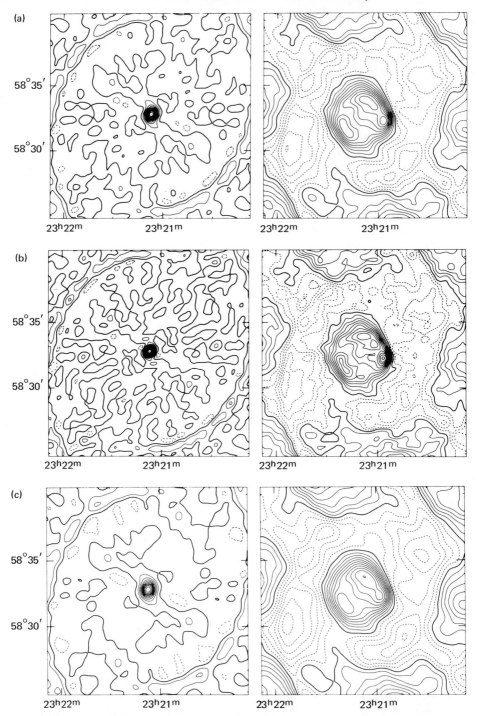

Figure 10.8 The synthesized beam and brightness distribution for CAS A using the $(u-v)$ coverage shown in Figure 10.7. Contours are at intervals of 10% of the peak value. Negative contours are shown as dashed lines and the zero-level contour is shown by the bold line. (a) Natural weighting, all points weighted equally. (b) The points weighted so as to produce a uniform transfer function. (c) Heavily tapered weighting giving a wide beam. The use of the different weighting functions affects the beam width and the sidelobes because of the edge-diffraction pattern of the finite $(u-v)$ coverage. The grating sidelobe response [due to the radical periodicity of the $(u-v)$ coverage] and the coherent, near-in sidelobes [due to the gap in the $(u-v)$ coverage] are unaffected by the different grading functions.

minimized. This is shown in Figure 10.8(c), where the outer spacings have been reduced by weighting. However, sidelobes caused by gaps in the (u,v) coverage (or feed support shadowing for a filled aperture) are not significantly effected by beam-weighting.

c) Grating Responses

The discreteness of the interferometric data manifests grating responses which have no filled-aperture analogue. Many arrays are built with a regular spacing between the elements. This is often called a stepping interval for a variable spacing interferometer (Figure 10.7). The regular stepping interval D causes large responses at angles of $n\lambda/D$ from the beam center, where n is a positive integer. These grating responses contain the same energy as in the central synthesized beam and they cannot be reduced by beam-weighting. For arrays which sample the (u,v) plane in two dimensions (Figure 10.7) the grating energy is distributed in an approximate ellipse, with an intensity peak of only $\sim 10\%$ for the first ring. The grating responses of the one-dimensional arrays, as in Figure 10.6, are approximately equal to the main response. The stepping interval can be made smaller than an antenna radius, thus moving the response outside of the antenna primary beam, $A(\boldsymbol{\sigma})$. Some of the closer spacings less than the element diameter cannot be measured, and this leads to curvature in the zero level of the synthesized map.

d) Bandwidth Correlation Area

Large-frequency bandwidths are often used for continuum measurements in order to increase the signal-to-noise ratio, which varies as the square root of the bandwidth. There are practical difficulties in using very wide bandwidths, however, since the characteristics of the RF, IF, and delay systems must be sufficiently frequency independent over the frequency range of interest.

Large bandwidths are also used to limit the coherence area of a synthesis array. Generally, a variable delay, τ_D, is inserted in the IF line in order to compensate for the time-variable geometric delay, $\tau(t)$, such that the delay error, $\Delta\tau = \tau_D - \tau(t)$, is less than

the reciprocal bandwidth (see Section 10.2.2.b). However, the geometric delay varies with position in the sky, and for large interferometer spacings the change of the instantaneous geometric delay across the reception area of the array (generally, the primary beam area of the individual elements) is larger than the reciprocal bandwidth. Thus, if delay is inserted to equalize the geometric delay at the center of the reception area, radiation off-axis will not add coherently over the entire frequency bandwidth. In this way the correlation area of the array can be made smaller than the primary response area. Low-frequency arrays often use large bandwidths to limit the effective reception area (Moseley *et al.*, 1970; Erickson and Fisher, 1973).

The details of the bandwidth correlation area can be illustrated in the following manner. The calculation of **b** for the synthesized beam or brightness distribution, given in Equations (10.29) and (10.30), has been implicitly determined using ω_o, the center frequency of the RF bandpass. At another frequency ω within the passband, the associated map will appear exactly the same except for the addition of a radial scale factor (ω_o/ω) applied to the map coordinates. The integration over bandwidth $\beta(\omega)$ will produce a radially smeared brightness distribution

$$I_\beta(\boldsymbol{\sigma}) = \int \beta(\omega)\, I\!\left(\frac{\omega_o}{\omega}\boldsymbol{\sigma}\right) d\omega \quad (10.31)$$

Thus, the result of an observation of a point source at the edge of the primary beam using a wide-band long-spacing array is a source extended in the radial direction at about the same centroid position. The total flux density of the source is unchanged but the peak brightness is lowered. In this way the response of the array to emission outside the primary beam area is rejected.

e) Other Effects

A detailed analysis of many perturbations affecting a synthesized map have been considered by Brouw (1971) and will only be listed here.

(1) The exact expansion of Equation (10.9) leads to an additional term to those in

Equations (10.13) and (10.14). Only for an east-west interferometer can the term be easily incorporated. The term is important for arrays larger than a few kilometers.

(2) Errors in the synthesized pattern are caused by a phase offset, a phase variation, a baseline error or a gain variation, delay stepping, frequency dependence of the system, and time-constant effects.

10.3.3 *Sensitivity*

a) *Random Noise*

The sensitivity of a radio-telescope is usually discussed in terms of the random noise fluctuations due to contributions from the sky brightness, ground radiation, and receivers. Other causes of nonrandom noise—variously described as instabilities, atmospheric attenuation, scintillation, and interference—that limit the performance of a telescope or an array are discussed later.

A radio-telescope attempts to measure the noise power received from a radio source in the presence of a much larger power from the receivers themselves. What is important is not the large power of the unwanted noise signal but rather the statistical fluctuations in that power. If we integrate noise power in a bandwidth $\Delta \nu$ Hz for a time t seconds, we have collected $\Delta \nu t$ independent samples of the noise power. Statistical theory shows that the r.m.s. error in the estimate of the noise power is proportional to $(\Delta \nu t)^{-1/2}$. The noise fluctuations can be expressed in units of temperature,

$$\Delta T_{\text{r.m.s.}} = M \frac{T_s}{\sqrt{\Delta \nu t}} \qquad (10.32)$$

where T_s is the system temperature and M is a factor ($M \geq 1$) which depends on the method of observing. The system temperature includes contributions from receiver noise (which is usually dominant at frequencies larger than 100 MHz), sky brightness, and ground radiation entering the telescope feed and losses in the feed itself. A complete description of the calculation of the system temperature is given by Christiansen and Högbom (1969).

In terms of a telescope with a geometric area Ag and efficiency η (typically ≈ 0.5), the flux density (Wm^{-2} Hz^{-1}) r.m.s. fluctuation $\Delta S_{\text{r.m.s.}}$ may be written as

$$\Delta S_{\text{r.m.s.}} = M \frac{2kT_s}{\sqrt{\Delta \nu t}} \frac{1}{\eta Ag} \qquad (10.33)$$

where k = Boltzmann's constant = 1.38×10^{-23} W Hz^{-1} k^{-1}. An extra factor of 2 arises with unpolarized radiation, because only one-half of the power is intercepted by a feed.

The term "sensitivity" for an array has two interpretations, depending on the nature of the experiment performed. For unresolved sources the quantity of interest in the r.m.s. flux density fluctuation $\Delta S_{\text{r.m.s.}}$ given by

$$\Delta S_{\text{r.m.s.}} = \sqrt{2} \frac{2kT_s}{\sqrt{\Delta \nu t}} \frac{1}{\eta Ai} \frac{1}{\sqrt{C}} \qquad (10.34)$$

which can be derived directly from Equation (10.33), with the following considerations:

(1) $M = \sqrt{2}$ for a correlation receiver.
(2) Ai = geometric aperture for an interferometer = $2(Ag_1 \cdot Ag_2)^{1/2}$, where $Ag_{1,2}$ is the geometric aperture for each element. For identical elements $Ai = 2Ag$ = total geometric area of the interferometer.
(3) C = number of correlators. The maximum number of correlators is $N(N-1)/2$ for N elements.

For an array with many identical elements the r.m.s. noise fluctuation is equal to that of a single aperture (with an ideal receiver $M = 1$) with a geometric area equal to that of the array. For calculating the minimum detectable flux density of a point source, the array configuration is inconsequential.

For extended sources of radio emission the r.m.s. noise of the brightness distribution $\Delta I_{\text{r.m.s.}}$ is a useful parameter. The quantity is equal to that obtained by blurring $S_{\text{r.m.s.}}$ over the synthesized beam.

$$\Delta I_{\text{r.m.s.}} = \frac{\Delta S_{\text{r.m.s.}}}{\Omega_{\text{syn}}} \qquad (10.35)$$

where Ω_{syn} is the equivalent solid angle of the synthesized beam. The brightness distribution of extended sources is commonly given in terms of brightness temperature $T_b = (\lambda^2/2k)$ I. Thus

$$\Delta T_{br.m.s.} = \sqrt{2}\,\frac{T_s}{\sqrt{\Delta\nu t}}\,\frac{1}{\eta A_i}\,\frac{1}{\sqrt{C}}\,\frac{\lambda^2}{\Omega_{syn}} \quad (10.36)$$

The brightness sensitivity *decreases* with *increasing* resolution of an array, and the sensitivity to extended sources correspondingly becomes less.

All of the sensitivity parameters have been given in terms of a root-mean-square value. The correct use of a sensitivity depends on the nature of the experiment and the characteristic of the fluctuations causing the sensitivity limit. A detection limit of five times the r.m.s. fluctuation is commonly used.

Equations (10.34) to (10.36) give a lower bound to the sensitivities, since the assumption has been made that all observations of a given signal-to-noise are given equal weight (*i.e.*, natural weighting). In general, the efficiency E of the array is equal to

$$E = \frac{\sum_{i=1}^{N} w_i}{\left(N\sum_{i=1}^{N} w_i^2\right)^{1/2}} \quad (10.37)$$

where w_i is the additional weight above natural weighting of the ith point. If $w_i = 1$ for all i, $E = 1$; otherwise $E < 1.0$.

An important result is the reduction of $\Delta I_{r.m.s.}$ or $\Delta T_{br.m.s.}$ by heavily tapering (weighting more heavily) the smaller spacings. Practical arrays undersample the longer baselines, and consequently further weighting the shorter spacings while decreasing the efficiency will nevertheless increase the sensitivity to an extended source. As an example, consider the set of observations shown in Figure 10.7. By disregarding data outside a radius R in the (u,v) plane, the synthesized beam area varies as R^{-2} but the efficiency $E \sim R^{1/2}$. Thus $\Delta I_{r.m.s.} \propto \Delta T_{r.m.s.} \propto R^{-3/2}$. For an extended source, optimal signal-to-noise will be achieved when the taper is such to produce a beam which matches in size the smallest

detectable source structure. For a point source, however, the peak brightness increases with resolution as R^2; hence the net signal-to-noise is proportional to $R^{1/2}$, the efficiency.

A limitation to useful sensitivity in many telescopes is confusion rather than noise. With present-day techniques a sensitivity limit of $10^{-29}\,\text{W Hz}^{-1}\,\text{m}^{-2}$ is obtainable. The density of sources is such that at frequencies less than 5000 MHz there are many sources above $\Delta S_{r.m.s.}$ in the primary beam of even the largest filled apertures. These weak sources produce fluctuations larger than $\Delta S_{r.m.s.}$ in the response of the telescope as it is moved. A useable sensitivity of $\Delta S_{r.m.s.}$ can be reached only with better resolution, *i.e.*, with an extended array.

b) Nonrandom Noise

i) Interference

One has achieved a great deal if observations are limited by the random noise fluctuations described above. Radio astronomy has certain protected bands in which communications broadcasts are not permitted, but man-made interference is an ever-present threat. The legal limits on the ignition system of a London taxi are insufficient to prevent interference at a distance of 20 miles. For this reason radio observatories tend to be located in remote locations and partially protected by mountain ranges. The continuous increase in communications, radar, and satellites makes the threat to radio astronomy very real.

Arrays have an intrinsic rejection against interference, in that the unwanted signal must correlate in the proper phase and delay between the elements to produce interference.

ii) Instabilities

Radio astronomy receivers have very high gains and small changes are amplified. Jumps in phase can occur due to a poor contact warming up, and slow drifts are not uncommon. Many of these instabilities can be calibrated out of the observations but often not perfectly.

iii) Atmosphere

At longer wavelengths interplanetary clouds of electrons produce scintillations of small-angular-diameter sources. This phenomenon is valuable for measuring the approximate angular size of sources. Ionospheric variations produce large distortions of signals at meter and decimeter wavelengths, as anyone who listens to shortwave radio broadcasts well knows. At centimeter and millimeter wavelengths water vapor variations can produce variations in path length, causing large fluctuations in the phase in an interferometer. The effect of these variations causes higher sidelobe levels (Hinder and Ryle, 1971).

10.4 Inversion Techniques

10.4.1 Normal Inversion Methods

a) Direct Inversion

The basic inversion equations have already been given in Section 10.1 (Equations 10.13 and 10.14). Expressed in the form of discrete observations, we get

$$I(x,y) = \sum_{k=1}^{K} w_k V(u_k,v_k) \exp\{-i2\pi(u_k x + v_k y)\}$$

$$(10.38)$$

where $V(u_k,v_k)$ is the visibility function of the kth point with projected spacing (u_k,v_k), (x,y) the sky coordinate, and w_k a weighting factor associated with each measured visibility function. The corresponding synthesized beam is given by

$$P(x,y) = \sum_{k=1}^{K} w_k \exp\{-i2\pi(u_k x + v_k y)\}$$

$$(10.39)$$

the response to a point source. Each observation corresponds to average values of the visibility function and (u,v) coordinates over a certain restricted interval discussed in Section 10.3.1. The beam shape can be adjusted through the weighting parameters w_k, and the effect of the various weighting on the sensitivity, beam size, and sidelobe levels was discussed in Section 10.3.2.a.

For about 2500 input (u,v) points and a 50×50 output (x,y) array (a modest amount of data), the computing time on a medium-sized computer is ~ 20 minutes of execution time. The total number of steps requires $\approx n^2 K \approx n^4$ basic operations where $n \times n$ is the size of the (x,y) array and K is the number of sampled points. Clearly, for a large number of points, faster techniques must be resorted to.

b) Fast Fourier Techniques

If the K input data points are interpolated into a rectangular $(L \times M)$ array, the inversion in Equations (10.38) and (10.39) may be computed in the u and v coordinates separately, saving computing time.

$$I(x,y) = \sum_{l=1}^{L} \exp\{-i2\pi u_l x\}$$

$$\times \sum_{m=1}^{M} V'(u_l,v_m) \exp\{-i2\pi v_m y\} \quad (10.40)$$

where (u_l,v_m) are the grid points and V' is the interpolated value of V (weighting included in V'). The number of calculation steps $\approx n^2 (M + L) \approx 2n^3$, which is less than for the direct inversion.

For interferometric data obtained with a linear array, the Fourier inversion can also be computed in polar coordinates, with the radial and azimuthal sums separately calculated. In this case no interpolation is necessary.

Finally, by using a fast Fourier transform algorithm developed by Cooley and Tukey (1965), the number of computational steps may be reduced to $\sim 2n^2 \log_2 n$. Both the (x,y) and (u,v) planes are gridded as follows:

$$
\begin{aligned}
u_l &= l \times \Delta u & l &= -n \leq l \leq n - 1 \\
v_m &= m \times \Delta v & m &= -n \leq m \leq n - 1 \\
x_j &= j/2n\Delta u & j &= -n \leq j \leq n - 1 \\
y_k &= k/2n\Delta v & k &= -n \leq k \leq n - 1
\end{aligned}
$$

then

$$I_{j,k} = \sum_{l=-n}^{-1} \sum_{m=-n}^{n-1} V'_{lm} \exp\left\{-i\pi \frac{jl + km}{n}\right\}$$

$$(10.41)$$

where $(\Delta u, \Delta v)$ are the increments in the (u,v) plane and $(1/2n\Delta u, 1/2n\Delta v)$ are the corresponding increments in the (x,y) plane. The area in the sky of $(1/\Delta u) \times (1/\Delta v)$ is called the "field of view."

The fast Fourier transform is a method used to conveniently compute the discrete Fourier transform of Equation (10.41). A readable description of the technique has been given in Cochran *et al.* (1967) and is not discussed in detail here. It is shown that the Fourier transform of N samples can be reduced to a simple sum of the Fourier transform of two samples, each with $N/2$ samples with a saving of computing time. This reduction may be continued as long as the number of samples is divisible by 2. Thus, in a sample of $N = 2^m$ points the original Fourier transform can be completely reduced to a number of computational steps much smaller than a straightforward application of Equation (10.41).

c) Gridding

The gridding of the data required by the fast Fourier transform involves two operations in the data. First, a convolution of the data is necessary to define the visibility function at the grid points. If $c(u,v)$ is this convolution function, then

$$V'(u,v) = \int du'dv' \, V(u',v') \, c(u - u', v - v')$$

$$(10.42)$$

and

$$I'(x,y) = I(x,y) \, C(x,y) \qquad (10.43)$$

where $I(x,y)$ is the actual brightness distribution, and $I'(x,y)$ is the brightness distribution after interpolation. The functions $C(x,y)$ and $c(u,v)$ are Fourier transform pairs.

Secondly, the sampling of the data at intervals $\Delta u, \Delta v$ causes aliasing in the map. That is, the gridded distribution I'' is given by

$$I''(x,y) = \sum_{M=-\infty}^{\infty} \sum_{N=-\infty}^{\infty} I'\left(x + \frac{N}{\Delta u}, y + \frac{M}{\Delta v}\right)$$

$$|x| < \frac{1}{\Delta u}, |y| < \frac{1}{\Delta v} \quad (10.44)$$

Thus, emission outside of the field of view will be reflected within the field of view. The primary response of the system and the factor $C(x,y)$ will both decrease the response of the aliases.

The aliasing due to gridding and the grating lobes of a map are both due to a stepping interval in the data. The grating lobes depend on the measurement of the data and cannot be substantially changed. The aliasing is an artifact of the fast Fourier technique and is not inherent to the map. With proper gridding the increase of inversion speed far outweighs the complications due to gridding.

The simplest type of gridding is to average the visibility data into (u,v) cells of size $\Delta u \times \Delta v$, with the summed visibility function becoming $V'(u,v)$. The convolving function in this case is $c(u,v)$, where $c(u,v) = 1$ for $|u| < \Delta u/2$ and $|v| < \Delta v/2$; $c(u,v) = 0$ otherwise. The convolution in the (u,v) plane tapers the brightness distribution by*

$$\frac{\sin(\pi\Delta ux)}{\pi\Delta ux} \times \frac{\sin(\pi\Delta vy)}{\pi\Delta vy}$$

If the $V'(u,v)$ is then sampled with interval $\Delta u \times \Delta v$, aliasing occurs, as defined by Equation (10.44). Although attenuated by the $|(\sin x/x)|$ terms, the aliasing can be very annoying in complicated brightness distributions. A simple solution is to enlarge the field of view and to significantly reduce the aliasing from adjacent images of the sky plane. The visibility function can also be convolved with a Gaussian function. This leads to a more desirable tapering of the brightness distribution than a simple averaging into cells, but is more time-consuming in a computer.

The choice of gridding also depends upon the type of observation. If a bright small-diameter source of extent Δx by Δy is synthesized, a grid size of $\Delta u \sim (1/\Delta x)$, $\Delta v \sim (1/\Delta y)$ (Bracewell, 1958) is acceptable. The aliasing of weak emission outside of the source

---

* A predictable relationship between the convolving and tapering functions exists only when the cells contain many data points randomly situated in the cell.

area is insignificant. At the other extreme, there is no need to grid more finely than about a half-diameter of the antenna size. The aliases are then well outside of the primary beam response and the grating response.

d) Analogue Methods

Apart from the above digital inversion techniques, which are the ones that have been used in practice, there exist in principle a number of analogue procedures based on the Fourier transform relationship in other realms of physics. For example, the diffraction pattern formed by an aperture illuminated by monochromatic light is the Fourier transform of the distribution of brightness across the aperture.

If this aperture were an analogue image of the (u,v) plane (say a photographic emulsion giving both amplitude and phase changes to the wavefront), then the diffraction pattern would be an image of the sky brightness distribution. Some of these analogue techniques have already been put into practice in speckle interferometry of starlight.

10.4.2 Other Inversion Methods

The Fourier inversion process discussed above typically leads to sidelobe levels of about 5%, even with good interferometric coverage in the (u,v) plane. Thus, faint features which are still more intense than the noise fluctuations can be obscured by the sidelobes of strong features. With incomplete (u,v) coverage, such as is obtained with a non-east-west line array, with sidelobe levels of perhaps 30%, the problem of recognizing low-intensity regions is even more difficult. Finally, interferometric data with inaccurate or undetermined phase information cannot be inverted using a Fourier analysis and other methods must be considered.

Many schemes have been developed which determine the brightness distribution from a given set of visibility functions. Most of the methods are not amenable to complete analysis because of their nonlinear character,

so derivations of uniqueness of solution or the errors of solution are difficult to ascertain. The schemes fall into two subclasses: model-fitting methods which use the measured visibility function directly to determine a compatible brightness distribution; and methods that use a Fourier inversion to obtain the brightness distribution which is then improved using the characteristics of the synthesized beam or some other criteria.

a) Model Fitting

Model fitting is an attempt to reproduce the observed visibility function using a simple brightness distribution which is composed of a collection of discrete components. The fitting procedure is usually an iterative one, in which the parameters describing the model (*i.e.*, number of components, specific values for component parameters) are adjusted until a satisfactory fit is obtained between the observed visibility function and the model visibility function (Fomalont, 1968). This type of analysis is one which minimizes an error difference, ΔV_i, by adjusting a model described by a set of parameters, P_i, such as

$$\Delta V_i = \sum_{k=1}^{K} w_k |V(u_k,v_k) - M(u_k,v_k; P_i)|^2$$

$$(10.45)$$

The summation is over the K measurement of the visibility function V. The weight of each point is w_k, and $M(u_k,v_k; P_i)$ is the model visibility function with a set of input parameters P_i. Gaussian-shaped components are most commonly used because of their convenient Fourier transform. In specific cases, other component shapes can be used. A uniformly illuminated circular disk, for example, is often used in analyzing high-resolution observations of planetary surfaces. Sample visibility functions for model sources are shown in the Appendix III.

Model fitting is most useful in analyzing interferometric data of strong, isolated small-diameter sources. For such data, which are completely defined by a small number of parameters (flux density, position and diameter of each component), a model-fitting technique satisfactorily determines accurate parameter values. The values may be difficult to obtain directly from a Fourier inversion because of the complication of the beam shape and sidelobes. Model fitting can be used for inverting complicated brightness distributions, but this requires a good method for minimizing ΔV and also leads to excessive computation time for a large amount of data.

Model fitting is a useful method for inverting data having very poor or no phase information. The error ΔV in Equation (10.45) can be defined in terms of the visibility amplitude instead of the visibility function, but the analysis technique is the same. For data with no phase information the position of the brightness distribution and its axial symmetry is arbitrary.

The solution of Equation (10.45) cannot be described in detail here. A least-square analysis usually leads to a nonlinear set of normal equations determined by $\partial V_i / \partial P_i$. The equations can be linearized in the neighborhood of an initial guess solution and small changes of the model parameters can be obtained. The improved solution is then used as an initial guess for the next iteration, etc. Convergence is not assured and is very slow for a complicated model. A more useful method, called a simplex analysis, is described by Nielsen (1964, p. 336).

Bates (1969) has discussed the interpretation of interferometric data with little or no phase information. He describes the square of the modulus of the observed visibility function in terms of the position of its zeroes in the complex (u,v) plane. The zeroes form conjugate pairs, and a complex visibility function whose modulus is equal to that observed can be generated by arbitrarily choosing one zero from each pair. If the modulus is defined by N zero pairs, then there are 2^N possible visibility functions. However, many of these solutions are not satisfactory because of significant negative response over part of the brightness distribution. The use of some phase information can further limit the acceptable choice of zero pairs.

b) Source Subtraction

Often maps of radio sources contain obvious sidelobe structure from several intense features in the field or other strong sources within the primary response and delay pattern. These sidelobes distort and hide weak and underlying structures, and it is difficult to disentangle by eye the real from the spurious features. A useful method of removing these sidelobes is to subtract the corresponding visibility function of the intense feature from the total visibility function. The inversion of the remainder will then be a map with the intense feature and all of its sidelobe removed. This process can be repeated as long as there are intense components which dominate the map. It is not necessary that the removed sources be within the field of view if a gridded inversion scheme is used after the subtraction.

The parameters of the source to be subtracted from the map should be precisely determined in order to remove completely the associated sidelobes. This can be done from measurements from the inverted map; however, the model-fitting technique described above can be used to better estimate exact parameters for the intense features. This is especially true if the intense features are somewhat larger than the synthesized beam.

The combination of source subtraction and model fitting is useful for incomplete aperture synthesis of relatively simple sources. Although the original map may contain sidelobe levels of 30%, after the subtraction of a reasonable model fit solution, the resulting sidelobe level of the remaining structure may be at only a 5% level of the original scale.

c) Cleaning

A useful method, called cleaning, corrects a map for the presence of sidelobes (Högbom, 1973). Cleaning is a type of band-limited deconvolution in which the brightness distribution is decomposed into a sum of beam patterns. Let $I_D(x,y)$ be the brightness distribution found by the inversion (dirty map) and $P_D(x,y)$ the corresponding beam (dirty beam). We wish to determine the set of numbers $A_i(x_i,y_i)$ such that

$$I_D(x,y) = \sum_i A_i P_D(x - x_i, y - y_i) + I_R(x,y) \tag{10.46}$$

where $I_R(x,y)$ is the residual brightness distribution after the decomposition. Normally, the decomposition is made from a gridded high-speed inversion for faster computing. The solution is considered satisfactory if $I_R(x,y)$ is of the order of the expected noise.

An important feature of cleaning is the ease with which the constraint of source size can be incorporated in the analysis. The field of search in the (x,y) plane can be arbitrarily limited to any region(s) by determining the values of A_i only in the region of the expected emission. If the region has been properly chosen, $I_R(x,y)$ should be small over the entire field of view. If there is emission outside the cleaning region, it will still appear in the residual term. Unlike source subtraction in the visibility plane, the region searched for sources must be well within the field of view if gridded inversion schemes are used.

The decomposition of Equation (10.46) cannot be done analytically. An iterative technique commonly used is:

(1) Determine dirty map, $I_D(x,y)$, and dirty beam, $P_D(x,y)$, using the same inversion methods.
(2) Find maximum value in the search area of $|I_D(x,y)|$; denoted by I_i at position (x_i,y_i).
(3) Subtract some fraction q of the dirty beam centered at (x_i,y_i) from the dirty map.

$$I_D'(x,y) = I_D(x,y) - q\,I_i\,P_D(x-x_i, y-y_i)$$

(4) Go back to step (2) and operate on I_D'.
(5) Terminate the iteration cycle when
  (a) I_i is less than a specified value
  (b) the number of iterations exceed a limit
  (c) I_i stops decreasing or begins to increase.

The technique yields a set of beam patterns of amplitude I_i and position (x_i,y_i), the sum of which approximately reproduces the original brightness distribution. During the iteration cycle the subtraction of a dirty beam at any point may occur several times. In practice the value of $q \approx 0.5$ gives fast and accurate convergence.

The cleaned map is obtained by the summation in Equation (10.46), with the dirty beam, P_D, being replaced by a clean beam, P_C. The clean beam is arbitrary but is most conveniently taken to be a Gaussian function with an elliptical beamwidth which matches that of the dirty beam.

Analysis of the cleaning procedure is difficult because of the iterative nature of the method. Practically the method works extremely well. If the cleaning area is well chosen, it is not uncommon to produce clean maps with no sidelobes above the noise level. Dirty maps with 30% sidelobe structures can be cleaned to a level of a few percent with adequate signal-to-noise. An example of cleaned maps is given in Figure 10.9. The corresponding dirty beam and dirty map are shown in Figure 10.8. Care must be taken in the inversion process so that the dirty map is, in fact, the convolution of the real brightness distribution and the dirty beam.

d) *Maximum Entropy Analysis*

A possible new method for determining the brightness distribution from a measured set of visibility functions has been developed by Burg (1973) for use in auto-correlation data; it has been generalized by Ables (1973) for interferometric data.

The usual inversion techniques (except for model fitting) implicitly assume a value of zero for the visibility function at unmeasured points in the (u,v) plane. This produces larger sidelobes and poorer resolution than necessary but, at least, defines a unique synthesized beam pattern—and improved maps can be obtained, as described in the previous section.

The maximum entropy analysis uses only those measured data and derives a brightness distribution which is the most random, *i.e.*, has the maximum entropy of any brightness distribution consistent with the measured data. The technique has been used success-

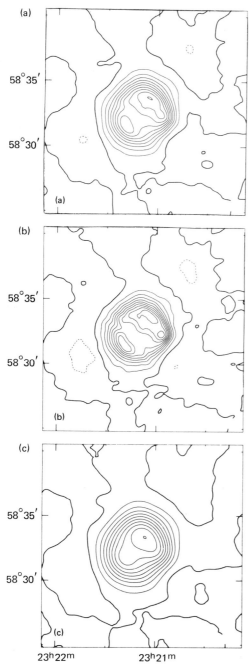

Figure 10.9 The results of using the cleaning inversion technique on CAS A. The dirty beam and dirty map have been taken from that in Figure 10.8(b).

fully on gridded one-dimensional data with improved resolution, virtually no sidelobes, and little extra cost in computational time.

However, the generalization to two-dimensional data may involve an excessive amount of computer time.

Appendix

I. Interferometric Parameters

Various quantities associated with interferometers are shown in Figure 10.A1. Given a separation of two antennas in kilometers and an observing frequency, the separation in wavelengths, the minimum fringe size, the maximum fringe rate due to the diurnal motion, and the maximum delay of a signal between antennas can be found. Lines corresponding to the radio frequencies 30 MHz, 300 MHz, 3 GHz, and 30 GHz as well as a typical optical frequency are given.

The determination of accurate pulsar positions using the change of pulsar phase with the Earth orbital motion is an interferometric technique. The frequency of the signal is equal to the pulsar repetition rate and the interferometer baseline is equal to the diameter of the orbit of the Earth. Positional accuracy to about 1/100 of a fringe size is obtained.

II. Fringe-Source Geometry

A) Baseline Coordinates

The equatorial coordinate system is generally used for describing the source and interferometer parameters. See Figure 10.A2.

$\hat{\mathbf{e}}_x$ toward the point $\delta = 0°, h = 0^h$ ⎫ left-
$\hat{\mathbf{e}}_y$ toward the point $\delta = 0°, h = 6^h$ ⎬ handed
$\hat{\mathbf{e}}_z$ toward the point $\delta = 90°$ ⎭ system

(1) Baseline vector **B** of length B (in any appropriate set of units) and direction defined from telescope 2 to telescope 1 with coordinates Declination D and hour angle H are given by

$$\begin{pmatrix} B_x \\ B_y \\ B_z \end{pmatrix} = \begin{pmatrix} B \cos D \cos H \\ B \cos D \sin H \\ B \sin D \end{pmatrix}$$

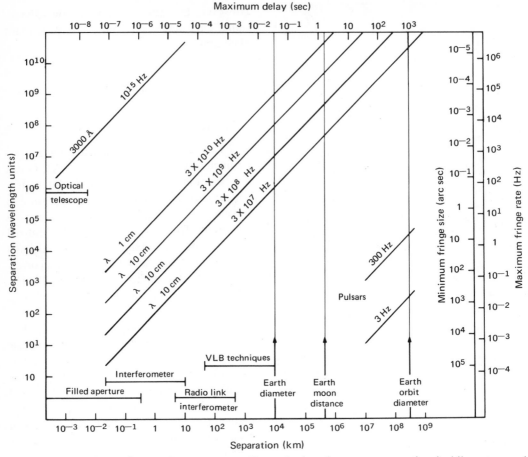

Figure 10.A1 Interferometric parameters. Given the interferometer separation in kilometers and the observing frequency, the separation in wavelengths, the maximum delay, the maximum fringe rate, and the minimum fringe size can be obtained. The range of separation used for the most common techniques of interferometry is shown at the lower part of the diagram.

(2) Source direction **s** (unit vector) given by declination δ and hour angle h is

$$\begin{pmatrix} s_x \\ s_y \\ s_z \end{pmatrix} = \begin{pmatrix} \cos \delta \cos h \\ \cos \delta \sin h \\ \sin \delta \end{pmatrix}$$

(3) For an azimuthal description of the baseline **B** in terms of B_{north} and B_{east} (the ground projections) and B_{elev} (the elevation difference),

$$\begin{pmatrix} B_x \\ B_y \\ B_z \end{pmatrix} = \begin{pmatrix} -\sin L & 0 & \cos L \\ 0 & -1 & 0 \\ \cos L & 0 & \sin L \end{pmatrix} \begin{pmatrix} B_{north} \\ B_{east} \\ B_{elev} \end{pmatrix}$$

where L is the latitude. This description is valid over only small distances, where the effect of the Earth's curvature is negligible.

(4) The delay is given by

$$\mathbf{B} \cdot \mathbf{s} = B_x \cos \delta \cos h + B_y \cos \delta \sin h + B_z \sin \delta$$
$$= B(\sin \delta \sin D + \cos \delta \cos D \cos(h - H))$$

in the same units as B.

(5) The angle θ between the interferometer pole and the source is

$$\cos \theta = \frac{\mathbf{B} \cdot \mathbf{s}}{|\mathbf{B}|} = \sin \delta \sin D + \cos \delta \cos D \cos (h - H)$$
$$= \cos (\delta - D) - \cos \delta \times \cos D (1 - \cos(h - H))$$

(6) The fringe frequency ν_f in hertz is given by

$$\nu_f = |\mathbf{B}| \cdot \frac{d(\cos \theta)}{dh} = -|\mathbf{B}| \cos \delta \cos D$$
$$\times \sin(h - H)$$
$$= \cos \delta \, [B_y \cos h$$
$$- B_x \sin h]$$

where B is in wavelengths.

B) Derived Quantities

The astrometric coordinate system is used for describing a tracking interferometer. See Figure 10.A2.

$\hat{\mathbf{e}}_u$ west to east as viewed from the source
$\hat{\mathbf{e}}_v$ south to north as viewed from the source
$\hat{\mathbf{e}}_w$ from source to observer.

(1) Baseline vector **B** becomes

$$\begin{pmatrix} B_u \equiv u \\ B_v \equiv v \\ B_w \equiv \mathbf{B} \cdot \mathbf{s} \end{pmatrix}$$

$$= \begin{pmatrix} \sin h & -\cos h & 0 \\ -\sin \delta \cos h & -\sin \delta \sin h & \cos \delta \\ \cos \delta \cos h & \cos \delta \sin h & \sin \delta \end{pmatrix} \begin{pmatrix} B_x \\ B_y \\ B_z \end{pmatrix}$$

$\mathbf{b} = \hat{\mathbf{e}}_u u + \hat{\mathbf{e}}_v v$ is the projected spacing and $B_w = \mathbf{B} \cdot \mathbf{s}$ is the delay.

(2) For a tracking interferometer the path of the projected baseline in the (u,v) plane over a 24-hour period is given by the ellipse

$$\frac{u^2}{a^2} + \frac{(v - v_o)^2}{b^2} = 1$$

where

$$a = \sqrt{B_x{}^2 + B_y{}^2} = B \cos D$$
$$b = a \sin \delta = B \cos D \sin \delta$$
$$v_o = B_z \cos \delta = B \sin D \cos \delta$$

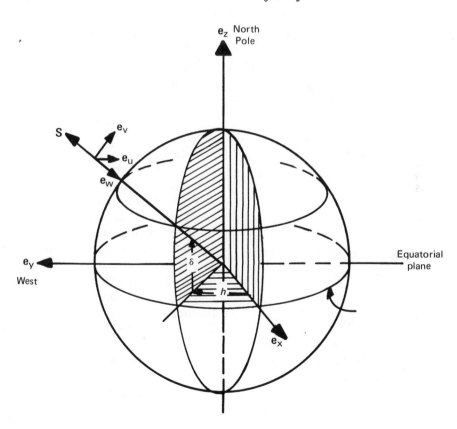

Figure 10.A2 Equatorial and astrometric coordinates used in interferometry. The unit vectors $\hat{\mathbf{e}}_x$, $\hat{\mathbf{e}}_y$, $\hat{\mathbf{e}}_z$ form the equatorial system. The unit vectors $\hat{\mathbf{e}}_u$, $\hat{\mathbf{e}}_v$, $\hat{\mathbf{e}}_w$ form the astrometric system. The direction of the source is given by **s**.

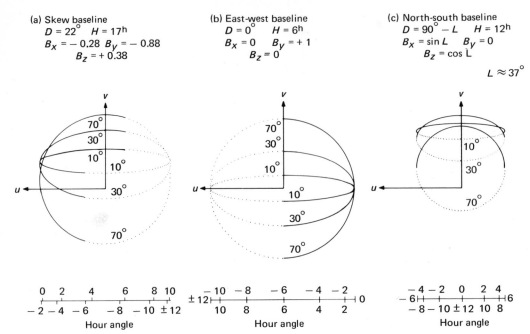

Figure 10.A3 The loci of points in the $(u–v)$ plane produced by a tracking interferometer for: (a) a skew baseline, (b) an east-west baseline, (c) a north-south baseline. The loci are drawn for declinations 70°, 30°, and 10°. The solid portion of each curve is given for the hour angle range -6^h to $+6^h$, the dotted portion for the hour angles $+6^h$ to $+18^h$. The hour angle scale, which is a function only of u, is given at the bottom of each diagram. For declinations south of the equator, use the curve with positive declination but flip the curves around $v = v_o$.

Examples are given in Figure 10.A3.

(3) The term σ, a displacement from the phase center direction **s**, may be written as

$$\begin{pmatrix} \sigma_u \\ \sigma_v \\ \sigma_w \end{pmatrix} = \begin{pmatrix} \Delta\alpha \cos \delta \\ \Delta\delta \\ 0 \end{pmatrix}$$

Neglecting the sky curvature one obtains,

$$\mathbf{B}\cdot\boldsymbol{\sigma} = u\Delta\alpha \cos \delta + v\Delta\delta = \mathbf{b}\cdot\boldsymbol{\sigma}$$

III. Visibility Functions Corresponding to Simple Brightness Models

The visibility function $V(u,v)$ can be obtained from the brightness distribution $I(x,y)$ using Equation (10.13):

$$V(u,v) = \int dx \int dy\, I(x,y) \exp\{i2\pi(ux + vy)\}$$

In Section 10.4 we discussed the general methods used to invert interferometric data. However, it is often useful to look at the general behavior of the visibility function in order to determine the approximate brightness distribution, which may then be helpful in choosing correct inversion parameters. For this reason model visibility functions are given in Figure 10.A4. For simplicity the visibility functions are plotted for an east-west projected spacing, except for the last plot. The visibility functions for any position angle can then be easily obtained. The visibility function amplitude (normalized to unity at zero spacing) is shown by the solid line; the visibility function phase is given by the dashed line in units of lobes (revolutions). All of the models can be scaled.

Details of each model are:

(a) The visibility function for a displaced point source. The displacement of the source along the angle of resolution produces a phase gradient. A displacement of 1 arc second gives a phase shift of 1 lobe at 206,265 wavelengths.

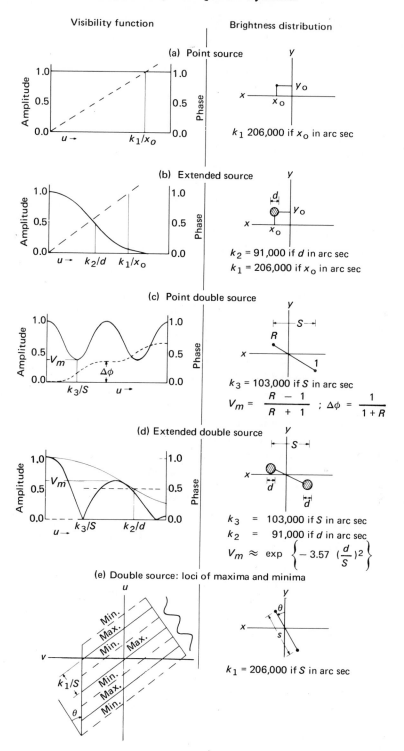

Figure 10.A4 The visibility functions for various brightness distribution models (see text).

(b) The visibility function for a displaced Gaussian source. The phase behavior is identical to the point source. The visibility amplitude is also Gaussian. A source of diameter 1 arc second (half-power diameter) yields a visibility, with half-amplitude at 91,000 wavelengths.

(c) The visibility function for a point double source. The minimum amplitude and the change of phase between successive maxima are both related to the intensity ratio, R, of the double. The period of the amplitude depends on the double separation. The sign of the phase change gives the direction of the stronger component, positive to east. The centroid of the double has been taken as the phase center.

(d) The visibility function for an equal double. The model is similar to that of an equal point double, with the visibility amplitude multiplied by an envelope equal to the visibility amplitude of the individual component. The axial ratio, s/d, of the double is given by the amplitude of the first maximum.

(e) The loci of visibility amplitude maxima and minima in the (u,v) plane for a double source. Lines of constant maxima are given by the solid lines, lines of constant minima by the dashed lines.

IV. Interferometric Polarimetry

A. *General Expression*

The output of a two-element interferometer given in Equation (10.11) can be generalized for arbitrary polarization and feed characteristics. If ϕ is the orientation of the feed and θ is the ellipticity, then the response function of the feed $\mathbf{G}$ is

$$\mathbf{G} = \hat{\mathbf{e}}_x (\cos\theta\cos\phi - i\sin\theta\sin\phi)$$
$$+ \hat{\mathbf{e}}_y (\cos\theta\sin\phi + i\sin\theta\cos\phi) \quad (10.A4.1)$$

and the voltage response to the generally polarized beam of Equation (10.17) is the complex dot product $\mathbf{G}\cdot\mathbf{E}$ for the polarized part of the radiation and $(1/\sqrt{2})|\mathbf{E}_u|$ of the unpolarized part. If $\mathbf{G}_1$ and $\mathbf{G}_2$ are the

characteristics of the two feeds, then the interferometer response is sensitive to the part of the radiation given by

$$(\mathbf{G}_1\cdot\mathbf{E}) \times (\mathbf{G}_2\cdot\mathbf{E})^* + \tfrac{1}{2}(\mathbf{G}_1\cdot\mathbf{G}_2)\,|E_u|^2$$
$$(10.A4.2)$$

which can be written in terms of $|Ex|^2$, $|Ey|^2$, $Ex\,Ey^*$, Ex^*Ey—the four polarization coherence functions. A more useful method is to write the response in terms of the four Stokes parameters, I, Q, U, and V.

From Equations (10.11), (10.17), (10.A4.1), and (10.A4.2) we get

$$R(t) = \exp\{i2\pi\mathbf{B}\cdot\mathbf{s}\} \int d\sigma\, F(\sigma)\, \exp\{i2\pi\mathbf{b}\cdot\mathbf{\sigma}\}$$
$$(10.A4.3)$$

with

$$\begin{aligned}
2F(\sigma) = {}& I(\sigma)\,\{\cos(\phi_1-\phi_2)\cos(\theta_1-\theta_2) \\
& \quad - i\sin(\phi_1-\phi_2)\sin(\theta_1+\theta_2)\} \\
& + Q(\sigma)\,\{\cos(\phi_1+\phi_2)\cos(\theta_1+\theta_2) \\
& \quad - i\sin(\phi_1+\phi_2)\sin(\theta_1-\theta_2)\} \\
& + U(\sigma)\,\{\sin(\phi_1+\phi_2)\cos(\theta_1+\theta_2) \\
& \quad + i\cos(\phi_1+\phi_2)\sin(\theta_1-\theta_2)\} \\
& + V(\sigma)\,\{\cos(\phi_1-\phi_2)\sin(\theta_1+\theta_2) \\
& \quad - i\sin(\phi_1-\phi_2)\cos(\theta_1-\theta_2)\}
\end{aligned}$$

where

(θ_1,ϕ_1) = ellipticity and orientation of feed 1

(θ_2,ϕ_2) = ellipticity and orientation of feed 2

$\mathbf{B}$ = baseline (defined from antenna 2 to antenna 1)

$\mathbf{b}$ = projected spacing

$\mathbf{s}$ = position of phase center

$\mathbf{\sigma}$ = angular coordinate

$I = E_o^2 + E_u^2$ = total intensity

$Q = E_o^2 \cos 2\beta \cos 2\chi$
$U = E_o^2 \cos 2\beta \sin 2\chi$ $\left. \begin{array}{l} \\ \\ \end{array} \right\} = \begin{cases} E_o^2\cos2\beta \\ \quad\times\exp(i2\chi) \\ = Q + iU \\ = P = \text{linear} \\ \quad\text{polarization} \end{cases}$

$V = E_o^2 \sin 2\beta$ = Circular polarization (left-hand positive)

B. Specific Examples

(a) Almost parallel feeds

$$\phi_2 = \phi_1 - \Delta\phi \qquad \theta_2 = \theta_1 - \Delta\theta$$

$$F_{||}(\sigma) = I(\sigma) + Q(\sigma) \cos 2\phi_1 \cos 2\theta_1$$
$$+ U(\sigma) \sin 2\phi_1 \cos 2\theta_1$$
$$+ V(\sigma) \sin 2\theta_1 - i I(\sigma) \Delta\phi \sin 2\theta_1$$
$$+ \text{ smaller terms if } I \gg Q,U,V$$

(b) Almost perpendicular feeds

$$\phi_2 = \phi_1 + \frac{\pi}{2} - \Delta\phi \qquad \theta_2 = -\theta_1 - \Delta\theta$$

$$F_\perp(\sigma) = Q(\sigma) [-\sin 2\phi_1$$
$$- i \cos 2\phi_1 \sin 2\theta_1] + U(\sigma) [\cos 2\phi_1$$
$$- i \sin 2\phi_1 \sin 2\theta_1] + V(\sigma) [i \cos 2\theta_1]$$
$$+ I(\sigma) [\Delta\phi \cos 2\theta_1 - i\Delta\theta]$$
$$+ \text{ smaller terms if } I \gg Q, U, V$$

(c) Linear polarized feeds

$\theta_1 = \theta_2 \doteq 0$, with the orientation designated by $F_{\phi_1,\phi_2}(\sigma)$

$$F_{0,0}(\sigma) \quad = I(\sigma) + Q(\sigma)$$
$$F_{90,90}(\sigma) \quad = I(\sigma) - Q(\sigma)$$
$$F_{45,45}(\sigma) \quad = I(\sigma) + U(\sigma)$$
$$F_{135,135}(\sigma) = I(\sigma) - U(\sigma)$$
$$F_{0,90}(\sigma) \quad = U(\sigma) + iV(\sigma)$$
$$F_{45,135}(\sigma) \quad = -Q(\sigma) + iV(\sigma)$$
$$F_{90,0}(\sigma) \quad = -F_{90,180}(\sigma) = U(\sigma) - iV(\sigma)$$
$$F_{135,45}(\sigma) = -F_{135,225}(\sigma)$$
$$= -Q(\sigma) - iV(\sigma)$$

(d) Circularly polarized feeds

$$\phi_1 + \phi_2 = \phi: \quad \text{arbitrary}$$

$\theta = \pi/4$ (left hand = L), $\theta = -\pi/4$ (right hand = R)

$$F_{LL}(\sigma) = [I(\sigma) + V(\sigma)] \exp(-i2\phi)$$
$$F_{RR}(\sigma) = [I(\sigma) - V(\sigma)] \exp(-i2\phi)$$
$$F_{LR}(\sigma) = [Q(\sigma) + iU(\sigma)] \exp\{-i2\phi\}$$
$$F_{RL}(\sigma) = [Q(\sigma) - iU(\sigma)] \exp\{-i2\phi\}$$

V. Bandwidth and Fringe-Washing Functions

(a) The monochromatic response at frequency ω is (see Figure 10.4)

$$R_\omega(t) = \cos \{\omega [(\tau(t) - \tau_D]$$
$$+ \omega_o\tau(t) - \phi(t)\}$$
$$= \text{Real} \{\exp [i\omega(\tau(t) - \tau_D)]$$
$$\times \exp [i\omega_o\tau(t) - i\phi(t)]\}$$

(b) The broad-band response is

$$R(t) = \int \alpha(\omega) R_\omega(t) d\omega$$

where $\alpha(\omega)$ is the bandpass of the system. This can be written as

$$R(t) = \exp i[\omega_o\tau(t) - \phi(t)] \int \alpha(\omega) \exp\{i\omega\Delta\tau\} d\omega$$

and using $\Delta\tau \equiv \tau(t) - \tau_D = $ delay offset becomes

$$R(t) = \beta(\Delta\tau) \exp i\{\omega_o\tau(t) - \theta(t)\}$$

where

$$\beta(\Delta\tau) \equiv \int \alpha(\omega) \exp\{i\omega\Delta\tau\} d\omega$$

is the fringe-washing function.
Examples of $\alpha(\omega) - \beta(\Delta\tau)$ pairs are given by Christiansen and Högbom (1969).

References

Ables, J. 1974. Submitted to *Astron. Astrophys.*, in press.

Baldwin, J. E., C. J. Field, P. J. Warner, and M. C. H. Wright. 1971. *Monthly Notices Roy. Astron. Soc.* 154:445.

Basart, J. P., G. K. Miley, and B. G. Clark. 1970. *IEEE Trans.*, *Ap-18*, 375.

Bates, R. H. T. 1969. *Monthly Notices Roy. Astron. Soc.* 142:413.

Bracewell, R. N. 1958. *Proc. IRE* 46:97.

Brouw, W. N. 1971. Doctoral dissertation. Leiden University.

Burg, J. P. 1973. In preparation.

Casse, J. L., and C. A. Muller. 1973. *Astron. Astrophys.* In press.

Chandrasekhar, S. 1950. *Radiative Transfer.* London: Oxford University Press, pp. 24–35.

Christiansen, N. N. 1953. *Nature* 171:831.

———, and J. A. Högbom. 1969 *Radiotelescopes.* London: Cambridge University Press.

Clark, B. G. 1968. *IEEE Trans.*, *Ap-16*, 143.

Cochran, W. T., J. W. Cooley, D. L. Favin, H. D. Helms, R. A. Kaenel, W. W. Lang, G. C.

Maling, Jr., D. E. Nelson, C. M. Rader, and P. D. Welch. 1967. *Proc. IEEE* 55:1664.

Cohen, M. H. 1958. *Proc. IRE* 46:172.

————. 1969. *Ann. Rev. Astron. Astrophys.* 7:619.

————, and D. B. Shaffer. 1971. *Astron. J.* 76:91.

Conway, R. G., and P. P. Kronberg. 1969. *Monthly Notices Roy. Astron. Soc.* 142:11.

Cooley, J. W., and J. W. Tukey. 1965. *Math. Computations* 19:297.

Elgaroy, D., D. Morris, and B. Rowson. 1962. *Monthly Notices Roy. Astron. Soc.* 124:395.

Erickson, W. C., and J. R. Fisher. 1973. In preparation.

Fomalont, E. B. 1968. *Astrophys. J. Suppl.* 15:203.

Hanbury Brown, R. 1968. *Ann. Rev. Astron. Astrophys.* 6:13.

Hinder, R., and Ryle, M. 1971. *Monthly Notices Roy. Astron. Soc.* 154:229.

Högbom, J. A. 1974. Submitted to *Astron. Astrophys.*, in press.

Kraus, J. D. 1966. *Radio Astronomy.* New York: McGraw-Hill.

MacPhie, R. H. 1966. *IEEE Trans.*, *Ap-14*, 369.

McCready, L. L., J. L. Pawsey, and R. Payne-Scott. 1947. *Proc. Roy. Soc.* A-190:357.

Mills, B. J., and A. G. Little. 1953. *Australian J. Phys.* 6:272.

Moffet, A. T. 1968. *IEEE Trans.*, *Ap-16*, 172.

Moran, J. M., B. M. Burke, A. H. Barrett, A. E. E. Rodgers, J. A. Ball, J. C. Carter, and D. D. Cudabeck. 1968. *Astrophys. J.* 152:L297.

Morris, D., V. Radhakrishnan, and G. A. Seielstad. 1964. *Astrophys. J.* 139:551.

Moseley, G. F., C. C. Brooks, and J. N. Douglas. 1970. *Astron. J.* 75:1015.

Nielsen, K. L. 1964. *Methods in Numerical Analysis.* New York: Macmillan.

Radhakrishnan, V., J. W. Brooks, W. M. Goss, J. D. Murray, and W. J. Schwarz. 1971. *Astrophys. J. Suppl.* 24:1.

Read, R. B. 1963. *Astrophys. J.* 138:1.

Rogstad, D. H., G. W. Rougoor, and J. B. Whiteoak. 1967. *Astrophys. J.* 150:9.

Ryle, M. 1952. *Proc. Roy. Soc.* A-211:351.

————. 1960. *J. IEE* 6:14.

————. 1962. *Nature* 194:517.

————, and A. Hewish. 1960. *Monthly Notices Roy. Astron. Soc.* 120:220.

————, and D. D. Vonberg. 1946. *Nature* 158:339.

Stanier, H. M. 1950. *Nature* 165:354.

Swenson, G. W. 1969. *Ann. Rev. Astron. Astrophys.* 7:353.

Wade, C. M. 1970. *Astrophys. J.* 162:381.

Weiler, K. W. 1973. *Astron Astrophys.* 26:403.

Wild, J. P. 1967. *Proc. Inst. Radio Electron. Engrs. Australia* 28:279.

CHAPTER 11

MAPPING NEUTRAL HYDROGEN IN EXTERNAL GALAXIES

Melvyn C. H. Wright

11.1 Introduction

11.1.1 Hydrogen Distribution and Kinematics

Hydrogen is the building material for stars and galaxies, and a study of its distribution and kinematics in external galaxies is essential to understanding the dynamics and evolution of these objects. Much can be learned by comparing the overall or integral properties of different galaxies, but an understanding of the detailed distribution of neutral hydrogen in a galaxy can be derived only from a high-resolution map.

The Magellanic Clouds were the first extra-galactic objects observed in the 21-cm line (Kerr, Hindman, and Robinson, 1954). The large angular extent of these objects enabled a detailed map to be made of the neutral hydrogen density and velocity distribution across the Clouds. The aim of observations today is to make similarly detailed maps for ordinary spiral and irregular galaxies.

The story of mapping neutral hydrogen in external galaxies is one of a continuing quest for angular resolution, but we shall first briefly review the classification and properties of the galaxies we will be talking about.

11.1.2 Classification Systems

Normal galaxies have been classified under several schemes. The Hubble classifi-

cation (Sandage, 1961), based on the morphological features visible on blue-sensitive plates, is perhaps the most widely used. This scheme, which divides spiral galaxies into three classes—Sa, Sb, Sc—depending on the openness of the spiral features, has been extended by Holmberg (1958) and the de Vaucouleurs (1964), and has much to recommend it, in that it would seem to represent a real separation in terms of the evolutionary parameters (e.g., fractional hydrogen content) of the galaxies. This is not to say that the series is an evolutionary sequence. Indeed, unless there can be considerable accretion of mass by a galaxy it would appear impossible for evolution to take place along the series from the irregular and Sc galaxies (which, on average have the highest fractional hydrogen content) to the more massive Sa spirals and ellipticals.

Equivalent classifications under these morphological systems are set out in the introduction to the de Vaucouleurs' (1964) *Reference Catalogue of Bright Galaxies*.

The Morgan (Yerkes) (1958, 1959) classification is a valuable complementary alternative to the Hubble classification. In this system the parameter *a-k* denotes increasing central concentration of light, while the luminosity is progressively from A- to K-type stars (Morgan and Mayall, 1957). Although the early-type (Sa) spirals tend to have a higher degree of central condensation and the Hubble type correlates well with the integrated color of galaxies (Holmberg, 1958), there is not a one-to-one correspondence between the Hubble and Morgan classifications. A further classification system for late-type spirals (Sc and Sb) is that of van den Bergh (1960a, 1960b), based on a strong correlation between the absolute luminosity and the degree of development of the spiral arms.

These classification systems are inter-related, but not exactly equivalent, and each is useful in discussing the properties of spiral galaxies.

11.2 Integral Properties of Galaxies

The overall characteristics of a galaxy are determined by its so-called integral properties, which include the morphological form under some classification system, color, total mass, luminosity, hydrogen mass, HII region abundance, radio continuum power, etc. By studying the correlations among these parameters we hope to learn something of the evolution of galaxies and the reasons for their different morphological forms.

Surveys of external galaxies in the 21-cm line have been made by a number of authors, and there are several excellent reviews of the integral properties of galaxies (*e.g.*, Roberts, 1967, 1969). We shall merely note here some of the established trends of the integral properties.

11.2.1 The Data

a) *Hydrogen Mass*

The mass of neutral hydrogen in a galaxy is derived from its 21-cm emission spectrum. The surface density (atom cm^{-2}) of hydrogen radiating in a 1 km sec^{-1} interval is given by

$$n_v = 1.823 \times 10^{18} T_s \tau_v \qquad (11.1)$$

where T_s is the spin temperature of the gas, and τ_v the optical depth (see, for example, Chapter 2).

T_s is assumed here to be constant along the column of radiating gas. If the gas is optically thin $(T_B \ll T_s)$, then the surface density is given by

$$n = 1.823 \times 10^{18} \int_{velocity} T_B \, dV \qquad (11.2)$$

and the total number of atoms, N, is obtained by integrating the surface density across the galaxy

$$N = 1.823 \times 10^{18} \int_{galaxy} dS \int_{velocity} T_B \, dV \qquad (11.3)$$

An element of surface area dS is related to the distance D by $dS = D^2 d\Omega$. Thus the number of atoms is proportional to D^2 and the integrated line flux is given by

$$N = 1.823 \times 10^{18} D^2 \int \int T_B \, dV \, d\Omega \qquad (11.4)$$

or in convenient units,

$$\frac{M}{M\odot} = 2.356 \times 10^5 D^2 \int S_v \, dV \qquad (11.5)$$

where D is in Mpc, S_v in flux units*, and dV in km sec^{-1}.

The assumption of small optical depth may be invalid, particularly where the velocity dispersion of the gas is small or the line-of-sight thickness is large owing to the high inclination of the galaxy. In this case the mass of neutral hydrogen derived above is a lower limit. Also, the variation of T_s (known to occur in our galaxy) makes the mass a lower limit from another point of view, as some of the hydrogen will be in cold clouds of substantial opacity. It is tacitly assumed that there is no interaction between radio continuum and line radiation. Epstein (1964) has considered these problems for different cases.

b) *Total Masses*

The masses of spiral galaxies are estimated by assuming that the galaxies are in rotational equilibrium.

The line-of-sight velocity of some component of the galaxy—usually neutral hydrogen or ionized hydrogen, oxygen, and nitrogen—can be measured by the Doppler shift of

* 1 flux unit = 10^{-26} Wm^{-2} Hz^{-1}.

its corresponding line emission. In the case of irregular galaxies (or for low-resolution neutral hydrogen measurements) the mass of the galaxy can be estimated (*e.g.*, Bottinelli *et al.*, 1968) directly from the measured velocity dispersion using the virial theorem (Chandrasekhar, 1942). More usually, for spiral galaxies a rotation velocity is estimated as a function of radius and a mass is derived from a model for the centrifugal equilibrium of the gas. The basic equation is of the same form in either case and the estimated mass scales linearly with the assumed distance to the galaxy. Assuming that the galaxy is a plane rotating disk in centrifugal equilibrium, a simple Keplerian estimate of the total mass is obtained as $M_T = (rv^2/G)$, where the maximum rotational velocity v occurs at a radius r. In convenient units this gives

$$M_T = 6.8 \times 10^4 \times r[\text{min arc}] \times D[\text{Mpc}]$$
$$\times v([\text{km sec}^{-1}])^2 \, M\odot \quad (11.6)$$

As discussed in more detail in Sections 11.5 and 11.6, the neutral hydrogen data usually have poor angular resolution and the total mass must be derived from a model for the rotation curve. The optical data, however, have high angular resolution and provide a detailed mass distribution, but only out to the limiting radius of the emission lines (*e.g.*, Burbidge, Burbidge, and Prendergast, 1963). There is some difficulty, then, in comparing masses derived from optical and radio data.

c) Luminosity

The measured luminosity of a galaxy increases with the area of the photographic image integrated. The Holmberg (1958) system of luminosities and diameters of galaxies provides a well-defined measurement down to a sky-brightness-limited isophote of $26^m.5$ per square second arc. The measured luminosities and colors on the UBV system can be corrected for an average galactic extinction by -0.25 cosec (latitude) and for inclination of the observed galaxy. The absolute luminosity of a galaxy is usually quoted in units of the solar luminosity and of course scales as the square of the assumed distance.

d) Selection Effects

Observational requirements force a selection of galaxies which may affect the correlations described below. The neutral hydrogen measurements are biased toward the late-type (Sc and Irr) galaxies and optical measurements of rotation curves tend to be of intrinsically bright galaxies. Upper limits for the HI-to-total mass ratio of $\approx 0.1\%$ have been established for several elliptical galaxies (Bottinelli *et al.* 1973), and we shall confine ourselves in what follows to the spiral and irregular galaxies.

11.2.2 Correlations

a) Mass Luminosity

The masses and luminosities of spiral and irregular galaxies cover about four orders of magnitude, and a plot of mass against luminosity (*e.g.*, Roberts, 1969) shows that the brighter galaxies are more massive. The mass-to-light ratio is approximately constant with one order of magnitude dispersion ($M/L \approx 4$ solar units $M\odot/L\odot$). There is no separation according to structural type in the mass-luminosity diagram, except in the case of the irregulars, which tend to be low-mass intrinsically faint systems. Without the irregulars the relationship between mass, luminosity, and morphological type is not obvious.

b) Hydrogen Mass Luminosity

The distance to an external galaxy is generally ill-determined and distance-independent parameters are of great value in establishing correlations between the integral properties of galaxies. The neutral hydrogen–to–luminosity ratio, M_H/L, is such a parameter and tends to be higher in the bluer later-type galaxies (*e.g.*, Epstein, 1964). The neutral hydrogen content is itself correlated with the luminosity, and these two correlations probably signify that star formation is proceeding rapidly in galaxies containing more hydrogen.

c) Fractional Hydrogen Content

A plot of neutral hydrogen mass versus total mass (*e.g.*, Roberts, 1969) again shows a correlation, but one which is highly dependent on the presence of the low-mass irregular systems. There is a clear relationship between the fractional hydrogen content and the structural type. The HI content ranges from 1 to 2% hydrogen in the Sa spirals to over 30% in the irregular systems. This is the most striking relationship in the integral properties of galaxies and probably indicates that the early-type spirals are more evolved than the late-type spirals and irregulars.

d) Ionized Hydrogen

Except for the nearest galaxies, ionized hydrogen has been discussed only in terms of large HII regions. Sersic (1960) has made a survey of HII regions in 66 galaxies and finds that the mean size of the three largest HII regions increases progressively with neutral hydrogen content from the earlier types to the Sc$^-$ galaxies and then decreases again in the dwarf Sc and irregulars. Hodge (1967) has cataloged the positions of HII regions in 66 galaxies and finds in an analysis of 25 of these (Hodge, 1969) that the distribution of HII regions within a galaxy tends to be a function of the morphological type, possibly indicating a difference in the state, or rate of evolution, of the galaxies.

e) Radio Continuum

Normal spiral galaxies are sources of nonthermal radio continuum, with flux densities of typically 0.1 to 1 flux unit, at 408 MHz. The optical and radio luminosities of spiral galaxies are closely correlated and recent data (Cameron, 1971) has shown that this correlation is even stronger for the restricted set of galaxies of Morgan type f. The radio continuum power is better correlated with the van den Bergh and Morgan classifications than with the Hubble type. The correlation with the van den Bergh classification and hence with spiral arm development is particularly interesting in the light of recent maps of spiral galaxies, *e.g.*, M31 (Pooley, 1969) and M51 (Mathewson, *et. al.*, 1972), showing that spiral arms are sources of radio continuum emission.

11.3 Mapping HI in Galaxies

11.3.1 Density Distribution

a) Extent of Distribution

The integral properties of galaxies have revealed several interesting correlations of total mass and hydrogen abundance with morphological type, but have not answered the question as to why the different morphological types exist. We can hope to gain insight into this, and the question of evolution and star formation in galaxies, only if we know more details of the structure and distribution of the different components of a galaxy. Particularly lacking in this respect is a detailed knowledge of the distribution of the neutral hydrogen, the most basic component of a galaxy. The mapping of neutral hydrogen in external galaxies allows us to find such things as:

(1) The overall form of the HI distribution, how this compares with that of the total mass distribution, and where the highest hydrogen concentrations are.
(2) The extent of the hydrogen distribution. Is it coextensive with the luminous population of a galaxy? How sharp are its boundaries?
(3) Neutral hydrogen wings, bridges between galaxies and companions or satellites with the same sort of relationship as the Magellanic Clouds are believed to have to the Milky Way.

b) Spiral Arms

The above questions can be answered with fairly limited resolution. With higher resolution we may comment further on whether the neutral hydrogen has the form of spiral arms and if these are coincident with the spiral arms defined by the bright HII regions or displaced from them.

A detailed correlation can be made of the surface densities of the young O and B stars, the HII regions they excite, and the neutral hydrogen from which they have been formed. This is a very important field, for in our galaxy the situation is confused by the complex velocity-distance relationship. In an external galaxy one has an overall view of the distribution of the different components.

11.3.2 Velocity Distribution

a) *Line-of-Sight Velocity*

Observations of neutral hydrogen are usually made with a multi-channel spectrometer so that the frequency or velocity distribution of the line radiation is also determined. Neutral hydrogen–line radiation arises from a forbidden transition, and the low-transition probability implies a small natural width of the line (5×10^{-16} Hz). The observed width of the line is therefore due to Doppler broadening (-4.74 kHz/km sec^{-1}) at the line rest frequency of 1420.4 MHz.

The line broadening is partly due to random and thermal motions in the gas, but is mainly due to larger-scale motions of the gas. In a spiral galaxy the dominant motion is one of rotation, although expansion motions and other "peculiar" velocities are of great interest. The line-of-sight velocity of the neutral hydrogen in any portion of the galaxy can be measured as a Doppler shift, and with sufficient angular and frequency resolution we observe different Doppler shifts across the face of the galaxy. The line-of-sight velocity is interpreted in terms of the rotation, expansion, and other velocities in the galaxy, which is assumed to have a plane rotating disk geometry.

b) *Isovelocity Contours*

The observed line-of-sight velocity $u(r,\phi)$ of hydrogen with azimuthal coordinates (r,ϕ) in a plane galaxy inclined at an angle i to the plane of the sky is given by

$$u(r,\phi) = v_o + v(r,\phi) \sin i \cos \phi$$
$$+ w(r,\phi) \sin i \sin \phi + z(r,\phi) \cos i \quad (11.7)$$

where v_o is the systemic velocity, $v(r,\phi)$ the tangential velocity (including both the rotation of the galaxy and "peculiar" streaming and random components), $w(r,\phi)$ a radial velocity in the plane of the galaxy, and $z(r,\phi)$ a velocity out of the plane of the galaxy. A complete analysis of the kinematics of a galaxy should include all these effects. If the motion of the hydrogen in a galaxy is one of pure rotation, then the lines of constant observed radial velocity will be the lines

$$v(r) \cos \phi = \text{constant} \quad (11.8)$$

A plot of $v(r)$ against radius r is called the rotation curve for the galaxy and typically has the form shown in Figure 11.1. For a

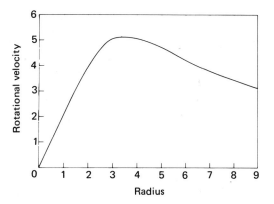

Figure 11.1 Model rotation curve showing (1) solid-body rotation out to 3 units radius, (2) constant velocity region between radii 3 and 4, (3) region of decreasing rotation velocity. (M. C. H. Wright.)

rotation curve of this form the lines of constant observed velocity (the *isovelocity contours*) will have the form shown in Figure 11.2. Where the rotation is solid-body, i.e., $v(r) \propto r$, the isovelocity contours are parallel to the minor axis of the galaxy. The region of solid-body rotation is shown enclosed by the dashed line in Figure 11.2. Other systematic motions of the galaxy have characteristic isovelocity contours; e.g., the addition of an expansion, $w(r) \propto r$, gives isovelocity contours in the solid-body region inclined at an angle to the minor axis. Closed isovelocity contours are produced by the region of the rotation curve where the rotation velocity is decreasing with radius.

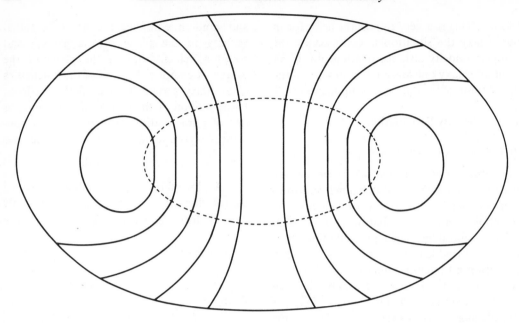

Figure 11.2 Isovelocity contours (lines of constant line-of-sight velocity) in a rotating galaxy. The contours are drawn at intervals of the units of velocity in Figure 11.1 The region of solid-body rotation is shown enclosed by the dashed line. [M. C. H. Wright, *Astrophys. J.* (1971) 166:455.]

c) Interpretation

From the velocity information we can investigate

(1) rotation of the galaxy and the shape of the rotation curve
(2) the total mass and a mass distribution for the galaxy
(3) angular momentum distribution
(4) velocity dispersion of the gas and hence an estimate of the thickness of the neutral hydrogen layer
(5) peculiar velocities, expansion motions
(6) perturbations due to spiral arms, streaming motions, and density waves.

The above list has been compiled in order of increasing observational difficulty. Observation of the spiral arms and their perturbations requires the highest resolution and sensitivity and is the subject of current research.

The highest physical resolution so far obtained has been the observations of the Magellanic Clouds with the 210-foot Parkes dish, where the half-power beamwidth is equivalent to about 200 pc at the distance of the clouds. In order to obtain comparable resolution in the nearest spiral galaxies an angular resolution of the order of 1 minute arc is required and we must resort to aperture synthesis techniques. Aperture synthesis in general is discussed in Chapter 10, and its application to spectral-line work is discussed in Section 11.6, but let us first see what has been achieved with single dish measurements.

11.4 Observations with a Single Dish

11.4.1 Observing Procedure

The radiometer system commonly used consists of a multi-channel (or swept-frequency) receiver, whose overall bandwidth covers the range of velocities to be found in the observed galaxy and is centered on or near the systemic velocity of the galaxy. The receiver is frequency-switched in order to establish a reference outside the hydrogen line, or observations are made on and off the source to find a frequency baseline for the spectrum. Scans are made across the galaxy, and the intensity in each frequency channel is recorded as a function of RA and declination. The sensitivity of the frequency channels may be calibrated by observing continuum sources or by injecting broad-band noise into the system. After calibrating the data, intensity-versus-frequency (or velocity) spectra may be drawn on a grid of points on the sky. These are the basic data. Integration under all the profiles yields a map of the hydrogen emission brightness temperature. Radio continuum emission from the galaxy may be subtracted, on the assumption of little interaction with the line radiation, by using channels outside the hydrogen-line emission or by using an off-frequency observation.

Because the resolution is limited, the frequency profiles are usually single-peaked, and it is an easy process to establish a velocity at the peak, median value, or leading edge of the profile. Thus a single velocity at each grid point can be derived and a map of iso-velocity contours of the line-of-sight velocity can be plotted.

Most galaxies subtend only a few beam areas, but for the nearest galaxies there are many beam areas and the maps derived give beam-smoothed estimates of the overall hydrogen density and velocity distribution.

11.4.2 Magellanic Clouds

As already mentioned, the observations of the Magellanic Clouds with the Parkes 210-foot dish have the highest physical resolution so far attained. The Magellanic Clouds cover some 800 square degrees of sky and, at a distance of ≈ 50 kpc, are the nearest extragalactic objects to our own galaxy. A low-resolution ($2°2$ beamwidth) HI survey (Hindman, Kerr, and McGee, 1963) showed that the Large and Small Clouds (LMC and SMC) are embedded in the same HI envelope. The total HI mass of the system is some 1.5×10^9 M$\odot$; 5.4×10^8 M$\odot$ is associated with the LMC and 4.8×10^8 M$\odot$ with the SMC.

The 210-foot survey of the LMC (McGee and Milton, 1966) revealed 52 HI complexes of mean HI mass 4×10^6 M$\odot$ and diameter 575 pc (≈ 1 HI atom/cc). The HI complexes are closely associated (in position and velocity) with HII regions and OB stars, but do not correlate with the stellar clusters.

Of 90 large HII regions cataloged by Henize (1956), 61 are closely associated with the HI complexes. The supergiant OB stars with measured velocities also have a high correlation in position and velocity with the HI complexes.

From an analysis of the rotation of the LMC and the position and velocity distribution of the population I objects, McGee and Milton (1966) proposed a spiral structure for the LMC. The total mass derived from rotational and random motions is greater than 6×10^9 M$\odot$, giving the LMC a fractional hydrogen content of 5 to 9%. The 1410-MHz radio continuum also correlates well with the integrated hydrogen contours.

The neutral hydrogen distribution in the SMC is rather smooth, with three major concentrations reaching peak brightness temperatures of 150, 110, and 100°K, merging into a high-level background. This smooth distribution is in striking contrast to the clumpy distribution in the LMC. The most interesting feature of the 210-foot survey (Hindman, 1967) is the presence of what would appear to be three massive (1 to 2×10^7 M$\odot$) expanding neutral hydrogen shells of diameter 1 to 2 kpc, which could be the result of supernova explosions. The SMC is also thought to be rotating, but is only slightly flattened. The estimated mass is 1.5×10^9 M$\odot$, with some 30% neutral hydrogen. As in the LMC, there is a correlation between the distribution of neutral hydrogen and the bright stars. About half of the cataloged HII regions (Henize, 1956) are located in the vicinity of the three major gas concentrations.

11.4.3 Nearby Spiral Galaxies

Although the Magellanic Clouds are closest and allow the most detailed comparison of optical and radio features, it is important to also observe the external spiral galaxies which more closely resemble the Milky Way. Our galaxy is thought to have a morphological form somewhere between that of M31 and M33 (see Burke, 1967). Both of these galaxies have been mapped with 10 minutes arc resolution, equivalent to 2 kpc at a distance of 680 kpc. At this resolution the most prominent feature of the HI distribution of M31 is a deficiency of hydrogen in the central regions, with the peak hydrogen distribution in the form of a broad ring at a radius of 50 minutes arc (Roberts, 1966; Gottesman and Davies, 1970). The OB star associations, HII regions, and radio continuum emission are closely correlated with the ridge of neutral hydrogen, all of which suggests that star formation is proceeding at maximum rate where the neutral hydrogen density is highest (see Section 11.7.5). There are several possibilities for the deficit of neutral hydrogen in the central regions of M31 (a feature it has in common with our galaxy and several other external galaxies; see Section 11.7.4).

A rotation curve has been derived out to 150 minutes arc from the center of M31. The computed mass distribution gives an almost constant M/L (mass-to-light) ratio (11.9 $M\odot/L\odot$) over the entire galaxy. The total mass within 150 minutes arc is some $2 \times 10^{11} M\odot$, giving M31 a fractional hydrogen content of 2%—typical for an Sb galaxy. There is a marked asymmetry in the isovelocity contours of adjacent quadrants of the galaxy, and the rotation curves derived separately for the north and south of the galaxy differ in the central regions by some 30 km sec^{-1}. These asymmetries have been interpreted as a tilt in the plane of the galaxy of some 10° and could be caused by the tidal effect of the companion galaxies M32 and NGC 205 in a similar way to the suggested influence of the Magellanic Clouds on the Milky Way (Avner and King, 1967).

M33 is an Sc galaxy and has the second largest angular size in the Northern Hemisphere. At an assumed distance of 690 kpc it lies some 190 kc from M31. Observations

with a 10-minutes-arc resolution (Gordon, 1971) reveal a neutral hydrogen distribution with the same overall dimensions as the Holmberg optical size (83 × 53 minutes arc). The neutral hydrogen distribution is asymmetric, with a major concentration in the south-preceding quadrant and a 10% central depression. At the extreme ends of the major axis are two companions, or wings, of the galaxy which have a very different position angle to that of the main body of the galaxy. These wings contain 10% of the total HI mass. The computed rotation curve is rather flat-topped, and the derived total mass within 60 minutes arc of the center is 2.3×10^{10} M⊙. The neutral hydrogen mass is some 7% of this total. The velocities in the wings do not follow those predicted by the rotation curve, and a consistent interpretation is that the wings are stable gaseous companions gravitationally bound to M33 rather like the Magellanic Clouds are to our galaxy. M33 is discussed in greater detail in Section 11.7.1.

11.4.4 Galaxies with Smaller Angular Diameters

With more distant galaxies, where the beam size is comparable with the angular diameter of the galaxy, it is still possible to measure the line profile and estimate the mass of neutral hydrogen. The velocity width of the profile allows an estimate of the rotation or random velocities within the galaxy and, with the usual assumption of gravitational equilibrium, the mass of the galaxy may be estimated.

When the hydrogen distribution subtends 2 or 3 beamwidths, an estimate of the large-scale HI distribution may be made and compared with the optical features. An example of this sort of observation is the work done by the Meudon group with the Nançay radio-telescope. This instrument has a beamwidth of 4 minutes arc E-W by ≳24 minutes arc N-S at 21-cm, which allows an estimate of the E-W hydrogen distribution. Using this instrument Bottinelli (1971) finds that the hydrogen distribution is asymmetrical with respect to the optical distribution in about 40 % of the galaxies observed. The ratio of the neutral hydrogen diameter to the optical diameter [measured on the Holmberg (1958) system] is observed to be a function of the morphological type, increasing toward the later-type galaxies. This characteristic gives a mean HI surface density independent of the galactic type. Details of the neutral hydrogen distribution may be deduced through a model-fitting procedure, and Bottinelli finds that a ring-like model for the HI distribution (as in M31) is consistent with about 30 % of the galaxies observed and that the neutral hydrogen seems to be more strongly concentrated toward the center in the early-type galaxies.

11.4.5 Model Fitting

a) HI Distribution and Rotation Curve

The technique of taking a model distribution for a source and smoothing it by the observing beam is well known. We can do rather better with line observations, as the distribution of HI in a particular velocity range is a function of both the density and velocity distribution of the gas. For a rotating galaxy with isovelocity contours, as in Figure 11.2, observations in the different velocity channels are essentially of the regions delimited by the isovelocity contours. The shape of these is determined by the rotation curve and other motions in the galaxy. Profiles of brightness temperature versus velocity may be generated from a model for the rotation curve and the neutral hydrogen distribution. The generation of model profiles is usually performed with a digital computer and many input models can be tried. The HI distribution and rotation curve, which give profiles that most resemble the observed profiles, are to be preferred. The chief drawback to this procedure is that there must always be two multi-parameter inputs: the rotation curve and the hydrogen distribution. These interact in the generation of the model profiles and it is usually possible to get a good fit to the observed profiles with more than one input model.

b) Minor Axis Profiles

The region radiating in a small velocity range about the systemic velocity of a rotating galaxy lies close to the minor axis (see Figure 11.2). In a direction orthogonal to the minor axis the region observed may be much narrower than the beamwidth. With limited resolution the shape of the hydrogen distribution in a small velocity range about the systemic velocity (*viz.*, the minor axis profile) is often observed to be double-peaked or flat-topped. This observation has led to the suggestion (Roberts, 1967) that the distribution of hydrogen along the minor axis is also double-peaked and that a ring-shaped distribution of hydrogen, as in M31, is a common feature of the HI distribution in external galaxies. Consideration of Figure 11.2 shows that the shape of the minor axis profile is a function of both the distribution of HI along the minor axis and the shape of the isovelocity contours defining the minor axis region observed.

In M31 the neutral hydrogen distribution along the minor axis is double-peaked in the integrated neutral hydrogen distribution as well as in a small velocity range centered on the systemic velocity, and the description of the overall distribution of HI as a ring is a good one. M33 has a similar double-peaked minor axis profile in a small velocity range, but this is due to the shape of the isovelocity contours, and the HI distribution is really rather flat-topped. It remains to be seen from higher-resolution observations whether a ring is a good general description of the neutral hydrogen distribution in external galaxies (see Section 7.3).

c) Mass Derivations

In many cases the internal motion of an external galaxy is well approximated by a rotation law, $v(r)$, and if we assume that the rotating galaxy is in dynamic equilibrium under self-gravitation, it is possible to derive a mass distribution from the rotation law.

Several schemes have been used for calculating a mass distribution from the rotation curve. Most mass derivations based on optically measured rotation curves have used a mass model of the form of concentric spheroids developed by Burbidge, Burbidge, and Prendergast (1963). The mass is calculated by fitting a polynomial to the rotation curve, and substituting this into the equation for the equilibrium of the concentric spheroids to derive the density as a function of radius. Use of a polynomial with more than five or six terms produces unrealistic oscillations in the rotation curve, and the calculated mass distribution is not particularly sensitive to the number of terms in the polynomial used.

Many spiral galaxies are highly flattened and a variable density disk is a good model. Model rotation curves developed by Brandt and Belton (1962) have been much used in neutral hydrogen work, as they may be inverted to give a mass distribution directly. The necessary functions are well tabulated (Brandt and Scheer, 1965). The Brandt curves are characterized by a maximum rotation velocity V_{max} at a radius R_{max}, also called the turnover radius. There is also a shape parameter n, which gives more sharply peaked rotation curves for larger values of n. The general equation is

$$v(R) = \frac{V_{max} \times R/R_{max}}{[1/3 + 2/3\,(R/R_{max})^n]^{3/2n}} \quad (11.9)$$

The only physical feature of the Brandt curve is that, at large radii, the galaxy must appear as a point mass and the rotation velocity is then Keplerian. There is very little evidence from observations as to the nature of actual rotation curves beyond the turnover radius, but the Brandt curve is often a good fit up to this point, and the derived mass within this radius compares well with masses derived by fitting concentric spheroids. The total mass derived by extrapolating the observed rotation curve along the best-fitting Brandt curve is, however, much larger, and we have the rather unsatisfactory result that much of the mass of the galaxy lies beyond the observed region. Fitting a rotation curve model to a number of

galaxies does offer a convenient and standard way of comparing the mass and derived quantities of these galaxies, but some care must be exercised in interpreting the results. The best procedure seems to be to quote a mass out to some standard radius such as the Holmberg (1958) radius. The angular momentum distribution may also be derived from the fitted density distribution, and this is of interest with respect to theories of galaxy formation from a condensing cloud of gas. The Brandt curve is characterized by only three parameters: R_{max}, V_{max}, and the shape parameter n. Since the angular momentum scales as $R_{max}^2\,V_{max}^3$ and the mass as $R_{max}\,V_{max}^2$, we must avoid comparing any two parameters such as mass and angular momentum derived by fitting the Brandt curve, as they will then be mathematically correlated, independent of any real physical correlation between mass and angular momentum. Indeed, in the absence of real correlation between mass and angular momentum a graph of $R_{max}^2\,V_{max}^3$ versus $R_{max}\,V_{max}^2$ has a slope determined by the relative dispersion in the distributions of R_{max} and V_{max}. Measurements of angular momentum are also rather unsatisfactory, as most of the angular momentum lies beyond the observed rotation curve.

For those 21-cm observations where there is insufficient angular resolution to measure the radius of maximum rotation velocity directly, a mass may still be obtained by estimating V_{max} from the width of the profile and R_{max} from the optical size of the galaxy. Figure 11.3(b) is a plot of R_{max}/a [where a is the Holmberg (1958) diameter] for 21 galaxies for which the rotation curves have been determined optically from long-slit spectroscopy of HII regions. The ratio is not a function of the morphological type, and the histogram is quite sharply peaked with $R_{max} \approx 0.1a$. If R_{max} cannot be measured, then we can estimate it as one-tenth a in order to obtain a mass estimate.

For irregular galaxies a mass may be estimated from the virial theorem (*e.g.,* Volders and Högbom, 1961); the mass will be

$$M = \frac{kaV_{\text{r.m.s.}}^2}{G} \qquad (11.10)$$

where a is the diameter of the galaxy, G is the gravitational constant, $V_{\text{r.m.s.}}$ is estimated from the width of the velocity profile, and k is a constant of order unity which depends upon an assumed model for the density and velocity distribution.

d) Noncircular Velocities

Noncircular velocities are apparent as departures of the isovelocity contours from symmetry about the major and minor axes. In particular we are interested in analyzing the isovelocity contours for expanding hydrogen and for streaming motions in the vicinity of spiral arms as predicted by the density wave theory (see Chapter 4 and also Lin, Yuan, and Shu, 1969). Analysis of these effects may be made by fitting the isovelocity contours with a model rotation curve, $V = V(r)$, and a set of parameters such as the systemic velocity, inclination, position angle, and rotation center.

The fitting for $V(r)$ can take place over the whole plane of the galaxy, with a higher weight given to points near the major axis. Having obtained the best-fit rotation curve, model isovelocity contours can then be subtracted from the observed isovelocity contours to give the residual velocity field. Examination of the residuals then shows more clearly the systematic noncircular or "peculiar" velocities. It should be noted that the isovelocity contours and the residual velocity field are velocities weighted by the hydrogen distribution within the beam area, and particular care must be taken in interpreting the results. The effect of beam smoothing is to

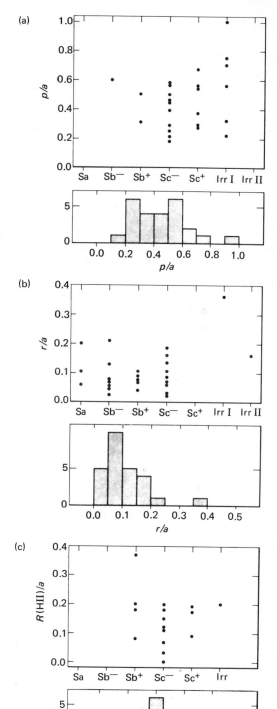

Figure 11.3 Plot against morphological type of galaxy and a histogram for
(a) turnover radius, ρ/a, for galaxies observed by Rogstad, Rougoor, and Whiteoak (1967)
(b) radius of maximum rotation velocity for optically derived rotation curves
(c) radius of maximum HII region count, $R(\text{HII})/a$, for galaxies observed by Paul Hodge. [M. C. H. Wright, *Astrophys. J.* (1971) 166:455.]

Figure 11.3

bias the measured velocity toward that of hydrogen concentrations within the beam. Two examples may be given:

(1) A galaxy which is a few beamwidths in diameter and which has a steep rotation curve toward the center has isovelocity contours drawn at intervals of the velocity resolution with a separation less than the observing beamwidth. Broad-frequency profiles will be observed, and the estimated rotation curve will have a smaller slope than the true curve.

(2) Suppose that the hydrogen distribution has the form of spiral arms with separation rather less than the observing beamwidth. Observations between the spiral arms where there is not much HI will give a beam-smoothed profile with a velocity biased toward that of the nearest spiral arm. There is usually a gradient in the rotation velocity, and observations on the inner and outer edges of the spiral hydrogen concentration will yield velocities biased in opposite directions. We may be looking for just this sort of velocity perturbation resulting from streaming motions near spiral arms! Clearly, caution must be exercised in drawing conclusions from observed peculiar velocities.

11.5 Interferometric Observations

11.5.1 Position Profiles

An interferometer with a multi-channel receiver is an inherently powerful tool for line observations, because the phase in each velocity channel contains information about the position of the source of radiation received in that velocity interval. A phase relationship over several channels can confirm a detection of a weak signal, while for stronger signals the phase gives the relative position in the sky of radiation received in each velocity channel. A plot of position versus velocity is often called a position profile.

11.5.2 Observations of External Galaxies

For a rotating galaxy, hydrogen radiating at different velocities will arise from regions of the galaxy (such as those in Figure 11.2) delimited by the isovelocity contours. If observations are made with an interferometer whose baseline has an orientation close to the position angle of the major axis of the galaxy, and whose length, D, is not so great that the hydrogen distribution is appreciably resolved (*i.e.*, $a > \lambda/D$, where a is the diameter of the galaxy), then the interferometer phase in each velocity channel indicates the distance along the major axis of hydrogen radiating in that velocity interval. The position profile is then a plot of the centroid positions of the regions in Figure 11.2 weighted by the hydrogen distribution (and smeared by random velocities, the observing beam, and the velocity filter). The amplitude in each velocity channel is roughly the product of the hydrogen density multiplied by the area of these regions. Observations of a number of galaxies have been made in this manner at the Owens Valley Radio Observatory (Rogstad, Rougoor, and Whiteoak, 1967). The amplitude-velocity and phase-velocity plots are a function of both the hydrogen distribution and the rotation curve, and the position profile obtained is subject to various interpretations.

11.5.3 Interpretation of the Data

a) Rotation Curves

For a solid-body rotation curve the isovelocity contours are everywhere parallel to the minor axis, and the position profile along the major axis reproduces the shape of the beam-smoothed rotation curve. For any other rotation curve, interpretation is more difficult and we must resort to model fitting. If the rotation velocity falls to small values at a large radius, as in Figure 11.1, then the isovelocity contours form small closed areas at the high velocities, as in Figure 11.2. Provided that these closed regions occur well within the HI distribution, then the amplitude decreases at the velocity corresponding to the appearance of small closed regions in the isovelocity contours. The position indicated by the interferometer phase at this velocity gives an estimate of the "turnover radius," where rotation velocity is a maximum. This was the original interpretation given to the position profiles by Rogstad *et al.* (1967). Through a model-fitting procedure they deduced the turnover radius of the rotation curve and hence together with the rotation velocity the masses of the galaxies.

b) Hydrogen Distribution

The turnover radius deduced by Rogstad *et al.* (1967) is, on the average, a factor of two larger than rotation maxima measured by Burbidge, Burbidge, and Prendergast (1963), and others, by means of optical long-slit spectroscopy of HII regions.

This leads us to consider an alternative interpretation of the position profile in terms of the extent of the hydrogen distribution (Wright, 1971a). If the rotation velocity does not decrease beyond the maximum, as in Figure 11.1, but instead remains at a high value, then the small closed regions do not appear in the isovelocity contours. The amplitude response then decreases at a velocity where there is not much hydrogen. The position profile at this velocity then tells us the extent of the hydrogen distribution.

c) Shape of HI Distribution and Rotation Curve

In either of the above interpretations the position profile always gives us a lower limit to the extent of the hydrogen distribution, and comparison of this radius with that of the total mass and HII region distribution shows that the HI distribution is much wider than either (see Figure 11.3).

We can use the disagreement between the optically derived rotation maximum and that deduced from the first interpretation to invert the argument as follows: If the first explanation is not correct, then the rotation velocity cannot fall to a low value as in Figure 11.1, but must remain at a high velocity within the extent of the hydrogen distribution. This is quite consistent with observed rotation curves (see Section 11.7), which are indeed rather flat-topped.

11.6 Aperture Synthesis Observations

11.6.1 Observational Requirements

a) Angular Resolution

Model fitting of low-resolution maps or of simple interferometer observations is clearly inadequate to analyze the details of the HI distribution and kinematics in spiral galaxies. Aspects concerning the spiral structure can be investigated only when we have obtained a map of the neutral hydrogen distribution with a resolution comparable to the angular separation of the luminous spiral arms. In the nearest galaxies this is of the order of 1 minute arc, and such resolution can at present be obtained only by the use of aperture synthesis techniques. Aperture synthesis observations of external galaxies are currently being made at four observatories. The instruments being used are the Cambridge Half-Mile Telescope in England (Baldwin *et al.*, 1971), the 12-element Westerbork Synthesis Radio Telescope (Brouw, 1971) in the Netherlands, the 2-element interferometer at the Owens Valley Radio Observatory, and the 3-element interferometer at the National Radio Astronomy Observatory (NRAO) in the United States.

Observations are usually made with the telescopes at each of several separations at which the telescopes track the source over a range of hour angle. Maps of the sky brightness are computed from a Fourier inversion of the recorded data. This is the method of Earth rotation synthesis (see Chapter 10). The maximum angular resolution is determined by the largest interferometer baseline, D, used in the aperture synthesis. The synthesized beam subtends a solid angle $(\lambda/D)^2$ cosec δ, where δ is the declination of the source.

If observations are made with interferometer baselines at intervals d to a maximum D, then a grating sidelobe response occurs at an angle λ/d in RA and $(\lambda/d) \times$ cosec δ in declination. The area of sky which can be mapped without ambiguity from grating side-lobes is $(\lambda/d)^2$ cosec δ. The baseline interval, d, should be sufficiently small that this area includes the extent of the neutral hydrogen radiating in any velocity channel; otherwise the synthesized map will be confused by grating sidelobes. There is no point, however, in making d much smaller than the diameter of the antennas, since the reception pattern of these then limits the area of sky which can be mapped.

b) Frequency Resolution

Observations of spiral galaxies show that the greatest range of velocities expected is about ± 300 km sec^{-1}, and it is desirable that a multi-channel receiver should cover the whole of this range. The maximum useful resolution in velocity is dependent on the angular size of the synthesized beam, since in general the line-of-sight radial velocity varies in a systematic way across the galaxy. Near the center of the galaxy the area of the synthesized beam intersects many isovelocity contours (see Figure 11.2); a large range of velocities is present within one beam area, and a high-velocity resolution is not required. Further out in the galaxy, however, there is a small range of rotation velocities within one beam area and a higher resolution might be useful. Dispersion in the gas along the line of sight might be of order 10 km sec^{-1} and this accordingly is a useful resolution. Thus a desirable minimum for a multi-channel receiver is 60 channels, each 10 km sec^{-1} wide, for each interferometer.

c) Sensitivity

Suppose that we spend equal times observing at each interferometer baseline at intervals d to a maximum D. The total integration time is proportional to D/d. The angular resolution obtained is proportional to $\theta = \lambda/D$ so that for a given stepping interval d (which is determined by the angular extent of the hydrogen) the fluctuations in aerial temperature, T_a, are proportional to $\theta^{1/2}$. As a result of the aperture synthesis, only flux collected by the *synthesized* beam contributes to the effective aerial temperature of the source. The aerial temperature, T_a, due to a

source of uniform brightness temperature, T_b, is obtained by multiplying T_b by the ratio of the synthesized beam area to the antenna beam area. Thus $T_a \propto T_b\theta^2$, and the sensitivity of the telescope to extended objects *decreases* with resolution as

$$T_a/\Delta T_a \propto T_b \; \theta^{3/2} \qquad (11.11)$$

This degrading of the signal-to-noise ratio with increasing resolution can be offset by spending a longer time at the larger aerial spacing, and the reason for this requirement may be understood in terms of the increased rate of sampling of the (u-v) plane at the longer baselines (see Chapter 10). It does, however, mean that the sensitivity requirements may limit the resolution rather than the available baseline.

11.6.2 Correlation Receivers

The requirement for a large number of velocity channels has favored the use of cross-correlation receivers. The principle on which the cross-correlation receiver operates is that, for two random time-varying signals, $V_1(t)$ and $V_2(t)$, the cross-correlation function

$$\sigma(\tau) = \int V_1(t) \, V_2(t - \tau) \, dt \quad (11.12)$$

is the Fourier transform of the visibility spectrum $V_1(\omega) \, V_2(\omega)$ of the two signals. Here the signals $V_1(t)$ and $V_2(t)$ are the voltages from the two telescopes forming the interferometer, and $V_1(\omega) \, V_2(\omega)$ is the cross-correlated spectrum at an angular frequency ω. The cross-correlation function is sampled over a range of delays $\pm \Delta T, \pm 2\Delta T, \ldots$ to a maximum delay $\pm T$ sec, and the visibility spectrum is obtained as the Fourier transform of the sampled cross-correlation function.

The maximum delay T determines the resolution of the synthesized frequency channel, $1/2T$ Hz, and the sampling interval ΔT produces a grating response in frequency at an interval $1/2\Delta T$. The exact shape of the equivalent frequency filter is the Fourier transform of the weighting applied to the cross-correlation function. If the latter is transformed with equal weight applied to each delay, then the equivalent frequency filters have a $(\sin \theta)/\theta$ response, with a half-width of $1.2/2T$ Hz and 22% sidelobes. Both positive and negative delays must be sampled to determine the amplitude and phase of the interferometer, and the correlation receiver is equivalent to a bank of $T/\Delta T$ adjacent frequency filters at intervals of $1/T$ Hz.

The cross-correlation may be achieved in practice either in an analogue device using physical delay steps or in a digital correlator. In the latter, simplified logic results if one-bit sampling of the correlation function is employed (so that only the sign of the sampled correlation function is recorded). This results in some loss in signal-to-noise ratio, but the visibility spectrum may be fully restored (Weinreb, 1963) through the Van Flyck correction. While increasing the complexity of the data processing, as an extra Fourier transform must be computed, the correlation receiver has a number of advantages over a conventional filter bank receiver in that the relative sensitivity of the frequency channels is easily calibrated. A digital correlator has good stability essential for a good synthesis, and the additional advantage that the bandwidth can be changed by simply changing the clock rate which determines the sampling interval.

11.6.3 Data Reduction

a) *Frequency Fourier Transform*

If a correlation receiver has been used, then the first stage is a Fourier transform to recover the visibility spectrum of the signal. It should be noted that the shape of the frequency filters is under our control as we can apply a weighting function to the correlation function before computing the Fourier transform (this is equivalent to convolving the spectrum after the Fourier transform). Thus we can reduce the 22% frequency sidelobes associated with a uniform weighting function, at the expense of an increased half-width. The frequency Fourier transform is often made in an on-line computer, which receives data directly from the correlator.

b) *Calibrations*

There are two basic calibrations in addition to the usual interferometer baseline and system gain-and-phase calibrations of a broad-band interferometer (see Chapter 10). These are:

(1) to determine the relative sensitivities of the frequency channels
(2) to determine the sensitivity and system phase of the telescope as a function of frequency.

c) *Spatial Fourier Transform*

Each frequency channel may be separately Fourier inverted as for a broad-band interferometer, as discussed in Chapter 10. The chief problem is the large quantity of data to be processed, and the computation may take several hours—even on the largest computers using fast Fourier transform techniques.

d) *Presentation of the Data*

The end-product of the above processing is a series of maps of the hydrogen distribution in the different frequency channels. These maps form the basic data, which can be considered as a three-dimensional array with x, y, and frequency coordinates. The maps of frequency channels containing line emission may be combined to produce a map of the integrated line emission and a map of frequency spectra at a grid of points on the area of sky studied. A velocity (or profile width) can then be fitted to the frequency spectra and a map of isovelocity contours drawn across the source. There is obviously a considerable data-handling problem associated with mapping a $1°$ square of sky, say, with an angular resolution of 1 minute arc at 60 different velocities, and the basic problem is presenting the information in a digestible form. It is to be expected that fairly exotic techniques will eventually be used in this final and most important stage of data reduction, namely, in the interface between the data and the astronomer.

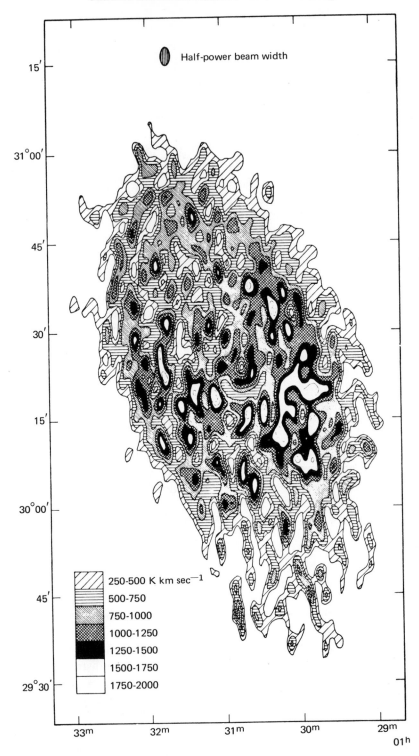

Figure 11.4 Integrated HI brightness in M33 to an angular resolution of 1.5 × 3 minutes arc. [Wright *et al.*, *Monthly Notices Roy. Astron. Soc.* (1972) 155:337.]

11.7 High-Resolution Maps

11.7.1 Observations of M33 with the Cambridge Half-Mile Telescope

a) Observations

The highest resolution observations of an external galaxy so far made using aperture synthesis techniques are those of M33, made with the Cambridge Half-Mile Telescope (Wright, Warner, and Baldwin, 1972). This telescope consists of two 9-m paraboloids on an east-west baseline and was designed as an aperture synthesis instrument for observing extended objects. The observations of M33 have an angular resolution of 1.5 minutes arc in RA and 3 minutes arc in declination, equivalent to a 300×600 pc area at the distance of M33 (690 kpc). (This compares with a linear resolution of 210 pc in the Magellanic Clouds using the Parkes 210-foot telescope.) The velocity resolution of these data is 39 km sec^{-1}, commensurate with the range of velocities expected within a 300×600 pc area over most of the galaxy. Observations were made at 59 interferometer baselines with telescope separations at 6m intervals to a maximum of 360m. The full baseline (720 m, $\approx$ a half-mile) was not used for reasons of sensitivity, as discussed in the previous section. The data were obtained using an analogue cross-correlation receiver and were processed much as described in the previous section. The basic data are in the form of nine maps of the HI distribution at 26 km sec^{-1} intervals. These nine maps cover the range of velocities found in the neutral hydrogen of M33.

b) Integrated HI Brightness Distribution

For these observations the half-width of a velocity channel is larger than the velocity interval between the channels, and a simple addition of these nine maps suffices to construct a map of the integrated hydrogen distribution in M33 (Figure 11.4). This map is of the surface brightness temperature of the hydrogen line integrated over the line profile. If the galaxy is everywhere optically thin

to the line radiation, then the map also represents the distribution of the HI surface density projected along the line of sight. The peak brightness temperature observed is 50°K but the distribution is in places unresolved in both angle and velocity so that the true brightness temperature may exceed 100°K and the line radiation may not be optically thin. Where the radiation is not optically thin, the brightness temperature gives only a lower limit to the surface density. There is no direct evidence of optically thick HI from absorption of continuum sources lying behind or in the disk of M33, and we can adopt as a working hypothesis that the line radiation is optically thin, so that the map of the integrated brightness temperature is also a map of the HI surface density.

c) Large-Scale Structure

The large-scale structure of the hydrogen distribution may be obtained with a higher signal-to-noise ratio on a lower-resolution map (which may be obtained in aperture synthesis observations by simply not including data from the larger interferometer spacings in the Fourier transform). A low-resolution map generated from the above data agrees well with the map obtained by Gordon (1971) with the NRAO 300-foot telescope and described in Section 11.4. Figure 11.5 is an integration in elliptical rings (circular in the galaxy plane) of the brightness temperatures of Figure 11.4, and shows that the average radial distribution is a plateau with a very sharp cut-off at the edges. The radial distribution in Figure 11.5 is not in good agreement with the suggestion by Roberts (1967) that the HI has a ring distribution as in M31. The average projected surface density is $\approx 3 \times 10^{21}$ atoms cm^{-2}, or 1.7×10^{21} cm^{-2} viewed normal to the plane of the galaxy. The sharp fall in density at the edges of the galaxy could be due to ionization by an inter-galactic flux of UV photons, as discussed by Sunyaev (1969). An alternative explanation is that the sharp gradients at the edges of the galaxy are associated with the warping of the plane of the HI disk indicated by the wings of the

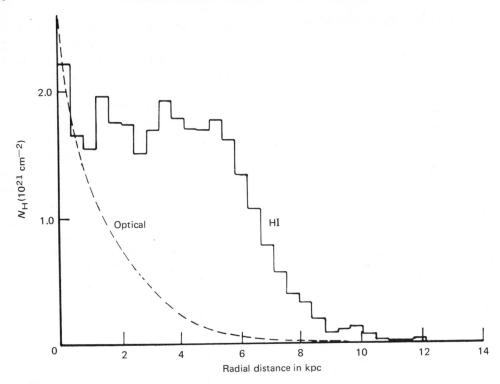

Figure 11.5 Integration in circular rings in the plane of M33 of the integrated brightness distribution. [Wright *et al.*, *Monthly Notices Roy. Astron. Soc.* (1972) 155:337.]

galaxy. A hat-brim model is envisaged with an increasing inclination of the plane of the galaxy to the line of sight along the edges of the galaxy at the ends of the minor axis.

There is a marked asymmetry in the HI distribution with a massive HI complex in the south-preceding quadrant of the galaxy.

d) Small-Scale Structure

The HI distribution is broken up into a large number of concentrations only partially resolved by the 1.5 × 3.0 minute arc beam. These concentrations have a typical peak surface density of 2.7×10^{21} cm^{-2} and a space density of ~ 1 to 2 atom cm^{-3}. They perhaps resemble the complexes discussed by McGee (1964) in the spiral arms of our galaxy, having sizes ≈ 500 to 2500 pc and densities ≈ 0.5 cm^{-3}, and those in the Large Magellanic Cloud having mean diameter 600 pc and density ≈ 1 cm^{-3}. A spiral arm structure can be seen in the inner regions of the

galaxy and is most evident in the trough running south from the galactic center (Figure 11.4). A best-fitting logarithmic spiral structure agrees with the optical spiral arms and the measured ratio of the average projected HI density in the arm and interarm regions is between 2 or 3 to 1. The troughs between HI concentrations are barely resolved by the beam, and the true density ratio may be as large as 6 to 1. An infinite contrast ratio is, however, ruled out by these observations.

e) Comparison with Optical Features

The extent of the HI distribution corresponds well with that of a well-exposed blueprint of the galaxy and the major and minor axis widths are close to the 83 × 53 minute arc given for the optical size by Holmberg (1958).

Figure 11.6 shows a superposition of the HI peaks onto a plate taken through a narrow-band red filter. It can be seen that the

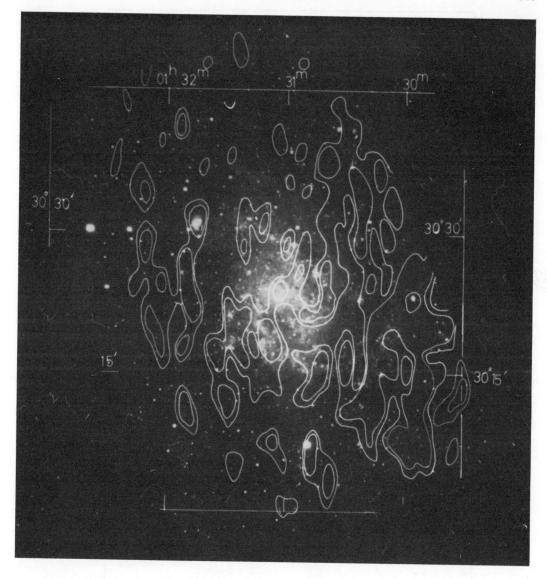

Figure 11.6 Superposition of the peaks of the HI distribution of M33 on a red print showing mainly HII regions.

HI concentrations follow the line of the optical spiral arms well in the south of the galaxy. The correlation is not so clear, and there are no strong HI concentrations on the northern spiral arm between the nucleus and NGC 604, where there is again a large concentration of HI. The contrast of the spiral arms is better in the composite HI + HII distribution in Figure 11.6 than in either the HI or HII regions separately, which indicates that the HI and HII are in some sense complementary.

f) *Rotation Curve and Total Mass*

The isovelocity contours (Figure 11.7) conform well to the pattern expected for a rotating galaxy and the rotation curve measured along the major axis is shown in Figure 11.8. The total mass can be derived by fitting a model rotation curve. The observed rotation

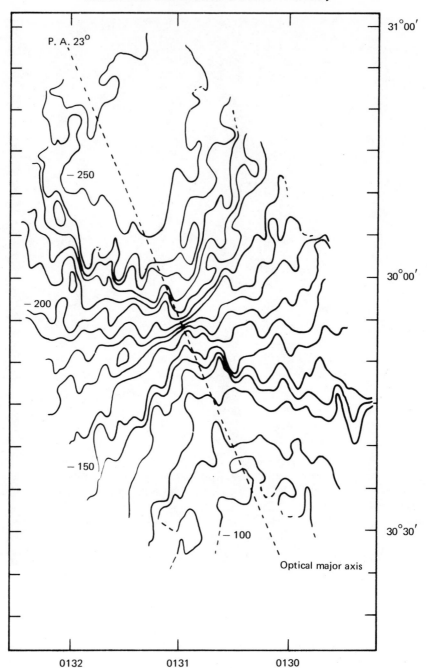

Figure 11.7 Isovelocity contours in M33 drawn at intervals of 10 km sec^{-1}.

curve has been fitted to three different types of rotation curve: a Brandt rotation curve with $n = 1.0$, a curve corresponding to an exponential distribution of mass (as discussed by Freeman, 1970), and an eighth-order polynomial. In all three cases the fitted curves agree with the observed rotation curve within 3 km sec^{-1}, and it will clearly be difficult to distinguish among them. The distributions of mass with radius deduced from these three

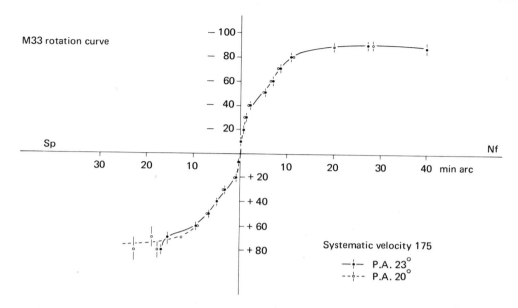

Figure 11.8 Rotation curve measured along the major axis of M33. The rotation velocities uncorrected for inclination are referred to a heliocentric systematic velocity of -175 km sec^{-1}.

fitted rotation curves are very similar within ≈ 20 minutes arc radius but diverge outside this radius. Using a Brandt curve with $R_{max} = 30$ minutes arc, $V_{max} = 100$ km sec^{-1}, and $n = 1.0$ gives a mass within 45 minutes arc of 1.7×10^{10} M$\odot$. The HI content is then some 9%, typical for an Sc galaxy. Because of the very flat rotation curve, the total mass of the galaxy extrapolated beyond the observed rotation curve is some 5×10^{10} M$\odot$, but this does not have much meaning.

g) Peculiar Velocities and Streaming Motions

It is clear from Figure 11.7 that there are local departures of the isovelocities from circular motion which exceed the noise level. It is essential, however, to consider the effect of beam averaging. A superposition of Figure 11.4 and Figure 11.7 shows that the deformations in the isovelocity contours often correspond to their crossing between HI peaks. The velocity in the interarm region is a beam average of the velocities of all HI concentrations within the beam at that time, and we may consequently discount many of the departures from smooth isovelocities. Some of the departures are real, however, and local

peculiar velocities can be 20 to 30 km sec^{-1}.

The line-of-sight velocity due to the rotation of the galaxy may be computed by selecting values for the rotation center, position angle, inclination, and rotation curve of the galaxy. If this model velocity field is subtracted from the observed velocity field, errors in the parameters selected show up in the residual velocity field with characteristic symmetries and enable best values for the rotation parameters to be determined. The residual velocity field may then be examined for systematic streaming motions predicted by the density wave theory of spiral arms. From the present observations of M33 it appears that such streaming motions are less than about 5 km sec^{-1}.

h) Comparison of Neutral and Ionized Hydrogen Velocities

In Figure 11.9 are plotted the velocities of the large HII regions measured by Mayall and Aller (1942) against the HI velocity at the HII region position. Because of the relatively large beamwidth (1.5 × 3 minutes arc), the HI velocities are best regarded as an average velocity of HI in the vicinity of the HII

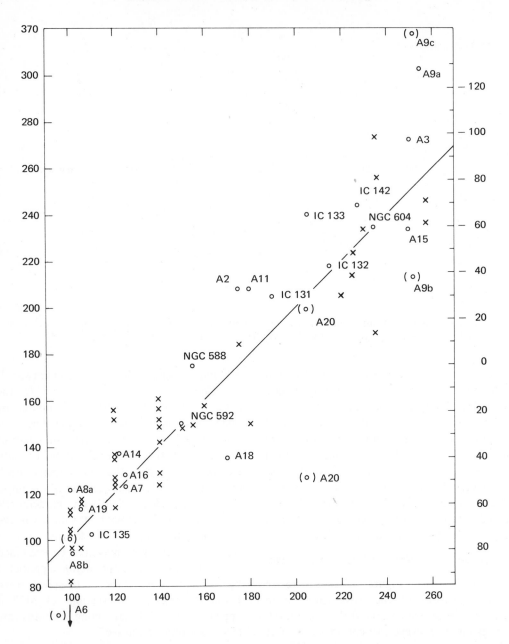

Figure 11.9 Velocities of HII regions (ordinate) plotted against the neutral hydrogen velocity at the position of the HII region. Open circles, *o*, are HII regions measured by Mayall and Aller (1942) with velocities by Brandt (1965); (*o*) velocities are measured by Mayall and Aller; (*x*) velocities are measured by Carranza *et al.* (1968). The left ordinate and abscissa scale are heliocentric, and the right ordinate is with respect to a systematic velocity of -175 km sec^{-1}. The line has a slope of 1.

region. The HII region velocities are local velocities of ionized gas within the HII region. The straight line has a slope of 1, showing that there is no systematic difference between the velocities of neutral and ionized gas. The vertical scatter in Figure 11.9 shows that the velocity of an HII region can differ by 20 to 30 km sec^{-1} from that of the neutral gas. Indeed, measurements within a single HII region can differ by the same magnitude. Estimates of the mass of gas in large HII regions, *e.g.*, in 30 Doradus in the LMC (Faulkner, 1967) and in NGC 604 in M33 (Wright, 1971b), show that velocity dispersions of this magnitude will disperse the HII region in $\approx 10^7$ years.

11.7.2 Observations at the Owens Valley Radio Observatory

a) *Observations*

The two-element interferometer at the Owens Valley Radio Observatory has been used to map a number of late-type galaxies to an angular resolution of 2 minutes arc and a velocity resolution of 21 km sec^{-1}. The Owens Valley interferometer is able to track sources over only a limited hour angle range but both E-W and N-S spacings are available so that good coverage of the (*u-v*) plane can still be obtained. Fixed observing stations are positioned at 100-foot intervals, giving a grating sidelobe response at 20 minutes arc radius and enabling sources smaller than this to be mapped without confusion.

For larger galaxies special techniques must be used to eliminate the sidelobe response. A powerful technique, first used in the

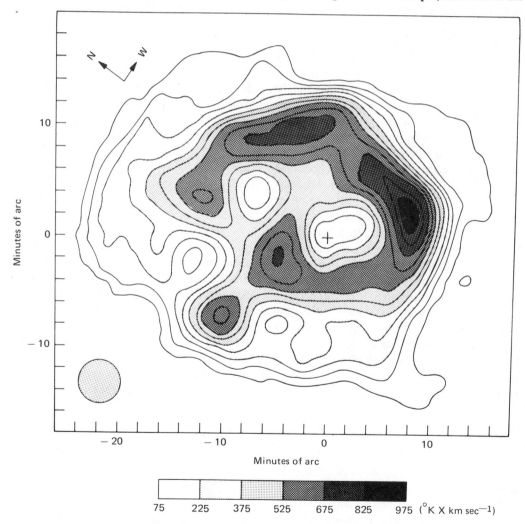

Figure 11.10 Integrated HI brightness in M101 to an angular resolution of 4 minutes arc.

case of M101 (Rogstad and Shostak, 1971), is to iteratively subtract the synthesized beam together with its sidelobes from the maximum of the source region of the synthesized map. The iteration stops when the sources being subtracted are a small percentage of the original maximum or when the residual map approaches the noise level. A map, corrected for the sidelobe response, is then reconstituted by convolving the subtracted sources with a clean sidelobe-free beam and adding them back into the residual map.

This procedure may not result in a unique map, but there are some tests on the validity of the procedure, which is discussed in more detail in Chapter 10.

b) *Integrated Hydrogen Distribution of M101*

Figure 11.10 shows the integrated HI brightness of M101 observed with a 4-minute-arc beam. Most remarkable in Figure 11.10 is the ring distribution and the marked asymmetry in the hydrogen distribution. If the gas is assumed to be optically thin, then the contour interval corresponds to an HI surface density of 1.4 atom cm^{-2}. For an assumed distance of 6.9 Mpc the integrated HI mass is 9×10^9 M$\odot$. If the HI distribution is really symmetric and the asymmetry in the integrated brightness is due to variations in the spin temperature of the gas, then this mass would be increased by $\approx 25\%$. Assuming that the gas is optically thin, the central depression in the integrated brightness represents a surface density of $\sim 3.5 \times 10^{20}$ atom cm^{-2}. Monnet (1971) has observed a weak background of Hα radiation in several galaxies. In M101 the estimated plasma density is nearly sufficient to account for the missing hydrogen, suggesting that the gas is highly ionized.

c) *Velocity Field and Streaming Motions in M101*

The velocity field of M101 (Figure 11.11) derived from the HI observations (Rogstad and Shostak, 1971) displays many deviations of $10 \sim 20$ km sec^{-1} from circular rotation of the gas. Many of these deviations are in the

form of ridges in the vicinity of the luminous spiral arms. Rogstad (1971) has interpreted this as evidence for the density wave theory of spiral structure (see Chapter 4).

There are further deviations in the iso-velocity contours near the nucleus of M101 which could indicate expansion motions ≈ 40 km sec^{-1} in the vicinity of an inner spiral arm and coincident with a central nonthermal radio source. The outflow of HI is sufficient to account for the even deeper hole in the HI distribution of the nuclear region.

11.7.3 Comparison of the HI Distributions of Late-Type Galaxies

a) *Observations*

Several other galaxies have recently been mapped at the Owens Valley. NGC 6946 and IC 342 (Rogstad *et al.*, 1973); NGC 2403, NGC 4236, (Shostak and Rogstad, 1973), and IC 10 (Shostak, 1974); M51, M81, and M82 (Weliachew and Gottesman, 1973); and Maffei 2 (Wright and Seielstad, 1973); IC 2574, and NGC 7640 (Seielstad and Wright, 1973). These observations have an angular resolution of 2 minutes arc and a velocity resolution of 10 or 21 km sec^{-1}.

NGC 2403 is an Sc$^-$ (Holmberg system) galaxy and is very similar in morphological form and stellar content to M33 [see the *Hubble Atlas of Galaxies* (Sandage, 1961)]. The integrated brightness distribution has a marked depression in the central region, similar to that in M101 but quite different from the HI distribution in M33. The three highest peaks in the integrated HI brightness distribution do not correspond well with the luminous spiral features, but rather lie in the dust lanes between the luminous spiral arms. A number of bright HII regions can be seen around the edges of these HI peaks. This can also be seen in places in M33. The rotation curve, as in M33, is very steep at the center and has a large constant-velocity region extending to the boundary of the detected hydrogen (as suggested in Section 11.5).

NGC 4236 is classified Sc$^+$ by Holmberg

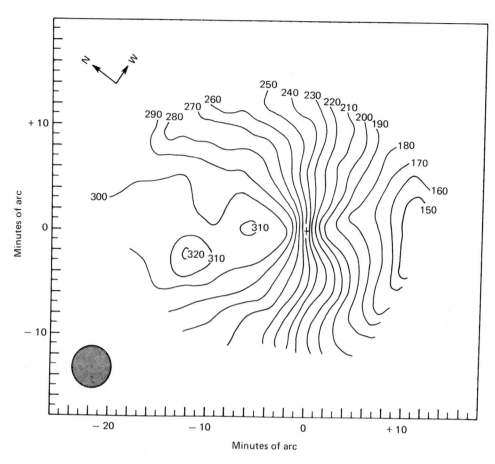

Figure 11.11 Isovelocity contours in M101.

(1958) and SB 7 by de Vaucouleurs (1964). A number of hydrogen concentrations are found, the largest peak in the north corresponding to a bright HII complex at the end of the "bar." The HII regions in the south, on the other hand, tend to lie around the HI peaks. All the rotation curves of NGC 4236, NGC 2403, and M33 have the same characteristic shape if scaled to common rotation curve maxima, R_{max} and V_{max}.

IC 10 is an irregular galaxy in a heavily obscured region of the sky. The integrated HI distribution shows a number of peaks and a large depression where the integrated brightness falls to zero. The peak brightnesses of all three galaxies are very similar ($\approx 1200°K$ km sec^{-1}), although the inclinations are different.

b) *Five Scd Galaxies*

In a comparison of the five Scd galaxies, M33, NGC 2403, IC 342, M101, and NGC 6946, Rogstad and Shostak (1972) find substantial similarities in the hydrogen distributions and rotation curves. The integral HI parameters are nearly constant or scale with the Holmberg (1958) diameter. Eighty percent of the observed HI mass lies within the Holmberg diameter for each galaxy. The mean hydrogen content is 11% of the total mass within this same boundary, and the central HI surface density is also nearly constant ($\simeq 8 \times 10^{20}$ atom cm^{-2}).

All the rotation curves are observed to be flat-topped, with a maximum velocity proportional to the Holmberg diameter. Because of the flat-topped rotation curves, the total masses extrapolated to infinite radius are indeterminate, but the total mass within the Holmberg radius is proportional to $R_{H_o} V_{max}^2$ and therefore scales as $R_{H_o}^3$ for these five galaxies. The constancy of the integral properties within one morphological class is an important justification for the use of the morphological classification system for spiral galaxies.

11.7.4 Central Hydrogen Depressions

As suggested by Roberts (1967) (see Section 11.4.5), it appears that a central depression in the HI surface brightness is a common feature in late-type spiral galaxies. M31 is the most obvious case where an HI "ring" is a good overall description. For the other galaxies so far observed the HI distribution is better described in terms of an asymmetry and a central depression. In no observed galaxy does the surface brightness resemble the overall light distribution, which is exponentially decreasing, or the total mass distribution (deduced from rotation curves or from an assumed mass-to-light ratio). The total mass and light distribution are centrally peaked; the HI surface brightness is rather flat or centrally depressed. There are several possible explanations:

(1) The usual assumption that the gas is everywhere optically thin may not be correct. It is almost certainly true that small cold optically thick clouds occur locally in every galaxy. If the spin temperature or the cloudiness of the HI varies systematically across the galaxy, then the surface brightness of HI line radiation may not resemble the HI surface density distribution. This explanation requires lower spin temperatures or very clumpy HI in the central regions of galaxies.

(2) The thickness of the HI layer may increase with radius, as observed in our galaxy (Kerr and Westerhout, 1965). The HI layer thickness in M33 can be estimated

from the measured velocity dispersion in the HI line profiles. If we assume that the vertical velocity dispersion (c^2) supports the gas against gravitational attraction, then the half-thickness ($\bar{z}$) of the gas layer is given by Mestel (1963) as

$$\bar{z} = \frac{c^2}{2\pi G \sigma} \qquad (11.13)$$

where σ is the total mass surface density in the plane of the galaxy. σ has been estimated from a mass-model fitted to the rotation curve and decreases with radius. To first order the observed velocity dispersion in M33 is constant; therefore the HI layer thickness increases with radius. In M33 the HI surface density is roughly constant with the radius (see Figure 11.5); therefore the HI volume density decreases with radius. This is consistent with the observed decrease in the rate of star formation with radius in M33 (Madore *et al.*, 1973). In M31 a ten-fold increase in layer thickness would make the HI volume density in the central regions equal to that at the position of the ring.

(3) The gas in the central regions may be ionized. Monnet (1971) has observed a weak background of Hα radiation in the central regions of several galaxies. In the case of M31, however, the emission measure was less than 10 cm^{-6} pc, which is too small by an order of magnitude to give a surface density comparable with that of the HI peak.

(4) The gas could be in molecular form without any observable consequences. Molecular hydrogen has been seen in the far-ultraviolet absorption spectra of the star ξ Persei, which has a strong ultraviolet continuum emission (Carruthers, 1970). However, observation of H_2 in external galaxies is far beyond present techniques.

(5) Neutral hydrogen is expanding away from the central regions of the Milky Way (see Oort, 1964; Kerr, 1967). The 3-kpc-arm feature of our galaxy is expanding at a velocity of some 53 km sec^{-1}, while asymmetry in the rotation curve for our

galaxy can be interpreted as a 7 km sec^{-1} expansion in the vicinity of the Sun. The outflow of neutral gas from the central regions (about 1 M$\odot$ per year) might be partially balanced by an inflow (under one interpretation of the high-velocity gas seen at high latitudes in the galaxy; see Oort, 1967), but is otherwise sufficient to deplete the central regions of HI in some 10^7 to 10^8 years.

Expansion motions have been observed in the central regions of some external galaxies (*e.g.*, M101) but in M31 they are less than 10 km sec^{-1} (Gottesman and Davies, 1970) and are too small to have removed a significant mass of HI away from the central regions.

(6) The gas in the central regions may have been used up by star formation. Star formation would initially be fastest in the central regions of galaxies where the gas density is highest. At the present epoch the rate of star formation in M31 is at its peak in the region of the ridge at 10-kpc radius.

The correct explanation of the deficit of HI in the central regions is probably a combination of explanations (2) and (6), although in some instances explanations (3) and (5) may be highly significant. Explanations (1) and (4) are unknown factors.

11.7.5 Rate of Star Formation

The rate of star formation is presumed to depend on the gas density. Attempts to find a relationship of the form

$$\frac{d\rho_*}{dt} = \text{constant} \times \rho_g{}^n \qquad (11.14)$$

[where ρ_* is the star (or HII region), and ρ_g the gas density] have yielded different values for the exponent n. Comparisons of HII region and bright star counts with HI surface densities have been made for the SMC, M31, and M33. For bright stars in the SMC Sanduleak (1969) obtained a value of $n = 1.84 \pm 0.14$. In M31 Hartwick (1971) obtained a value of $n = 3.50 \pm 0.12$ by comparing HII

region counts with Robert's (1966) HI density.

A comparison of the HI surface density in M33 (Wright *et al.*, 1972) with HII region and star counts in M33 (Madore, 1971) finds a value of $n = 1.72 \pm 0.01$ for stars and $n = 3.15 \pm 0.18$ for HII regions. Most of the stars counted in M33 are the less massive O and B stars, whilst only large HII regions (excited by massive O stars) are included. The different values of n obtained in M33 tend to support the suggestion by Schmidt (1963) that n is the largest for the most massive stars. The values of n obtained for HII regions are in good agreement in M31 and M33.

11.8 Conclusions

The observations presented in the previous section represent the present state of the art of mapping neutral hydrogen in external galaxies. Aperture synthesis observations of a number of nearby late-type spiral galaxies are in progress at the present time, and we should soon be able to see if the HI distributions of, say, Sb and Sc galaxies are characteristically different.

Maps made with the Westerbork telescope to an angular resolution of 23 seconds arc should soon be available, allowing a detailed comparison of HI structure in many more galaxies. Details of the HI density and velocity distribution in the luminous spiral arms should make possible a much better understanding of star formation and of the dynamics of the spiral patterns.

References

Avner, E. S., and I. R. King. 1967. *Astron. J.* 72: 650.

Baldwin, J. E., C. Field, P. J. Warner, and M. C. H. Wright. 1971. *Monthly Notices Roy. Astron. Soc.* 154:445.

Bergh, S. van den. 1960a. *Astrophys. J.* 131:215.

———. 1960b. *Astrophys. J.* 131:558.

Bottinelli, L. 1971. *Astron. Astrophys.* 10:437.

———, L. Gougenheim, J. Heidmann, and N. Heidmann. 1968. *Ann. Astrophys.* 31:205.

———, L. Gougenheim, and J. Heidmann. 1973. *Astron. Astrophys.* 25:451.

Brandt, J. C. 1965. *Monthly Notices Roy. Astron. Soc.* 129:309.

——, and J. J. Belton. 1962. *Astrophys. J.* 136:352.

——, and L. S. Scheer. 1965. *Astrophys. J.* 70:471.

Brouw, W. N. 1971. *Data Processing for the Westerbork Synthesis Radio Telescope.* Dissertation, Leiden University.

Burbidge, E. M., G. R. Burbidge, and K. H. Prendergast. 1963. *Astrophys. J.* 138:375.

Burke, B. F. 1967. *I.A.U. Symposium No. 31, Paper 31.* New York: Academic Press.

Cameron, M. J. 1971. *Monthly Notices Roy. Astron. Soc.* 152:403.

Carruthers, G. R. 1970. *Astrophys. J.* 161:L81.

Chandrasekhar, S. 1942. *Principles of Stellar Dynamics.* Chicago: University of Chicago Press.

Epstein, E. E. 1964. *Astrophys. J.* 69:521.

Faulkner, D. J. 1967. *Monthly Notices Roy. Astron. Soc.* 135:401.

Freeman, K. C. 1970. *Astrophys. J.* 160:811.

Gordon, K. J. 1971. *Astrophys. J.* 169:235.

Gottesman, S. T., and R. D. Davies. 1970. *Monthly Notices Roy. Astron. Soc.* 149:262.

Hartwick, F. D. A. 1971. *Astrophys. J.* 163:431.

Henize, K. G. 1956. *Astrophys. J., Supp. Ser.* 2:315.

Hindman, J. V. 1967. *Australian J. Phys.* 20:147.

——, F. J. Kerr, and R. X. McGee. 1963. *Australian J. Phys.* 16:570.

Hodge, P. W. 1967. *Astron. J.* 72:129.

——. 1969. *Astrophys. J.* 155:417.

Holmberg, E. 1958. *Medd. Lunds, Astr. Obs. Ser. 2,* No. 136.

Kerr, F. J. 1967. *I.A.U. Symposium No. 31.* New York: Academic Press.

——. 1968. In *Stars and Stellar Systems.* Vol. 7. "Nebulae and Interstellar Matter." Chicago: University of Chicago Press.

——, J. V. Hindman, and B. J. Robinson. 1954. *Australian J. Phys.* 7:297.

——, and G. Westerhout. 1965. In *Stars and Stellar Systems.* Vol. 5. "Galactic Structure." A. Blaauw and M. Schmidt, eds. Chicago: University of Chicago Press.

Lin, C. C., C. Yuan, and F. H. Shu. 1969. *Astrophys. J.* 155:721.

Madore, B. F. 1971. Unpublished masters thesis. University of Toronto.

——, S. van den Bergh, M. C. H. Wright, and J. E. Baldwin. 1973. In preparation.

Mathewson, D. S., P. C. van der Kruit, and W. N. Brouw. 1972. *Astron. Astrophys.* 17:468.

Mayall, N. U., and L. H. Aller. 1942. *Astrophys. J.* 95:4.

Mestel, L. 1963. *Monthly Notices Roy. Astron. Soc.* 126:553.

McGee, R. X. 1964. *I.A.U. Symposium No. 20.* "The Galaxy and the Magellanic Clouds." F. J. Kerr and A. W. Rodgers, eds. Canberra: Australian Academy of Science, p. 126.

——, and J. A. Milton. 1966. *Australian J. Phys.* 19:343.

Monnet, G. 1971. *Astron. Astrophys.* 12:379.

Morgan, W. W. 1958. *Publ. Astron. Soc. Pacific* 70:364.

——. 1959. *Publ. Astron. Soc. Pacific* 71:394.

——, and N. U. Mayall. 1957. *Publ. Astron. Soc. Pacific* 69:291.

Oort, J. H. 1964. *I.A.U. Symposium No. 20.* "The Galaxy and the Magellanic Clouds." Canberra: Australian Academy of Science, p. 1.

——. 1967. *I.A.U. Symposium No. 31.* "Radio Astronomy and the Galactic System." New York: Academic Press.

Pooley, G. G. 1969. *Monthly Notices Roy. Astron. Soc.* 144:101.

Roberts, M. S. 1966. *Astrophys. J.* 144:639.

——. 1967. *I.A.U. Symposium No. 31,* Paper 32. New York: Academic Press.

——. 1969. *Astron. J.* 74:859.

Robinson, B. J., and J. A. Koehler. 1965. *Nature* 208:993.

Rogstad, D. H. 1971. *Astron. Astrophys.* 13:108.

——, G. W. Rougoor, and J. B. Whiteoak. 1967. *Astrophys. J.* 150:9.

——, and G. S. Shostak. 1971. *Astron. Astrophys.* 13:99.

——, and G. S. Shostak. 1972. *Astrophys. J.* 176:315.

——, G. S. Shostak, and A. H. Rots. 1973. *Astron. Astrophys.* 22:111.

Sandage, A. 1961. *Hubble Atlas of Galaxies.* Carnegie Institution of Washington.

Sanduleak, N. 1969. *Astrophys. J.* 74:47.

Schmidt, M. 1963. *Astrophys. J.* 137:758.

Seielstad, G. A., and M. C. H. Wright. 1973. *Astrophys. J.* 184:343.

Sersic, J. L. 1960. *Z. Astrophys.* 50:158.

Shostak, G. S. 1973. *Astron. Astrophys.* 24:411.

——. 1974. *Astron. Astrophys.* In press.

——, and D. H. Rogstad. 1973. *Astron. Astrophys.* 24:405.

Sunyaev, R. A. 1969. *Astrophys. Lett.* 3:33.

Vaucouleurs, G. de, and A. de Vaucouleurs. 1964. *Reference Catalogue of Bright Galaxies.* Austin: University of Texas Press.

Volders, L., and J. A. Högbom. 1961. *Bull. Astron. Inst. Neth.* 15:307.

Weinreb, S. 1963. *Technical Report No. 412.* Massachusetts Institute of Technology, Research Laboratory of Electronics.

Weliachew, L., and S. T. Gottesman. 1973. *Astron. Astrophys.* 24:59.

Wright, M. C. H. 1971a. *Astrophys. J.* 166:455.

———. 1971b. *Astrophys. Lett.* 7:209.

———, P. J. Warner, and J. E. Baldwin. 1972. *Monthly Notices Roy. Astron. Soc.* 155:337.

———, and G. A. Seielstad. 1973. *Astrophys. Lett.* 13:1.

CHAPTER 12

RADIO GALAXIES AND QUASARS

Kenneth I. Kellermann

12.1 Introduction

The discrete sources of radio emission were first distinguished from the general background radiation during the 1940's as a result of their rapid amplitude scintillations; and initially, it was thought that the scintillations were due to fluctuations in the intrinsic intensity of the discrete sources. Assuming that the dimensions of the sources could not greatly exceed the distance traveled by light during a typical fluctuation period of about 1 minute, it was concluded that the discrete sources were galactic stars located at relatively small distances from our solar system. Thus the term "radio star" was often used in referring to these sources.

The identification of two of the strongest sources, Virgo A and Centaurus A, with the nearby galaxies M87 and NGC 5128 by Bolton, Stanley, and Slee (1949) made it clear that at least some of the discrete sources were of extra-galactic origin. In 1954 an accurate position measurement of the strong source Cygnus A led to the identification of this source with a relatively faint 15th magnitude galaxy having a red shift $z \simeq 0.06$ (Baade and Minkowski, 1954), and the extra-galactic nature of the discrete sources was generally recognized.

Although a few other radio sources were identified with galaxies during the 1950's, progress was slow because of the low accuracy of the radio source positions. By 1960, however, the positions of most of the strongest sources had been determined with an accuracy

of about 10 seconds of arc and many were identified with various galaxy types. These galaxies, which are identified with strong radio sources, are generally referred to as "radio galaxies."

Most of the radio galaxies have bright emission lines, and so their red shift may be relatively easily determined. The faintest and most distant identified radio galaxy is the strong radio source 3C 295, which has a red shift of 0.46 and an apparent magnitude of about −21. This identification, which was made in 1960, was the result of accurate radio positions determined at Caltech and at Cambridge; it culminated a long search for distant galaxies and stimulated the search for galaxies of even higher red shift.

Continued efforts to identify distant galaxies were concentrated toward sources of small diameter and high surface brightness on the reasonable assumption that these were most easily observed at a large distance. A primary candidate was 3C 48, which had an angular size less than 1 second of arc, and was at the time the smallest strong source known. Accurate position measurements made in 1961 resulted in what appeared to be a unique identification with a 16th-magnitude *stellar* object having a faint red wisp extending away from it. The absence of any other optical visible object near the radio source and the later discovery of significant night-to-night variations in light intensity led to the reasonable conclusion that 3C 48, unlike other radio sources, was a true radio star. Soon two other relatively strong sources, 3C 286 and 3C 196, were also identified with "stars," and it appeared that more than 20% of all sources were of this class. The optical and radio properties were surprisingly dissimilar for the three objects, and there were no unique radio properties to separate them from radio galaxies.

Early efforts at interpreting the emission-line spectrum of 3C 48 were relatively unsuccessful, although the possibility of a large red shift was apparently considered. By 1962 most of the lines were thought to be identified with highly excited states of rare elements.

The identification of 3C 273 with a similar stellar object, however, again cast doubt on the galactic interpretation which by 1963 was widely accepted. 3C 273 was tentatively identified in Australia with a 13th-magnitude star from a moderately accurate position determined with the 210-foot telescope. The position and identification were confirmed by a series of lunar occultations. These showed that the radio source was double, one component being within 1 second of the optical image, and the other component being elongated and coincident with a jet-like extension to the star. The identification was beyond question, although one wonders why it had not been made much earlier, as 3C 273 was the brightest then unidentified source, of small angular size and located in an unconfused region of the sky near the galactic pole.

The optical spectrum of 3C 273 showed a series of bright emission lines which could be identified only with the Hydrogen Balmer series, but with a red shift of 0.16. This red shift was confirmed when the H_γ line was found near the predicted wavelength of 7590 Å in the near infrared. Adopting this red shift of 0.16 then led to the identification of the MgII lines appearing at 3239 Å.

A re-inspection of the 3C 48 spectrum gave a red shift of 0.37 if a strong feature at 3832 Å was identified with MgII. Other lines could then be identified with OII, NeIII, and NeV. Additional spectra taken of other similar sources led to the identification of CIII (1909 Å), CIV (1550 Å) and finally Ly-α, permitting red shifts as great as 3.5 to be measured.

Radio sources in this hitherto unrecognized class are usually referred to as quasi-stellar radio sources, QSS, or quasars. Following the identification of the first quasars it was realized that they all had a strong UV excess, and the search for further quasar identifications was simplified by looking for a very blue object located at the position of radio sources. In fact, many such objects were found which optically appear to be quasars, but are radio-quiet. These were originally referred to as Blue Stellar Objects (BSO), Quasi-Stellar Galaxies (QSG), or Quasi-Stellar Objects (QSO). Today the word "quasar" is generally used to refer to the entire class of stellar-type objects with large apparent red shifts, while QSO and QSS refer to radio-quiet and radio-active quasars, respectively.

It is now widely accepted that the radio emission from both galaxies (including our own Galaxy) and quasars is due to synchrotron emission from relativistic particles moving in weak magnetic fields. The amount of energy required in the form of relativistic particles is, however, very great, and the source of energy and its conversion into relativistic particles has been one of the outstanding problems of modern astrophysics. The remainder of this chapter is devoted to a description of the observed properties of radio galaxies and quasars, and how these are interpreted in terms of the synchrotron mechanism.

12.2 Radio Surveys

Catalogs of discrete sources have been prepared from extensive surveys using instruments especially designed for this purpose. Initially these surveys were made at relatively long wavelengths near one meter, but as techniques have improved at the shorter wavelengths, the surveys have been extended to wavelengths as short as a few centimeters.

In order to isolate the discrete sources from the intense emission observed from the galactic background, most of the earlier surveys were made with interferometer systems, which are relatively insensitive to the distributed background emission (see Chapter 10).

Today, catalogs of sources based on surveys made between 10 MHz and 5 GHz are available. Some of the surveys, particularly those made at the longer wavelengths, cover essentially the entire observable sky down to source densities of about 10^3 ster^{-1}. Other instruments, intended mainly for cosmological studies, have reached source densities of about 10^5 ster^{-1} over very restricted parts of the sky. New instruments just coming into operation will reach densities of 10^6 ster^{-1}.

Generally the surveys have produced approximate values for the position and flux density for the cataloged sources. These catalogs have then been used as the basis for subsequent more accurate measurements over a wide range of wavelengths of properties such as

(1) the angular position in the sky
(2) the radio brightness distribution
(3) the radio-frequency spectrum
(4) the amount and direction of any polarization and its angular distribution
(5) the time-dependence of the radio emission.

A useful summary of radio source surveys has been compiled by Dixon (1970), and contains the results of many separate surveys. The two most widely used catalogs are based on the Cambridge 3C and 3CR surveys and the Parkes survey, which contain between them the great majority of sources which have been studied in detail and for which optical data are available.

12.2.1 3C and 3CR Surveys

The 3CR catalog (Bennett, 1962) is based on the original 3C survey (Edge *et al.*, 1960) made at Cambridge at 159 MHz using a complex interferometer system. This survey was preceded by the 2C survey made with the same instrument at 81 MHz with a resolution two times poorer in each coordinate. The 2C catalog contained 1936 sources, but owing to the poor resolving power, it became

clear at an early stage that many of these sources were not real, and were due to blends of two or more sources in the primary antenna beam. Moreover, except for the strongest sources the determination of the flux density and angular coordinates was poor. The 3C survey with twice the resolution contained only 471 sources and was considerably more reliable. Nevertheless, because of the relatively poor primary resolving power, there were still large errors in the positions and flux densities. In particular, it was frequently uncertain in which lobe of the interference pattern a particular source was located, and this introduced large positional uncertainties.

In order to reduce these uncertainties an additional survey was made at 178 MHz using a large parabolic cylinder antenna. The narrow E-W beam of this antenna eliminated nearly all of the lobe ambiguities of the original 3C catalog. The data from the two surveys were combined to form the then most reliable radio source catalog—the *Revised 3C or 3CR Catalogue*.

The same parabolic cylinder antenna was later used together with a smaller moveable antenna as an aperture synthesis instrument (see Chapter 10) to make a complete high-resolution survey of the northern sky (the 4C survey), which contains about 2000 sources.

12.2.2 Parkes Survey

The most extensive survey of the Southern Hemisphere radio sources has been made with the 210-foot radio-telescope near Parkes, Australia. This survey was initially made at 408 MHz (75 cm), except for the region $20° > \delta < 27°$, which was surveyed at 635 MHz (50 cm). Each source detected in the 50- or 75-cm finding survey was re-observed at 20 and 11 cm to obtain more accurate positions and flux densities at these wavelengths. The Parkes survey essentially replaces an earlier survey made at 86 MHz using a Mills Cross–type antenna with a resolution of about 48 minutes of arc. Unlike the Cambridge 2C survey, made about the same time and at nearly the same wavelength, the Mills

Cross survey was mostly limited—not by inadequate resolution to separate nearby sources, but by low sensitivity. Nevertheless, later work has shown that this survey was surprisingly accurate for its time, at least for the sources not near the limit of detection.

12.2.3 Nomenclature

Most of the earlier surveys used a variety of complex schemes for naming and identifying sources. Unfortunately, the use of many names for a single source has led to needless confusion among workers. The recognition of this problem has led to the growing use of a system first used in the Parkes catalog. In this system each source is identified by a name of the form HHMM ± DD, where HHMM represents the hours and minutes of right ascension, and ± DD the degrees of declination prefixed by the sign. In order to preserve the identity of the observatory, the catalog name is frequently preceded by a symbol identifying the observatory. Some prefixes in common use are PKS (Parkes), DW (Dwingeloo), B (Bologna), G (Green Bank), A (Arecibo). Often additional symbols are used to identify a particular survey.

12.3 Optical Identifications

Of the thousands of sources which have been cataloged only a few hundred have been reliably optically identified with galaxies or with quasars. For a considerable number of sources accurate positions have been measured, but no optical object is found above the plate limit of the Palomar Sky Survey. For a much larger number of sources the position accuracy is not sufficient to distinguish between the two or more objects lying within the error rectangle.

Optical identifications are important for two reasons.

(1) It is not possible from radio measurements alone to determine the distance to a radio source. Thus, only if the radio source is identified with a galaxy or quasar is it possible to measure the red shift and thus deduce the distance from the Hubble law. Distances are, of course, required to estimate the absolute radio luminosity and linear dimensions from measurements of radio flux density and angular size.

(2) Optical studies of radio galaxies and quasars may give some insight into the problem of the origin of the intense radio emission.

For these reasons much of the earlier work on the discrete radio sources was concentrated on the determination of accurate positions to permit unambiguous optical identifications. Today, coordinates of at least the stronger sources may be routinely determined by interferometry with an accuracy of the order of 1 second of arc. Nevertheless, the identifications are difficult. Firstly, often there is no apparent optical counterpart of the radio position, so that any associated optical object is either subluminous or at such a great distance that it is not visible, even to large optical telescopes. Secondly, many radio sources have dimensions of the order of 1 minute of arc or more and a complex distribution of radio brightness so that more than one galaxy or quasar is found within the area covered by the source. Often, but not always, the identified galaxy lies near the centroid of radio emission, but it may also be coincident with one of the individual radio components, so that unambiguous identifications are difficult.

12.3.1 Types of Identifications

Identifications of extra-galactic radio sources are with a variety of objects, for example, as defined by Matthews, Morgan, and Schmidt (1964), with

(1) normal spiral galaxies
(2) Seyfert galaxies
(3) E, or elliptical, galaxies which are often the brightest member of a cluster
(4) D galaxies, which are similar to elliptical galaxies, but contain an extended halo

(5) dB or Dumbell galaxies, which contain a double nucleus imbedded in a common halo

(6) N galaxies, which contain a compact bright nucleus with a strong emission-line spectrum, superimposed on a faint nebulous envelope

(7) QSO's or quasi-stellar objects (quasars).

The above sequence is very roughly in increasing order of absolute radio luminosity, although there is a wide spread of luminosity within each class. In general the spiral galaxies are relatively weak radio emitters, while the various types of ellipticals (E, D, dB, N) are considerably more intense. Some of the giant ellipticals, however, contain a very weak radio source coincident with its nucleus (Heeschen, 1970).

In recent years there has been a growing realization that the distinction between the various optical categories is largely subjective and may vary depending on the observer, the size of the telescope, and the distance of the object. For example, the prototype quasar, 3C 48, is now sometimes classed as an N galaxy (*e.g.*, Morgan, 1972). In another case the optical jets from the quasar 3C 273 and the giant elliptical galaxy M87 show surprising similarity.

12.3.2 Radio Galaxies

Essentially all of the identified radio galaxies show moderate to strong narrow emission lines such as OII at 3727 Å in their optical spectra, and this property has been used to confirm preliminary optical identifications based on the positional agreement of the radio source and the galaxy. Most of the identified galaxies are classed as giant ellipticals which have a surprisingly narrow dispersion in absolute optical magnitude of -20.8 ± 0.6 (H = 100 km/sec/Mpc). Initially most of the identifications were with galaxies which showed some sort of optical peculiarity. For example, M87 (Virgo A) has a well-known jet extending from its nucleus; NGC 5128 (Centaurus A) contains a conspicuous dark band across the galaxy; Cygnus A has a double nucleus with intense emission lines, which were the basis for the idea that the radio emission was the result of collisions between two galaxies. Other identifications were with Seyfert galaxies such as NGC 1086 and NGC 1275.

Although more recently radio galaxies which show no obvious optical peculiarity have been identified, there has been some degree of bias toward accepting identifications with galaxies which show some abnormality. Since the identification process is a very subjective one, involving not only positional coincidence but the size and structure of the radio source, as well as the presence of strong emission lines and optical features such as jets or dust, the connection between these phenomena and the radio emission is not clear unless the difficult task of obtaining optical identifications and red shifts of a complete sample of radio sources is achieved.

Because two of the early radio source identifications were with the Seyfert galaxies NGC 1068 and NGC 1275, it has been widely thought that the Seyfert phenomenon is associated with intense radio emission. In fact, this is not the case, since none of the other well-known Seyfert galaxies show radio emission much greater than the normal spiral galaxies. Moreover, even NGC 1068 has a luminosity of only 10^{40} ergs/sec—only slightly greater than other spiral galaxies—and the classification of NGC 1275 as a Seyfert is now questioned by many astronomers. Although a few other radio sources have been identified with galaxies that were later classed as Seyferts, it appears in general that the radio luminosity of Seyfert galaxies does not significantly exceed that of normal spirals, and the fraction of Seyfert galaxies which are strong radio sources appears to be comparable with that of giant ellipticals.

Similarly, attempts to detect radio emission from other "peculiar" galaxies such as the Zwicky compact galaxies, Markarian galaxies, or the interacting galaxies cataloged by Arp and Vorontzov-Velyaminov have been for the most part unsuccessful, and contrary to a widely held belief, there is no evidence that these "exotic" galaxies are more likely to be strong radio sources than galaxies chosen at random.

Most Seyfert and Seyfert-type galaxies do contain a relatively weak small source at their nuclei. These nuclei also show surprisingly strong infrared emission, with an intensity roughly proportional to the radio flux (van der Kruit, 1971).

Red shifts are available for only about 100 of the identified radio galaxies. These have absolute radio luminosities which range from about 10^{40} to 10^{45} ergs/sec. In contrast to the radio galaxies which were all identified as galaxies that are coincident with cataloged radio sources, are the so-called normal galaxies. These are the optically bright galaxies from which weak radio emission has been detected as a result of a special search. The absolute radio luminosity of the "normal" galaxies is of the order of 10^{37} to 10^{38} ergs/sec, which is comparable to the power radiated from our own Galaxy.

It is not completely clear to what extent the normal galaxies are a separate class or just an extension of the radio galaxy phenomena. Originally there was some evidence of a relative deficiency of sources with intermediate luminosities in the range 10^{39-40} ergs/sec. However, it now appears that this was largely a manifestation of the different techniques used to investigate "radio" galaxies and "normal" galaxies and that the luminosity function is continuous in the range $10^{38} < L < 10^{45}$ ergs/sec.

The identified quasars are all very strong radio sources with luminosities comparable with the most luminous radio galaxies. Of the quasars which were first located from optical measurements, only a few percent have been detected as radio sources. But considering the large red shifts, the upper limit to the radio luminosity of the undetected quasars is still large, and comparable with that of the strongest radio galaxies.

Searches for radio emission from rich clusters of galaxies have been more fruitful, although it appears that the radio emission originates in the brightest cluster member, and that cluster membership does not affect the probability of radio emission. About 5 to 10 % of the giant elliptical galaxies which are the brightest cluster members show detectable radio emission (Rogstad and Ekers, 1969).

There is some evidence that the radio luminosity depends weakly on the optical luminosity for both galaxies and quasars. In the case of galaxies almost all the strong radio sources show bright emission lines.

12.3.3 Quasars

The optical properties of quasars may be summarized as follows:

(1) They appear stellar on direct photographs, although in some cases there is a faint jet or wisp extending from the stellar object.
(2) They have large red shifts ranging up to $z = 3.5$.
(3) Assuming that the red shifts are cosmological and that the distance is given by the Hubble law with a Hubble constant, H, equal to 100 km/sec/Mpc, then the absolute optical magnitudes range from -22 to -26 or more, so that they are up to 100 times brighter than the most luminous galaxies at optical wavelengths.
(4) Often the optical emission is variable on time scales ranging from a few hours to a few years.
(5) The luminosity rises sharply toward the infrared, where most the radiated energy lies. There is also an excess of UV emission compared with galaxies, so that the presence of a large red shift causes the quasar to appear blue when measured by UBV photometry or when the color is estimated from the "red" and "blue" plates of the Palomar Sky Survey. For very large red shifts ($z \gtrsim 3$), however, the color may appear neutral or even "red."
(6) The spectra show intense broad emission lines, with line widths corresponding to velocities up to 4000 km/sec. The most commonly observed lines are those of Lyα ($\lambda1216$), CIV ($\lambda1549$), CIII ($\lambda1909$), MgII ($\lambda2798$), OIII ($\lambda4363$, $\lambda4959$, $\lambda5007$), and the Hydrogen Balmer series.

(7) Some quasars show narrow absorption lines, often with several sets of red shifts which are usually smaller than the emission-line red shift.

12.4 Radio Properties

The extra-galactic sources can be conveniently divided into two groups: the *extended* or *transparent* sources, and the *compact* or *opaque* sources. The observed properties of the two groups are discussed in more detail in the following pages.

Quite surprisingly there is no simple relation between the dimensions of the radio-emitting region and the dimensions of the optical galaxy or QSO, and it must be emphasized that the division into *compact* and *extended* radio sources in no way separates the quasars from the radio galaxies. In fact,

insofar as we know for any individual source, the quasars are indistinguishable from the radio galaxies on the basis of their radio properties alone. There are, however, clear statistical differences; the majority of the compact sources are identified with quasars or with galaxies that have bright nuclei, such as N-type or Seyfert galaxies. However, many also appear in normal-looking elliptical galaxies as well. Likewise the extended radio sources are not identified only with galaxies, but are frequently associated with quasars showing no visible optical extent. Because the compact sources are all affected by self-absorption (see Section 12.5.4), their spectra are flat and they are therefore most easily detected by radio surveys made at short wavelengths.

The total power radiated at radio frequencies extends from about 10^{38} ergs/sec from so-called normal galaxies such as our

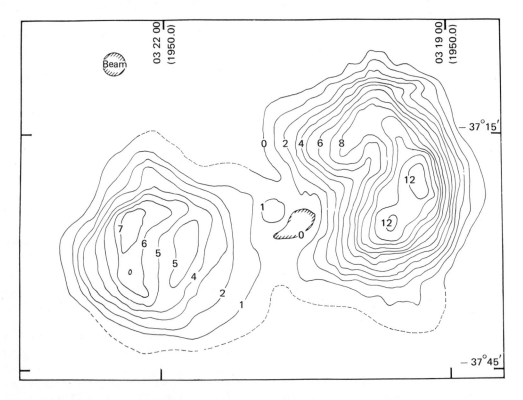

Figure 12.1 Brightness distribution of Fornax A at 75 cm, mapped with a beamwidth of about 2.8 of arc beam, using the Molonglo Cross radio telescope in Australia. Contour unit is 100°K. [Taken from Cameron (1969), *Proc. Astron. Soc. Australia* 1:229].

own Galaxy and other spirals, to 10^{41} ergs/sec for the weaker radio elliptical galaxies such as M87 or Centaurus A, and up to 10^{45} ergs/sec for the most luminous radio galaxies such as Cygnus A and 3C 295 and many quasars. In many numerical examples Cygnus A is often discussed as a "typical source," whereas in fact it is a very outstanding source since less than one source in 10 million in any given volume of space approaches this luminosity.

12.4.1 Brightness Distribution

The three-dimensional structure of extra-galactic sources is usually inferred from the observations of the angular size and brightness distribution projected onto the plane of the sky. Often only the brightness distribution in one dimension is determined. The data on brightness distributions are obtained by one of four common procedures.

a) Pencil Beam Observations

Only a few of the extra-galactic sources are sufficiently large for their structure to be studied by even the largest pencil beam radio telescopes. Moreover, these sources are all relatively nearby, of low absolute luminosity and very low surface brightness, and therefore not representative of typical extra-galactic sources found in radio source surveys. One very extended source which has been mapped with pencil beam telescopes is Fornax A (shown in Figure 12.1), which consists of two extended components, with little or no fine structure, located symmetrically on either side of a bright elliptical galaxy.

b) Lunar Occultations

For a number of years the highest resolutions were obtained by observing the diffraction pattern of a radio source as it was occulted by the Moon. Most of the analysis was based on a technique described by Scheuer (1962) and later extended by von Hoerner (1964) to restore the true brightness distribution from the observed Fresnel diffraction pattern. The maximum resolution is generally limited by the sensitivity of the telescope.

Although a radio-telescope especially designed for this purpose has recently been completed in India, the technique has not enjoyed widespread use for several reasons:

(1) High resolution is obtained only for very strong sources or when using very large antennas, such as the 1000-foot Arecibo spherical reflector. The integration time is limited to about one second by the passage of the Moon across the source, and very short receiver time constants are required to obtain good resolution.

(2) Each occultation gives only a one-dimensional "strip" distribution. Several occultations are required to reconstruct the two-dimensional structure.

(3) At the shorter wavelengths, the Moon is an intense source of thermal radio emission and the small tracking irregularities present in radio-telescopes may completely mask the occultation of the much weaker extra-galactic sources.

(4) Interference from terrestrial radio emission reflected from the Moon is often a serious problem.

c) Interplanetary Scintillations (IPS)

In 1962 a group working at the Cavendish Laboratory in England discovered that radio sources with structure of the order of a second of arc or less showed rapid scintillations when observed through the solar corona. This effect was used extensively for several years to study small-scale-structure radio sources (*e.g.*, Cohen, Gundermann, and Harris, 1967; Little and Hewish, 1966). However, in recent years the high-resolution interferometers now available have been used for this kind of work, and IPS are now rather used to study the interplanetary medium. The reader who is interested in further details about the application of lunar occultations and interplanetary scintillations is referred to the review by Cohen (1969).

d) Interferometry and Aperture Synthesis

During recent years the techniques of interferometry and aperture synthesis have

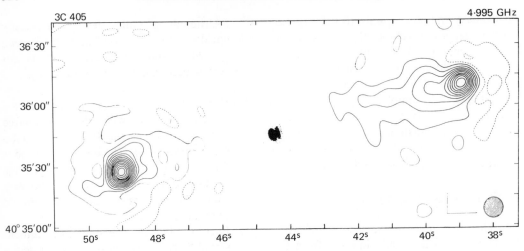

Figure 12.2 Brightness distribution of the intense radio galaxy Cygnus A at 6 cm observed with the 6-second-of-arc beam of the Cambridge 1-mile aperture synthesis radio-telescope. [Taken from Mitton and Ryle (1969), *Monthly Notice Roy. Astron. Soc.* 146:221].

been greatly improved and are now providing accurate and detailed information on the structure of extra-galactic sources. Because of the importance of these techniques, a whole chapter of this book has been devoted to describing the methods in considerable detail.

Most of the sources that have been studied have angular dimensions of less than a few minutes of arc, and about half of all sources are less than 15 seconds of arc in size. For the resolved sources a simple single-component structure is very rare, and most sources show a surprising amount of structure. Often the source is extended along a single axis, and the most common configuration is the double structure where most of the emission comes from two well-separated components. Frequently the two components are of approximately equal size, and with luminosity as illustrated by the map of Cygnus A shown in Figure 12.2. Typically the overall dimensions are about three times the size of the individual components, which are located symmetrically about the galaxy or QSO. But in some cases the ratio of component separation to component size may be very great. Often an extended low surface brightness component is located between the two primary components, or extends as a tail away from a single bright region, as shown by

the radio photograph of the radio galaxies 3C 129 and 3C 129.1 illustrated in Figure 12.3.

Other sources have been observed to show more complex structures containing three or more widely separated components which may also be aligned along a single axis (Macdonald *et al.*, 1968; Fomalont, 1969). In all of these extended sources the radio emission comes primarily from regions well removed from the optical galaxy or quasar. High-resolution studies of these extended clouds often show considerable structure, with one or more highly condensed regions existing many kiloparsecs from what is presumed to be the parent galaxy or quasi-stellar object. Typical linear dimensions for these extended sources range from about 10 kpc to several hundred kiloparsecs.

Following the early detection of apparent extensive halos around two nearby galaxies, M31 and M33, and the speculation of our own Galactic halo (Chapter 1), it was widely supposed that the radio emission from "normal" galaxies originated in a much larger volume of space than the optical emission. This is not the case, however, and in fact in general the "normal" galaxy emission is often highly concentrated to the galactic nucleus.

In many objects, such as quasars or

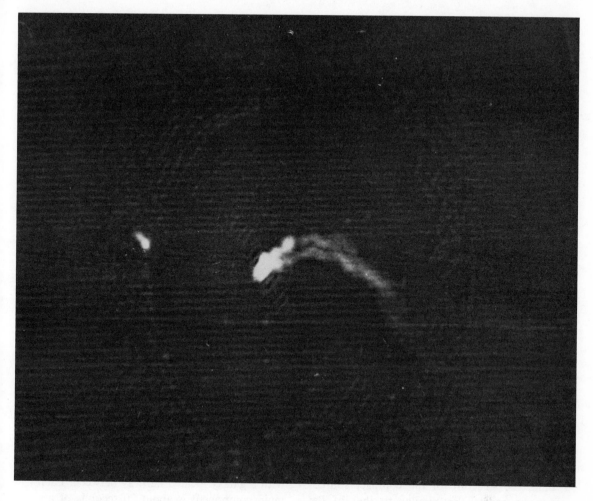

Figure 12.3 Radio photograph of the sources 3C129 and 3C129.1 obtained with the 21-second-of-arc beam of the Westerbork Synthesis Radio Telescope at 21 cm. (Miley, private communication.)

Seyfert and N galaxies, but also including some elliptical galaxies of normal appearance, there are one or more very compact radio sources coincident with the region of brightest optical luminosity. These compact sources also have complex structures with component sizes ranging from about 0.1 arc to well under 0.001 arc. Often compact and extended components exist simultaneously. The extended components may appear as (1) a "halo" surrounding a compact "core" component, as in the galaxy M87, (2) a jet extending away from the compact component as in 3C 273, or (3) a pair of unconnected extended components lying on either side of

the compact central sources, as in 3C 111. Generally the sources in categories (1) and (3) are identified with E galaxies and those in (2) with N or Seyfert galaxies, or quasars.

The observations of the sub-structure of the compact sources are made with independent oscillator tape-recording interferometers using telescopes widely spaced around the world with base-lines up to 80% of the Earth's diameter. Since intermediate baselines are insufficiently sampled, the data are not adequate to completely reconstruct the brightness distribution and it is necessary to resort to model fitting. To the extent that the brightness distribution can be inferred from the limited

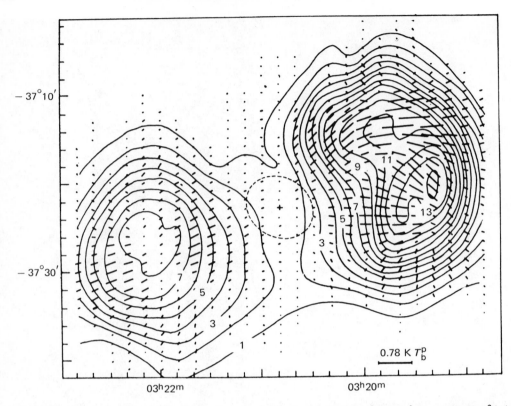

Figure 12.4 Polarization vectors of Fornax A observed at 6 cm superimposed on contours of total brightness temperature. The resolution was 4 minutes of arc beam using the Parkes 210-foot radio-telescope. In the eastern component the polarization is very low at the intensity maximum, while in the other it is much greater. [Taken from Gardner and Whiteoak (1971), *Australian J. Phys.* 29:899].

data, the structure of the compact sources appears remarkably similar to that of the extended sources, in the sense that in general they do not show circular symmetry, but consist of two or more well-separated components, lying along a single axis. Thus over a range of angular (and linear) dimension of about a factor of 10^5, the radio sources show essentially similar structure—only the scale size varies.

The smallest linear size which has been directly measured is the compact source located in the nucleus of the nearby galaxy M87, which contains about 1% of the total flux density and is only about 0.1 pc across in extent.

One of the best-studied sources is the intense radio galaxy Cygnus A, shown in Figure 12.2, which contains two major components separated by about 2 minutes of arc,

with a galaxy located halfway between. Each main component is about 20 seconds of arc in size, and contains a faint tail somewhat elongated along the line joining the components. Near the outer edge of each of the main components is an intense bright core. High-resolution observations of the western core show that the core itself contains a double source with a separation of about 5 arc seconds along a position angle about 20° from the line joining the two main components. The eastern component is also a double with a separation of a few arc seconds along a line nearly perpendicular to the line joining main components. Each of the subcomponents has an angular size of the order of one arc second (Miley and Wade, 1971). Also there is a weak, compact component coincident with the optically identified galaxy.

Some sources, particularly those located

in dense clusters, have an extended region of low surface brightness emission. It has been postulated that these are trails of relativistic electrons left by the motion of a radio galaxy through a gaseous medium. The radio photograph of 3C 129 and 3C 129.1 shown in Figure 12.3 clearly shows a double-helix trail suggesting rotation of the particle-ejecting region of the galaxy (Miley *et al.*, 1972).

12.4.2 Polarization

Nearly all of the radio galaxies and quasars have some degree of linear polarization ranging from integrated values of a few tenths of a percent to several percent, with the greatest value about 20%. At least for the extended sources, the integrated polarization is generally greatest at the shorter wavelengths, and the greatest polarization is found in the low surface brightness objects. In general the plane of polarization rotates at a rate approximately proportional to λ^2 and it is generally considered that this is due to Faraday rotation. Since the amplitude and sign of rotation appear to depend on galactic coordinates, it is thought that much, but not necessarily all, of the rotation occurs within the Galaxy. The degree of depolarization at longer wavelengths may also depend on galactic coordinates, but this is not clearly established.

Observations have also been made to map the distribution of polarized emission. In some cases the observed polarization reaches a degree of polarization comparable to that expected from a uniform magnetic field, indicating remarkably aligned magnetic fields over large volumes of space. Usually the regions of lowest surface brightness show the greatest polarization. Although there is no simple general relationship between the source geometry and the polarization direction, in many sources the polarization appears to be either parallel or perpendicular to the direction of elongation. In several sources the polarization is radial suggesting a circumferential magnetic field (Fomalont 1973). Figure 12.4 shows the polarization vectors observed

for Fornax A at a 6-cm wavelength.

Recently, several observers have detected small amounts of circular polarization in a few compact sources.

12.4.3 Spectra

With the exception of the 21-cm line of neutral hydrogen found only in relatively nearby galaxies, there are no sharp features in the radio spectra of galaxies and quasars, and the observations are confined to measurements of the continuous spectra. Since, unlike optical telescopes, radio-telescopes generally operate only over a limited range of wavelengths, the determination of spectra over a wide range of wavelengths requires combining data obtained by many observers using widely different types of telescopes. Because radio-telescopes may differ widely in their characteristics, each antenna and radiometer system must be separately calibrated at every wavelength where observations are made. Generally, this is done by observing one or more sources whose intensity had previously been determined on an "absolute" scale. Until recently, the problem of obtaining an "absolute" calibration of these primary standards was a formidable one, and the experimental discrepancies discouragingly large. Today, however, the situation is vastly improved and standard sources calibrated with an absolute accuracy of 3 to 10% are available over a wide range of wavelengths. The determination of relative intensities is much easier and is routinely done to an accuracy of a few percent, at least at the shorter wavelengths where confusion from the galactic background is less important.

Hundreds of extra-galactic sources have now been observed over a range of wavelengths extending from a few centimeters to a few meters, and a smaller number over the wider range from a few millimeters to a few tens of meters (10 MHz to 100 GHz). This range of 10^4 to 1 in wavelength may be compared with the range of only about 2:1 available for ordinary optical spectra. Although a wide variety of spectral shapes are

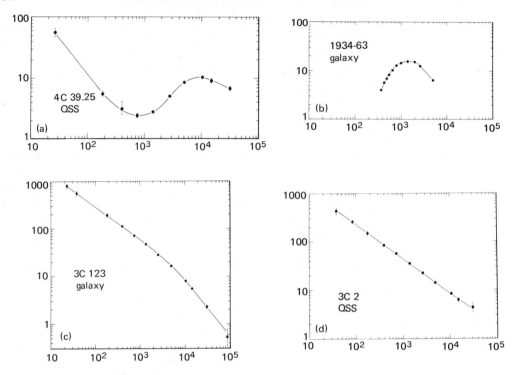

Figure 12.5 Radio-frequency spectra of four sources: (a) 4C39.25, Class **CPX**; (b) 1934-63, Class C_{max}; (c) 3C123, Class **C–**; (d) 3C2, Class **S**.

found, no unambiguous distinction exists between the spectra of radio galaxies and those of quasars.

Radio spectra are usually displayed in the form of a logarithmic plot of flux density vs. frequency. Sources with simple power law spectra are then represented by a straight line. The spectral index, α, is defined by the relation* (flux density) $\propto$ (frequency)$^\alpha$.

Although the radio spectra of only a few sources follow a simple power law accurately, a spectral index may be defined at any frequency as the tangent to the curve on a log S—log ν plot, or by the measurement of flux density at two arbitrarily selected frequencies. The observed spectra are conveniently divided into three groups (Figure

12.5) defined by their appearance on a log S— log ν plot.

(1) Class **S**. Straight spectra. Sources with Class **S** spectra generally have indices near -0.8 with nearly all sources in the range $-1.3 < \alpha < -0.6$. There are no Class **S** spectra with indices flatter than -0.5.

(2) Class **C**. Curved spectra. Sources with Class **C** spectra may be further subdivided into three groups:

(a) Class **C–**. Class **C–** spectra have a negative second derivative (defined by the second derivative of the log S—log ν relation), so that the spectrum is steeper at short wavelengths. Typically the region of curvature extends over a decade or so of wavelength, while at wavelengths significantly removed from the maximum curvature there are two well-defined power laws with indices differing by about 0.5.

---

* Since for most of the earlier known sources, the flux density decreased with wavelength, the spectral index was sometimes defined by (flux density) $\propto$ frequency $^{-\alpha}$, so that α was always a positive number. Now that indices of both signs are known to occur, the definition given in the text is being more widely used.

(b) Class C_{max}. Class C_{max} spectra have a power law (Class **S**) or dual power law (Class **C**−) spectrum at short wavelengths, but a sharp cut-off at long wavelengths, which is probably due to self-absorption.

(c) Class **C**+. Class **C**+ spectra have positive curvature and can usually be clearly separated into two groups:

   (i) Class C_l+. Class C_l+ have enhanced spectra at long wavelengths.

   (ii) Class C_s+. Class C_s+ have enhanced spectra at short wavelengths.

(3) Class **CPX**. Class **CPX** spectra are complex and show one or more relative minima. **CPX** spectra are generally believed to be composed of two or more Class C_{max} spectra, plus in some cases a Class **S** spectrum, or one Class C_{max} together with a Class **S** spectrum.

The Class **S** and Class **C**− spectra are mostly found in the extended radio sources, which are transparent at radio frequencies, while the C_{max}, C_s+, and **CPX** spectra generally indicate a compact opaque component. C_l+ spectra are typically found in sources located in rich clusters of galaxies.

The histogram of the distribution of spectral indices shows three distinct populations. One, which predominates in the surveys made at relatively long wavelengths, contains class **S** and **C**− spectra, with a narrow distribution of indices about a median value near −0.8 and a dispersion of 0.15. The steepest value of the index which is observed for these sources is about −2.0. Very few sources, however, have indices steeper than −1.3. The sources shown in the histogram of Figure 12.6 are mostly of the class **S** or **C**−.

A second population, of very steep spectra with $\alpha \sim -2$ appears in some data obtained at decameter wavelengths. Because of the experimental difficulties in measuring accurate flux densities at long wavelengths, the fraction of these C_l+ spectra which show this anomalous increase in the decameter flux is not well determined. Only a single spectrum of this type is found among the 3CR sources illustrated in Figure 12.6.

The third population is found mostly in surveys made at shorter wavelengths and consists of sources having flat spectra, and a much broader dispersion in spectral index about a median value near zero. These are all of the type C_{max}, C_s+, or **CPX** associated with compact sources. The preponderance of indices near zero appears to be due to the fact that the form of the spectra in these sources results from the superposition of a number of Class C_{max} components, with spectral peaks extending over a range of frequencies. Only a few sources of this type are in the 3C catalog and, as shown in Figure 12.6, these are mostly quasars.

For most of the extended radio sources where the radio brightness distribution has been mapped at several wavelengths, the structure is found to be essentially independent of wavelength. In other words, the spectral index is constant throughout the source. An exception to this, however, is the so-called core-halo source, in which the compact "core" component generally has a very flat spectrum due to synchrotron self-absorption.

Although it is not possible in any individual case to distinguish between radio galaxies and quasars on the basis of their spectra, there do appear to be statistical differences in the spectral distributions for the two classes of identifications. The radio galaxies show mostly class **S** spectra and, as shown in Figure 12.6, have a narrow dispersion of indices near a median of −0.8. The small "tail" in the distribution toward flat spectra represents sources that are mostly identified with galaxies having prominent optical nuclei, such as Seyfert or N galaxies, but several otherwise normal-looking E galaxies are also included. In the case of the quasars, both spectral populations are represented, but the "flat spectra" population is considerably more prominent than in the case of the galaxies.

The unidentified sources selected from long-wavelength surveys show a spectral

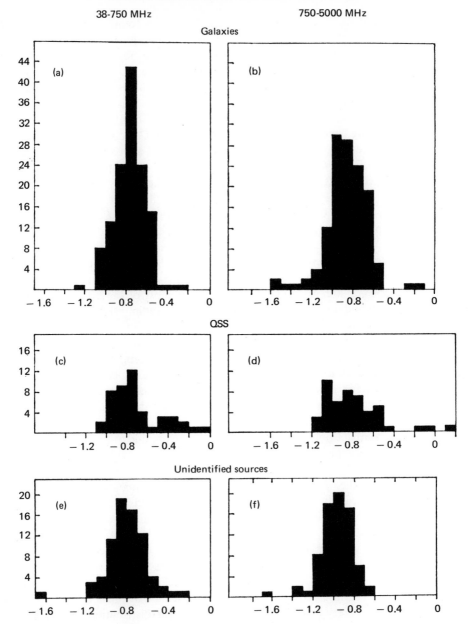

Figure 12.6 Histogram showing the distribution of spectral indices for (a) radio galaxies between 38 and 750 MHz, (b) radio galaxies between 750 and 5000 MHz, (c) quasars between 38 and 750 MHz, (d) quasars between 750 and 5000 MHz, (e) the unidentified sources with $|b''| \geq 10°$ between 38 and 750 MHz, and (f) the unidentified sources between 750 and 5000 MHz. [Taken from Kellermann, Pauliny-Toth, and Williams (1969), *Astrophys. J.* 157:1].

distribution which is similar to that of the radio galaxies, although the mean index is somewhat steeper. It is thus possible to interpret these unidentified sources as radio galaxies which are beyond the plate limit of the Palomar Sky Survey. The slightly

steeper mean index observed for these sources may be explained if they have dual power law spectra, and the large red shift has moved the high-frequency (steeper) part of the spectrum into the observed frequency range. Detailed investigation of the form of radio source spectra of unidentified radio galaxies, however, indicates that the spectra do not steepen sufficiently at high frequencies for the unidentified sources to be interpreted in this way.

An alternative explanation is that if the identified sources are associated with radio galaxies beyond the plate limit, they must be relatively distant, and thus have a high absolute radio luminosity. The steep spectra observed for the unidentified sources could then reflect a relation between high luminosity and steep spectra (cf. Section 12.4.5).

Because the sources with flat or inverted spectra are relatively weak at long wavelengths, they were not detected in the earlier surveys, which were made at meter wavelengths. As techniques have been pushed toward shorter wavelengths, the fraction of flat spectra sources detected by surveys has increased. Near 6-cm wavelength the observed fraction of flat spectra (opaque) sources and steep spectra (transparent) sources is about equal.

The expected dependence of the spectral index distribution on observing wavelength may be easily calculated. Given the normalized spectral index distribution, $P(\alpha)$, at any wavelength v_1 for sources with flux density stronger than S_1, and provided that $P(\alpha)$ is independent of S, the spectral distribution, $Q(\alpha)$, for all sources stronger than S_2 at v_2 is then

$$Q(\alpha) = \frac{N[(v_2/v_1)^{-\alpha} S_2, v_1]}{N(S_2, v_2)} P(\alpha) \quad (12.1)$$

where $N(S, v)$ is the integral number flux density relation (see Chapter 13).

If $N(S) \propto KS^x$ where K and x are constants, then

$$Q(\alpha) = \frac{N(S_2, v_1)}{N(S_2, v_2)} \left(\frac{v_2}{v_1}\right)^{-\alpha x} P(\alpha) \quad (12.2)$$

where the first term is simply a normalization constant.

In the special case where $P(\alpha)$ is a Gaussian with a mean index α_o and dispersion σ, then $Q(\alpha)$ is also Gaussian with the same dispersion σ but with a mean index displaced by the amount

$$\Delta\alpha = -x\sigma^2 \ln v_1/v_2 \quad (12.3)$$

Because of the factor $(v_2/v_1)^{-\alpha x}$ in Equation (12.2), it is clear that the spectral index distribution is very frequency-dependent and that sources with flat spectra dominate short-wavelength surveys in the same way that steep spectra dominate the long-wavelength surveys. For example, if the slope, x, of the log N—log S relation is taken as 1.5, then the ratio of flat spectra ($\alpha \sim 0$) to steep spectra ($\alpha \sim -1$) sources for two surveys made at frequencies one decade apart is $10^{1.5} \sim 32$.

This change in the spectral index distribution with sample frequency should not be confused with the observed change in the index distribution when a given sample of sources is observed at different wavelengths. The brightest **n** sources which determine $P(\alpha)$ at v_1 are not the same **n** sources which determine $Q(\alpha)$ at v_2. The change in spectral distribution of a given sample observed at different wavelengths is due to spectral curvature.

Comparisons of surveys made at different wavelengths, and of the dependence of the spectral index distribution on wavelength, are in reasonable agreement with that predicted by the expressions above. This means that the full range of spectral population is represented at each observing frequency, although the relative proportion of the two spectral classes may vary considerably.

12.4.4 Intensity Variations

Many of the compact sources show pronounced intensity variations on time scales from less than a week to a few years. There is no simple pattern to the observed variations; in particular, there is no evidence for any periodic phenomena. Rather the variations appear as bursts, first at short wavelengths, and then at longer wavelengths with reduced amplitudes, with the duration of each burst being longer at the longer wavelengths. Below some critical wavelength, the amplitude, shape, and time of occurrence of the burst are independent of wavelength. Often the duration of a single outburst is comparable

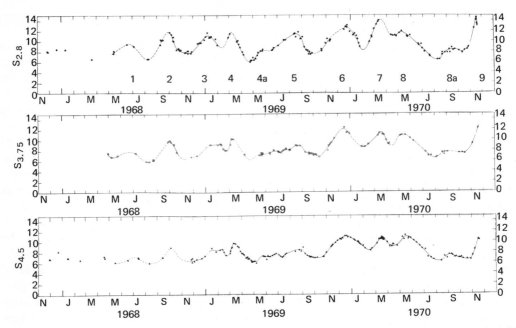

Figure 12.7 Variation in the radio intensity of BL Lacerte (VRO 42.22.01) observed at 2.8, 3.75, and 4.5 cm. [Taken from MacLeod, Andrew, Medd, and Olsen (1971), *Astrophys. Lett.* 9:19.]

to the time between outbursts so that the individual events are not resolved. Figure 12.7 shows the relatively rapid intensity variations observed at three wavelengths in the source BL Lac (VRO 42.22.01). Slower, more typical variations are found in the radio galaxy 3C 120, shown in Figure 12.8.

Quasars and the nuclei of some galaxies also vary at optical wavelengths. Although there appears to be no detailed relation between the intensity variations seen at radio and optical wavelengths, those sources which are most active at radio frequencies are also usually prominent optical variables.

There is no obvious difference in the pattern of the radio intensity variations seen in radio galaxies and quasars. The change in radiated power observed in the radio galaxies is typically of the order of 10^{42} ergs/sec/yr. For the quasars, if they are at cosmological distances, the change is very much greater and may be as much as 10^{45} ergs/sec/yr, *e.g.*, comparable with the total radio luminosity of the strongest radio galaxies such as Cygnus A.

Variations are also observed in the pola-

rization of the compact extra-galactic sources. These measurements are very difficult since the observed polarization is typically only a few percent, and so the experimental uncertainties are large. The limited data indicate that the most rapid changes in polarization occur when the total flux is increasing or is near a maximum (*e.g.*, Aller, 1970).

12.4.5 Empirical Relations

The establishment of relations between the various observed radio source properties is clearly important to the understanding of the origin and evolution of extra-galactic sources. Some of the better-determined relations are summarized below. Some of these are well understood in terms of current theory, while others are not.

(1) Spectral index—angular size. All of the sources with flat or inverted spectra ($\alpha \gtrsim -0.5$) have components with small angular dimensions ($\theta \ll 1''$). This is well understood as the effect of self-absorption.

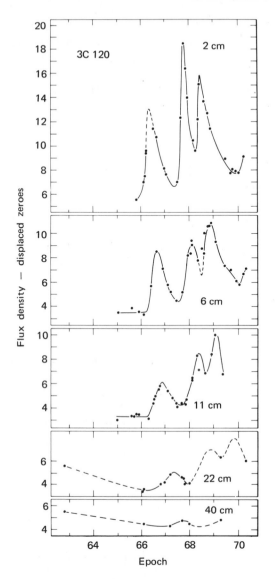

Figure 12.8 Variations in the intensity of the radio galaxy 3C 120 observed at 2, 6, 11, 22, and 40 cm. Three distinct outbursts are apparent in the 2-cm data. These bursts occur at later times and with reduced amplitude at the longer wavelengths.

(2) Spectral-index variability. All of the sources which show flux density variations have flat ($\alpha \gtrsim -0.5$) or CPX spectra in the spectral region where the intensity varies. This is a reflection of the fact that only the opaque or partially opaque sources vary. Sources which have

a single sharp low-frequency cut-off generally do not vary as much, at least over the time scales for which accurate observations are available.

(3) Radio-optical variability. Sources which show large radio flux density variations also frequently show large variations at optical wavelengths as well, but there is no simple one-to-one relationship.

(4) Variability-wavelength. The largest and most rapid variations generally occur at the shorter wavelengths.

(5) Spectral-index-luminosity. Among the extended transparent radio galaxies there is a tendency for the most luminous sources to have the steepest spectra (Heeschen, 1960), particularly for sources with Class S spectra. This relationship does not seem to hold, however, for the quasars which all have a very high radio luminosity.

(6) Luminosity-structure. Sources with relatively simple brightness distributions are of relatively low luminosity.

(7) Luminosity-brightness. Sources with a high surface brightness have a high radio luminosity. Two sequences are apparent, as shown in Figure 13.9. The apparent continuity of the quasars and radio galaxies in this Heeschen diagram, if the quasar red shifts are assumed cosmological, is often taken as evidence of the cosmological origin of the red shifts. However, if the red shifts are local, then the quasars fall in the same high brightness—low luminosity part of the diagram as the cores of E galaxies such as M87, NGC 1052, and the Seyfert galaxy nuclei, so that the argument can go either way.

(8) Polarization-wavelength. For the transparent sources (Class S spectra) the observed polarization is greatest at shorter wavelengths. In the opaque sources the reverse is true.

(9) Circular polarization is found only in some compact sources.

(10) The greatest linear polarization is found in the sources with low surface bright-

ness, or in low surface brightness regions of resolved sources.

(11) Sources found in clusters tend to have smaller dimensions.

(12) The strong radio galaxies nearly all show intense emission lines.

(13) The two most rapidly varying sources, BL Lac and OJ287, appear to be quasars, but their optical spectra do not show the emission lines common to other quasars.

12.5 Theories of Radio Sources

Theories describing the source of energy in radio galaxies and quasars have been as numerous and varied as the authors proposing them. These theories which, in general, make no attempt to interpret the growing detailed observational data include:

(1) the collisions of stars or galaxies

(2) the collapse of stars, superstars, galaxies, or intergalactic matter

(3) the explosion of stars, superstars, or galaxies, including chain reactions

(4) matter—anti-matter annihilation, creation of matter

(5) quark interactions.

In recent years, however, theoretical efforts have concentrated on interpreting the observed spectra, polarization, structure, and time variations in terms of the synchrotron hypothesis, rather than on exotic mechanisms for producing energy.

For many years nearly all of the theoretical effort in this direction was made in the Soviet Union (*e.g.*, Ginzburg, 1951; and Shklovsky, 1952). Today, however, the synchrotron model is widely accepted, and it is to be hoped that an increased understanding of the observational material in terms of synchrotron radiation will lead to a better understanding of the source of energy. In particular, the exciting discovery by Dent in 1965 of rapid time variations in the radio emission of some quasars and galaxies has opened the possibility of observing the synchrotron

emission from relativistic particles within a few months to a few years of the time they are accelerated. This offers a previously unexpected opportunity to study the source at a very early epoch and may ultimately specify the initial conditions in radio sources, and hence the source of energy.

In the remainder of this section we summarize briefly the basic results of the synchrotron theory as they apply to radio astronomy and the theory as applied to the data in an attempt to understand the origin and evolution of extra-galactic radio sources. A thorough review of the basic synchrotron process is given in the book by Pacholczyk (1970).

12.5.1 Synchrotron Radiation

A single electron of energy E spiraling in a magnetic field at ultra-relativistic velocities $[(1 - v^2/c^2) \ll 1]$ has its radiation concentrated in a cone of half angle $\phi \sim E/mc^2$. An observer sees a short burst of emission lasting only during the time, Δt, that the cone is pointed toward the observer. The radiation is concentrated in the high-order harmonics, $\eta = (E/mc^2)^2$, of the classical gyrofrequency $v_g = eB/m$. The frequency distribution of the radiation is given by a complex expression conveniently represented by

$$P(v, E, \theta) = 2.3$$
$$\times 10^{-29} B_\perp \left(\frac{v}{v_c}\right) F\left(\frac{v}{v_c}\right) \text{ W Hz}^{-1} \quad (12.4)$$

where

$$F\left(\frac{v}{v_c}\right) = \int_{v/v_c}^{\infty} K_{5/3}(\eta)\, d\eta$$

and where $B_\perp = B \sin \theta$ is the component of the magnetic field perpendicular to the line of sight; $K_{5/3}(\eta)$ is a modified Bessel function; θ is the angle between the electron trajectory and the magnetic field (pitch angle); and the critical frequency v_c is given by

$$v_c = cB_\perp E_{\text{Gev}}^2 = (1.6 \times 10^4) B_\perp E_{\text{Gev}}^2 \text{ GHz}$$
$$(12.5)$$

The spectrum of the observed radiation depends on the angle ϕ between the line of sight and the electron trajectory and on the plane of polarization. In the remainder of this chapter the subpscript "$\perp$" is dropped and the symbol B is understood to represent the perpendicular component of the magnetic field.

The total power radiated by each electron is given by

$$\frac{dE}{dt} = AB^2E^2 = (6.08 \times 10^{-9})\,B^2 E_{\text{Gev}}{}^2\,\text{ergs/sec}$$

$$(12.6)$$

where $A = 6.08 \times 10^{-9}$ and $c = 1.6 \times 10^7$.

The distribution given by equation (12.4) has a broad peak near $v \sim 0.28 v_c$. For $(v/v_c) < 0.3$, $P(v) \propto v^{1/3}$. For $(v/v_c) > 3$, $P(v) \propto (v/v_c)^{1/2}\,e^{-v/v_c}$ and the radiation falls off rapidly with increasing frequency.

For an assembly of electrons with a number density $N(E)dE$ between E_1 and E_2, Equation (12.4) can be integrated to find the total radiation at any frequency from all electrons. Using Equation (12.5) and making a change of variable this becomes

$$P(v\theta) = 4.1 \times 10^{-29}\,B^2\,v^{1/2}$$

$$\times \int_{v/v_1}^{v/v_2} (v/v_c)^{-3/2}\,N(v/v_c)F\,(v/v_c)\,d(v/v_c) \quad (12.7)$$

where v_1 and v_2 are the critical frequencies defined by Equation (12.6) corresponding to E_1 and E_2.

In the special case where the electron energy distribution is a power law, that is, $N(E)\,dE = KE^{-v}\,dE$, Equation (12.7) becomes

$$P(v,\theta) \propto B^{(\gamma+1)/2}\,v^{-(\gamma-1)/2}$$

$$\times \int_{v/v_1}^{v/v_z} (v/v_c)^{-(3-\gamma)/2}\,F(v/v_c)\,d(v/v_c) \quad (12.8)$$

For $\gamma \gtrsim 1$ the major contribution to the integral is when $(v/v_c) \sim 1$ so that the limits of integration may be extended from zero to infinity without introducing significant error. The integral is then essentially constant when $3v_1 \lesssim v \lesssim 10v_2$. The radio spectrum then is a power law,

$$S \propto v^\alpha \quad (12.9)$$

with a spectral index $\alpha = -(\gamma - 1)2$.

It must be emphasized that this approximation is valid only when $\gamma \gtrsim 1$; and in particular that no form of energy distribution can give a spectrum that rises faster than the low-frequency asymptotic limit of $v^{1/3}$ for a single electron.

As described in Section 12.4.3, many sources show nearly power law radio-frequency spectra, with a common spectral index $\alpha \sim -0.8$ corresponding to an electron energy distribution index, $\gamma \sim 2.6$. Deviations from a constant radio spectral index may be explained as being due to (1) variations in γ as a function of energy which may exist either in the initial electron energy distribution or occur as a result of differential energy loss in an initial power law distribution; (2) self-absorption in the relativistic electron gas; (3) absorption in a cold HII region between us and the source; (4) the effect of a dispersive medium in which the electrons are radiating.

12.5.2 Effect of Energy Losses

Even if relativistic electrons are initially produced with a power law distribution, differential energy losses can alter the energy spectrum. Relativistic electrons lose energy by synchrotron radiation and by the inverse Compton effect, which are both proportional to the square of the energy; by ordinary bremsstrahlung and adiabatic expansion, which are directly proportional to the energy; and by ionization, which is approximately proportional to the logarithm of the energy. Approximating the logarithmic term by a constant, the rate of energy loss may be written

$$\frac{dE}{dt} = aE^2 + bE + c \quad (12.10)$$

If electrons are being supplied to the source at a rate $N(E,t)$, then the equation of continuity describing the time dependence of the energy distribution $N(E,t)$ is

$$\frac{\partial N(E,t)}{\partial t} = \frac{\partial}{\partial E} \frac{dE}{dt} N(E,t) + N(E,t) \quad (12.11)$$

If at $t = 0$

$$N(E) = \begin{cases} K E^{-\gamma} & E_1 < E < E_2 \\ 0 & E < E_1, E > E_2 \end{cases} \quad (12.12)$$

and if synchrotron and inverse Compton losses dominate and there is no injection of new particles [$N(E,t) = 0$], then

$$N(E,t) = \begin{cases} \dfrac{K E^{-\gamma}}{(1 - AB^2 Et)^{2-\gamma}} & E_1' < E < E_2' \\ 0 & E < E_1', E > E_2' \end{cases}$$

$$(12.13)$$

where

$$E' = \frac{E}{1 + AB^2 Et}$$

Thus, even with an initial energy distribution extending to unlimited energy, there will be a cut-off at

$$E_c = \frac{1.64 \times 10^8}{B^2 t} \quad (12.14)$$

and a corresponding cut-off in the synchrotron radiation spectrum. In the special case where $\gamma \sim 2$, $N(E,t)$ can become very large for energies slightly less than E_c because of the piling up near E_c of electrons with large initial energies as the result of their more rapid rate of energy loss. In this case, if E_2 is sufficiently large so that $(E_2'/E_c) \sim 1$, then the radiation spectrum will become flat just below the upper cut-off frequency, $\nu_c = c BE_c^2$. Above ν_c the spectrum sharply decreases rapidly for all values of γ.

If the distribution of electron pitch angles is random, then the cut-off frequency for each pitch angle differs. At low frequencies where energy losses are not important, the spectral index, α, remains equal to its initial value $\alpha_o = (1 - \gamma)/2$. But at higher frequencies if the pitch angle of each electron remains constant (Kardashev, 1962), $\alpha = (\frac{4}{3}\alpha_o - 1)$. If on the other hand the pitch angle distribution is continuously made random, e.g., by

irregularities in the magnetic field, then all the electrons see the same effective magnetic field and the spectrum shows the same sharp cut-off observed with a single pitch angle. The frequency at which synchrotron radiation loss is important is given by

$$\nu_b \sim B^{-3} t_{yr}^{-2} \text{ GHz} \quad (12.15)$$

If the relativistic electrons are continuously injected with $Q(E) = K E^{-\gamma_0}$, then for $\nu < \nu_b$ the spectral index remains constant with $\alpha = \alpha_o$. But at higher frequencies where the rate of energy loss is balanced by the injection of new particles, the equilibrium solution of Equation (12.11) with $(\partial N/\partial t) = 0$ gives $\alpha = (\alpha_o - \frac{1}{2})$.

12.5.3 Interpretation of Spectral Data

The main features of the radio spectra which must be explained by any theory of the generation of the relativistic particles are:

(1) the relatively sharp concentration of the energy index near $\gamma \sim 2.6$
(2) the extreme values of the index for the Class **S** spectra of $\alpha \sim -0.5$ and $\alpha \sim 2.0$, and the inclusion of nearly all sources in the range $-0.5 > \alpha > -1.3$
(3) the absence of any sharp high-frequency cut-off in the radio spectra expected from synchrotron radiation losses
(4) the detailed form of the spectra of transparent sources
(5) the observed index of about $-\frac{1}{4}$ for the transparent part of the spectrum of variable sources.

It is tempting to interpret the change in spectral index of ~ 0.5 observed in the Class **C**− spectra as the effect of synchrotron radiation losses balanced by continuous injection. However, in this case we might expect that the index most commonly observed for the Class **S** spectra would correspond either to

(1) the low-frequency asymptotic value of the **C**− spectra if the **S** spectra refer to young

sources where ν_b is greater than the maximum observed frequency

(2) the high-frequency asymptotic value of the **C**− spectra if the **S** spectra refer to old sources where ν_b is less than the minimum observed frequency.

In fact the median value of α for the **S** spectra of about -0.8 is intermediate between the extremes of the **C**− spectra, which are about -0.65 and -1.15, so the interpretation is not clear.

It is also possible that the range in spectral index observed in the **C**− spectra is due to partial self-absorption (see Section 12.5.5) in dense regions of the source which become opaque at relatively short wavelengths. This interpretation is suggested by the fact that a larger fraction of sources with **C**− spectra contains compact components (as determined, for example, from observations of interplanetary scintillations) than is found for sources with **S** spectra (Kellermann, Pauliny-Toth, and Williams, 1969).

Finally, it must be remarked that the most recent absolute calibrations indicate that at wavelengths of about one meter and longer most published flux densities are systematically low, so that the true spectral curvature may in fact be much less than has previously been thought.

The value of approximately $-\frac{1}{4}$ observed for the spectral index of variable sources in the transparent part of the spectrum suggests that this is the initial injection spectrum (so that $\gamma_o \sim 1.5$). The steeper spectra observed for the more extended sources are then somehow the result of synchrotron radiation losses.

The high probability of finding an observed index steeper than α_o by about 0.5 suggests the continuous injection model. But the limiting value of -1.3 suggests instantaneous injection with $\alpha = \frac{4}{3}[(-\frac{1}{4})-1] = -1.33$. Both of these can be accounted for by assuming that there is a repeated generation of particles with a characteristic period T. Then from Equation (12.15) the spectral index after an elapsed time $t \gg T$ have three distinct spectral regions defined by

(1) $\quad \alpha = \alpha_o \qquad\qquad \nu < B^{-3}\,t^{-2}$

(2) $\quad \alpha = \alpha_o - \frac{1}{2} \qquad B^{-3}\,t^{-2} < \nu < B^{-3}\,T^{-2}$

(3) $\quad \alpha = \frac{4}{3}\,\alpha_o - 1 \qquad \nu > B^{-3}\,T^{-2}$

In case (1) the effect of synchrotron radiation losses is not important, and the index remains equal to its initial value. In case (2) the time scale for radiation loss is longer than the period T between bursts, the injection can be considered quasi-continuous, and the spectrum is in equilibrium since radiation losses are balanced by the injection of new particles. In case (3) synchrotron losses dominate, and the spectrum steepens. The observed curvature in the spectrum of many sources is also consistent with this model.

The problem with the recurring injection model is that the high-frequency part of the spectrum, $\alpha = (\frac{4}{3}\,\alpha_o - 1)$, can be understood only in the case where the electron pitch angles are conserved for times greater than the radiation lifetime. For $B = 10^{-4}$ and $\nu = 1$ GHz, this is 10^6 years. Otherwise, the spectrum must show a sharp cut-off at high frequencies, unless, in fact, the relativistic electrons lose energy not by synchrotron radiation, but by some other mechanism which preserves the power law form of the energy distribution ($dE/dt \propto E$). The absence of a clear spectral cut-off may also be due to the superposition of a range of cut-offs, such as would occur, for example, if the radio emission came from a number of discrete regions having very different magnetic field strengths.

The observed spectral data are also consistent with a source where multiple injection and radiation losses are important during the initial stages, but followed by a longer period of adiabatic expansion when the spectral shape is preserved. But this interpretation poses serious energy problems, as discussed in Section 12.5.7.

It will be important to determine from future observations at shorter wavelengths at what wavelength the expected spectral cut-off occurs (if anywhere) and so help to specify the mechanism or mechanisms by which the electrons lose energy, and thus place further observational constraints on the radio source lifetimes.

12.5.4 Effect of Absorption by Ionized Hydrogen

The observed radio spectrum may differ from the radiated spectrum owing to the influence of the medium between the source and the observer. If a cold cloud of ionized gas is located in front of the source, then the observed flux density will fall off sharply below the frequency, v_o, where the optical depth is unity. For an electron temperature T_e,

$$v_o \sim 3.6 \times 10^5 \, T_e^{-3/2} \, \varepsilon \text{ MHz} \quad (12.16)$$

where $\varepsilon = \int n_e \, dl$ is the emission measure, and n_e the density of thermal electrons. The observed spectrum is then

$$S \propto v^\alpha \, e^{-(v_0/v)^2} \quad (12.17)$$

If the ionized medium is mixed with the synchrotron source, then for $v \ll v_o$

$$S \propto v^{(\alpha+2)} \quad (12.18)$$

If the density of thermal electrons is sufficiently great, then at frequencies where the index of refraction, η, becomes less than unity, the form of the spectrum will differ from that *in vacuo*. When $\eta < 1$, the velocity of a relativistic electron is less than the phase velocity of light in the medium; the radiation is no longer so highly concentrated along the electron trajectory, and the energy no longer appears in the high-order harmonics of the gyrofrequency. This is commonly called the Razin, or Tsytovich, effect and is important below a frequency v_r given by

$$v_r \sim 20 \frac{n_e}{B} \text{ MHz} \quad (12.19)$$

For $v < v_r$ the spectrum cuts off very sharply (see Chapter 3).

12.5.5 Synchrotron Self-Absorption

In Sections 12.5.2 and 12.5.3 we have assumed that the flux from a group of relativistic electrons is merely the arithmetical sum of the radiation from the individual electrons, *i.e.*, the electron gas is assumed to be transparent. If, however, the apparent brightness temperature of the source approaches the equivalent kinetic temperature of the electrons, then self-absorption will become important and part of the radiation is absorbed. The precise form of the radiation spectrum is complex, but can be calculated from the emission and absorption coefficients of relativistic electrons in a magnetic field. The parameters depend on the electron energy and pitch-angle distributions and can be determined only from numerical integrations. The form of the spectrum in the limiting case of a completely opaque source may be derived quite straightforwardly, however. Consider the radiation from an optically thick blackbody of solid angle Ω at temperature T. The observed flux density is

$$S = \frac{2k}{c^2} v^2 \, T \, \Omega \quad (12.20)$$

If the optically thick relativistic electron gas is described as a blackbody whose temperature is given by the equivalent kinetic temperature of the electrons, $E = kT$, then using Equation (12.16), we can write

$$S \propto B^{-1/2} \, \theta_{\text{sec}}^2 \, v^{2.5} \quad (12.21)$$

In other words, the source may be thought of as a blackbody ($S \propto v^2$) whose temperature (energy) depends on the square root of the frequency ($E \propto v_c^{1/2}$).

Rewriting Equation (12.21) and using a more precise analysis, including the small dependence on the index γ and the effect of the red shift z, the magnetic field is given by

$$B \sim 2.3 \times 10^{-5} \, (S_m/\theta^2)^{-2} \, v_m^5 \, (1 + z)^{-1}$$
$$\text{for } (\gamma = 2.5) \quad (12.22)$$

where S_m is the maximum flux density at the self-absorption cut-off frequency, v_m is in GHz, and θ is in milliarc seconds.

Although the apparent effect of synchrotron self-absorption is evident in many sources where indices as steep as $+1$ are often observed at long wavelengths, there has not been any direct observation of the theoretically expected value of $+2.5$. This has led some to question whether or not the observed low-frequency cut-offs are in fact due to self-absorption. However, this argument is probably irrelevant for the same reason that the theoretical index of $+2.0$ is never observed in the thermal emission spectra of the familiar HII regions. The explanation in both cases is that there is a wide range of opacities in these sources which cause different parts of the source to become opaque at different wavelengths, producing a gradual, rather than a sharp, transition from the transparent to the opaque case.

12.5.6 Inverse Compton Scattering

The maximum brightness temperature of any opaque synchrotron source is limited by inverse Compton scattering to about 10^{12} °K. This is the brightness temperature corresponding to the case where the energy loss by synchrotron radiation is equal to the energy loss by inverse Compton scattering and may be derived as follows (Kellermann and Pauliny-Toth, 1969).

For a homogeneous isotropic source

$$\frac{L_c}{L_s} = \frac{U_{\rm rad}}{U_B} = \frac{6L}{\rho^2 B^2 c} = \frac{{\rm constant}\,(S_m\,\nu_c\,R^2)}{R^2\,\theta^2\,B^2\,c} \quad (12.23)$$

where L_c = power radiated by inverse Compton scattering, L_s = radio power radiated by synchrotron emission, $4\pi r^2 \int_{\nu_m}^{\nu_c} S\,d\nu \sim 4\pi r^2 S_m \nu_c$, $U_{\rm rad} = 3L/4\pi r^2 c$ = energy density of the radiation field, $U_B = B^2/8\pi$ = energy density of the magnetic field, R = the distance to the source, θ = angular size, and the radius $\rho = R\theta/2$. Then using Equation (12.22) and recognizing that $S_m/\theta^2\nu^2$ is proportional to the peak brightness temperature, T_m, and including the effect of second-order scattering, we have

$$\frac{L_c}{L_s} \sim \frac{1}{2}\left(\frac{T_m}{10^{12}}\right)^5 \nu_c \left[1 + \frac{1}{2}\left(\frac{T_m}{10^{12}}\right)^5 \nu_c\right] \quad (12.24)$$

where ν_c is the upper cut-off frequency in MHz. Taking $\nu_c \sim 100$ GHz, then for $T_m <$

10^{11}°K, $L_c/L_s \ll 1$ and inverse Compton scattering is not important; but for $T_m > 10^{12}$°K, the second-order term becomes important, $L_c/L_s \sim (T_m/10^{12})^{10}$, and the inverse Compton losses become catastrophic. The exact value of T_m corresponding to $L_c/L_s = 1$ is somewhat dependent on the specific geometry, the value of γ, and the spectral cut-off frequency ν_c, but the strong dependence of L_c/L_s on T_m implies that T_m cannot significantly exceed 10^{12}°K, *independent* of wavelength. This places a lower limit to the angular size of

$$\theta \gtrsim 1.3 \times 10^{-3}\,S^{1/2}\,\nu_m^{-1} \quad (12.25)$$

If the compact sources expand with conservation of magnetic flux, then T_m varies with radius ρ as $T_m \propto \rho^{-(\nu-1)/(\nu+4)}$, so that for $\gamma \sim 1$, T_m remains constant and otherwise depends only weakly on ρ.

12.5.7 Energy Considerations

The problem of the origin and evolution of extra-galactic radio sources has been a formidable one; in particular the source of energy needed to account for the large power output and the manner in which this energy is converted to relativistic particles and magnetic flux is still a mystery. Assuming only that synchrotron radiation from ultra-relativistic electrons is responsible for the observed radiation, the necessary energy requirements may be estimated in a straight-forward way.

If the relativistic particles have a power law distribution with an index γ between E_1 and E_2, then for $\gamma \neq 2$, the energy contained in relativistic electrons is

$$E_e = \int_{E_1}^{E_2} E\,N(E)\,dE$$

$$= \frac{K}{(2-\gamma)}\,[E_2^{(2-\gamma)} - E_1^{(2-\gamma)}] \quad (12.26)$$

The constant K can be evaluated if the distance to the source is known; then the total luminosity L of the source may be estimated by integrating the observed spectrum, giving ($\gamma \neq 3$)

$$L = \int_{E_1}^{E_2} N(E) \frac{dE}{dt} dE = \int_{E_1}^{E_2} AH^2 E^{(2-\gamma)} dE$$

$$= A \frac{KB^2}{(3-\gamma)} [E_2^{(3-\gamma)} - E_1^{(3-\gamma)}] \quad (12.27)$$

or, for $\gamma = 2.5$, $H = 100$ km/sec/Mpc, and $q_o = +1$,

$$L \sim 10^{44} z^2 S \quad (12.28)$$

where S = flux density at 1 GHz.

Eliminating K we have

$$E_e = \frac{(3-\gamma)}{(2-\gamma)} \frac{E_2^{(2-\gamma)} - E_1^{(2-\gamma)}}{E_2^{(3-\gamma)} - E_1^{(3-\gamma)}} \frac{L}{A} \quad (12.29)$$

Using Equation (12.5) to relate E_2 and E_1 to the cut-off frequency, and grouping all the constant terms together,

$$E_e = C_e L B^{-3/2} \quad (12.30)$$

The magnetic energy is just

$$E_m = \int \frac{B^2}{8\pi} dV = C_m B^2 V \quad (12.31)$$

The total energy in fields and particles ($E_e + E_m$) is minimized when

$$\frac{\partial E}{\partial t} = 0 \quad (12.32)$$

or when

$$B = \left[\frac{3}{4} \frac{C_e}{C_m} \frac{L}{V} \right]^{2/7}$$

$$\sim 1.5 \times 10^{-4} \theta^{-6/7} z^{-2/7} S^{2/7} \text{ gauss}$$
$$(\gamma = 2.5) \quad (12.33)$$

The value of B estimated in this way depends almost entirely on the angular size, θ, and is relatively insensitive to the flux density or distance.

From Equations (12.30), (12.31), and (12.33), if θ is expressed in arc seconds,

$$E_e = \tfrac{4}{3} E_m \sim 10^{59} \theta^{9/7} z^{17/7} S^{4/7} \text{ ergs}$$
$$(\gamma = 2.5) \quad (12.34)$$

That is, the energy is nearly equally distrib-

uted between relativistic particles and the magnetic field.

Somewhat surprisingly there is little relation between the minimum energy computed in this way and the total radio luminosity. Typically, the total energy contained in the extended sources estimated is in the range 10^{57} to 10^{61} ergs and the magnetic field between 10^{-5} and 10^{-4} gauss. It is largely because of this apparent very great energy requirement (up to 0.01 % of the rest energy of an entire galaxy) that theoretical efforts to explain the origin of radio galaxies have been for the most part unsuccessful.

One interesting result is that if $E_e \sim E_m$ the total energy strongly depends on the size of the source ($E \propto r^{9/7}$). This gives the curious situation that the larger sources with low surface brightness and low luminosity, such as Centaurus A, appear to contain almost as much energy as the smaller high surface brightness objects such as Cygnus A. This is not, of course, what would be expected if, as generally assumed, the larger sources were older; this has led to the interesting suggestion that sources may collapse rather than expand. Another way out of this situation which also reduces the energy requirements on the larger sources is that if, as recent observations suggest, sources break up into a number of small components, or if the emission comes from only a thin shell, only a small fraction, Φ, of the projected volume of a source actually has particles and a magnetic field. The minimum total energy is then multiplied by a factor of $\Phi^{3/7}$, and the corresponding magnetic field is increased by the factor $\Phi^{-2/7}$. Finally, of course, we remark that there is no direct evidence that these minimum energy calculations are at all relevant. The true conditions may show considerable departure from equipartition; however, this greatly amplifies the energy requirements.

For some years it was widely thought that the relativistic electrons were secondary particles produced as the result of collisions between high-energy protons. If the ratio of energy in protons to that in electrons is k, then the minimum total energy is increased by a factor of $(1 + k)^{4/7}$ and the magnetic field by $(1 + k)^{2/7}$. Estimates of the value of k were about 100, so the energy requirements were about an order of magnitude greater. However, the discovery of rapid time variations in many sources, and its implications for the rapid production of particles, suggests that the secondary production mechanism is probably not relevant, and unnecessarily exaggerates the energy requirements. This elimination of the factor k, and inclusion of the fill-in-factor, Φ, can easily reduce the energy estimates by two or more orders of magnitude.

A characteristic lifetime for radio sources may be estimated from the relation $t \sim E/(dE/dt)$. Lifetimes of radio sources determined in this way are very long. For $E \sim 10^{61}$ and $(dE/dt) \sim 10^{45}$ ergs/sec the lifetime is 10^8 to 10^9 years. Similar ages are obtained from the fraction ($\sim 10\%$) of giant elliptical galaxies that are found to be strong radio sources, and an estimated age of 10^{10} years for the age of elliptical galaxies.

Equation (12.15) shows that in a 10^{-4} Gauss field, electrons radiating at $\nu > 1$ GHz are expected to decay in about 10^6 years. Thus the absence of a spectral cut-off even at $\nu \gtrsim 10$ GHz suggests a continued or multiple injection of relativistic particles (*e.g.*, van der Laan and Perola, 1969), or a very short lifetime.

For the compact opaque sources magnetic field strength is given directly by the measured peak surface brightness (S_{max}/θ^2), the frequency of maximum flux density, ν_m, and Equation (12.22). For the relatively nearby radio galaxies with small red shifts the magnetic field derived in this way is independent of the red shift, and in any case depends only weakly on the red shift. If the distance is known, then the total energy in the form of relativistic particles, E_p, is given by

$$E_p \sim 3.1 \times 10^{62} \, S_{max}{}^{5.5} \, \theta^{-9} \, \nu_m{}^{-10.5} \, z^2 \text{ ergs}$$
(12.35)

and the magnetic energy, ε_m, by

$$E_m \sim 2.5 \times 10^{48} \, S_{max}{}^{-4} \, \theta^{11} \, \nu_m{}^{10} \, z^3 \text{ ergs}$$
(12.36)

while the ratio of the two quantities is given by

$$\frac{E_p}{E_m} \sim 10^{14} \, S_{max}{}^{9.5} \, \theta^{-20} \, \nu_m{}^{-20.5} \, z^{-1} \quad (12.37)$$

In Equations (12.35) to (12.37) θ is in milliarc seconds, S in f.u., ν_m in GHz, $H = 100$ km/sec/Mpc, and $\alpha = -0.75$.

Although the energies calculated in this way are very sensitive to the observed size and self-absorption cut-off frequencies, estimates of the energy content can be made at least for those sources where there is accurate data. For the relatively nearby compact radio galaxies, such as NGC 1275, the energy content is $\sim 10^{52}$ ergs. If the compact quasars are at cosmological distances, their energies are considerably greater and are about $10^{55 \pm 2}$ ergs.

The energies derived for the compact sources are very much less than the minimum energy of the extended sources, so that a single compact source does not simply evolve by expansion into an extended source. The relation between the compact and extended sources is particularly unclear, since both the luminosity, L, and energy content, E, decrease with expansion. As discussed in Section 12.5.9, $L \propto r^{-2\gamma}$, and of course $E \propto r^{-1}$. Thus, taking 100 arc sec and 0.001 arc sec as dimensions of typical extended and compact sources and $\alpha = -0.75$ ($\gamma = 2.5$), upon expansion the energy and luminosity are decreased by a factor of 10^5 and 10^{25}, respectively, so that even a multiple explosion of compact sources does not appear adequate to explain the extended sources.

It is clear therefore that some continuing energy supply must be available. A possible mechanism for this has been suggested by Rees (1971), who has proposed that the relativistic particles are accelerated by low-frequency electromagnetic waves generated by the release of rotational energy of collapsing stars at the galactic nucleus. The subsequent motion of the particles in the electromagnetic field then produces a "synchro-Compton" radiation, similar in many ways to the usual synchrotron emission of electrons in a magnetic field. One particular attraction for this model is that it avoids the problem of generating a large magnetic flux.

12.5.8 Polarization

The synchrotron radiation from a single electron is elliptically polarized, and the degree of polarization is a function of ν/ν_c. In a uniform magnetic field the linear polarization of an ensemble of electrons with a power law index γ, the polarization is perpendicular to the magnetic field in the transparent part of the spectrum and is given by

$$P(\gamma) = \frac{3\gamma + 3}{3\gamma + 7} \qquad (12.38)$$

and is of the order of 70 % for typical values of γ. In the opaque part of the spectrum, the polarization is parallel to the magnetic field and is given by

$$P(\gamma) = \frac{3}{6\gamma + 13} \qquad (12.39)$$

so that P is typically only about 10 %.

Since the observed polarization in the transparent sources is typically only a few percent, it may be concluded that the magnetic fields are generally tangled, and so the observed polarization integrated over the source is greatly reduced. This is confirmed by the observations of polarization that indeed approaches the theoretical value in limited regions of some sources, although it is somewhat remarkable that such highly ordered fields can exist over regions extending up to 10 or more kpc.

In the elongated sources the orientation of the polarization vectors indicate that the magnetic field is often aligned perpendicular or parallel to the direction of elongation. In some sources there appears to be a radial magnetic field.

If the pitch-angle distribution is non-isotropic, then there is a net circular polarization since the circularly polarized components of the radiation from the individual electrons do not completely cancel. Even if the distribution is isotropic, there will be a small net

circular polarization, since there are more electrons in the solid angle defined by $\theta + d\theta$ than in the one defined by $\theta - d\theta$. This effect is particularly important if the cone of radiation of a single electron ($\theta \sim E/m_c{}^2$) is large, which will occur at very low frequencies or in regions of high magnetic field strength.

In a uniform magnetic field of B Gauss, and isotropic distribution of electron pitch angles, the integrated circular polarization is $\sim 100 \ (3B/\nu)^{1/2}$ percent at a frequency ν (Sciama and Rees, 1967). In a few sources the degree of circular polarization has been measured to be ~ 0.01 to 0.1% near 1 GHz. This corresponds to magnetic fields $\sim 3 \times 10^{-5 \pm 1}$ gauss—in good apparent agreement with the values derived from the synchrotron self-absorption cut-off frequency and the angular size.

12.5.9 Interpretation of Time Variations

The observations of time variations provide direct evidence in some sources of repeated energetic events which may provide a nearly continuous input of energy necessary to account for the observed energy requirements of the extended sources.

The form of the observed intensity variations is most simply interpreted in terms of a cloud of relativistic particles which is initially opaque out to short wavelengths, but which, due to expansion, becomes optically thin at successively longer wavelengths. In its simplest form the model assumes that the relativistic particles initially have a power law spectrum, that they are produced in a very short time in a small space, that the subsequent expansion occurs at a constant velocity, and that during the expansion the magnetic flux is conserved. Thus

$$N(E,t,r) = K\,E^{-\gamma}\,\delta(\theta)\,\delta(t) \quad (12.40a)$$

$$\frac{\theta_2}{\theta_1} = \frac{t_2}{t_1} \qquad (12.40b)$$

$$\frac{B_2}{B_1} = \left(\frac{\theta_1}{\theta_2}\right)^2 = \left(\frac{t_1}{t_2}\right)^2 \quad (12.40c)$$

where θ is the angular dimensions, t the

elapsed time since the outburst, B the magnetic field, and the subscripts 1 and 2 refer to measurements made at two epochs t_1 and t_2. A more detailed mathematical description of the model has been given by van der Laan for the nonrelativistic (1966) and the relativistic (1971) case. The discussion below follows that of Kellermann and Pauliny-Toth (1968).

The observed flux density as a function of frequency, ν, and time, t, is given by

$$\frac{S(\nu,t)}{S_{m_1}} = \left(\frac{\nu}{\nu_{m_1}}\right)^{5/2} \left(\frac{t}{t_1}\right)^3$$

$$\times \left\{ \frac{1\ \exp(-\tau(\nu/\nu_{m_1})^{-(\gamma+4)/2}\,(t/t_1)^{-(2\gamma+3)})}{1 - \exp(-\tau)} \right\}$$

(12.41)

where S_{m_1} is the maximum flux reached at frequency ν_{m_1} at time t_1.

If the optical depth is taken as the value of τ at the frequency, ν_m, at which the flux density is a maximum, then it is given by the solution of

$$e^{\tau_\nu} - \left(\frac{\gamma + 4}{5}\right)\tau_\nu - 1 = 0 \quad (12.42)$$

The maximum flux density at a given frequency as a function of time occurs at a different optical depth, τ_t, given by the solution of

$$e^{\tau_t} - \frac{2\gamma + 3}{3} - 1 = 0 \quad (12.43)$$

In the region of the spectrum where the source is opaque ($\tau \gg 1$), the flux density increases with time as

$$\frac{S_2}{S_1} = \left(\frac{t_2}{t_1}\right)^3 \quad (12.44)$$

Where it is transparent ($\tau \ll 1$), the flux density decreases as

$$\frac{S_2}{S_1} = \left(\frac{t_2}{t_1}\right)^{-2\gamma} \quad (12.45)$$

The wavelength, λ_m, at which the intensity is a maximum is given by

$$\frac{\lambda_{m2}}{\lambda_{m_1}} = \left(\frac{t_2}{t_1}\right)^{(4\gamma+6)/(\gamma+4)} \quad (12.46)$$

and the maximum flux density, S_m, at that wavelength is given by

$$\frac{S_{m2}}{S_{m_1}} = \left(\frac{\lambda_{m2}}{\lambda_{m_1}}\right)^{-(7\gamma+3)/(4\gamma+6)} \quad (12.47)$$

In most variable sources the outbursts occur so rapidly that the emissions from different outbursts overlap both in frequency and time, and so a detailed quantitative analysis is difficult. As pointed out by van der Laan (1966), the spectra of individual bursts are cumulative, suggesting spacially separated outbursts. If the different events occur in the same volume of space, the number of relativistic particles would be cumulative, rather than the spectra.

To the extent that it has been possible to separate events in some sources, the individual outbursts seem to follow surprisingly well the simple model of a uniformly expanding cloud of relativistic particles. The data relating S_m, t_m, and ν_m (Equations 12.46 and 12.47) indicate that the initial value of γ is in the range 1 to 1.5. This agrees with the spectral index of $\sim -\frac{1}{4}$ initially observed in the optically thin region of the spectrum. At least for one year following an outburst the expansion appears to continue at a constant rate, and the value of γ is unchanged by radiation losses or by inverse Compton scattering at least for $\nu < 10$ GHz. From Equation (12.15) this places a limit on the magnetic field of $B_o \lesssim 1$ Gauss. From the requirement that $T < 10^{12}$ and Equations (12.22) and (12.25), we have $B_o \gtrsim 0.1$ Gauss. Thus we conclude that $B_o \sim 1$ Gauss and in those sources where there are good data the magnetic flux seems to be approximately conserved, at least during the initial phases of the expansion. But because the data from long-baseline interferometer observations when used in Equation (12.22) indicate that $B \sim 10^{-4}$ gauss over a wide range of dimensions for both variable and nonvariable sources, and since this is also the value of the field estimated from minimum energy arguments, it appears that the flux is conserved for only a limited time, after which the relativistic particles diffuse through a fixed magnetic field of about 10^{-4} Gauss. In this way many repeated outbursts may provide the particles in the extended sources, although as explained earlier, this presents formidable energy problems unless energy is continuously supplied.

In the case of the variable radio galaxies, whose distance can be determined from their red shift, the initial dimensions appear to be well under one light year and the initial particle energy in a single outburst about 10^{52} ergs. Repeated explosions over a period of 10^8 years at a rate of one per year are required to account for the minimum total energy in the extended sources, but even this falls short by a factor of about 10^5 if account is taken of energy lost during the expansion.

The direct measurement of the angular size and expansion rate of variable sources using long-baseline interferometry is now possible, and can be used to determine uniquely the magnetic field (Equation 12.18), and when the distance is known the total energy involved in each outburst (Equations 12.25 and 12.27).

The model of a uniform isotropic homogeneous instantaneously generated sphere of relativistic electrons, which expands with a uniform and constant velocity, where magnetic flux is conserved, and where the only energy loss is due to expansion, is mathematically simple. Clearly, such sources are not expected to exist in the real world, and it is indeed remarkable that the observed variations follow even approximately the predicted variations. A more realistic model must take into account nonconstant expansion rates, the nonconservation of magnetic flux, changes in γ, the finite acceleration time for the relativistic particles, and the initial finite dimensions. But these are relatively minor modifications, and the observed departures from the predictions of the simple model should not, as is sometimes done, be used to infer that the general class of expanding source models is not relevant to the variable source phenomena. Rather the departures

from the simple mathematical model can be used to derive further information about the nature of the source.

In the case of the continued production of relativistic particles, or where the initial volume of the source is not infinitely small, the initial spectrum is not opaque out to very short wavelengths, and the source is always transparent at frequencies higher than some critical frequency, ν_o. In the transparent region of the spectrum the flux variations occur simultaneously and reflect only the rate of particle production and/or decay due to synchrotron and inverse Compton radiation.

The experimental determination of ν_o may be used to estimate the initial size of the source. Characteristically $\nu_o \sim 10$ to 30 GHz, corresponding to initial dimensions of about 10^{-3} arc second for $B \sim 1$ Gauss. For typical radio sources with $0.1 < z < 1$, the initial size derived in this way is from 1 to 10 light years. This is roughly consistent with the direct determination of the angular sizes made by long-baseline interferometry, but it must be emphasized that so far these measurements have not been made in sufficient detail to permit a detailed comparison, or to estimate from Equation (12.21) the initial magnetic field.

In those sources where good data exist in the spectral region $\nu > \nu_o$, the observed variations occur simultaneously as expected from the model, and with equal amplitude, indicating an initial spectral index $\alpha \sim 0$, or $\gamma \sim 1$, in good agreement with the value of γ derived from Equations (12.46) and (12.45).

In the spectral region $\nu > \nu_o$, the observed flux variations depend on the total number of relativistic particles, their energy distribution, and the magnetic field. Thus observations in this part of the spectrum reflect the rate of generation of relativistic particles more closely than observations in the opaque part of the spectrum.

In some sources ν_o occurs at relatively low frequencies of 1 or 2 GHz. This poses a serious problem, for the following reason. If variations occur on a time scale of the order of τ, then it is commonly assumed that the dimension of the emitting region, l, is less than $c\tau$, since otherwise the light travel time from different parts of the source to the observer would "blur" any variations which occur. Using the distance obtained from the red shift, a limit to the angular size, θ, may be calculated, and from Equation (12.22) an upper limit to the magnetic field strength is obtained.

For a typical quasar, such as 3C 454.3, $t \sim 1$ yr, $z \sim 1$, $\theta \lesssim 10^{-4}$ arc second, and $B \lesssim 10^{-5}$ Gauss. With such weak fields the energy required in relativistic particles is very high and is $\gtrsim 10^{58}$ ergs, and the repeated generation of such enormous energies in times of the order of one year or less is a formidable problem. Also the limit to the angular size deduced from the light travel time argument often results in a peak brightness temperature which may exceed the expected maximum value of 10^{12} K (Equation 12.24). For these reasons it has been questioned by some whether or not in fact the quasars are at the large distances indicated by their red shifts (*e.g.*, Hoyle, Burbidge, and Sargent, 1969), or whether they do indeed radiate by the ordinary synchrotron process.

One way in which the theoretical brightness temperature limit may be exceeded is if the relativistic electrons are radiating coherently. Stimulated emission or negative absorption leading to coherent radiation is possible in opaque synchrotron sources, if the relativistic electrons are moving in a dispersive medium where the index of refraction is less than unity.

However, other than the seemingly excessive brightness temperature implied by some of the variable source observations, the expanding source model and the ordinary incoherent synchrotron process appear to be adequate to explain all of the observed phenomena.

Another way to explain the rapid variations was pointed out by Rees (1967), who showed that if the source is expanding at a velocity $v \sim c$, then the differential light travel time between the approaching and receding parts of the source can cause the *illu-sion* of an angular expansion rate corresponding to an apparent linear velocity $v > c$. In this case the angular size and peak brightness temperature are larger than suggested by the observations; and from Equation (12.36), which depends on a high power of θ, the required particle energy is greatly reduced. However, there is a limit to the extent that the total energy requirements can be reduced by this "super-light" expansion theory, since as the particle energy is decreased when θ is increased, the magnetic energy is increased. The minimum value of the total energy occurs when the two are approximately equal, and for the typical quasar it is $\sim 10^{55}$ ergs (*e.g.*, van der Laan, 1971).

Unfortunately, the variation in total intensity for the relativistically expanding source is very similar to that for the nonrelativistic model, so that they cannot be easily distinguished merely from observations of the intensity variations. The direct observations of the variations in angular size likewise do not distinguish between "superlight" velocities at cosmological distances and nonrelativistic velocities in a "local" model for quasars.

An interesting variation on the expanding source model has been suggested by the Russian astrophysicists Ozernoy and Sazonov (1969), who propose that two or more discrete components are "flying apart" at relativistic velocities, while at the same time expanding. Evidence for relativistic component velocities has been obtained from long-baseline interferometer observations, but with the meager data so far available it has not been possible to uniquely distinguish between actual component motions and properly phased intensity variations in stationary components.

It may be expected, however, that future observations of intensity variations as a function of wavelength, when combined with the direct observation of the variations in angular size, not only will uniquely determine the dynamics and energetics of the radio outbursts, but also will specify the initial conditions of the outburst with sufficient accuracy to limit the range of theoretical speculation concerning the source of energy and its conversion to relativistic particles. In particular, there must be increased emphasis on observations made at the shortest possible wavelengths, since these most nearly reflect the conditions during the time just following the outburst (Equation 12.44).

12.5.10 Evolution of the Radio Sources

As a result of the discovery and continued observations of the intensity variations in radio galaxies and quasars, it is becoming increasingly clear that, on a cosmic time scale, the generation of particles occurs during very short times and is presumably the result of repeated violent events in the nuclei of galaxies and in quasars. This essentially precludes any statistical process such as Fermi acceleration, whose time scale is measured in millions of years. The problem of the energy source is made even more difficult because of the large energy loss expected when the cloud of relativistic particles expands.

However, even aside from the question of the source of energy and its conversion to relativistic particles, there is the problem of the formation and evolution of the extended radio sources. In particular, how are the clouds of relativistic particles confined, and what determines the characteristic double or multiple shape?

One particular problem has been the need to explain the extremely fine structure found in the outer parts of some extended sources such as Cygnus A, or the existence of sources with very high ratio of component separation to component size. If there is a cold dense plasma to contain the relativistic particles, then the minimum energy requirements are magnified as a result of the necessary increase in the kinetic energy. On the other hand, the initial presence of a large kinetic energy may provide a source of continuous energy input to balance that lost by expansion. However, it is difficult to see any easy way for the kinetic energy of the cold plasma to be transformed into energy accelerating the relativistic particles.

One interesting suggestion has been that

rather than accelerating the relativistic particles themselves, the galactic nuclei and quasars expel massive coherent bodies which then explode at some distance from the origin and produce the particles *in situ* (*e.g.*, Burbidge, 1972). But there is no experimental evidence that this occurs since all of the observed very compact opaque sources, and all of the variable sources, are coincident with quasars or nuclei. Thus, it appears that unless the number of coherent ejections is large ($N > 1000$) so that individual events are not seen, the relativistic particles in each source are in fact generated at a common point and diffuse out to form two or more extended clouds of particles. It has therefore been widely assumed that the individual components are confined by the inter-galactic gas or intergalactic magnetic fields.

The density of any intergalactic gas deduced from measurements of the X-ray background is too small for confinement by static pressure, but De Young and Axford (1967) have suggested that the internal pressure of the ejected "plasmons" can be balanced by the ram pressure of the intergalactic gas. Wardle (1971) has noted that the ram pressure model provided a natural explanation for the power–surface brightness diagram discussed by Heeschen (1966). The required expansion velocities are 5 to 10% that of light, so that the characteristic ages are only 10^5 to 10^6 years. This is much less than deduced from the energy arguments (see Section 12.5.7). This shorter time scale would also make it easier to understand the lack of any high-frequency cut-off in the radio source spectra discussed in Section 12.5.3.

Since the density of any intergalactic gas is expected to vary as $(1 + z)^3$, this model also provides a natural explanation for the small size found by Miley (1971) for double radio sources at large red shifts. The presence of extended tails observed for several radio galaxies (Miley *et al.*, 1972) is also predicted by the model (De Young, 1971). The rather sharp outer boundaries found in many of the resolved sources lend support to the ram pressure model, although the required gas density or magnetic field strengths are uncomfortably large (*e.g.*, Burbidge, 1972). However, the possibility of sufficient gas being present in clusters or being ejected from the galaxy itself does not seem unlikely, and the observation that double radio sources found in clusters tend to be smaller than those not in clusters provides indirect evidence for intergalactic matter in clusters (De Young, 1972).

A model describing the evolution of double radio sources has been developed by Ryle and Longair (1967). They explain the ratio of observed intensity and separation from the parent object as the effect of the differential light travel time and other relativistic effects on two *identical* objects expanding with highly relativistic velocities.

12.6 Summary

There is convincing quantitative evidence that all of the extra-galactic radio sources radiate by the commonly accepted incoherent synchrotron process, or a mathematically similar mechanism such as synchro-Compton radiation. This includes:

(1) The shapes of the spectra of the extended (transparent) sources are power law or dual power, and their detailed shapes are in agreement with synchrotron models where the relativistic particles both gain and lose energy.
(2) In the compact sources the spectral peak occurs at shorter wavelengths in the smaller sources, as predicted by the synchrotron model, and the measured angular sizes are in good agreement with those estimated from the observed self-absorption cut-off wavelength.
(3) The maximum observed brightness temperature is $\sim 10^{12}\,°K$, as is expected from an incoherent synchrotron source, which is "cooled" by inverse Compton scattering.
(4) The variations in intensity and polarization and their dependence on wavelength and time are in good agreement with those expected from an expanding cloud of relativistic particles.

The outstanding problems in the understanding of the radio galaxies and quasars are:

(1) the source of energy and its conversion to relativistic particles
(2) the connection, if any, between the compact opaque sources and the extended transparent one
(3) the evolution and confinement of the extended sources.

The third and possibly the second problem present challenging complex problems in magneto-hydrodynamics and plasma physics, while the first may well require fundamentally new physics before it can be understood.

Bibliography

Aller, H. D. 1970. *Astrophys. J.* 161:19.
Baade, W., and R. Minkowski. 1954. *Astrophys. J.* 119:206.
Bennett, A. 1962. *Monthly Notices Roy. Astron. Soc.* 68:165.
Bolton, J. G., G. Stanley, and B. Slee. 1949. *Nature* 164:101.
Burbidge, G. 1972. *Proc. I.A.U. Symposium No. 44*, D. S. Evans, ed. New York: Springer-Verlag, p. 492.
Cameron, M. J. 1969. *Proc. Astron. Soc. Australia*, Vol. 1, p. 229.
Cohen, M. 1969. *Ann. Rev. Astron. Astrophys.* 7:619.
——, E. J. Gundermann, and D. E. Harris. 1967. *Astrophys. J.* 150:767.
De Young, D. S. 1971. *Astrophys. J.* 167:541.
——. 1972. *Astrophys. J.* 173:L7.
——, and W. I. Axford. 1967. *Nature* 216:129.
Dixon, R. S. 1970. *Astrophys. J. Suppl.* 20:1.
Edge, D. O., J. R. Shakeshaft, W. B. McAdam, J. E. Baldwin, and S. Archer. 1960. *Mem. Roy. Astron. Soc.* 68:37.
Fomalont, E. 1969. *Astrophys. J.* 157:1027.
——. 1973. *Astrophys. Lett.* 12:187.
Gardner, F. F., and W. B. Whiteoak. 1971. *Australian J. Phys.* 29:899.
Ginzburg, V. L. 1951. *Dokl. Acad. Nauk* 29:418.

Ginzburg, V. L., and S. I. Syrovotskii. 1964. *The Origin of Cosmic Rays.* New York: Pergamon Press.

Heeschen, D. S. 1960. *Publ. Astron. Soc. Pacific* 72:368.

———. 1966. *Astrophys. J.* 146:517.

———. 1970. *Astrophys. Lett.* 6:49.

von Hoerner, S. 1964. *Astrophys. J.* 140:65.

Hoyle, F., G. Burbidge, and W. L. Sargent. 1969. *Nature* 209:751.

Kardashev, N. 1962. *Astron. Zh.* 39:393. English trans. *Soviet Astron.* 6:317.

Kellermann, K. I., and I. I. K. Pauliny-Toth. 1968. *Ann. Rev. Astron. Astrophys.* 6:417.

———, and I. I. K. Pauliny-Toth. 1969. *Astrophys. J.* 155:L31.

———, I. I. K. Pauliny-Toth, and P. J. Williams. 1969. *Astrophys. J.* 157:1.

Little, L. T., and A. Hewish. 1966. *Monthly Notices Roy. Astron. Soc.* 134:221.

Macdonald, G. H., S. Kenderine, and A. C. Neville. 1968. *Monthly Notices Roy. Astron. Soc.* 138:259.

MacLeod, J. M., B. H. Andrew, W. J. Medd, and E. J. Olsen. 1971. *Astrophys. Lett.* 9:19.

Matthews, T., W. Morgan, and M. Schmidt. 1964. *Astrophys. J.* 140:35.

Miley, G. K. 1971. *Monthly Notices Roy. Astron. Soc.* 152:477.

———, and C. M. Wade. 1971. *Astrophys. Lett.* 8:11.

———, G. C. Perola, P. C. van der Kruit, and H. van der Laan. 1972. *Nature* 237:269.

Mitton, S., and M. Ryle. 1969. *Monthly Notices Roy. Astron. Soc.* 146:221.

Morgan, W. W. 1972. *Proc. I.A.U. Symposium No.* 44. D. S. Evans, ed. New York: Springer-Verlag, p. 97.

Ozernoy, L. M., and V. N. Sazonov. 1969. *Astrophys. Space Sci.* 3:395.

Pacholczyk, A. 1970. *Radio Astrophysics.* San Francisco: Freeman and Co.

Rees, M. 1967. *Monthly Notices Roy. Astron. Soc.* 135:345.

———. 1971. *Nature* 229:312.

Rogstad, P., and R. Ekers. 1969. *Astrophys. J.* 157:481.

Ryle, M., and M. Longair. 1967. *Monthly Notices Roy. Astron. Soc.* 136:123.

Scheuer, P. A. G. 1962. *Australian J. Phys.* 15:333.

Sciama, D. W., and M. J. Rees. 1967. *Nature* 216:147.

Shklovsky, I. S. 1952. *Astron. Zh.* 29:418.

van der Kruit, P. C. 1971. *Astron. Astrophys.* 15:110.

van der Laan, H. 1966. *Nature* 211:1131.

———. 1971. *Nuclei of Galaxies.* D. J. K. O'Connell, ed. Amsterdam: North Holland.

———, and G. C. Perola. 1969. *Astron. Astrophys.* 3:468.

Wardle, J. 1971. *Astrophys. Lett.* 8:221.

General Bibliography

Annual Review Articles

A. T. Moffet. "The Structure of Radio Galaxies," *Ann. Rev. Astron. Astrophys.* 4:145, 1966.

E. M. Burbidge. "Quasi-Stellar Objects," *Ann. Rev. Astron. Astrophys.* 5:399, 1967.

P. A. G. Scheuer, and P. J. S. Williams. "Radio Spectra," *Ann. Rev. Astron. Astrophys.* 6:321, 1968.

K. I. Kellermann, and I. I. K. Pauliny-Toth. "Variable Radio Sources," *Ann. Rev. Astron. Astrophys.* 6:417, 1968.

M. Schmidt. "Quasi-Stellar Objects," *Ann. Rev. Astron. Astrophys.* 7:527, 1969.

M. H. Cohen. "High-Resolution Observations of Radio Sources," *Ann. Rev. Astron. Astrophys.* 7:619, 1969.

G. R. Burbidge. "The Nuclei of Galaxies," *Ann. Rev. Astron. Astrophys.* 8:369, 1970.

Book

G. R. Burbidge, and E. M. Burbidge. *Quasi-Stellar Objects.* New York: Freeman. 1967.

Symposia Proceedings

I. Robinson, A. Schild, and E. L. Schücking, eds. "Quasi Stellar Sources and Gravitational Collapse." *Proceedings of the First Texas Symposium on Relativistic Astrophysics.* Chicago: University of Chicago Press. 1965.

K. N. Douglas, I. Robinson, A. Schild, E. L. Schücking, J. A. Wheeler, and N. J. Woolf, eds. "Quasars and High-Energy Astronomy." *Proceedings of the Second Texas Symposium on Relativistic Astrophysics.* New York: Gordon and Breach. 1969.

D. J. K. O'Connell, ed. *Nuclei of Galaxies.* Amsterdam; North Holland. 1971.

D. S. Evans, ed. 1972. "External Galaxies and Quasi-Stellar Objects." *Proc. I.A.U. Symposium* 44.

CHAPTER 13

COSMOLOGY

Sebastian von Hoerner

Introduction

Cosmology tries to describe the universe as a whole (assuming this to be a meaningful concept). The single objects, from atoms to stars to galaxies and their clusters, are used as sources of observational information, but cosmology must also provide the proper frame which enables their formation and evolution. The main emphasis in cosmology is mostly not on the objects. It is on the metric of space and time; and on the average density of matter, radiation, and energy, on its change with time and perhaps its spatial fluctuations.

Unfortunately, cosmology has up to now been concerned mostly with theory and only very little with observational information; many radio observations obtain their cosmological relevance only in connection with some optical observations; and any "latest news" in cosmology has invariably turned out to be wrong. These three facts are reflected in the contents of this chapter.

Much more emphasis than usual will be put on problems, oddities, and uncertainties, since these seem to be, after all, very essential features of this fascinating field of study.

13.1 General Problems

13.1.1 Limited Experience

We want to describe the whole universe, but our range of experience is badly limited to the following:

(1) Our telescopes are able to reach only to a certain *distance*.

(2) The human *time* scale is very short as compared to cosmological changes.

(3) Our *laws* of physics are derived from moderate ranges of density and temperature, whereas all big-bang models of the universe begin with a singularity.

(4) The following *objects* are known and studied: galaxies for about 40 years, clusters of galaxies about 30, super-clusters still undecided; quasars 10 years (but distance still uncertain); and background radiation for 7 years. Theoreticians suggest the existence of "black holes" (remnants from gravitational collapse) and "white holes" (delayed little-bangs). Anti-matter should be just as frequent as matter but is not seen. Finally, the "hidden mass" problem (Section 13.4.1) indicates that all visible matter is perhaps only 1/100 of the total. The question is: How complete, or at least how representative and informative, is this list of known objects?

(5) A similar question concerns the *observables*: We have observed light for thousands of years, but radio waves only 40 years; we just started with X-rays and γ-rays, and maybe we observe neutrinos and gravitational waves—but what else are we missing?

(6) In addition, most world models have a *horizon* (Section 13.2.3), a principal limit to any observation.

13.1.2 Entanglement

a) The Problem

We would like to deal separately with questions concerning *space* and *time* of world models, and *evolution* of the observed objects. We need "evolution-free model tests" and "model-free evolution tests"; the former to be divided into measurements of spatial curvature and isotropy, and independent measurements of time-dependent things like expansion and acceleration. In the actual observations, however, all three items are completely tangled up; disentangling them is most urgent and difficult (and completely unsolved in most cases).

b) Distance = Past

The further we look out into space, the further we look back in time, because of the finite speed of light; only in steady-state theory is this of no concern. But in all bigbang models we see the more distant parts of the universe in earlier phases, all the way back to time zero if we could look out to infinite red shift [see Figure 13.1, calculated with $H_o = 100$ (km/sec)/Mpc].

c) Objects versus World

We see only objects, but neither space nor time. Objects are formed and evolve, they have a history of their own. We must distinguish between their individual evolution and class evolution (Table 13.1): Class evolution means that certain average properties of the objects (*i.e.*, creation frequency, luminosity, diameter, lifetime) may be functions of time-dependent parameters (*i.e.*, surrounding density, chemical composition, radiation temperature).

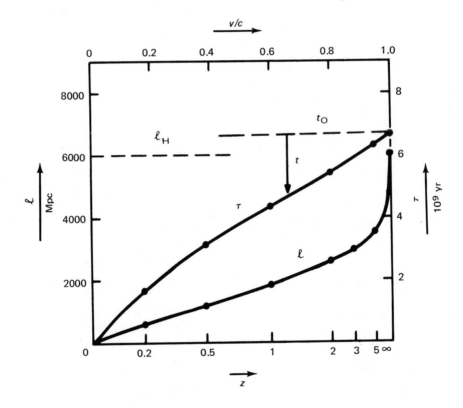

Figure 13.1 Distance and time; Einstein–de Sitter model. z = red shift; v = expansion speed at emission; τ = light travel time; t = world age at emission; l = present proper distance (radar) of object; l_H = present distance of horizon; t_o = present world age.

Table 13.1

Evolution	Matters if objects have a	Which holds for
Individual	lifetime $> 10^{10}$ yr	optical galaxies
Class	lifetime $\ll 10^{10}$ yr	quasars and radio sources

For disentangling, we need a theory of the evolution of the objects. This does exist for the individual evolution of optical galaxies (approximate at present, but improvable); it is completely missing for the class evolution of quasars and radio galaxies. If such theory is missing, then the observational data must be solved for an additional number of unknowns (evolution parameters in addition to model parameters). With enough evolution parameters, any set of observations then can give a good fit to any given world model; this, for example, is our present situation with number counts.

13.1.3 Observables and Their Standards

Most observables are useful for cosmology only if we know standards. For using the observed radio flux, S, or optical magnitude, m, we must know the absolute luminosity, L, of the source ("standard candle"); for using the angular diameter, θ, we need the linear diameter, D ("standard rod"); and for any number count, $n(S)$ or $n(z)$, we need the luminosity function, $\phi(L)$.

Optical and radio luminosities—of galaxies as well as quasars—have a very large range of more than a factor of 1000 in L. Fortunately, the optical luminosity function of galaxies drops very steeply at the bright end, and many galaxies occur in clusters; thus, the brightest galaxy in a rich cluster seems to be a fairly good standard candle, with a scatter in L of about a factor of 1.3 ($\pm .25$ mag). But this is not so for quasars, nor radio galaxies, where we are left with the full range of L.

Optical diameters of rich clusters may prove useful in the near future. Radio diam-

eters of galaxies and quasars (or separation between doubles) have a tremendous range, a factor of 10^7 in D, but it seems that their upper limits of about 300 kpc can be used. Our knowledge of the radio luminosity function is also very poor (see Figure 13.8), with an uncertainty of $\phi(L)$ of at least a factor of 4 (up and down) over most of its range, and at least a factor of 10 at its ends.

13.1.4 How Distant Are Quasars?

Appreciable differences between world models show up only for red shifts $z \geq 1$. The observed galaxies and clusters have $z \leq .25$, with only one exception at $z = .46$; quasars are observed within $.16 \leq z \leq 2.88$. About half of the identified radio sources are quasars. Thus, quasars are the most (maybe the only) promising objects of cosmology; if they are not at cosmological distance, then even the $N(S)$ plot of radio sources seems useless, since half are quasars. But whether or not they are distant is still unknown. For summaries, see M. Burbidge (1967) and Schmidt (1969, 1971) in general, and Cohen (1969) for structure. Some arguments, against and in favor of their cosmological distance (CD), are discussed below.

a) Against CD

The first objection raised regarded the large amount of energy (up to 10^{61} ergs) and mass (up to $10^8 \, M_\odot$) following from CD, confined to an extremely small volume (.01 pc = 1 light-week diameter) following from the fast variability of many quasars. However, this argument has lost importance, since massive small objects can be obtained by gravitational collapse, or by stellar-dynamical evolution up to stellar collisions (von Hoerner, 1968), or a combination of both (Spitzer and Saslaw, 1966).

Second, the red shift distribution $n(z)$ showed a high and narrow maximum at $z = 1.95$ (G. Burbidge, 1967), and some periodicities (Burbidge, 1968; Cowan, 1968). But both effects have completely disappeared with a larger number of data (208 quasars) (Wills, 1971).

Third, one might expect quasars (just as radio galaxies) to be a certain type or a special phase of galaxies. Optical and radio galaxies occur preferentially in clusters, but quasars do not. Five cases were once claimed by Bahcall *et al.* (1969), but these have been questioned by Arp (1970).

Fourth, Arp (1967, 1968, 1971) and Weedman (1970) find several cases of close companionship with large differences Δz in red shift; like a quasar or Seyfert nucleus sitting in the spiral arm of a nearby galaxy, or connected to it by a bridge. The observed examples seem to be too numerous and too striking for just a chance projection. If confirmed and real, they would give evidence for non-CD large red shifts of unexplained origin (which could affect galaxies as well as quasars).

Fifth, very-long-baseline observations show fast lateral expansion for two quasars (Whitney *et al.*, 1971; Cohen *et al.*, 1971; Gubbay *et al.*, 1969; Moffet *et al.*, 1971). If at CD, the apparent lateral velocity V of expansion (or separation of doubles) would exceed the speed of light: $V/c \approx 2$ for 3C 273, and $V/c \approx 3$ for 3C 279. Either we drop CD or we choose between several possible but somewhat artificial geometrical explanations and a relativistic explanation (Rees, 1967): If a source shoots off a companion almost at us (angle β to line of sight) with a velocity, v, almost the speed of light, then the apparent lateral speed, V, is maximum for $\cos \beta = v/c$, in which case $V = \gamma v$, with $\gamma = (1 - v^2/c^2)^{-1/2}$. For example, $v/c = .95$ needs $\beta = 18°$, and yields $V/c = 3.05$. This explanation seems to save CD and might even help for some companions with large differential

red shifts. The ejection of a cloud of rest mass m with speed $v \approx c$, however, needs an energy of

$$E = (\gamma - 1)\, mc^2 \qquad (13.1)$$

or $E = 2.2\, mc^2$ in our example. Since even nuclear fusion would give only 1% of mc^2, we must burn up a very large mass, of 220 m, and funnel all that energy just into the kinetic energy of the small cloud m, without destroying the cloud. This sounds very difficult.

Sixth, some quasars show very narrow absorption lines at various very different red shifts. This seems a great difficulty (but not only for CD).

b) *In Favor of CD*

First of all, there is wishful thinking, of course, since little observational cosmology is left without the quasars. Second, this has been done all the time for all distant galaxies; should the whole Hubble relation be rediscussed? Third, nothing else seems to work: (1) Gravitational red shifts would give much broader lines (or ridiculously small distances). (2) To overcome this difficulty, Hoyle and Fowler (1967) suggested a cluster of many collapsed objects, with an emitting cloud at its center; but Zapolski (1968) found that this gives too short a lifetime as well as other difficulties. (3) Shooting off clouds at high speed from many galaxies should give blue shifts as well (Faulkner *et al.*, 1966). (4) Terrell (1967) thus assumes our Galaxy as the sole origin of these explosions; but this needs 10^{63} ergs of kinetic energy (or burning up to $10^{11}\, M\odot$) for 10^6 quasars, and each single shot would also lead to the problem mentioned with Equation (13.1).

Fourth, there is a nice continuity (and some overlap) between radio galaxies and quasars. Heeschen (1966) plotted radio luminosity versus surface brightness, which was extended and confirmed by Braccesi and Erculiani (1967). Meanwhile, many other plots of various quantities gave similar results—including the radio luminosity function (Section 13.5.1). Most of this would be mere coincidence without assuming CD. A strong similarity between radio galaxies and quasars is also shown by spectra, structure, and variability. Fifth, without CD it would again be mere coincidence that the stellar collision model (Dyson, 1968; von Hoerner, 1968) gives about the right values for mass, radius, luminosity, and variability.

Sixth, the angular sizes of quasars as a function of red shift not only continue nicely the sizes of the radio galaxies, but fall off with $\theta \sim 1/z$, just as they should (Legg, 1970; Miley, 1971). Some consider this the strongest argument for CD; but actually it is very odd that $\theta \sim 1/z$ continues further down than any world model would allow (Section 13.5.2). The $m(z)$ and $S(z)$ relations are mostly, but not completely, blurred by the large scatter of L, and are discussed in Section 13.5.2.

In summary, quasars are the most important but most uncertain objects for cosmology. The most promising possibilities for future investigation seem to be:

(1) More interferometer and very-long-baseline work
   (a) Fast geometrical changes (against CD)
   (b) Continue $\theta(z)$ relation (in favor of CD)
   (c) Proper motions? (300 km/sec at 6 Mpc distance gives 10^{-4} arc seconds in 10 years)
(2) Further examples and details about close odd companions of galaxies
(3) A medium-sized optical telescope in space
   (a) Are quasars galactic nuclei?

(b) Do they occur in clusters of galaxies?
(c) Optical diameters
(d) Find Lyman α absorption in galaxies

13.2 Basic Theory

This section treats concepts and formulas which are more basic and general than the various theories and models treated later.

13.2.1 Space, Time, References

Time is usually considered as just a fourth coordinate. This may be used as a convenience, but there is a fundamental difference; as C. F. von Weizsäcker said: "The past is factual, the future is possible," meaning that time has an arrow while space has not. *Space* has three dimensions, and in our normal experience space is "flat" or Euclidean, meaning that parallel lines keep their distance constant. But space may also be curved, as already discussed by Gauss, and only observations can tell.

(a) Absolute versus Relative

An absolute frame of rest and even its origin were defined for the ancient Greeks by the center of the Earth (see Table 13.2). Galilei and Newton "relativated" location and velocity, whereas unaccelerated (inertial) motion and the absence of rotation still kept an "absolute" meaning. Mach's principle, from about 1893, postulates everything to be relative—or more exactly, to be (somehow)

Table 13.2 Various Degrees of Relativity*

	Ancient Greeks (Earth)	General relativity, any curved empty space	Galilei, Newton, special relativity	Mach's principle
Location	x (origin)	$\dot{x}$ (rest)	$\ddot{x}$ (acceleration)	
Angle	α (direction)	$\dot{\alpha}$ (rotation)	$\ddot{\alpha}$ (torque)	

*The vertical line connected with each theory separates the relative quantities (left) from those which have absolute standards (right).

defined by the total masses of the universe.

Special relativity draws the line where Newton did. With general relativity, Einstein wanted originally to go further and to fulfill (and specify) Mach's principle, but actually went one step back by permitting a curved empty space, which defines a frame of rest as can be shown (von Hoerner, 1973a).

13.2.2 Metric

A "metric" is the generalization of Pythagoras' law, including time, and with general metric coefficients $g_{\mu\nu}$, but restricted to small distances ds:

$$ds^2 = \sum_{\mu=1}^{4} \sum_{\nu=1}^{4} g_{\mu\nu} dx^\mu dx^\nu \quad (13.2)$$

A metric is said to be Riemannian if it has the quadratic form of Equation (13.2) and if the coefficients depend on coordinates only (space and time) but not, for example, on their derivatives.

The universe is usually imagined as being filled with a "substratum" (evenly smeared-out matter and radiation) expanding with the universe but without peculiar motion. A "fundamental observer" is at rest in the substratum. The expansion is included in the $g_{\mu\nu}$, which makes the space coordinates "co-moving." For simplification and in accordance with our limited observations, the universe is usually postulated to be homogeneous and isotropic. Schur's theorem says that isotropic means always homogeneous, too, but not vice-versa. Homogeneous plus isotropic is frequently called "uniform."

Weyl's postulate says "world lines of fundamental observers do not intersect (except maybe at the origin)." If it holds, time in Equation (13.2) is orthogonal on space, and a "cosmic time" can be established, the same for all fundamental observers. (Counterexample: two satellites in different orbits, meeting each other sometimes, are not fundamental; they do not fulfill the postulate and generally keep different times.)

Under the three assumptions of Riemannian metric, Weyl's postulate, and uniformity, Equation (13.2) reduces to the Robertson-Walker metric:

$$ds^2 = c^2 dt^2 - R^2(t) \frac{dx^2 + dy^2 + dz^2}{(1 + kr^2/4)^2},$$

$$\text{with } r^2 = x^2 + y^2 + z^2 \quad (13.3)$$

Here, t = cosmic time, $R(t)$ = radius of curvature of three-space if $k \neq 0$, and $R(t)$ = distance between any two fundamental observers if $k = 0$; space may be closed ($k = +1$), flat ($k = 0$), or hyperbolic ($k = -1$). The x, y, z are *some* co-moving metric space coordinates but can be transformed into many other forms. For example, the transformation $\bar{r} = r/(1 + kr^2/4)$, with polar coordinates, yields

$$ds^2 = c^2 dt^2 - R^2(t) \left\{ \frac{d\bar{r}^2}{1 - k\bar{r}^2} + \bar{r}^2(d\theta^2 + \sin^2 \theta d\phi^2) \right\} \quad (13.4)$$

But both r and $\bar{r}$ are somewhat confusing (see Table 13.3). A better metric distance, called u by Sandage (1961a) and ω by McVittie (1952, 1965), is obtained by the transformation $u = \arcsin \bar{r} = 2 \arctan(r/2)$ for $k = +1$, $u = \bar{r} = r$ for $k = 0$, and $u = \text{arcsinh } \bar{r} = 2 \text{arctanh}(r/2)$ for $k = -1$. This leads directly

Table 13.3 Three Types of Metric Distances

Metric distance	$k = +1$		$k = -1$	
	Equator	Antipole	"Equator"	$l = \infty$
r	2	∞	1.312	2
$\bar{r}$	1	0	2.30	∞
u	$\pi/2$	π	$\pi/2$	∞

to a physically useful distance *l*:

$$l = uR = \text{proper distance}$$
$$= \text{rigid-rod distance}$$
$$= \text{radar distance} \quad (13.5)$$

Defining a function

$$\mathscr{S}(u) = \bar{r} = \begin{cases} \sin u & \text{for } k = +1 \\ u & 0 \\ \sinh u & -1 \end{cases} \quad (13.6)$$

then yields the metric as

$$ds^2 = c^2 \, dt^2 - R^2(t)$$
$$\times \{du^2 + \mathscr{S}^2(u)\,[d\theta^2 + \sin^2\theta \, d\phi^2]\} \quad (13.7)$$

Table 13.2 compares r, $\bar{r}$, and u (for $k = -1$, an "equator" has been formally defined by $u = \pi/2$, too). The behavior at the antipole and at infinity shows clearly why we called u better, and less confusing, than r or $\bar{r}$.

The following lists a few properties of a sphere of radius $l = uR$ in curved uniform three-space, centered at the observer; derivations are omitted here but can easily be obtained from metric Equation (13.8)

	Element	Whole sphere	
Circum-ference	$dC = R\mathscr{S}d\phi;$	$C = 2\pi R\mathscr{S}$	(13.8)
Surface area	$dA = R^2\,\mathscr{S}^2$ $\times \sin\theta \, d\theta \, d\phi;$	$A = 4\pi R^2 \mathscr{S}^2$	(13.9)
Volume	$dV = 4\pi R^3$ $\times \mathscr{S}^2 \, du;$	$V = \dfrac{4\pi}{3} R^3 \, \mathscr{R}$	(13.10)

with

$$\mathscr{R}(u) = 3 \int_0^u \mathscr{S}^2(u) \, du$$

$$= \begin{cases} \frac{3}{2}\,[u - \frac{1}{2}\sin(2u)] \\ \quad = u^3(1 - u^2/5 \pm \ldots) & k = +1 \\ u^3 & 0 \\ \frac{3}{2}[\frac{1}{2}\sinh(2u) - u] \\ \quad = u^3(1 + u^2/5 \pm \ldots) & -1 \end{cases}$$
$$(13.11)$$

Finally, for $k = +1$ the *whole universe* has the following *total* values:

Circumference (origin-antipole-origin) $\quad C = 2\pi R \quad (13.12)$

Area of plane (to anti-pole, all directions) $\quad A = 4\pi R^2 \quad (13.13)$

Volume (total of all space) $\quad V = 2\pi^2 R^3 \quad (13.14)$

13.2.3 Horizons

The *particle horizon* (or just "horizon") is a sphere in three-space, of metric radius u_{ph}, where objects would be seen with infinite red shift and at age zero. Particles within this sphere are observable; those outside are not. Since $u_{ph}(t)$ increases monotonously, more particles are observable all the time, and none of them can ever leave the horizon again (see Rindler, 1956b).

A particle horizon means that the whole *history* of the universe is observable, back to age zero, but only a limited part of all space (6000 Mpc in Figure 13.1). A given world model has a particle horizon if the integral exists:

$$u_{ph} = c \int_0^t \frac{dt}{R(t)} \quad (13.15)$$

If Einstein's cosmological constant $\Lambda > 0$, the expansion of the universe may finally be so much accelerated that some distant photon coming our way will never reach us. Photons just reaching us at $t = \infty$ define the *event horizon*; it sets an upper limit to the age at which we can see a given object, and it exists if the integral exists:

$$u_{eh} = c \int_t^\infty \frac{dt}{R(t)} \quad (13.16)$$

If *both horizons* exist, some distant object is first not observable. Second, it enters our particle horizon at some time given by Equation (13.15) with infinite red shift and age zero. Third, it gets older and its red shift decreases. Fourth, the red shift goes through a minimum and increases again, while the object seems to age more slowly. Fifth, if we observe it infinitely long, the red shift goes to infinity, but the age at which we see the object approaches a finite age given by Equation (13.16).

In four-dimensional space-time both horizons are light cones. Our particle horizon is our forward light cone emitted by us at $t = 0$; our event horizon is our backward light cone reaching us at $t = \infty$.

13.2.4 Observable Properties

The following formulae are derived from Equations (13.7) to (13.11). They assume nothing else except Riemannian metric, Weyl's postulate, and uniformity; $R(t)$ and k are left unspecified. Any special cosmological theory then will provide the dynamics, meaning a differential equation for $R(t)$, mostly in the form $\dot{R} = \dot{R}(R)$. And a special world model then will have selected values for constants of integration, k, and other parameters.

The indices mean: o = present, r = received, e = emitted, bol = bolometric, v = certain frequency and limited bandwidth. The spectral index α is defined by $L_v \sim v^{+\alpha}$; spectral curvature is neglected. The first and second derivatives of $R(t)$ are frequently used as

Hubble parameter $\qquad H_o = \dfrac{\dot{R}_o}{R_o}$ (13.17)

Deceleration parameter $\qquad q_o = \dfrac{-R_o \ddot{R}_o}{\dot{R}_o{}^2}$

(13.18)

a) Red Shift z

For the red shift one can derive

$$1 + z = \frac{R_o}{R_e}$$

$$z = \tau H_o + \left(1 + \frac{q_o}{2}\right)\tau^2 H_o{}^2 \pm \dots \quad (13.19)$$

where τ = light travel time = $t_o - t_e$.

The metric distance u is derived from Equation (13.7), with $ds = d\theta = d\phi = 0$, as

$$u = c \int_{t_e}^{t_o} \frac{dt}{R(t)} = c \int_{R_e}^{R_o} \frac{dR}{R\,\dot{R}(R)} \quad (13.20)$$

With Equation (13.19) we then obtain the following formula which describes how a special theory, via dynamics, enters the formulae connecting observables and red shift:

$$u(z) = c \int_{R_0/(1+z)}^{R_o} \frac{dR}{R\,\dot{R}(R)} = c \int_0^z \frac{dz}{(1+z)\,\dot{R}(z)} \quad (13.21)$$

or approximately

$$u(z) = \frac{cz}{R_o H_o}\left\{1 - \frac{q_o + 1}{2}z \pm \dots\right\} \quad (13.22)$$

Sandage (1962) has discussed the possibility of a true "evolution-free model test": a distant object is observed over a long time, to determine its change of red shift. Unfortunately, dz/dt is of the order of H_o; measuring z with an accuracy of 10^{-4}, say, then needs observations over 10^6 years.

b) Flux S (or Magnitude m)

The flux can be written as

$$S_{bol} = \frac{L_{bol}}{4\pi l_{bol}{}^2} \quad \text{or} \quad S_v = \frac{L_v}{4\pi l_v{}^2} \quad (13.23)$$

with a luminosity distance

$$l_{bol} = R_o\,\mathscr{S}(u)\,(1 + z) \quad (13.24)$$

The flux observed at frequency v in a given bandwidth is emitted at a higher frequency and within a broader band, which together gives a factor $(1 + z)^{1+\alpha}$ for the observed flux, or its root for the luminosity distance. Thus

$$l_v = R_o\,\mathscr{S}(u)\,(1+z)^{(1-\alpha)/2} = l_{bol}(1+z)^{-(1+\alpha)/2} \quad (13.25)$$

and approximately

$$S_v = \frac{L_v}{4\pi}\left(\frac{H_o}{cz}\right)^2\{1 + (q_o + \alpha)z \pm \dots\} \quad (13.26)$$

In flat space, $l_{bol} = (1 + z)l$, which means that $S \sim l^{-2}$ not in "Euclidean space" (as sometimes stated) but only in "static Euclidean space," where all $z = 0$.

In optical astronomy the transition from Equation (13.24) to Equation (13.25) is much more complicated (strong lines, curved spectra, wide band) and is called the K-correction. The conversion from optical magnitude, m, into flux, S, in flux units, is done by

$$S(m) = 10^{a - 0.4m} \quad \text{with } a = \begin{cases} 3.258 & \text{for } U \\ -3.621 & B \\ 3.584 & V \end{cases}$$

(13.27)

c) *Angular Size* θ (D = *Linear Size*)

$$\theta = \frac{D(1 + z)}{R_o \mathscr{S}(u)} = \frac{D(1 + z)^2}{l_{bol}}$$ (13.28)

d) *Surface Brightness b* (B = *nearby value*)

From Equations (13.23) and (13.28) we find

$$b_{bol} = \frac{B_{bol}}{(1 + z)^4}$$ (13.29a)

and

$$b_v = \frac{B_v}{(1 + z)^{3 - \alpha}}$$ (13.29b)

These formulae do not depend on world models; if they are not fulfilled, the reason can only be evolution (class or individual, of the sources). Thus Equations (13.29a) and (13.29b) are a true "model-free evolution test"; or would be, if we had a standard for B—which we don't. But with more data, the following limit could be used. Kellermann and Pauliny-Toth (1969) give a theoretical upper limit for the brightness temperature, $T \leq 10^{12}{}^\circ$K, close to which are several of the maxima of variable quasars and galactic nuclei. Since $\alpha = 0$ at the maximum, and $T \sim b\lambda^2$, which means $T_r \sim T_e(1 + z)^2$, Equation (13.29b) predicts an observed z-dependence of

$$T_{max} = \frac{10^{12}{}^\circ\text{K}}{(1 + z)}$$ (13.30)

e) *Parallax Distance*

Call α and γ the angles from the end points of a (perpendicular) baseline D to some distant object; then its parallax distance can be defined as $l_{par} = D/\beta$, with $\beta = \pi - (\alpha + \gamma)$. Call θ the angle under which D is seen from the object; then in flat space $\beta = \theta$. But in general

$$l_{par} = \frac{R_o \mathscr{S}(u)}{\sqrt{1 - k \mathscr{S}^2(u)}} = \begin{cases} R_o \tan u & k = +1 \\ R_o u & 0 \\ R_o \tanh u & -1 \end{cases}$$

(13.31)

and

$$\frac{\beta}{\theta} = \sqrt{1 - k \mathscr{S}^2(u)} = \begin{cases} \cos u & k = +1 \\ 1 & 0 \\ \cosh u & -1 \end{cases}$$

(13.32)

With $u(z)$ from Equation (13.21), Equation (13.31) then gives $l_{par} = l_{par}(z)$, which is a true "evolution-free model test," or would be, if we could measure angles with an accuracy of 10^{-10} arc second; for D = Earth orbit and $H_o = 100$, we have approximately

$$\beta = 3 \times 10^{-10} \text{ arcsec } \frac{1}{z} \left(1 - \frac{q_o + 1}{2} z \pm \ldots \right)$$

(13.33)

Weinberg (1970) introduced the parallax distance for a different purpose:

$$\frac{k}{R_o^2} = (1 + z)^2 l_{bol}^{-2} - l_{par}^{-2}$$ (13.34)

yields a direct measure of the curvature and thus is a "dynamic-free curvature test" but unfortunately contains evolution (in L_{bol} needed for l_{bol}); whereas Equation (13.31) contains dynamics, $u(z)$, but no evolution.

f) *Number Counts*

With all sources of the same L, the cumulative count would give

$$N(S) = \frac{1}{4\pi} Q \mathscr{R}(u)$$

= number/steradian with flux $\geq S$ (13.35)

where $Q = (4\pi/3) \rho_o R_o^3 = (4\pi/3)\rho R^3$ = constant, neglecting source evolution, and for all theories except steady-state; ρ is the number of sources/volume; $\mathscr{R}(u)$ is defined in Equation (13.11); $u(S)$ is obtained from Equations (13.23) and (13.25), with the dynamics equation, (13.21), for a special theory. Approximately,

$$N(S) = \frac{1}{3} \rho_o \left(\frac{L}{4\pi S}\right)^{3/2}$$

$$\times \left\{1 - \frac{3}{2}(1-\alpha)\frac{H_o}{c}\left(\frac{L}{4\pi S}\right)^{1/2} \pm \ldots\right\} \quad (13.36)$$

and the famous slope of the log N − log S plot is

$$\beta = \frac{d \log N}{d \log S} = -\frac{3}{2} + \frac{3}{4}(1-\alpha)$$

$$\times \frac{H_o}{c}\left(\frac{L}{4\pi S}\right)^{1/2} \pm \ldots \quad (13.37)$$

with

$$\frac{H_o}{c}\left(\frac{L}{4\pi S}\right)^{1/2} \approx z \quad (13.38)$$

Thus, the slope is appreciably less steep than −3/2, even for small z. For $\alpha = -0.8$ and $\alpha = 0$, for example, we have

z	.05	.10	.15	.20
$-\beta$ ($\alpha = -0.8$)	1.43	1.36	1.30	1.23
$-\beta$ ($\alpha = 0$)	1.46	1.43	1.39	1.35

$$(13.39)$$

Observationally, the slope β should be calculated from the data by the maximum-likelihood method of Crawford *et al.* (1970).

Instead of the cumulative count $N(S)$, one should plot the differential count $n(S) = -dN/dS$, where $n(S)dS$ is the number/steradian within $S \ldots S + dS$, because:

(1) The $n(S)$ points are statistically independent of each other and give an honest picture; whereas the $N(S)$ points contain the same (strong) sources again and again, feigning more accuracy than they have (Jauncey, 1967).

(2) Any details are shown sharper in the $n(S)$ plot; they are more smeared-out and propagated to fainter fluxes in the $N(S)$ plot. For good examples, see Bridle *et al.* (1972) and Kellermann *et al.* (1972).

Approximately,

$$n(S) = \frac{1}{2} \rho_o \left(\frac{L}{4\pi}\right)^{3/2} S^{-5/2}\left\{1 - 2(1-\alpha)\right.$$

$$\times \underbrace{\frac{H_o}{c}\left(\frac{L}{4\pi S}\right)^{1/2}}_{\approx z} \pm \ldots\left.\right\} \quad (13.40)$$

Because of the wide spread of L, one must know (or pretend to know) the luminosity function $\phi(L)$ when evaluating the observed counts. The previous formulae should then be integrated, $\int \ldots \phi(L)dL$; or one may introduce the red shift z and count $n(S,z)\,dS\,dz$, with

$$n(S,z) = 3\,Qc\,\frac{R_o{}^2\,\mathscr{S}^4(u)\,\phi(L)}{\dot{R}(z)\,(1+z)^\alpha} \quad (13.41)$$

where all functions on the right-hand side can be expressed in terms of S and z by using Equations (13.23) and (13.25), and the dynamics equation, (13.21). In addition, it seems that evolution must be introduced, too, which will be discussed in connection with the observations (Section 13.5.4.c).

Equations (13.36), (13.37), and (13.40) show that the flatness of the bright end, as calculated in Equation (13.39), is the *same* for *all* models (except steady-state, where it is still flatter). It depends only on the spectral index α, and probably on evolution, but not on model parameters like q_o or k. Differences between world models appear at the fainter part of the plot, from terms of higher order. Thus, the brightest part of the log N/log S plot can yield only a model-free evolution test.

13.2.5 Matter, Anti-matter, and Radiation

For the dynamics, one needs an equation of state, $p = p(\rho, T)$. But the pressure is significant only in the early phases of big-bang models; thus the following applies only to those models. On the other side, all big-bang models become more and more similar to each other the further we go back in time; thus the following applies to all big-bang models in about the same way (almost model-independent).

a) Comparison

In general, we have

$$p = p_m + p_r \qquad (13.42)$$

$$\rho = \rho_m + \rho_r \qquad (13.43)$$

where m = matter (nucleons, electrons), r = radiation (photons, neutrinos), and o = present value. At present,

$$\frac{3p_{mo}}{c^2} = \rho_{mo}\left(\frac{w}{c}\right)^2$$

$$= 10^{-6}\, \rho_{mo} = 3 \times 10^{-36 \pm 1}\ \text{g/cm}^3$$

$$(w \approx 300\ \text{km/sec}) \quad (13.44)$$

$$\frac{3p_{ro}}{c^2} = 6.8 \times 10^{-34}\ \text{g/cm}^3$$

$$(3^\circ\text{K background radiation}) \quad (13.45)$$

$$\rho_{mo} = 3 \times 10^{-30 \pm 1}\ \text{g/cm}^3 \qquad \text{(visible versus}$$

hidden matter, Section 13.4.1) (13.46)

$$\rho_{ro} = \frac{3p_{ro}}{c^2} = 6.8 \times 10^{-34}\ \text{g/cm}^3 \quad (13.47)$$

Thus at present, to a good approximation,

$$p = 0 \qquad \rho = \rho_m \qquad (13.48)$$

Going backward in time, we have, if matter and radiation do not interfere (decoupled),

$$\rho_m \sim R^{-3}; \qquad \text{conservation of mass;}$$

$$d(\rho R^3) = 0 \quad (13.49)$$

$$\rho_r \sim R^{-4}; \qquad \text{conservation of energy;}$$

$$dE + pdV = 0; \ T_r \sim R^{-1} \quad (13.50)$$

The densities of matter and of radiation then were equal when

$$\rho_r = \rho_m \ \text{when}\ 1 + z = \frac{R_o}{R} = 5 \times 10^2 \ldots 5 \times 10^4$$

$$(13.51)$$

and $\qquad T_r = 10^3 \ldots 10^5\ {}^\circ\text{K}$

$$(13.52)$$

for $\qquad\qquad$ visible ... hidden matter

b) Equation of State

Instead of deriving $p(\rho,T)$ from physics, one usually just defines $\varepsilon = p/(\rho c^2)$ and makes simplifying assumptions about $\varepsilon(t)$; see Chernin (1966), McIntosh (1968), Zeldovich (1970). The range is $0 \le \varepsilon \le \frac{1}{3}$, between a dust universe (matter only) and a universe containing radiation only. In general, from $dE + pdV = 0$,

$$\rho \sim R^{-3(1 + \varepsilon)} \qquad (13.53)$$

The very early phase of a big-bang universe is the *hadronic state* (Hagedorn, 1965, 1970; Moellenhoff, 1970). All surplus energy goes into pair-creation of heavy and super-heavy particles (hadrons), without further increase of the temperature. This leads for $t \to 0$ to $\rho \to \infty$ and $p \to \infty$, but $\varepsilon \to 0$ and $T \to T_h$; densities are above 10^{15} g/cm^3, and

$$T \le T_h = 1.86 \times 10^{12}\,{}^\circ\text{K} = 160\ \text{MeV}$$

$$(13.54)$$

This is followed by a phase of dominating radiation, up to the limit of Equation (13.51), followed by our present phase of dominating matter. Since the dominance is always strong, except for short transitions, a fairly good approximation for $\varepsilon(t)$ is just a step-function:

1. Hadronic state	$\varepsilon = 0$	for	$T \ge 10^{11}\,{}^\circ\text{K}$, $t \le 10^{-4}$ sec
2. Radiation universe	$\varepsilon = 1/3$	before limit (13.51)	(13.55)
3. Dust universe	$\varepsilon = 0$	after limit (13.51)	

How far back in time may we trust our physics? Except for a more general feeling of distrusting all extremes, nobody has come up with any well-founded limitation. Frantschi *et al.* (1971) find that quarks will be present in ultra-dense matter, but will not change the equation of state; Misner (1969a) finds that quantization gives no change for at least

$$R \ge \left(\frac{Gh}{c^3}\right)^{1/2} = 10^{-33}\ \text{cm} \quad (13.56)$$

c) *Decoupling, Viscosity, Relics*

Some agent is said to decouple (from the rest of the world) when its collision time becomes larger than the Hubble time (H_o^{-1}), or when its mean free path becomes larger than the particle horizon—whichever comes first. Some equilibrium then is terminated. Two such agents are important: neutrinos and photons.

Surrounding the time of decoupling, this agent may yield a large viscosity; whereas earlier the range of interaction was too small, and later there is no interaction. A large viscosity may smear out primeval finite-size inhomogeneities and anisotropies (fluctuations, turbulence, condensations); it increases the uniformity or keeps it up.

Decoupling also leaves (nonequilibrium) *relics*. Neutrino decoupling at $10^{10}\,^\circ$K means the termination of nucleon pair-creation, which means a frozen-in neutron/proton ratio, which finally defines the *helium*/hydrogen ratio, Y. The helium is made at 10^8 to $10^9\,^\circ$K, when deuterium no longer gets thermally disintegrated while neutrons are still not decayed. One finds $Y = .30$, almost model-independent; except that large fluctuations of T would decrease Y (Silk and Shapiro, 1971).

The observed $3\,^\circ$K *background* radiation is (most probably) the relic of photon decoupling which happened at about $3000\,^\circ$K when hydrogen recombination suddenly decreased the opacity. This was predicted by Alpher and Herman (1948); Gamow (1956), forgotten and repredicted by Dicke *et. al.* (1965); and observed independently by Penzias and Wilson (1965). With expansion, $T_r \sim R^{-1}$, which means we see these photons now with a red shift $z \approx 1000$.

Photon decoupling is probably followed by a time of Jeans instabilities ($z \approx 100$), leading to *condensations* of matter, decoupling from each other, with galaxies and clusters as relics. But there are some serious problems, and it seems we do not yet have a satisfactory theory of galaxy formation, which certainly is a severe drawback.

d) *Matter and Anti-matter*

In our experiments there is always pair-creation and pair-annihilation—for heavy particles (conservation of baryon number) as well as for light ones (conservation of lepton number). And in our theories this particle-antiparticle symmetry is one of the "cornerstones" of quantum physics. The meaning of this conservation law is direct and exact (as opposed to statistical): one particle and one antiparticle of a pair are created at the same instant and the same spot.

Thus the creation of matter, either 10^{10} years ago in a big-bang or all the time in steady state, should give equal amounts of matter and anti-matter, well mixed. But we see no anti-matter nor any sign of annihilation. For a summary of this problem see Steigman (1969, 1971).

A possible solution is given by Omnes (1969, 1970), and is supported and explained by Kundt (1971) in a good summary of the early phases. At the end of the hadron state there is a phase transition with *demixing*, yielding one-kind droplets of 10^5 g, stopped by lack of time from world expansion. Next follows a state of *diffusion* and annihilation along the droplet boundaries, stopped again by lack of time. Finally comes a state of *coalescence*, where surface tension makes the droplets merge into larger and larger ones, stopped by condensation of matter. The largest droplet size then is about 10^{46} g, corresponding to clusters of galaxies.

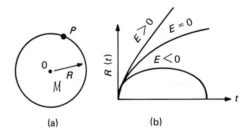

Figure 13.2 Newtonian cosmology. (a) Attraction of arbitrary particle, P, to arbitrary origin, O, by mass, M. (b) Three resulting types of expansion: Elliptic ($E < 0$), parabolic ($E = 0$), and hyperbolic ($E > 0$). Identical with general relativity for $p = \Lambda = 0$.

This theory works for big-bang models only (if it works at all). It results in a universe divided into alternating cells of cluster size, of matter or of anti-matter only (which, at present, is neither contradicted nor supported by any observation). It could be decided in the future if a γ-ray background would be observed, of the right intensity and spectrum, as predicted from the annihilation along the droplet boundaries.

13.3 World Models

13.3.1 Newtonian Cosmology

Between 1874 and 1896 Neumann and von Seliger applied Newton's law of gravitation to an infinite, Euclidean, uniform universe. They found no static universe, which was considered an obvious demand at that time. They solved the problem by inventing a repulsive force increasing with distance, very similar to the cosmological constant Λ of Einstein. But this did not find much favor and was forgotten.

After general relativity was introduced by Einstein in 1915, and the expansion of the universe found by Hubble in 1929, Milne and McCrea showed in 1934 the striking similarity between Newtonian and relativistic cosmologies. See Heckmann (1968), Bondi (1960), and McVittie (1965).

Newtonian world *dynamics* can be properly derived (see the last quotations). For a sloppy derivation, see Figure 13.2(a). Select an arbitrary origin and an arbitrary particle at some distance, R, and consider the particle as being attracted to the origin by the gravitation from the sphere of radius R about the origin. Call

$$\mathcal{M} = \frac{4\pi}{3} R_o{}^3 \rho_o = \frac{4\pi}{3} R^3 \rho = \text{constant}$$

The differential equation of the dynamics then is

$$\ddot{R} = -\frac{G\mathcal{M}}{R^2} \qquad (13.57)$$

or, integrated once, and representing the

conservation of energy,

$$\dot{R}^2 = \frac{2G\mathcal{M}}{R} + 2E \qquad (13.58)$$

where E = constant of integration = total energy per mass (kinetic plus potential). Equation (13.58) can be integrated analytically yielding $t(R)$, while $R(t)$ cannot be written analytically except for $E = 0$, where

$$R(t) = \left(\frac{9}{2} G\mathcal{M}\right)^{1/3} t^{2/3} = R_o (6\pi G \rho_o)^{1/3} t^{2/3}$$
$$\text{for } E = 0 \quad (13.59)$$

There are three types of expansion (see Figure 13.2b). They merge together at the beginning:

$$R(t) \sim t^{2/3} \qquad \text{for } t \to 0 \qquad \text{for any } E \quad (13.60)$$

The three items of Newtonian cosmology, dynamics $R(t)$, world age $t_o(H_o, q_o)$, and light travel time $\tau(z)$, are all identical with those of general relativity for $p = \Lambda = 0$. Most observables, however, depend on space curvature and are identical in both cosmologies for only the parabolic case $E = 0$ ($k = 0$ and $q_o = 1/2$).

The elliptical case, $E < 0$, of Newtonian as well as of relativistic cosmology, is frequently called an oscillating universe, although Hawking and Ellis (1968) have proved that no "bouncing" is possible. This is one of the completely unsolved problems, regarding universes as well as any massive black hole, for a comoving observer. It can be stated as:

What comes after a gravitational collapse? (13.61)

13.3.2 General Relativity (GR)

a) Early History

Special relativity was founded by Einstein in 1905, but did not contain accelerations or gravitation. GR followed in 1915; it is basically a theory of gravitation, while cosmology is just one of its fields of application. At first, Einstein was interested only in solutions describing a static universe.

In 1922 Friedmann suggested pressure-free dynamic big-bang solutions, which were justified in 1929 when Hubble observed the expansion of the universe. Lemaitre further studied the big-bang models in 1931, also with pressure, as well as dynamical models evolving out of an almost static case or staying close to one for a long time.

Between 1948 and 1953, Gamow, Alpher, Herman, Hayashi, and others predicted from big-bang models a present background radiation of 4 to 6°K, and a primordial helium abundance of $Y = .29$. For a good and critical summary about optical observables and our instrumental limits of observation, see Sandage (1961a, 1961b).

b) Basic Concepts

(1) *Mass = energy.* $(E = mc^2)$. A system of restmass m_o has the total (inertial = gravitational) mass

$$m = \frac{E}{c^2} = m_o + \frac{(E_{\text{kin}} + E_{\text{pot}} + E_{\text{rad}} + \ldots)}{c^2}$$
$$(13.62)$$

Photons and neutrinos have energy and thus have mass, but do not have any restmass.

(2) *Principle of covariance.* Laws of nature are independent of our choice of coordinates, including curved ones (tensor calculus). Additional demand: space-time metric shall be Riemannian, Equation (13.2).

(3) *Principle of equivalence.* There is no difference, locally, between a free fall in a gravitational potential and an unaccelerated flight. Potentials can be reduced to zero by transformations to proper coordinates, leading to curved space-time (sometimes called "geometrization of physics"). Then:

(a) Trajectories of $s = \min$ (geo-
    particles   desics, short-
                     est way)

(b) Trajectories of $s = 0$ (null-
    photons   geodesics)

(4) *Field equations.* These equations express the equivalence, in covariant form, by equating the (physical) energy-momentum tensor with some (geometric) tensor built from the metric $g_{\mu\nu}$ and its first and second derivatives. Symmetry leaves 10 independent equations. For *uniform* models (homogeneous and isotropic), this reduces to only two differential equations for $R(t)$:

$$8\pi G \,\rho = -\Lambda + 3\frac{kc^2}{R^2} + 3\left(\frac{\dot{R}}{R}\right)^2 \quad (13.63)$$

$$\frac{8\pi G}{c^2}\, p = +\Lambda - \frac{kc^2}{R^2} - \left(\frac{\dot{R}}{R}\right)^2 - 2\frac{\ddot{R}}{R}$$
$$(13.64)$$

This yields the following combination, identical with the Newtonian equation, (13.57), for $p = \Lambda = 0$.

$$\ddot{R} = -\frac{4\pi G}{3}\left(\rho + \frac{3p}{c^2}\right) R + \frac{1}{3}\Lambda R \quad (13.65)$$

Most models start with a *singularity* at $t = 0$, with $\rho = \infty$:

Big-bang, $R \sim$ $\begin{cases} t^{1/2} \text{ radiation only, } \varepsilon = \frac{1}{3} \\ t^{2/3} \text{ matter only, } \quad \varepsilon = 0 \end{cases}$
$$(13.66)$$

(5) *Cosmological constant* Λ. See McCrea (1971) for a good discussion. In the literature one finds three main versions:

(a) The geometric tensor of the field equations must have zero divergence for yielding conservation laws. This is a differential equation of first order, having Λ as its constant of integration, whose value must be found from observation.

(b) Einstein introduced Λ for enabling a static universe. After Hubble's discovery, Λ was not needed ($\Lambda = 0$).

(c) Einstein introduced Λ for fulfilling Mach's principle in a finite static world, but abandoned it ($\Lambda = 0$) when DeSitter showed that non-Machian empty universes still are possible.

It seems that Λ must be found from *observation*. Since the only theoretical reason for $\Lambda = 0$ is simplicity, one could as well demand $E = 0$ ($k = 0$, $q_o = 1/2$) for simplicity and forget about observation altogether. (Furthermore, the simplest universe is an empty one!)

c) Tests of General Relativity

(1) *Gravitational red shift*. From a stellar surface to infinity, $z = GM/rc^2$; or $cz = 0.635$ km/sec for the Sun and about 50 km/sec for white dwarfs. Observations agree with theory, within their mean errors of $\pm 5\%$ for the Sun (Brault, 1963), and $\pm 15\%$ for white dwarfs (Greenstein and Trimble, 1967).

Two clocks at different height, h, in the same building keep different time, with $z = gh/c^2 = 1.09 \times 10^{-16}\ h$/meter. This was measured with the Mössbauer effect by Pound and Snider (1965), who found agreement within a small mean error of $\pm 1.0\%$.

(2) *Perihelion advance*. For Mercury the observed advance is 5596 seconds/century. From this we must subtract 5025 for precession, and 528 for perturbations from other planets. There remains a residual of 43 seconds/century, which was first found by Leverrier in 1859. General relativity predicts a value of 43.03 and the best observations today give 43.1 ± 0.5. Agreement is also obtained for Venus, Earth, and Icarus (Shapiro *et al.*, 1968a), but with larger errors. But according to Dicke, about 20% of Mercury's observed advance is due to an oblateness of the Sun. A future decision is possible with artificial solar satellites of different eccentricities.

(3) *Light deflection at the Sun's rim*. General relativity demands exactly 1.75 arc seconds, Brans-Dicke only about 1.63 arc seconds. *Optical* measurements, from eclipses during the last 50 years, are summarized by von Kluber (1960) and are about 30% too high, but the uncertainties are large. *Radio* interferometers give

Seielstad *et al.* (1970) 1.77 ± 0.20 arcsec
Muhleman *et al.* (1970) 1.82 ± 0.26 arcsec
Hill (1971) 1.87 ± 0.33 arcsec
Sramek (1971) 1.57 ± 0.08 arcsec

$$(13.67)$$

(4) *Light delay at the Sun's rim*. This so-called fourth test was suggested by Shapiro in 1964: for radar reflected by Mercury or Venus beyond Sun, GR demands a delay of 0.2 msec. This can easily be measured, but the orbits are not well enough known, and one must solve for a total of 24 parameters. Measurements (Shapiro *et al.*, 1968b) agree within errors of $\pm 20\%$.

(5) *PPN-approximation* (parameterized post-Newtonian; Thorne and Will, 1971; Will, 1971). A minimum of theoretical assumptions gives nine free parameters to be found by observation. Two solar satellites with gyros, enclosed in self-correcting spheres for shielding against radiation pressure and plasma drag, may in some years enable us to determine most of these parameters.

13.3.3 General Relativity, Pressure-Free Uniform Models

a) Formulae, Calculations

In addition to H_o and q_o from Equations (13.17) and (13.18), we define three dimensionless parameters:

Density parameter

$$\sigma_o = 4\pi G\rho_o/3H_o^2 \qquad (13.68)$$

Cosmological constant, normalized

$$\lambda_o = \Lambda/3H_o^2 \qquad (13.69)$$

Curvature parameter, normalized

$$\kappa_o = k(c/H_oR_o)^2 = k\,(c/\dot{R}_o)^2 \quad (13.70)$$

The two differential equations of GR, (13.63) and (13.64), then yield for the present and $p = 0$,

$$\lambda_o = \sigma_o - q_o \qquad (13.71)$$

$$\kappa_o = 3\sigma_o - q_o - 1 \qquad (13.72)$$

This would be the way to obtain Λ, R_o, and k from observation, if the problem of the hidden mass could be settled (Section 13.4.1.b).

The pressure-free uniform models are, basically, a two-parameter family; once σ_o and q_o are chosen, λ_o and κ_o follow from Equations (13.71) and (13.72), while H_o does not describe a model but tells only its present age.

The differential equation for $R(t)$ is Equation (13.63), with $\rho = \rho_o(R/R_o)^3$:

$$\dot{R}^2 = 2G\left(\frac{4\pi}{3}\rho_o R_o^3\right)\frac{1}{R} - kc^2 + \frac{1}{3}\Lambda R^3$$

$$(13.73)$$

to be compared with the Newtonian equation, (13.58). Equation (13.73) is easily solved for $\rho_o = 0$ and/or $\lambda_o = 0$, and some results are given in Table 13.4. Refsdal, Stabell, and de Lange (1967) have calculated numerically 101 different more general models, printed practically all needed properties as functions of z, and included a large number of useful graphs.

For connecting the red shift via dynamics with the other observables, we need $\dot{R}(z)$, explicitly in Equation (13.41), and implicitly for $u(z)$ in Equation (13.21). Using the parameters defined above we have

$$\left(\frac{\dot{R}}{R_o H_o}\right)^2 = 2(1 + z)\sigma_o - \kappa_o + (1 + z)^{-2}\lambda_o$$

$$(13.74)$$

b) *Special Features*

There is some confusion regarding the words "elliptical" and "hyperbolical." The same word applies to both space curvature and expansion type only for $\lambda_o = 0$ and a small surrounding. Most models have elliptical (closed) space but hyperbolical (never-stopping) expansion, or hyperbolical (open) space but elliptical (through maximum to collapse) expansion. Note: Euclidean (flat) space also is "open."

With increasing distance, z goes through a minimum for some models, S has a minimum for some more, and angles θ have a

minimum for most models. In some models, sources close to the antipole would show two images, separated by 180°.

The simplest (nonempty) world model is the *Einstein–de Sitter* model, $q_o = \sigma_o = 1/2$, where both cosmological constant and space curvature are zero, $\Lambda = k = 0$. Its Newtonian analogy is the model with zero energy, $E = E_{\text{pot}} + E_{\text{kin}} = 0$. Furthermore, this is the only model with complete agreement between Einstein and Newton, regarding the dynamics as well as all observables.

c) *Classification*

(See Stabell and Refsdal, 1966.) A short summary is given in Figure 13.3 and Table 13.5. There, "de Sitter" means $R(t) = R_o \exp(tH_o \sqrt{\lambda_o})$; "Einstein–de Sitter" is $R(t) = R_o(\frac{3}{2} tH_o)^{2/3}$ = Newton; "Milne" is $R(t) = ct$; "static" is the original Einstein solution with $R = R_E = c^2/\Lambda$, where $\Lambda = 4\pi G\rho_o$. The big-bang singularity ($R = 0$) is either "strong" ($\dot{R} = \infty$) or "mild" ($\dot{R} = c$).

13.3.4 General Relativity, Early Phases of Big-Bang Models

a) *Uniform Models*

A good summary is given by Kundt (1971), from which Figure 13.4 is taken. The model used is $q_o = \sigma_o = 1/2$ ($k = \Lambda = 0$), with approximation (13.55), but the early phases are almost model-independent.

b) *Fluctuations*

Primordial fluctuations (of temperature, velocity, density, etc.) may decrease the helium production considerably (see Silk and Shapiro, 1971); for example an r.m.s. $(\Delta T/T) = 0.5$ reduces the helium fraction by a factor 0.1 in the hot spots and by a factor 0.6 in the average. Fluctuations probably play a crucial role in the formation of galaxies (which we omit because of too many unsettled problems). Small primordial anisotropies will be smoothed out to $= 0.03\%$ by the high viscosities from neutrino and photon decoupling (Misner, 1968) but not the larger anisotropies (Stewart, 1968).

Table 13.4 Various Distances: In General and in Four Simple World Models

Name	Defined	Calculated	Approximation
Metric distance	$u = \displaystyle\int_0^r \frac{d\bar{r}}{\sqrt{1-k\bar{r}^2}}$	$u = c\displaystyle\int_{t_e} \frac{dt}{R(t)}$	$= \dfrac{cz}{R_o H_o}\left(1 - \dfrac{1+q_o}{2}z \pm \cdots\right)$
Radar distance = rigid-rod distance	Light travel time $= l_{\text{rad}}/c$	$l_{\text{rad}} = R_o u$	$= \dfrac{cz}{H_o}\left(1 - \dfrac{1+q_o}{2}z \pm \cdots\right)$
Luminosity distance (bolometric)	$S_{\text{bol}} = \dfrac{L_{\text{bol}}}{4\pi l_{\text{bol}}^2}$	$l_{\text{bol}} = R_o \mathscr{S}(u)(1+z)$	$= \dfrac{cz}{H_o}\left(1 + \dfrac{1-q_o}{2}z \pm \cdots\right)$
Luminosity distance (monochromatic)	$S_\nu = \dfrac{L_\nu}{4\pi l_\nu^2}$	$l_\nu = l_{\text{bol}}(1+z)^{-(1+\alpha)/2}$	$= \dfrac{cz}{H_o}\left(1 - \dfrac{\alpha+q_o}{2}z \pm \cdots\right)$
Parallax distance	$\Delta\phi = (\text{baseline})/l_{\text{par}}$	$l_{\text{par}} = \dfrac{R_o \mathscr{S}}{\sqrt{1-k\mathscr{S}^2}}$	$= \dfrac{cz}{H_o}\left(1 - \dfrac{1+q_o}{2}z \pm \cdots\right)$
Volume distance	$V = \dfrac{4\pi}{3}l_{\text{vol}}^3$	$l_{\text{vol}} = R_o\left\{3\displaystyle\int_0^u \mathscr{S}^2(u)\,du\right\}^{1/3}$	$= \dfrac{cz}{H_o}\left(1 - \dfrac{1+q_o}{2}z \pm \cdots\right)$
Diameter distance	$\phi = D/l_\phi$	$l_\phi = l_{\text{bol}}(1+z)^{-2}$	$= \dfrac{cz}{H_o}\left(1 - \dfrac{3+q_o}{2}z \pm \cdots\right)$

q_o	σ_o	k	λ_o	Name	$\dfrac{H_o R_o}{c}$	$t_o H_o$	$\dfrac{H_o}{c}l_{\text{rad}}$	$\dfrac{H_o}{c}l_{\text{bol}}$	$\dfrac{H_o}{c}l_{\text{par}}$	$\dfrac{H_o}{c}l_{\text{vol}}$
1	1	$+1$	0	/	1	.571	$\arcsin\dfrac{z}{1+z}$	z	$\dfrac{z}{\sqrt{1+2z}}$	$\left[\dfrac{3}{2}\left\{\arcsin\dfrac{z}{1+z} - \dfrac{z}{1+z}\sqrt{1-\left(\dfrac{z}{1+z}\right)^2}\right\}\right]^{1/3}$
$\frac{1}{2}$	$\frac{1}{2}$	0	0	Einstein–de Sitter	/	.667	$2\left\{1-\dfrac{1}{\sqrt{1+z}}\right\}$	$2\{1+z-\sqrt{1+z}\}$	$2\left\{1-\dfrac{1}{\sqrt{1+z}}\right\}$	$2\left\{1-\dfrac{1}{\sqrt{1+z}}\right\}$
0	0	-1	0	Milne	1	1	$\ln(1+z)$	$z(1+\tfrac{1}{2}z)$	$z\dfrac{1+\frac{1}{2}z}{1+z+\frac{1}{2}z^2}$	$[\tfrac{3}{4}\{(1+z)^2 - (1+z)^{-2} - 2\ln(1+z)\}]^{1/3}$
-1	0	0	1	de Sitter (steady state)	/	∞	z	$z(1+z)$	z	z

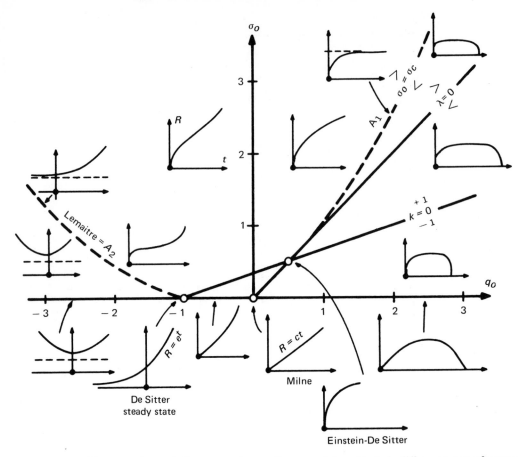

Figure 13.3 The q_o, σ_o plane of all pressure-free uniform models, with their different types of expansion, $R(t)$.

The present fluctuations (galaxies, clusters) or any larger anisotropies cause a *distortion* of light-rays, resulting in (1) apparent ellipticity (Kristian and Sachs, 1966; Kristian, 1967); (2) errors of angular measurements (Gunn, 1967) and of magnitude (Kantowski, 1969); and (3) splitting up of a strong source into many faint ones (Refsdal, 1970). All these effects are negligible for $z \leq 1$, but some may be large for $z > 2$.

c) *The Mixmaster Problem*

All nonempty big-bang models have a strong singularity and thus a particle horizon. From its definition in Equation (13.15), it follows that $u_{ph} \to 0$ for $t \to 0$. This means there was no interaction in the beginning. How can the universe then look as

homogeneous as we see it? If the last chance for homogenizing was at photon decoupling, $z \approx 1000$, we may calculate the particle horizon for that time, and we then would expect large inhomogeneities and anisotropies beyond this horizon, which means today for distances ≥ 100 Mpc and for angles $\geq 3°$, which is not the case. But the problem is more basic than that, and I think it is much more severe than most people realize (or admit); we can state:

All nonempty uniform big-bang models assert a "common but   (13.75) unrelated" origin of all things.

As a solution, Misner (1969b) suggested the "mixmaster universe" with slightly anisotropic expansion. For $t \to 0$ it has an infinite

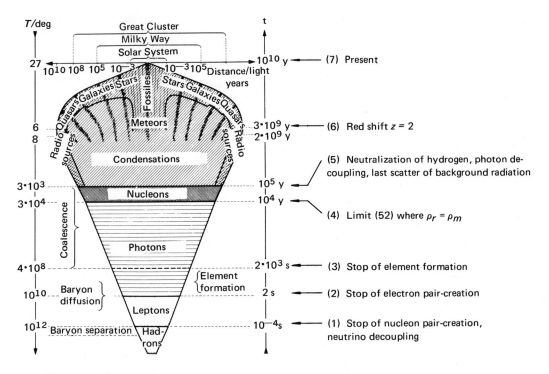

Figure 13.4 The early phases. The outer boundary is our past light cone. [Kundt (1971), *Springer Tracts in Mod. Phys.* 58:1.]

series of partial singularities, leaving always one nonsingular direction for mixing, with changing directions, which hopefully would give enough primordial mixing for the uniformity observed today. There are three objections: (1) It is very complicated. (2) According to some, the mixing occurs only in thin tubes of decreasing length and thus covers only a small fraction of space. (3) This mixmaster phenomenon occurs even in empty space, thus once more emphasizing the

"physical reality of empty space," which sounds very odd.

Being unsatisfied and going back to the roots, we find that it all comes from the fact that general relativity asserts velocities $>c$; actually $v \to \infty$ with $\dot{R} \to \infty$ for $t \to 0$. (In GR special relativity holds only locally but breaks down globally.) The mixmaster problem does not occur in Milne's universe. Maybe we need some change of GR which prevents any v or $\dot{R} \geq c$.

Table 13.5 Expansion Types, Singularities, and Horizons

Figure 13.3; q_o, σ_o plane	Expansion type	Singularity	Horizon
Left of A$_2$	Reversing	None	Both
On A$_2$	From static to de Sitter	None	Both
Between A$_2$ and A$_1$	Big-bang to de Sitter	$\sigma = 0$: mild	Event
		$\sigma > 0$: strong	Both
On A$_1$	Big-bang to static	Strong	Particle
Right of A$_1$	Big-bang to collapse	$\sigma = 0$: mild	None
		$\sigma > 0$: strong	Particle

13.3.5 Other Theories

a) *Steady-State Theory*

The steady-state theory was introduced in 1948 by Bondi, Gold, and Hoyle (see Bondi, 1960). The old (weak) cosmological principle of uniformity, where all fundamental observers see the same picture at any *place* and in any *direction*, is now extrapolated to the perfect (strong) cosmological principle, including "at any *time*."

It follows that $H = \dot{R}/R$ = constant. Thus the expansion is exponential, $R(t) = R_o e^{Ht}$, where R is an arbitrary scale factor since the curvature is zero, $k = 0$; further $q_o = -1$ and $\lambda_o = \sigma_o = +1$ (no free parameters). Expansion plus steady density needs continuous creation of matter, of $3\rho_o H_o = 1$ atom $(yr)^{-1} (km)^{-3}$. Metric, dynamics, and observables are all identical with de Sitter, except for number counts (flatter than any *GR*). The average age of matter is only $1/(3H_o)$.

Originally, it was clearly said that the steady-state theory is easy to disprove because it has no free parameters and no evolution. Then came two objections: the N/S plot was too steep, demanding evolution, and the 3°K background suggests a dense hot beginning. This disproves the original theory; Hoyle and Narlikar have introduced a fluctuating steady state with evolving irregularities, maybe local little-bangs, where the universe is steady only over very large ranges of space and time. In this way the theory can (just barely) be saved, but it has lost all its beauty. For comparisons with observations see Burbidge (1971) and Brecher, Burbidge, and Strittmatter (1971).

The steady-state theory needs continuous creation of matter (just as unexplained as a primordial creation of the universe); it would need additional creation of 3°K background radiation, and maybe that of helium (again unexplained, while big-bang predicts both); a steep source count would need a local hole or local evolution (a nuisance but possible); finally, it violates the conservation of baryon and lepton number (if Omnes' theory could be confirmed, this would be the strongest argument for big-bang).

b) *Scalar-Tensor Theory*

Brans and Dicke (1961) add, to the tensor field of GR, a scalar field, $\phi(r,t)$, and a coupling constant, ω, with a nonconstant $G(r,t)$ of gravitation. This is basically a theory of gravitation, and will best be checked by local PPN-tests, but does not make much difference for cosmology (Greenstein, 1968; Roeder, 1967; Dicke, 1968).

c) *Hierarchical Cosmology*

Hierarchy is a very attractive idea and was suggested by Charlier in 1908 (for avoiding Olbers' paradox, which now is irrelevant). It was splendidly revived by de Vaucouleurs (1970): Atoms are clustered in stars, stars in stellar clusters; these are clustered in galaxies, followed by clusters of galaxies, clusters of clusters, and so on to infinity (or to $2\pi R_o$ if $k = +1$). For simplification, assume each supercluster consisting of N clusters occupying the fraction β of its volume; then the density $\rho(r)$, averaged over a sphere of radius r, is

$$\rho(r) \sim r^{-\theta}, \quad \text{with } \theta = \frac{3}{1 + \dfrac{\ln N}{\ln(1/\beta)}} \quad (13.76)$$

and $\rho \to 0$ for $r \to \infty$ (if $\theta > 0$). A large-scale hierarchy of objects could well be the result and the left-over from a primordial hierarchical turbulence. And maybe it could even come close enough to an empty universe to avoid a strong singularity and the mixmaster problem, if $k \leq 0$.

d) *Kinematic Relativity*

On kinematic relativity, see Milne (1948). For a more critical summary see Heckmann (1968); for a more positive one see Bondi (1960). Before any laws of physics are introduced, Milne postulates exact uniformity; from this he derives the Lorentz transformation, to be valid not only locally (as in *GR*) but also globally. As for gravitation, he first found $G(t)$ but later tried to keep G = constant.

The connection to *GR* was worked out by Robertson and Walker from 1935 to 1937 (Rindler, 1956a). The metric and dynamic, $R(t) = ct$, are identical with the pressure-free *GR* model of $q_o = \Lambda = 0$, $k = -1$, which in GR is empty but now contains matter. Remarks: (1) it looks very promising because of avoiding the mixmaster problem without being empty; (2) how on Earth can a universe contain matter without having any deceleration, $q_o = q(t) = 0$, for all time?

e) Dirac-Jordan Cosmology

Dirac and Jordan claim that the three dimensionless large numbers, Coulomb/ gravity $\approx R_o$/electron radius $\approx$ (total particle number)$^{1/2} \approx 10^{40}$ are identical and constant. It follows that $G(t) \sim t^{-1}$, $R \sim t^{1/3}$, $q_o = \sigma_o = 2$, $\Lambda = 0$. This has not found much favor. *Alfven and Klein* suggest a reversing model with strong annihilation at the minimum of R (Alfven, 1965, 1971).

13.4 Optical Observations

13.4.1 Hubble Parameter, Density, and Age

a) Hubble Parameter

The Hubble parameter, H_o, gives the linear increase of velocity (red shift) with distance, $v = cz = H_o l$. Since $H = \dot{R}/R = H(t)$, it is not constant, and H_o is just its present value. Distance determinations are still uncertain (see Sandage, 1970). Hubble's original value in 1936 was $H_o = 560$ (km/sec)/Mpc; Baade reduced it in 1950 to $H_o = 290$; the present range of uncertainty is 50 to 130, and for simplicity one generally uses

$$H_o = 100 \text{ (km/sec)/Mpc}; \qquad H_o^{-1} = 10^{10} \text{ yr}$$
$$(13.77)$$

b) Density

The density, ρ_o, is extremely uncertain because of the "hidden mass" problem of groups and clusters of galaxies. First, one obtains the visible mass $\mathcal{M}_g$ of all galaxies in a cluster from their number and type (single masses calibrated with nearby galaxies from rotation curves); second, one obtains the virial mass $\mathcal{M}_v$ needed to keep the cluster gravitationally stable against the measured scatter of velocities; then one should have $\mu = \mathcal{M}_v/\mathcal{M}_g = 1$, but actually finds μ up to 2000 with a median of $\mu = 30$ (Rood, Rothman, and Turnrose, 1970). Thus only 1/30 of the matter is visible; the rest is hidden and might be left over from galaxy formation (Oort, 1970). It is a severe problem that we do not observe this hidden mass or its radiation, although many estimates say that we should (Turnrose and Rood, 1970). Ambartsumian has frequently suggested that most of the clusters and groups with $\mu \gg 1$ actually *are* unstable and flying apart, but this would give a very young age for most of the objects.

The visible matter of the universe was estimated as 3×10^{-31} g/cm^3 by Oort (1958) and 6×10^{-31} by van den Bergh (1961). If the hidden matter were stars, Peebles and Partridge (1967) find an upper limit of 4×10^{-30} g/cm^3 from measuring the background sky brightness and subtracting zodiacal light and faint stars. And for Equation (13.68) we have

$$\sigma_o = \frac{\rho_o}{\rho_c} \qquad \text{with a density unit of}$$

$$\rho_c = \frac{3H_o^2}{4\pi G} = 4 \times 10^{-29} \text{ g/cm}^3 \quad (13.78)$$

In summary:

Visible matter	$\sigma_o = .01$
Sky background, stars	$\sigma_o \leq .10$
Hidden mass with $\mu = 30$	$\sigma_o = .30$
Einstein–de Sitter, Newton	$\sigma_o = .50$

(13.79)

The close agreement between visible plus hidden mass and the simplest of all nonempty world models ($k = \Lambda = 0$, zero energy) is somewhat amazing. Even the visible matter alone is not off by a large factor. There is no *a priori* reason why ρ_o should be comparable to H_o^2/G.

c) Age

The age, t_o, of the universe seemed for a long time to be less (factor 2 to 3) than that of the oldest objects. Present values give good agreement; but see de Vaucouleurs (1970) for some nicely formulated doubt about the finality of our present values.

Almost all big-bang models give ages somewhat less than H_o^{-1} (except very close to the Lemaitre limit A_2 in Figure 13.3); for $\Lambda = 0$, we have $t_o H_o = 0.571$ for $q_o = 1$, and $t_o H_o = 2/3$ for $q_o = 1/2$. The last one then

gives $t_o = (7 \pm 2) \times 10^9$ years, with $H_o = (100 \pm 25)$ (km/sec)/Mpc. As for the objects, the oldest globular clusters give $t_o = (9 \pm 3) \times 10^9$ years, according to Iben and Faulkner (1968); and the age of radioactive elements such as uranium can be found from the estimated original, and the observed present abundance ratios with $t_o = (7.0 \pm .7) \times 10^9$

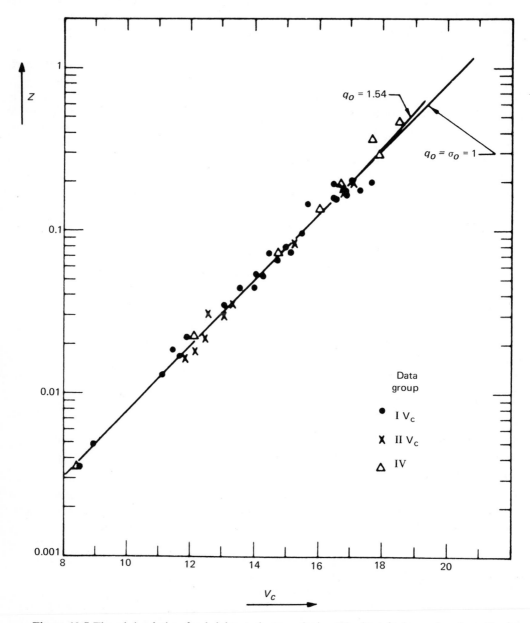

Figure 13.5 The $z(m)$ relation for brightest cluster galaxies. The best fit (assuming $\Lambda = 0$) gives $q_o = 1.54$; the straight line, $q_o = 1$, is drawn for comparison. [Peach (1970), *Astrophys. J.* 159:753.]

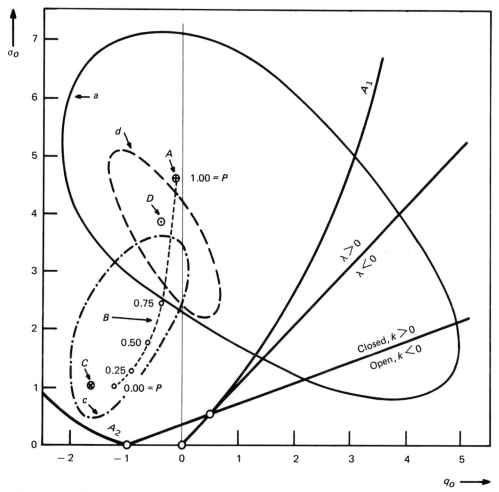

Figure 13.6 The σ_o, q_o plane of world models, with results from $z(m)$ tests. *A, B, C, D*: best-fitting values; *a, c, d*: 64 % confidence $\approx$ probable error. *A, a*: 20 brightest cluster galaxies (Solheim, 1966). *B*: shift of point *A* for evolution correction, with P = fraction of light from unevolving stars (Solheim, 1966). *C, c*: 15 quasars (Solheim, 1966). *D, d*: 46 brightest cluster galaxies (Peach, 1970).

years according to Dicke (1969). In summary:

Globular clusters	$t_o = (9 \pm 3) \times 10^9$ yr
Radioactive elements	$t_o = 7 \pm 1$
Einstein–de Sitter, Newton	$t_o = 7 \pm 2$ (13.80)

13.4.2 Red Shift—Magnitude Relation

Since the red shift increases with distance (Hubble), the flux of a source is $S \sim z^{-2}$ for small z and depends on world models for large z; see Equations (13.23) and (13.27), and Sandage (1961a). As a standard candle one mostly uses the brightest galaxy in rich

clusters; quasars are visible at much larger red shifts, but their luminosities scatter too much (Solheim, 1966).

There are several corrections to the luminosity. (1) The K-correction discussed before Equation (13.27); see Solheim (1966) and Oke and Sandage (1968). (2) A richness correction, if the brightest galaxy of a rich cluster is brighter than that of a poor one. This can be neglected if $N \geq 30$ (Sandage, 1961a), but see also Abell (1972). (3) Evolution of luminosity, plus light travel time (see Sandage, 1961b, 1968; Solheim, 1966; Tinsley, 1968; Peach, 1970.) This is still very uncertain, but improvable. (4) Thomson scatter from inter-galactic electrons may increase q_o by 15% (Peach, 1970). (5) Distortion effect from local inhomogeneities (Kantowski, 1969) may increase q_o from 1.5 to 2.7 (Peach, 1970). Figures 13.5 and 13.6 show some recent results, and illustrate their uncertainties.

13.4.3 Number Counts

(1) $n(z,m)$ of quasars, or luminosity-volume test, is discussed in Section 13.5.3.
(2) $n(z)$ of quasars, the odd distribution which has now disappeared, has been treated in Section 13.1.4.
(3) $N(m)$ of galaxies is of no use according to Sandage (1961a)—but might be from outside our atmosphere.
(4) $N(m)$ of optical QSO (see Braccesi and Formiggini, 1969). 300 objects were selected optically for UV and IR excess, of which 195 objects are complete to $m_b = 19.4$. These give an $N(S)$ slope of $\beta = -1.74$. A fraction of 20% are estimated to be white dwarfs, which gives a correction resulting in

$$N(S) \text{ slope} \qquad \beta = -1.80 \pm 0.15 \quad (13.81)$$

Spectra are known for 27 of these objects, with red shifts between 0.5 and 2.1. For the model $q_o = \sigma_o = 1$, the slope should be $\beta = -1.1$ without evolution, but $\beta = -2.0$ with the evolution $(1 + z)^5$ found by Schmidt (1968). The authors conclude that evolution is definitely needed and that optical and radio evolution are about the same. The latter agrees with Golden (1971), but not with Arakelian (1970).

13.4.4 Angular Diameters

In all nonempty expanding models the angular diameter, $\theta(z)$, drops to a minimum and then increases again for increasing z. This makes the diameter a very promising observable; values for the minimum are shown in Figure 13.7. They require $z > 1$.

Single galaxies are just marginal, reaching only to $z \leq 0.5$, where all reasonable models are still too similar and, for dimensions of 10 kpc, give $\theta = 2$ to 3 seconds of arc, which is too small for accurate measurements, except possibly from outside the atmosphere. Peach and Beard (1969) investigated 646 Abell clusters with diameters from Zwicky but, unfortunately, found some very strong systematic effect. If it could be removed, the uncertainty in q_o would be ± 0.2.

From the diameter and mass of a cluster, one can calculate its relaxation time, t_r. For those clusters where $3t_r \leq 10^{10}$ years, we do not expect (much) evolution of the linear diameter, and angular diameters thus could yield an (almost) evolution-free model test.

13.5 Radio Observations

In this whole section we make the (helpful but unproven) assumption that quasars are at cosmological distance; see Section 13.1.4.

13.5.1 Radio Sources for Cosmology

a) Types and Numbers

In the 3CR catalog there are about 40% galaxies, 30% quasars, and 30% empty fields (unidentified but tried). Surveys at shorter wavelengths have more quasars per galaxy; deeper surveys have much more empty fields (up to 70%), but also more quasars/galaxy. Table 13.6 gives some data about luminosity and frequency of occurrence. From the latter, one derives lifetimes between 10^2 years for the brightest and 10^9 years for the faint radio galaxies, assuming that each large elliptical (plus some other) galaxy goes once through an active phase; quasars then would give lifetimes between 0.1 and 10^3 years. These are only lower limits, since lifetimes are longer if the active phase occurs in some very special types of galaxy only. From estimated energies, divided by the luminosities, one derives

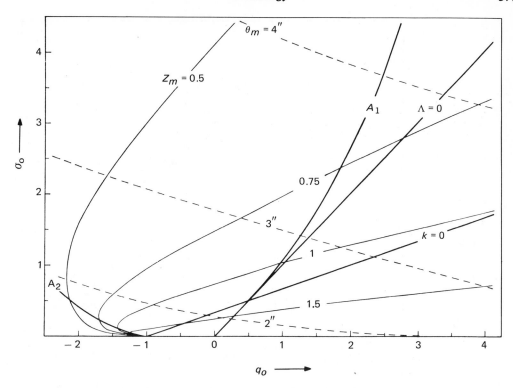

Figure 13.7 In nonempty expanding models the angular diameter $\theta(z)$ has a minimum, θ_m, at a certain red shift, z_m. A linear diameter of 10 kpc is used. [Refsdal, Stabell, and de Lange (1967), *Mem. Roy. Astron. Soc.* 71:143.]

Table 13.6 Luminosity L, Spectral Index α, and Space Density ρ of Radio Sources

	L_{rad} (178 MHz) W/Hz	L_{rad}/L_{opt}	$\bar{\alpha}$(rad/opt)	ρ (Mpc)$^{-3}$
Opt $\begin{cases} \text{all galaxies} \\ \text{bright ellipticals} \end{cases}$				4×10^{-2} 2×10^{-3}
Normal galaxies	10^{17} to 10^{23}	0.1 to 10	0.0	2×10^{-2}
Radio galaxies	10^{23} to 10^{28}	100 to 10,000	-0.4	2×10^{-4}
Optically selected quasars	$\leq 10^{22}$	≤ 0.001	$\geq +0.5$	3×10^{-7}
Radio selected quasars	10^{27} to 10^{29}	100 to 10,000	-0.4	1×10^{-9}

lifetimes of 10^4 to 10^8 years. In the following we always assume lifetimes $\ll 10^{10}$ years, which means *class* evolution only.

For the radio sources the ratio L_{rad}/L_{opt} goes up to 10^4, which certainly is an advantage; but it shows no correlation with the radio index α nor any other observable; there are some extremely luminous sources, but L_{rad} varies over 12 powers of 10 with (almost) no luminosity indicator or standard candle; VLB experiments give extremely high resolution, but the linear sizes go from 10^{-3} to

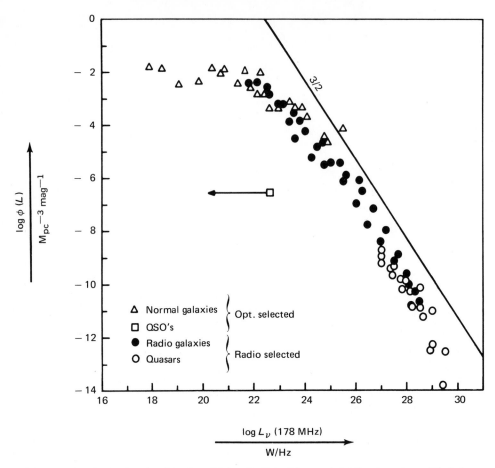

Figure 13.8 Radio luminosity function $\Phi(L)$; all data. The straight line has the critical slopes.

10^5 pc, over 8 powers of 10, with (almost) no diameter indicator or standard rod. Thus, radio astronomy reaches extremely far out into space but yields (almost) no information.

b) The Radio Luminosity Function

The radio luminosity function, $\phi(L)$, is derived from the $n(S,z)$ counts by the luminosity-volume method (Section 13.5.3) or a similar one. It is needed for evaluating the $N(s)$ counts regarding world models and evolution. Figure 13.8 is a compilation of available data, which shows a large uncertainty, especially at both ends. Data at the bright end can be obtained only using assumptions about world models and evolution, and the

latter may give factors up to 300 (and just as much uncertainty).

Another problem, not seen clearly by some authors, is the existence and relevance of several critical slopes of $\phi(L)$. For simplicity, we discuss the static Euclidean case. From Equation (13.40), and with $\phi(L)$, we find for the number of sources with luminosity L, and observed at flux S,

$$n(S,L)\, dS\, dL = \text{constant}\, \frac{dS}{S^{5/2}} L^{3/2} \phi(L)\, dL$$

$$(13.82)$$

Since S and L are separated, we see at any S the same distribution of sources, $L^{3/2}\phi(L)$. Furthermore, the sources most probably seen, at any S, are those where $L^{3/2}\phi(L) = $ max; which means those sources where the slope

$$\gamma = \frac{d\log\phi}{d\log L} = -3/2 \qquad (13.83)$$

In a well-behaved luminosity function these would be the ones to be called standard candles. But Figure 13.8 shows that almost the whole range of radio galaxies and quasars has a slope of $-3/2$; thus, the sources seen at any S may have any L (which means any distance or any z). Actually, the "half-probability" width of the distribution $L^{3/2}\phi(L)$ in Figure 13.8 is five powers of 10 in L, from 10^{24} to 10^{29} W/Hz. This indeed explains why the $z(S)$ relation just looks like a scatter diagram; it will be a decent Hubble diagram only if the range of observed S is much larger than the width of $L^{3/2}\phi(L)$ or if $S_{max}/S_{min} \gg 10^5$.

But for obtaining $n(S)dS$, the slope must be steeper than $-5/2$ in order to make Equation (13.83) integrable. With Equation (13.38) the slope must be $< -6/2$ for obtaining a Hubble relation $\bar{z}(S)$, and for giving it any accuracy we even need

$$\gamma < -7/2 \qquad \text{for finite r.m.s. } (z - \bar{z}) \quad (13.84)$$

Since the bright end does not look steep enough, it seems that the observed red shifts have been kept finite only by the grace of expansion, model, and evolution effects, entering approximation Equation (13.82) as terms of second and higher order.

It must be emphasized that a luminosity function as bad as Figure 13.8, and extrapolated in several ways within the large range of our uncertainty, is what should be used for evaluating the $N(S)$ counts when checking models and evolution.

There is one more problem. Like any decent distribution function, the luminosity function should be used normalized, with $\int \phi(L)dL = 1$. But Figure 13.8 shows that this is clearly impossible at the faint end. Other normalizations may be used—or none—but then the distinction between density evolution and luminosity evolution becomes

problematic and, indeed, is a mess: many authors criticizing each other for not having done it properly (von Hoerner, 1973b).

c) Intrinsic Correlations

Our lack of luminosity and distance indicators results from the absence of strong narrow correlations of L and D with distance-free observables like spectral index α, or surface brightness B. Also, for the theories of sources we would appreciate some strong correlations between intrinsic source properties. Only few correlations were found and they have a large scatter. First, Heeschen (1966) plotted L over B (Figure 13.9), which shows a clear correlation but a scatter of ± 0.6 in $\log L$ for 90%, plus 10% outsiders far away. This was confirmed by Braccesi and Erculiani (1967) as a correlation of L and D, and by Longair and Macdonald (1969) with a total of 150 sources at 178 MHz, giving a larger scatter of ± 1.0 in $\log L$. The smooth continuity in Heeschen's plot, from 12 normal galaxies over 28 radio galaxies to 14 quasars, has been used as an argument for the cosmological distance of quasars (Section 13.1.4).

Second, a weak correlation between L and α was claimed by Heeschen (1960), Braccesi and Erculiani (1967), Bash (1968), and Kellermann, Pauliny-Toth, and Williams (1969). It shows better for radio galaxies and looks more doubtful for quasars. The latter can be improved if only very straight-lined spectra are selected.

13.5.2 Correlations Involving Distance

a) The z(S) Relation for Quasars

The $z(S)$ relation for quasars is shown in Figure 13.10. It just looks like a scatter diagram, without a Hubble law; Hoyle and Burbidge (1966) thus concluded that quasars are not at cosmological distance; but a large scatter of $L^{3/2}\phi(L)$ explains it just as well.

Furthermore, in critical cases one should use the median and not the average. At the bright end of the luminosity function we need

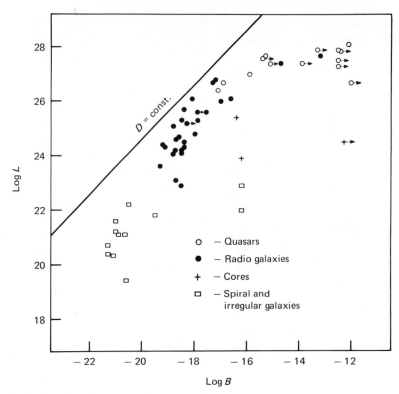

Figure 13.9 Luminosity and surface brightness, for radio galaxies and quasars, at 1400 MHz. [Heeschen (1966), *Astrophys. J.* 146:517.]

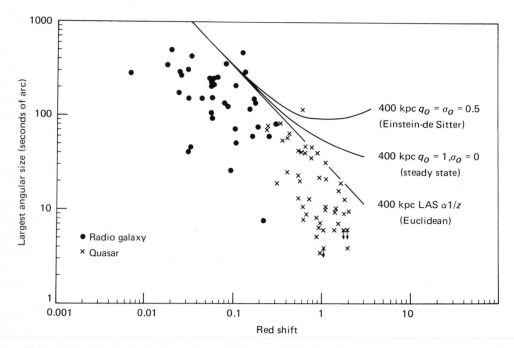

Figure 13.10 Angular size versus red shift. [Miley (1971), *Monthly Notices Roy. Astron. Soc.* 152: 477.]

$\gamma < -3$ for obtaining $\bar{z}$, and $\gamma < -7/2$ for its accuracy, the r.m.s. $(z - \bar{z})$; whereas the median and its accuracy, the quartiles, both require only $\gamma < -5/2$. Indeed, the median z_m in Figure 13.11 shows a fairly good correlation with S. Checking world models, however, would need a luminosity indicator for reducing the scatter.

b) Angular Size

Legg (1970) collected data for 32 radio galaxies and 25 quasars having double structure and known red shifts. He found a good correlation, with a well-defined upper envelope of

$$\theta(z) \sim \frac{1}{z} \quad \text{and} \quad D_{max} = 400 \text{ kpc} \quad (13.85)$$

Miley (1971) used the largest angular size, *LAS* (diameter of singles, separation of doubles), of 39 radio galaxies and 47 quasars (see Figure 13.11) with the same results. Most puzzling in Figure 13.11 is the fact that

$$\theta(z) \sim \frac{1}{z} \quad (13.86)$$

continues beyond any possible world model. From Figure 13.6 one can show that the steady-state or de Sitter model ($q_o = -1$, $\sigma_o = 0$) gives of all possible models the smallest θ for large z; whereas $\theta \sim 1/z$, called "Euclidean" in Figure 13.11, is actually "static Euclidean," which is not possible for an expanding universe. Both Legg and Miley conclude that they need a "diameter evolution" for explaining the small θ; Legg suggests $D_{max} \sim (1 + z)^{-3/2} = (R/R_o)^{3/2}$, in agreement with a theoretical estimate of Christiansen (1969).

About the opposite type of deviation is found by Longair and Pooley (1969); comparing the diameter distributions $n(\theta,S)$, of the 3CR and 5C surveys, they find too many large diameters for small S.

But there is a strong selection effect, since at large z we see only large L, while L and D are correlated. Also a correction for this effect must be worked out and applied

before checking models with diameters. Second, even if Figure 13.11 could be explained in this way, it still remains a puzzle why any correction should just yield the (impossible) static Euclidean continuation of $\theta \sim 1/z$.

c) Other Correlations

Other correlations have been tried in considerable numbers, *e.g.*, by Bash (1968), but without much success. Hogg (1969) finds a good correlation between index α and size (steep ones being larger) because small sources are opaque.

13.5.3 The $n(z,S)$ Counts and Luminosity-Volume Test

One would like to get a large sample $n(z,S)$, complete down to some faint S_o, and to derive from it the luminosity function, the source evolution, and finally the world model. Unfortunately, the data depend (as usual) much more on the first two than on the last one. And for quasars, a sample is limited in two ways: radio (detection) and optically (red shifts); this would be better for optically selected QSO's, having one limit only.

The $n(z,S)$ plot, Figure 13.11(a), yields luminosity function, $N(z)$ and $N(S)$ counts, by summations along different lines; it yields the $z(S)$ relation by taking the median. Strictly speaking, there are two types of luminosity function: first, the directly obtained one, which prevails along our past light cone and thus contains different (and unknown) amounts of evolution for different L; second, if possible, one would like to obtain the full time-dependent luminosity function, $\phi(L,t)$. A glance at Figure 13.11 shows immediately that this cannot be properly done because the range of observed S is much too small for splitting up the data into several groups of z. We badly need larger samples, complete to fainter limits.

The luminosity-volume test was suggested by Schmidt (1968) and Rowan-Robinson (1968); see also Arakelian (1970). For critical discussion, summaries, and new data, see

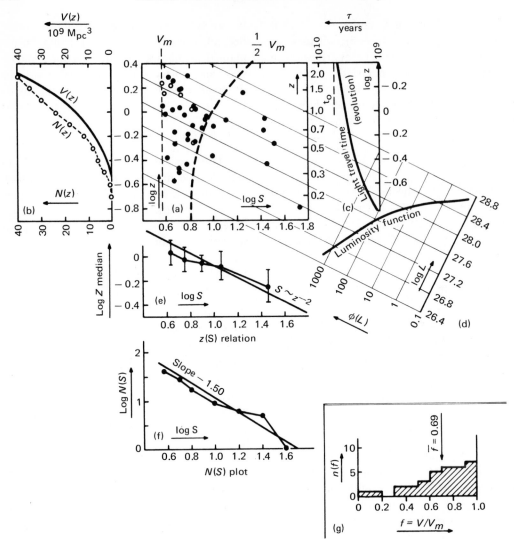

Figure 13.11 The $n(z,S)$ count and its derivatives. (a) Radio-complete sample of 40 quasars (Schmidt, 1968). Open circles: beyond optical completeness limit. Right and left of "½V_m" should be about equal numbers. (b) Volume and cumulative numbers, $N(z)$. (c) Light travel time, $\tau(z)$, used for evolution. (d) Luminosity function, number per volume. (e) The $z(S)$ relation, using the median and not the average. (f) The cumulative $N(S)$ plot. (g) Luminosity-volume test; without evolution, $n(f) =$ constant and $\bar{f} = 0.5$.

Longair and Scheuer (1970), Schmidt (1970), Rees and Schmidt (1971), Davidson, Davies and Cox (1971), and Rowan-Robinson (1971). The basic procedures:

(1) Estimate completeness limits of sample (radio and optical), omit all sources beyond.

(2) Apply the model-independent part of a red shift correction (K-correction) to S:

$$S_{cor} = \frac{S_{obs}}{(1 + z)^{1+\alpha}} \qquad (13.87)$$

(3) Adopt a world model or two. Calculate distance l and luminosity L for each source, and find the distance l_m where a source of

this L would just be, if it were near one or both completeness limits.

(4) Calculate volume V of sphere with radius l, and V_m with l_m. Take ratio $f = V/V_m$. Get distribution $n(f)$ and average $\bar{f}$.

(5) $1/V_m$ is the contribution of this source to the luminosity function $\phi(L)$, which then is obtained as the sum of all $1/V_m$ in each group of L.

(6) For a uniform world without evolution, we should have $n(f)$ = constant and $\bar{f} = 0.5$. A result like Figure 13.11(g) then means that there were more sources in the past.

(7) The slope β of the $N(S)$ plot is related to $\bar{f}$ as shown by Longair and Scheuer. In the static Euclidean approximation,

$$\bar{f} = \frac{-\beta}{(-\beta + 3/2)} \qquad (13.88)$$

All users of the luminosity-volume test agree that *evolution* is definitely needed, and that the co-moving density of sources (or their luminosity) was higher in the past by a factor $(1 + z)^n$, with $n = 3 \ldots 14$. This leads to the problem of very short evolutionary time scales, of only 10^9 years (Rowan-Robinson, 1971). And Schmidt's data (1968) show large f already at small z (see Figure 13.11a), which looks more like a local hole than evolution.

13.5.4 The $N(S)$ Counts

a) Observations

Just to count sources down to various flux limits sounds very easy but actually isn't.

The first surveys were all badly resolution-limited. If errors from noise plus resolution are larger than the statistical error, one needs an additional "spill-over" correction, since the more numerous faint sources will spill over, via errors, to the fewer bright sources, more frequently than vice-versa. These corrections can best be done by a Monte Carlo method, adding some known artificial sources to the record. The radio equivalent of a K-correction, Equation (13.87), cannot be applied to the data since the red shifts are not known, but it is taken care of on the model side when models are compared with the data. The faulty error bars of $N(S)$ plots, and the preference for differential counts, $n(S)dS$, were discussed in Section 13.2.4.f.

Table 13.7 shows the negative $N(S)$ slope, $-\beta$, for the bright end. We see that Jauncey's proper maximum-likelihood method yields smaller values and larger error limits than does the $N(S)$ plot and its error bars as used by some authors.

The full range of the counts is shown in Figure 13.12 for $n(S)$, and in Figure 13.13 for $N(S)$; correcting for various wavelengths gives good agreement. We see the famous steep slope at the bright end, and the well-pronounced flattening at the faint end, where we finally must have $\beta > -1$ for avoiding Olbers' paradox ($\beta = -0.8$ is reached already), and $\beta = 0$ for a finite total number.

b) Results and Interpretations

Four points must be emphasized. First, a slope of $\beta = -1.50$ does not necessarily

Table 13.7 Slope of $N(S)$ Counts, $-\beta$
(ci = very certain identifications only)

Type of source	3CR, 178 MHz			6-cm survey Pauliny-Toth and Kellermann (1972)
	Veron (1966)	Jauncey (1967)	ci	
Total	$1.85 \pm .05$	$1.78 \pm .12$		$1.76 \pm .11$ ($n = 271$)
Radio galaxies	$1.55 \pm .05$	$1.58 \pm .14$	$1.26 \pm .13$	$1.54 \pm .18$ ($n = 103$)
Quasars	$2.20 \pm .10$	$2.00 \pm .29$	$1.56 \pm .26$	$1.54 \pm .17$ ($n = 96$)
Empty fields	($\approx$ quasars)	$2.50 \pm .45$		$2.07 \pm .33$ ($n = 67$)

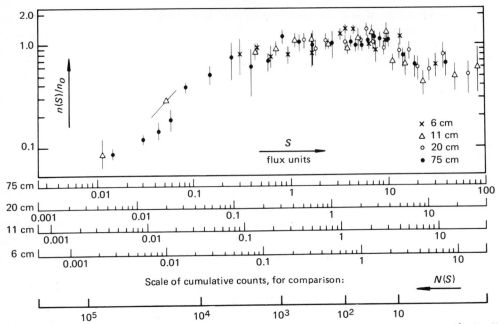

Figure 13.12 The differential counts, $n(S)$, normalized with static Euclidean $n_o \sim S^{-5/2}$. [Kellermann (1972), Warren Lecture 1971, *Astron. J.* 77:531.]

mean "no evolution," as sometimes stated; the numbers given in Equation (13.39) show that the bright part of $N(S)$, with an average red shift of 0.20, say, yields $\beta = -1.23$ without evolution (see also Figure 13.14). Second, the flatness of the bright part is model-independent, and depends only on spectral index, luminosity function, and evolution (Section 13.2.4.f); but evolution at the bright part would mean evolution in the *recent* past. Third, in case of large-scale clustering (de Vaucouleurs, 1970) there is nothing wrong with a *local hole* for explaining the flatness of the bright part. Actually, we should not sit at the average density $\bar{\rho}$, but at $\bar{\rho} \pm$ r.m.s. $(\rho - \bar{\rho})$. Fourth, a slope of $\beta = -1.50$ may be explained by a local hole or evolution, but it would still impose the same type of puzzle as Equation (13.86) does; Kellermann (1972) lists four such puzzles or paradoxa.

Maybe the puzzle can be solved by considering a luminosity function with a near-critical slope, where S shows only very little correlation with z, which may give $\beta = -1.50$ for the bright part, but it would not help for a steeper slope (see von Hoerner 1973b).

In Table 13.7, only the first line (total) is actually relevant. That the empty fields show a steeper slope seems trivial: since radio and optical fluxes are correlated, the radio-faint sources will be optically undetectable more often than the bright ones. Similarly the identified sources then must have a flatter slope. Their "true" slope could be obtained only with due regard to the optical detection limit, which then would spoil the basic idea of the $N(S)$ counts as a simple radio device. Once we need optical identification, we'd better go one step further and get red shifts, too, and then work with all tests indicated in Figure 13.11. Either the $N(S)$ count contains unidentified sources—then the slopes of "galaxies" and "quasars" (and their difference) are not meaningful; or all sources of a sample are identified—then the single slopes can be used, but they do not contain all available information and are confined to a small sample only.

c) Evolution

Detailed calculations and checks have been done by many authors. All of them

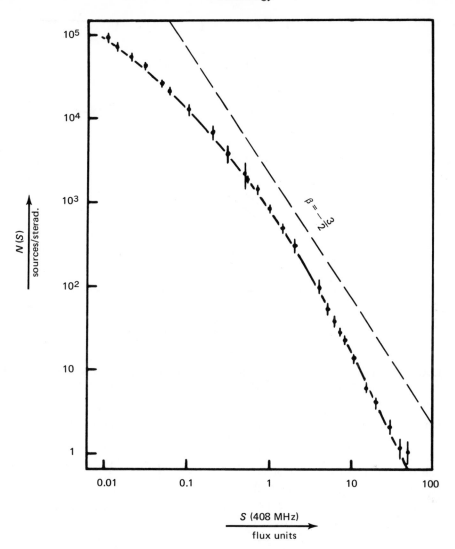

Figure 13.13 The cumulative counts, $N(S)$. [Ryle (1968), *Am. Rev. Astron. Astrophys.* 6:249.]

agree that evolution is needed for explaining the steep slope of the bright part. Longair (1966) finds that only strong sources evolve, but Rowan-Robinson (1967) includes weak sources, too. Schmidt (1968) supports density evolution where the number per co-moving volume is $\rho_c \sim (1 + z)^n$, with a constant luminosity distribution; but a luminosity evolution is claimed by Rowan-Robinson (1970) and by Davidson, Davies, and Cox (1971). Longair and Scheuer (1970) say it does not matter whether in the past the density of sources was higher or their luminosity.

Figure 13.8 shows that the luminosity function cannot be normalized unambiguously. Most of it follows a straight line of critical slope, and the uncertainty is large. This means that a clear distinction between evolution types is hopeless. The easiest then is density evolution with about

$$\rho_c \sim (1 + z)^5 \qquad (13.89)$$

Since this diverges for large z, while actually the counts get flatter at the faint end ($\beta = -0.8$), one needs a strong reduction for large z, and the easiest is a cut-off at some z^*

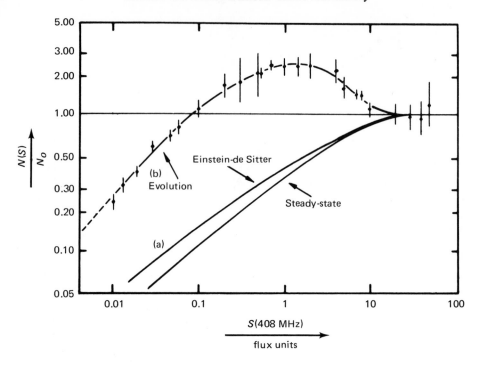

Figure 13.14 The cumulative counts, $N(S)$, normalized with static Euclidean $N \sim S^{-3/2}$. (a) Without evolution; (b) with evolution. [Ryle (1968), *Am. Rev. Astron. Astrophys.* 6:249.]

beyond which there are no sources, of about

$$z \leq z^* = 5 \qquad (13.90)$$

This approach has only two free parameters, n and z^*. If that does not fit the data well enough, one needs more parameters; 3 are used by Longair (1966), and 5 by Davidson *et al.* (1971); both achieve rather good fits (see Figure 13.14).

A different approach (to be preferred if we had more and better data) is the one of Ringenberg and McVittie (1969, 1970). Instead of assuming a steep increase, as in Equation (13.89), and a sharp cut-off, as in Equation (13.90), with maybe some more free parameters, they introduce a free evolution function to be determined numerically from the data. The result is a steep increase again, much steeper for bright than for faint L, but a more gradual decrease for large z.

A cut-off is certainly needed: If it takes $t_g = 6 \times 10^8$ years (three rotations, say) to make a galaxy and let it get an explosive core, or to make strong sources and quasars

in any other way, then there are no sources beyond

$$z^* = \left(\frac{\frac{2}{3} H_o^{-1}}{t_g} \right)^{2/3} - 1 = 4.0 \quad (13.91)$$

for the Einstein–de Sitter model with $H_o = 100$.

13.5.5 The 3°K Background Radiation

A thermal background radiation was predicted for all models with a hot and dense beginning (see Section 13.2.5.c). It was found independently by Penzias and Wilson (1965); they observed a total of 6.7°K at $\lambda = 7.3$ cm, of which 2.3°K was attributed to the atmosphere, about one degree to spill-over, and the remaining $T = 3.5 \pm 1.0$°K to a cosmic origin. Measurements at other wavelengths give about the same temperature.

This can be regarded as an argument for big-bang and against steady-state theory. Therefore some people tried whether a background of many faint discrete sources could

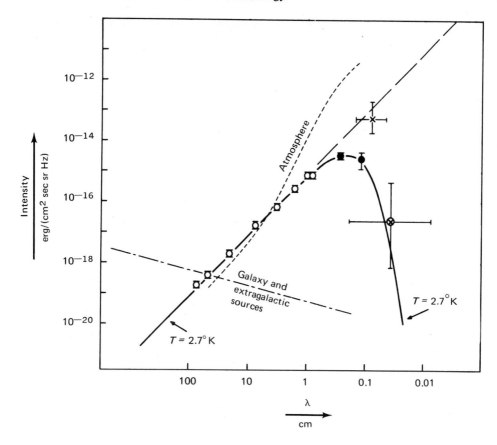

Figure 13.15 The "3°K background" radiation.

○ Radio measurements, quoted by Wolfe and Burbidge (1969).
● Rotational *CN* absorption (Heygi, Traub, and Carleton, 1972).

× Several older rocket measurements.
⊗ New rocket and balloon measurements (see Blair *et al.*, 1971; Beckman *et al.*, 1972).

explain the observations, too (see Wolfe and Burbidge, 1969). The observed spectrum can be explained if a new type of source is postulated with the right kind of spectrum. But the observed isotropy (lack of bumpiness) would demand a space density of these sources much larger than that of galaxies, and the idea now is mostly dropped (see Penzias, Schraml, and Wilson, 1969; Hazard and Salpeter, 1969).

Is it a thermal blackbody spectrum? The observable range is limited by galactic and atmospheric radiation (see Figure 13.15). The older measurements covered only the Rayleigh-Jeans part, which seemed to continue at too high a frequency. But the most recent observations yield a good confirmation of the

thermal spectrum, with a value of

$$T = (2.7 \pm 0.1)°K \qquad (13.92)$$

The observed isotropy is remarkable (see Partridge, 1969; Wolfe, 1969; Pariiskii and Pyatunina, 1971). No deviation ΔI of the intensity I was found:

Down to	5 arcmin	2 arcmin	10 arcsec	Polarization
$\Delta I/I \leq$	0.03%	1%	15%	1%

$$(13.93)$$

This radiation defines a very distant frame of rest (surface of last scattering). If we have a velocity v, we observe a small daily

Table 13.8 Solar Motion from 3°K Background Radiation, and from
Red Shifts of Galaxies in Local Supercluster

	Anisotropy of 3°K	Red shifts of galaxies	Unit
Frame of reference	Surface of last scatter	Local supercluster	—
Solar velocity	320 ± 80	400 ± 200	km/sec
Right ascension	10.5 ± 4	14 ± 2	hours
Declination	-30 ± 25	-20 ± 20	degrees

anisotropy of $T(\theta) = T_o[1 + (v/c)\cos\theta]$, where $v = 100$ km/sec yields $\Delta T = T_o(v/c) = 1$ mK (millidegree). This was actually observed by Conklin (1969) and Henry (1971). The resulting motion agrees so well with estimates from the red shifts of surrounding galaxies that our local supercluster can have only a small or no peculiar velocity (≤ 200 km/sec, say); see Table 13.8.

13.6 Summary

a) Oddities of General Relativity

First, empty space has an amazing degree of physical reality (see Table 13.1); actually GR is "less relative" than Newtonian physics. The model $\Lambda = \sigma = 0$ has space curvature ($k = -1$) and thus a well-defined absolute frame of rest; it also starts with a singularity (infinite curvature), and is completely empty and force-free.

Second, at the singularities (big-bang and collapse) the velocities $v = u\dot{R}$ become infinite; $v > c$ then leads to a particle horizon and thus to the rather serious mixmaster problem, Equation (13.75). A good theory should avoid $v > c$, if necessary by postulate. Third, GR seems to be too similar to Newtonian cosmology. Equation (13.57) is identical with Equation (13.65) for $p = \Lambda = 0$; in this case ρ is the matter density only. Fourth, our physics cannot continue before a big-bang and after a collapse (no bouncing) (see Equation 13.61).

Fifth, in Equation (13.65) we have $p = 0$ for a matter universe, and $\rho = 3p/c^2$ for a radiation universe. Thus, for equal total energy ρ, radiation is "twice as attractive" as matter.

b) Disentangling

Unfortunately, most observations have turned out to be "model-free evolution tests"

Table 13.9 Several Models and Observational Ranges

q_o	σ_o	k	λ_o	(1) $z(\phi_{min})$	(2) $z(u = \pi/2)$	(3) $u(z = 2)/u_{ph}$	(4) $u(z = 2)/\pi$
-1	.3	$+1$	$+1.3$	1.10	3.50	.445	.397
-1	1	$+1$	$+2$	.75	1.61	.522	.546
0	0	-1	0	∞	3.80	0	.350
0	.3	0	$+.3$	1.40	(∞)	.397	0
$+1$	0	-1	-1	∞	3.2	0	.396
$+1$	.3	-1	$-.7$	1.65	12	.373	.273
$+1$	1	$+1$	0	1.00	∞	.465	.232
$+.5$	.5	0	0	1.30	(∞)	.423	0

or nearly so. Truly model-free is the surface brightness [$b(z)$ (see Equation (13.29)], and almost so is the slope of the bright end of the $N(S)$ counts [see Equation (13.41)].

Evolution-free model tests are hard to find, because of the rarity of evolution-free observables. Since the red shift is one, Sandage suggested $\dot{z}(z)$ but found it needs over a million years (Section 13.2.4.a). Another test is the parallax distance $l_{par}(z)$ [see Equation (13.31)], but it would need an accuracy better than 10^{-10} arcsec.

If no evolution-free model test can be found, one needs a theory of source evolution, meaning class evolution for radio sources, which does not exist at present.

c) More Unsolved Problems

First: Are quasars at cosmological distance? Second: Do we have clusters of antimatter? Third, there is the problem of the hidden mass. Fourth, we badly need indicators of luminosity and linear size, for standards. Fifth, we need some test for evolution versus local hole. Sixth, one should develop a hierarchical cosmology.

d) Most Urgent Needs

For various models, Table 13.9 shows:

(1) the red shift z, where angular size ϕ has its minimum
(2) the red shift z, where source is at $u = \pi/2$, which is at the equator for closed models
(3) u of $z = 2$, divided by u of particle horizon
(4) u of $z = 2$, divided by $u = \pi$, which is at antipole for closed models.

The columns of Table 13.9 have the following meaning:

(1) This should be reached for model tests working with angular size or separation.
(2) To be reached for model tests working with m, S, N, or n. All these quantities depend mainly on space curvature; we must see a good part of the total curvature before model differences become really large. Compare (1) and (2): *angles* are the most sensitive observable.
(3) At present, the observational limit is about $z = 2$ [Wills (1971) finds 19 of 208 QSO's have $z \geq 2$].
Column (3) says: We see now already 1/2 of the distance which can possibly be seen.
(4) We see now about 1/3 of the total curvature.

From (3) and (4) we find: Most important for observational cosmology is not to extend our range to very large z, but to obtain:

Higher accuracy
Complete samples to fainter limits
Theory of sources, intrinsic correlations
Distance of quasars.

e) Our Meager Results

First, the 3°K background seems now well established (see Figure 13.15). It follows naturally from a hot dense early phase, and the big-bang seems the only way to get one. This excludes the lower left-hand part of Figure 13.3, and with the asymptotic models A_2. (In steady state it would mean that not only matter is created all the time but radiation, too.)

Second, the authors evaluating $N(S)$ counts and luminosity-volume tests agree that evolution is definitely needed but do not agree on its details. The simplest evolution is about as $(1 + z)^5$, for density and/or luminosity of sources, with a cut-off at about $z^* = 5$. The latter is nicely explained if it takes a few galactic (rotation) time scales for making sources after the big-bang. It seems to me that the whole curve of Figure 13.14 (not just its bright end) speaks fairly strongly against steady state.

Third, we see no antipole-images up to $z = 2$, say. This excludes a somewhat larger left-hand part of Figure 13.3, starting somewhat above A_2. Fourth, the density is somewhere between $\sigma_o = 0.01$ for visible galaxies and $\sigma_o = 0.30$ for hidden mass according to virial equilibrium of clusters, both limits being rather uncertain. Fifth, the optical $z(m)$ relation of cluster galaxies (Figure 13.6), yields about $-2 \leq q_o \leq +1$ and $0.5 \leq \sigma_o \leq 4$, with very uncertain limits, as indicated by Figure 13.5.

In summary, the universe started most probably with a big-bang about 10^{10} years ago, and radio sources appeared about 5×10^8 years thereafter. Either these sources were then much more numerous (and/or brighter) than they are today, or we happen to sit in a local hole of a spatially fluctuating density. Both q_o and σ_o (and thus λ_o, too) are of the order of unity, but the sign of q_o and λ_o is not known. The following big questions are still unsolved:

Is three-space closed, flat, or hyperbolic?
Is the expansion monotonic or oscillating?
Is the cosmological constant zero or not?

$$(13.94)$$

But our inability to answer these questions implies a positive and significant statement, too, for which no *a priori* reason is known:

The universe is not very different from the *simplest* nonempty model, the Einstein–de Sitter model with

$$q_o = \sigma_o = 1/2 \; (k = \Lambda = 0) \quad (13.95)$$

the only model with full agreement between Newton and Einstein.

References

Abell, G. O. 1972. *Proc. I.A.U. Symposium.* 44: 341.

Alfven, H. 1965. *Rev. Mod. Phys.* 37:652.

———. 1971. *Nature* 229:184.

Alpher, R. A., and R. C. Herman. 1948. *Nature* 142:419.

Arakelian, M. A. 1970. *Nature* 225:358.

Arp, H. 1967. *Astrophys. J.* 148:321.

———. 1968. *Publ. Astron. Soc. Pacific* 80:129.

———. 1970. *Astrophys. J.* 162:811.

———. 1971. *Astrophys. Lett.* 7:221.

Bahcall, J. N., M. Schmidt, and J. E. Gunn. 1969. *Astrophys. J.* 157:L77.

Bash, F. N. 1968. *Astrophys. J.* 152:375.

Beckman, J. E., P. A. R. Ade, J. S. Huizinga, E. I. Robson, and D. G. Vickers. 1972. *Nature.* 237:154.

Bergh, S., van den. 1961. *Z. Astrophys.* 53:219.

Blair, A. G., J. G. Beery, F. Edeskuty, R. D. Hiebert, J. P. Shipley, and K. D. Williamson, Jr. 1971. *Phys. Rev. Lett.* 27:1154.

Braccesi, A., and L. Erculiani. 1967. *Nuovo Cimento* (X) 50:398.

———, and L. Formiggini. 1969. *Astron. Astrophys.* 3:364.

Brans, C., and R. H. Dicke. 1961. *Phys. Rev.* 124:925.

Brault, J. 1963. *Bull. Amer. Phys. Soc.* 8:28.

Brecher, K., G. Burbidge, and P. Strittmatter. 1971. *Astrophys. Space Sci.* 3:99.

Bridle, A. H., M. M. Davis, E. B. Fomalont, and J. Lequeux. 1972. *Nature-Phys. Sci.* 235:123.

Burbidge, M. 1967. *Ann. Rev. Astron. Astrophys.* 5:399.

Burbidge, G. R. 1967: *Astrophys. J.* 147:851.

———. 1968. *Astrophys. J.* 154:L41.

———. 1971. *Nature* 233:36.

Chernin, A. D. 1966. *Soviet Astron.* 9:871.

Christiansen, W. 1969. *Month Nat.* 145:327.

Cohen, M. H. 1969. *Ann. Rev. Astron. Astrophys.* 7:619.

———, W. Cannon, G. H. Purcell, D. B. Shaffer, J. J. Broderick, K. I. Kellermann, and D. L. Jauncey. 1971. *Astrophys. J.* 170:207.

Conklin, E. K. 1969. *Nature* 222:971.

Cowan, C. L. 1968. *Astrophys. J.* 154:L5.

Crawford, D. F., D. L. Jauncey, and M. S. Murdoch. 1970. *Astrophys. J.* 162:405.

Davidson, W., M. Davies, and B. G. Cox. 1971. *Australian J. Phys.* 24:403.

Dicke, R. H. 1968. *Astrophys. J.* 152:1.

———. 1969. *Astrophys. J.* 155:123.

———, P. J. Peebles, D. T. Roll, and D. T. Wilkinson. 1965. *Astrophys. J.* 142:414.

Dyson, F. J. 1968. *Astrophys. J.* 154:L37.

Faulkner, J., J. E. Gunn, and B. A. Petersen. 1966. *Nature* 211:502.

Frantschi, S., J. N. Bahcall, G. Steigman, J. C. Wheller. 1971. *Comments Astrophys. Space Sci.* 3:121.

Gamow, G. 1956. *Vistas Astron.* 2:1726.

Golden, L. M. 1971. *Nature* 234:103.

Greenstein, G. S. 1968. *Astrophys. Lett.* 1:139.

Greenstein, J. L., and V. L. Trimble. 1967. *Astrophys. J.* 149:283.

Gubbay, J., A. J. Legg, D. S. Robertson, A. T. Moffet, R. D. Ekers, and B. Seidel. 1969. *Nature* 224:1094.

Gunn, J. E. 1967. *Astrophys. J.* 150:737.

Hagedorn, R. 1965. *Suppl. Nuovo Cimento* 3:147.

———. 1970. *Astron. Astrophys.* 5:184.

Hawking, S., and G. F. Ellis. 1968. *Astrophys. J.* 151:25.

Hazard, C., and E. E. Salpeter. 1969. *Astrophys. J.* 157:L87.

Heeschen, D. S. 1960. *Publ. Astron. Soc. Pacific* 72:368.

———. 1966. *Astrophys. J.* 146:517.

Henry, P. S. 1971. *Nature* 231:516.

Heygi, D. J., W. A. Traub, and N. P. Carleton. 1972. *Phys. Rev. Lett.* 28:1541.

Hill, J. M. 1971. *Monthly Notices Roy. Astron. Soc.* 153.

Hoerner, S. von. 1968. Paris Symposium. "N-Body Problems," *Bull Astron. Ser. 3*, 3:147.

———. 1973a. *Astrophys. J.* 181:261.

———. 1973b. *Astrophys. J.* 186:741.

Hogg, D. E. 1969. *Astrophys. J.* 155:1099.

Hoyle, F., and G. R. Burbidge. 1966. *Nature* 210:1346.

———, and W. A. Fowler. 1967. *Nature* 213:373.

Iben, I., and J. Faulkner, 1968. *Astrophys. J.* 153:101.

Jauncey, D. L. 1967. *Nature* 216:877.

Kantowski, R. 1969. *Astrophys. J.* 155:89.

Kellermann, K. I. 1972. *Astron. J.* 77:531.

———, and I. I. K. Pauliny-Toth. 1969. *Astrophys. J.* 155:L71.

———, I. I. K. Pauliny-Toth, and P. Williams. 1969. *Astrophys. J.* 157:1.

———, M. M. Davis, and I. I. K. Pauliny-Toth. 1972. *Astrophys. J.* 170:L1.

Klüber, H. von. 1960. *Vistas Astron.* 3:47.

Kristian, J. 1967. *Astrophys. J.* 147:864.

———, and R. K. Sachs. 1966. *Astrophys. J.* 413:379.

Legg, T. H. 1970. *Nature* 226:65.

Longair, M. S. 1966. *Monthly Notices Roy. Astron. Soc.* 133:421.

———, and G. H. Macdonald. 1969. *Monthly Notices Roy. Astron. Soc.* 145:309.

———, and G. G. Pooley. 1969. *Monthly Notices Roy. Astron. Soc.* 145:121.

———, and P. A. Scheuer. 1970. *Monthly Notices Roy. Astron. Soc.* 151:45.

McCrea, W. 1971. *Quart. J. Roy. Astron. Soc.* 12:140.

McIntosh, C. B. 1968. *Monthly Notices Roy. Astron. Soc.* 140:461.

Miley, G. K. 1971. *Monthly Notices Roy. Astron. Soc.* 152:477.

Misner, C. W. 1968. *Astrophys. J.* 151:431.

———. 1969a. *Phys. Rev.* 186:1319.

———. 1969b. *Phys. Rev. Lett.* 22:L1071.

Moellenhoff, C. 1970. *Astron. Astrophys.* 7:488.

Moffet, A. T., J. Gubby, D. S. Robertson, and A. J. Legg. 1971. *Proc. I.A.U. Symposium No. 44*, Dordrecht: Reidel, p. 228.

Muhleman, D. O., R. D. Ekers, and E. B. Fomalont. 1970. *Phys. Rev. Lett.* 24:1377.

Oke, J. B., and A. Sandage. 1968. *Astrophys. J.* 154:21.

Omnes, R. 1969. *Phys. Rev. Lett.* 23:38.

———. 1970. *Astron. Astrophys.* 10:228.

Oort, J. H. 1958. *Inst. Physique Solvay, Brussels* 11:163.

———. 1970. *Astron. Astrophys.* 7:405.

Pariiskii, Y. N., and T. B. Pyatunina. 1971. *Soviet Astron.* 14:1067.

Partridge, R. B. 1969. *Am. Sci.* 57:37.

Pauliny-Toth, I., and K. I. Kellermann. 1972. *Astron. J.* 77:797.

Peach, J. V. 1970. *Astrophys. J.* 159:753.

———, and J. M. Beard. 1969. *Astrophys. Lett.* 4:205.

Peebles, P. J., and R. B. Partridge. 1967. *Astrophys. J.* 148:713.

Penzias, A. A., J. Schraml, and R. W. Wilson. 1969. *Astrophys. J.* 157:L49.

———, and R. W. Wilson. 1965. *Astrophys. J.* 142:419.

Pound, R. V., and J. L. Snider. 1965. *Phys. Rev.* 140:B788.

Rees, M. J. 1967. *Monthly Notices Roy. Astron. Soc.* 135:354.

———, and M. Schmidt. 1971. *Monthly Notices Roy. Astron. Soc.* 154:1.

Refsdal, S. 1970. *Astrophys. J.* 159:357.

———, R. Stabell, and F. G. de Lange. 1967. *Mem. Roy. Astron. Soc.* 71:143.

Rindler, W. 1956a. *Monthly Notices Roy. Astron. Soc.* 116:353.

———. 1956b. *Monthly Notices Roy. Astron. Soc.* 116:662.

Ringenberg, R., and G. C. McVittie. 1969. *Monthly Notices Roy. Astron. Soc.* 142:1.

———, and G. C. McVittie. 1970. *Monthly Notices Roy. Astron. Soc.* 149:341.

Roeder, R. C. 1967. *Astrophys. J.* 142:131.

Rood, H. J., V. C. Rothman, and B. E. Turnrose. 1970. *Astrophys. J.* 162:411.

Rowan-Robinson, M. 1967. *Nature* 216:1289.

———. 1968. *Monthly Notices Roy. Astron. Soc.* 138:445.

————. 1970. *Monthly Notices Roy. Astron. Soc.* 149:365.

Ryle, M. 1968. *Ann. Rev. Astron. Astrophys.* 6:249.

————. 1971. *Nature* 229:388.

Sandage, A. 1961a. *Astrophys. J.* 133:355.

————. 1961b. *Astrophys. J.* 134:916.

————. 1962. *Astrophys. J.* 136:319.

————. 1968. *Observatory* 88:91.

————. 1970. *Phys. Today* 23:34.

Schmidt, M. 1968. *Astrophys. J.* 151:393.

————. 1969. *Ann. Rev. Astron. Astrophys.* 7:529.

————. 1970. *Astrophys. J.* 162:371.

————. 1971. *Observatory* 91:209.

Seielstad, G. A., R. A. Sramek, and K. W. Weiler. 1970. *Phys. Rev. Lett.* 24:1373.

Shapiro, I. I., M. E. Ash, and W. B. Smith. 1968a. *Phys. Rev. Lett.* 20:1517.

————, M. E. Ash, and W. B. Smith. 1968b. *Phys. Rev. Lett.* 20:1265.

Silk, J., and S. L. Shapiro. 1971. *Astrophys. J.* 166:249.

Solheim, J. E. 1966. *Monthly Notices Roy. Astron. Soc.* 133:321.

Spitzer, K., and W. Saslaw. 1966. *Astrophys. J.* 143:400.

Sramek, R. A. 1971. *Astrophys. J.* 167:L55.

Stabell, R., and S. Refsdal. 1966. *Monthly Notices Roy. Astron. Soc.* 132:379.

Steigman, G. 1969. *Nature* 224:477.

————. 1971. *General Relativity and Cosmology.* New York: Academic Press.

Stewart, J. M. 1968. *Astrophys. Lett.* 2:133.

Terrell, J. 1967. *Astrophys. J.* 147:832.

Thorne, K. S., and C. M. Will, 1971. *Astrophys. J.* 163:595.

Tinsley, B. M. 1968. *Astrophys. J.* 151:547.

Turnrose, B. E., and H. J. Rood. 1970. *Astrophys. J.* 159:773.

Vaucouleurs, G. de. 1970. *Science* 167:1209.

Veron, P. 1966. *Nature* 211:724.

Weedman, D. W. 1970. *Astrophys. J.* 161:L113.

Weinberg, S. 1970. *Astrophys. J.* 161:L233.

Whitney, A. R., I. I. Shapiro, A. E. E. Rogers, D. S. Robertson, C. A. Knight, T. A. Clark, R. M. Goldstein, G. E. Marandino, and N. R. Vondenberg. 1971. *Science* 173:225.

Will, C. M. 1971. *Astrophys. J.* 169:125, 141.

Wills, D. 1971. *Science* 234:168.

Wolfe, A. M. 1969. *Astrophys. J.* 159:L61.

————, and G. R. Burbidge. 1969. *Astrophys. J.* 156:345.

Zapolski, H. S. 1968. *Astrophys. J.* 153:L163.

Zeldovich, Y. B. 1970. *Comments Astrophys. Space Sci.* 2:12.

Literature

The following general literature on cosmology is recommended:

Some Textbooks

R. Tolman. 1950. *Relativity, Thermodynamics and Cosmology.* London: Oxford University Press. Classical introduction to relativity, detailed formulae.

O. Heckmann. 1968. *Theorien der Kosmologie.* Heidelberg: Springer.

E. A. Milne. 1948. Kinematic Relativity. London: Oxford University Press.

G. C. McVittie. 1952. *Cosmological Theory.* London: Methuen.

G. C. McVittie. 1961. *Fact and Theory in Cosmology.* New York: Macmillan.

H. Bondi. 1960. *Cosmology.* London: Cambridge University Press. Discussion of various basic philosophies.

G. C. McVittie. 1965. *General Relativity and Cosmology.* Urbana: University of Illinois. Frequently quoted for formulae, short derivations.

W. Rindler. 1969. *Essential Relativity.* New York: Van Nostrand Reinhold.

Summaries, Lectures

Brandeis Summer Institute. 1965. *General Relativity.* (Contr.: Trautman, Pirani, Bondi.) Englewood Cliffs, N.J.: Prentice-Hall. Differential geometry.

Brandeis Summer Institute. 1969. *Astrophysics and General Relativity.* (Contr.: Field, Greenstein, Greisen, Lin, Layzer, Lynden-Bell, Misner, Moffet, Sachs.) New York: Gordon and Breach.

H. Y. Chiu. 1967. Lecture Notes Relativity Theory and Astrophysics. Greenbelt, Md.: Goddard Space Flight Center. NASA.

H. Y. Chiu. 1969. "Cosmology of our Universe." *Sci. J.* Good short summary, history, tables, figures.

W. Kundt. 1971. *Survey of Cosmology.* Springer Tracts in Mod. Phys. *58.* Especially informative on early phases.

Enrico Fermi School. 1971. *General Relativity and Cosmology.* (Contr.: Sachs, Ehlers, Geroch, Ellis, Sciama, Thorne, Burbidge, Rees, Bertotti, Ipser, Heinzmann, Borner, Kundt, Steigman.) New York: Academic Press.

INDEX